HISTORICAL
GEOLOGY

HISTORICAL GEOLOGY

Third Edition

Carl O. Dunbar

PROFESSOR EMERITUS OF PALEONTOLOGY
AND STRATIGRAPHY, YALE UNIVERSITY

Karl M. Waage

PROFESSOR OF GEOLOGY
YALE UNIVERSITY

JOHN WILEY & SONS, INC.

New York / London / Sydney / Toronto

Frontispiece. Detail of Paleozoic strata lying above Pre-cambrian rocks in the Grand Canyon.

Library of Congress Catalogue Card Number: 72–89681

SBN 471 22507 X

Printed in the United States of America

Copyright, 1915 by Louis V. Pirsson and Charles Schuchert under the title
A Textbook of Geology. Part II—Historical Geology

Copyright renewed, 1943 by Eliza B. Pirsson

Copyright, 1924 by Charles Schuchert under the title
A Textbook of Geology. Part II—Historical Geology

Copyright, 1933, 1941 by Charles Schuchert and Carl O. Dunbar under the title
A Textbook of Geology. Part II—Historical Geology

Copyright, 1949 © 1960 by John Wiley & Sons, Inc.
Historical Geology by Carl O. Dunbar

PREFACE

This volume is a successor to *Historical Geology* by Carl O. Dunbar, and preserves the same general framework; but, in view of the quite extraordinary increase in knowledge during the last two decades, nearly all the text has been rewritten and two new chapters have been added.

Few subjects involve evidence from such diverse fields of science. Astronomy, space exploration, oceanography, geophysics, geochemistry, biology, and all the classical branches of geology contribute to the history of the earth. Since the student may find such a range of technical information formidable, we have endeavored to explain each observation as it is introduced and to show its relevance. Believing that it is more important for the student to learn *how* a geologist thinks about the earth then *what* he thinks about any particular detail, we prefer to emphasize principles rather than to catalogue facts, so as to appeal more to understanding than to memory. Pertinent facts are to be remembered, of course, for they are the steppingstones to knowledge. It would be as impossible to think or reason without facts as it would be to breathe in a vacuum.

But we remember the admonition of Plutarch, "The mind is not a vessel to be filled but a fire to be lighted." And surely the history of our world is a subject to fire the imagination. It is a drama in which the actors were real and live, and the stage was the whole wide world. This viewpoint has controlled both the selection and the treatment of the subject matter of this book. No effort has been spared in the selection of illustrations to make the story vivid, and no extraneous material has been allowed to break its continuity, from the fiery birth of the planet to the unfolding of our modern world.

The Prologue is intended to set the stage and insure understanding of the principles used in interpreting the historic record in the rocks. The Appendix is provided to orient students who have enrolled in geology with inadequate training in biology. For such students the history of life can have little meaning until they have learned something of the structure and relationships of at least the major groups of animals and plants. We recommend the study of the Appendix as an outside assignment to be completed before Chapter 8 is reached.

This book is intended primarily for American students, and for several reasons the physical history of the North American continent is given major prominence. Physical events such as the rise of mountains, episodes of volcanism, and the ebb and flow of epicontinental seas are regional or local in extent and differ in detail from continent to continent. Any attempt to cover them all within a single volume would amount to little more than a catalogue of observations without meaningful discussion. Fortunately the history of North America is rather better known than that of any other continent, and it illustrates all the principles involved in the physical history of the earth.

In contrast with the physical changes which were regional and reversible, the progress of life has been worldwide and irreversible; therefore it is treated as a global phenomenon.

The paleogeographic maps have been redrawn since the last edition in view of our present understanding of continental drift and of the spreading of the sea floors. None of these maps older than the Cretaceous is carried beyond the present shorelines. The entire set of paleogeographic maps has been grouped together and is situated at the beginning of the survey of the geologic record (pages 119–132). This has been done to give the student an opportunity to view the changing areas of inundation over time.

This revision is essentially a synthesis of the work of many specialists to whom we are indebted. Since much of it is relatively new we have documented the text in many places with bibliographic references that will enable both the teacher and the interested student to examine the sources of our data. For our illustrations, many of which are new, we are indebted to many individuals and institutions, each of which is acknowledged. Special thanks should go to Peabody Museum of Natural History at Yale, to the American Museum in New York, the Field Museum in Chicago, the U. S. National Museum, the U. S. Geological Survey, the National Geographic Society, and to *Life Magazine*. It is a pleasure also to acknowledge our indebtedness to Rudolph F. Zallinger, whose fine murals grace the halls of Peabody Museum and whose etching of the Bereskova Mammoth was a gift created especially for the earlier edition of *Historical Geology.* The drawings of invertebrate fossils for Plates 1 to 17 are the work of L. S. Douglass and those for Plate 18 were drawn by Shirley G. Hartman. Solene Roming helpfully assisted with the line drawings and with the manuscript in general. Photographs of the Peabody Museum specimens are the work of Richard Smith and Percy Morris. To all these people, and to our friends at John Wiley, go our cordial thanks.

November, 1969

CARL O. DUNBAR
KARL M. WAAGE

PREFACE TO THE SECOND EDITION

Geology is still a rapidly growing science, and in many respects it has been stimulated in recent years by the epoch-making discoveries in atomic physics. In the single decade since publication of the first edition of *Historical Geology*, for example, more has been learned than was previously known about the use of radioactive isotopes in the dating of ancient rocks. Dating of late Pleistocene objects by radiocarbon was just beginning when the first edition was in press. Since the age of Precambrian rocks in many parts of the world have been determined, it is now evident that attempts at their classification on more than a regional basis is not yet feasable, and such terms as Archeozoic and Proterozoic eras are no longer useful. In the meantime advances in both geochemistry and astronomy have produced revolutionary changes in conceptions of the origin and cosmic history of the Earth. Growing knowledge and more mature deliberation have helped also to clarify the geologic history of mankind.

In attempting to bring the subject up to date, I have maintained the general organization of the first edition, but several of the chapters have been completely rewritten and most of the text has been recast in order to include new information and to make a more effective presentation. Discussion of the cosmic history of the Earth has been shifted to the Prologue as Chapter 4, preceding the chapter on evolution, which is introduced by a brief epitome of current speculation on the origin of life on Earth.

Once again I am deeply indebted to my colleagues with whom problems have been discussed: to Karl Waage for help at many points, especially in the Mesozoic chapters; to Joseph Gregory for information about recent discoveries in vertebrate paleontology; to Horace Winchell for Figures 10 and 11; and to Karl Turekian for help with Chapters 2 and 4 in which geochemistry plays an important role. Since neither of these gentlemen has seen the manuscript in its final form, however, I alone must bear the entire responsibility for any errors or shortcomings. Richard Foster Flint's recent volume, *Glacial and Pleistocene Geology*, has been an indispensable source of information for Chapter 18, and A. S. Romer's *Vertebrate Paleontology* has been an equally important sourcebook for data on the history of vertebrate animals. For problems of correlation the *Correlation Charts* prepared by the Committee on Stratigraphy of the National Research Council have been freely used.

It is a pleasure to express my thanks to correspondents who have supplied information and illustrations for the book. Dr. Lauge Koch advised me, in advance of publication, about the new discoveries of the Devonian tetrapods in East Greenland, and, at his suggestion, Dr. Erik Jarvik rushed to me a copy of his fine paper and with permission to reproduce the illustrations in Figure 162; Professor A. C. Blanc of the University of Rome sent me data about the Neanderthal remains of Grotto San Felice at Monte Circeo and a picture of the fine skull reproduced as Figure 388; Dr. V. A. Orlov of the Academy of Sciences in Moscow kindly sent me original copy for Figures 230 and 231; Sir Wilbert Le Gros Clark of London has permitted me to reproduce figures from his fine volume, *The Fossil Evidence for Human Evolution;* Dr. Thomas Hendricks supplied information concerning the age of the Stanley and Jackfork formations in Arkansas and Oklahoma; Dr. Russell K. Grater sent me the fine photograph reproduced as Figure 28; Dr. A. R. Palmer supplied photographs of the amazing insect reproduced in Figure 23; and Professor Sherwood D. Tuttle wrote me numerous thoughtful suggestions as to the contents of the book and, among other things, stimulated me to include the discussion of the origin of life.

Many of the illustrations are new, and they come from many sources, each of which is gratefully acknowledged in the figure legends. The new format of the book has made it possible to use larger illustrations than in the previous edition. Special thanks go to Time and Life, Inc., for permission to reproduce portions of the mural, *The Age of Mammals,* which was painted by Rudolph F. Zallinger, under my supervision, for Life's *The World We Live In,* and for Figure 390, which is from Life's *Epic of Man.*

My cordial thanks go also to Rudolph F. Zallinger for the beautiful etching of the Beresovka mammoth which introduces Chapter 3, and to Shirley Glaser who retouched a number of the photographs and made several of the drawings, including the figures of Plate 18. I am again indebted to Percy A. Morris who made all the photographs of fossils used from the collections of Yale's Peabody Museum. All the line drawings and diagrams are mine.

New Haven, Connecticut   Carl O. Dunbar
October, 1959

PREFACE TO THE FIRST EDITION

The history of the earth is a drama in which the actors are all real, and the stage is the whole wide world. The student must sense the action and feel the essence of high adventure in the *march of time* as shifting scenes unfold and living actors cross the stage. This viewpoint has controlled both the selection and the treatment of subject matter in this volume. No effort has been spared in preparing the illustrations to make the story vivid, and no extraneous material has been allowed to break its continuity, from the fiery birth of the planet to the unfolding of our modern world. The *prologue* is intended to set the stage and to insure understanding of the principles used in interpreting the Earth's history.

The *Appendix* was prepared for students who enroll in geology without previous training in biology. For such beginners the history of life on the Earth can have little meaning until they have learned something of the structure and relationships of at least the major groups of animals and plants. For several years Yale students have been required to study the material embraced in the Appendix as an outside assignment during the first weeks of the course. Upon beginning the study of the Paleozoic Era, each is required to pass a sight test showing that he can recognize the major groups of animals and plants. This treatment has been eminently satisfactory.

The subject matter of historical geology is inherently diversified, involving as it does certain aspects of astronomy, anthropology, and biology, as well as geology. The danger exists, therefore, that the beginner will feel bewildered by the mass of unfamiliar facts drawn from such widely different fields and, in the welter of details, will lose sight of the grand conceptions. For this reason we have tried to group all details about great principles. Believing it to be more important for the student to learn *how* a geologist thinks about the Earth than *what* he thinks

about any particular detail, we have taken pains to emphasize principles of interpretation rather than to catalogue facts about the history of the Earth, appealing thus to understanding rather than to memory.

Stratified rocks with their entombed fossils form a manuscript in stone and are the source of much of our knowledge of the past history of the world. It must be confessed, however, that stratigraphic descriptions are dull and detailed unless related to the physical history they record. The late Ordovician formations in New York State, for example, might appear to have no more than purely local interest; but, when they are viewed as parts of a piedmont and coastal plain that was growing westward into an inland sea while the eastern border of the continent was rising into mountains, they take on significance. Therefore, in treating of each period of geologic history, we have tried at the outset to help the student visualize the physical geography of the time and understand the major physical changes our continent was undergoing. The panels of paleogeographic maps were designed as an aid to this end.

To avoid the monotony inherent in the systematic account of period after period, emphasis is varied from chapter to chapter. In the discussion of the Cambrian, for example, considerable space is devoted to paleogeography and to the bases for subdividing the rocks into series and formations; in the Ordovician chapter, the Taconian orogeny is discussed at length because it illustrates the principles used in recognizing and dating all later orogenic disturbances. In this way, general principles of interpretation are developed one after another in the early chapters, leaving room in later ones for greater detail. This we believe to be fitting, since human interest in the rocks and the fossils increases as we approach the modern world.

The facts of historical geology are drawn from many sources, and most of them are common knowledge. We have made no attempt to give credit for such general information except to cite works on subjects that are controversial and others so new that they may not be generally known to teachers of geology. Such references bear exponents in the text, referring to numbered citations at the end of the chapter.

This volume is a successor to, and an outgrowth of, the *Textbook of Historical Geology* by Schuchert and Dunbar. It preserves the same point of view and the same general organization except for the introductory chapters which are arranged so as to bring the geologic time scale near the front. The text of the previous work has been largely recast to take account of advances in knowledge or to make a more effective presentation. Special care has been given to the illustrations, many of which are new. The "bleed cuts" will speak for themselves.

In the preparation of this volume friendly assistance has been received from many sources. It is not possible to mention them all specifically, but my thanks are none the less real. Among those to whom I am most particularly indebted are my colleagues, Chester R. Longwell, Richard F. Flint, Adolph Knopf, Joseph T. Gregory, and John Rodgers, whom I have consulted on various problems in their several fields; G. Edward Lewis of the United States Geological Survey, who helped with the chapter on Mammals; Cornelius Osgood, chairman of the Department of Anthropology at Yale, who read and criticized the chapter on Man as it stood in the previous volume by Schuchert and Dunbar; W. W. Rubey, with whom I discussed plans for the book during the memorable days we spent aboard the U.S.S. *Panamint* on the way to Bikini. Others too numerous to mention have written to offer suggestions or to reply to inquiries about specific details. It is a pleasure to acknowledge all this friendly help, but, since none of those mentioned has read the manuscript in its final form, the writer alone must assume the responsibility for any shortcomings or mistakes.

The illustrations are from many sources, all of which are gratefully acknowledged in the credit lines attached to individual figures. The frontispiece and Figs. 192, 226, 232, and 262 are portions of a great mural in Peabody Museum, painted by Rudolph Zallinger under the direction of the scientific staff. My cordial thanks go also to other members of the museum staff: particularly, to Shirley P. Glaser, who drew several of the new text figures; to Percy A. Morris, who made nearly all the photographs of fossils in the museum; to Sally H. Donahue, for her faithful help in the long and tedious preparation of the manuscript; and to Clara M. LeVene, for her indispensable aid in the final editing of the work and the preparation of the index.

To Charles Schuchert my obligation is unbounded. At his feet I learned much of what appears here as my own. His association was a constant stimulus, and his memory is an abiding inspiration.

New Haven, Connecticut   CARL O. DUNBAR
October, 1948

CONTENTS

HISTORICAL
GEOLOGY

PART I

PART I

PROLOGUE

Figure 1-1 A manuscript in stone. Wall of the Grand Canyon from Sublime Point on the north rim. Here the layers of bedded rock form a manuscript in stone more than a mile thick, recording more than 100,000,000 years of earth's history. (Joseph Muench.)

CHAPTER **1**   *Records in Stone*

And some rin up hill and down dale
knapping the chucky stanes to pieces
wi' hammers, like sae many road
makers run daft. They say it is to
see how the world was made.
                    SIR WALTER SCOTT

The layers of stratified rock are tablets of stone bearing a record, in code, of the long history of our world and of the dynasties of living creatures that preceded the advent of man (Fig. 1-1). To break that code and discover its secrets is a challenge that stirs the imagination. Some of its characters are now well understood and many of the larger features of the earth's history are clear, but there is still much to learn as we discover how better to interpret the more subtle meanings of the record. We begin here by examining some of the general principles of interpretation.

EVIDENCE OF ANCIENT LANDS AND SEAS

Plotting the Ancient Seaways. If North America were depressed 600 feet, the Missis-

sippi Valley would be a great inland sea, the present Coastal Plain a shallow sea floor, and Florida a submarine bank (Fig. 1-2). Over these submerged areas sediment would spread, burying seashells, sharks teeth, and occasional bones of porpoises and whales, while beach gravel and barnacles and oyster banks would accumulate along the shore zone.

If later the continent were uplifted, the sea would of course disappear, but these marine deposits would form a telltale record of the submergence, and by plotting them on a map we could restore more or less accurately the limits of the vanished seaways; Fig. 1-2 depicts such a hypothetical submergence of North America. Figure 1-3 is an actual map of this type for Malaysia, which was long a part of the Asiatic mainland but in recent geologic time has been submerged about 200 feet, flooding the lower parts to form a wide shelf

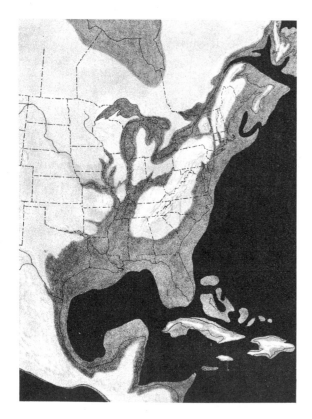

Figure 1-2 Southeastern part of North America as it would appear if depressed 600 feet. Shallow sea shaded; deep sea black.

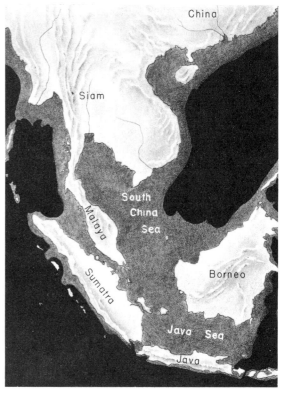

Figure 1-3 Malaysia, a portion of southeast Asia as it exists today, having been partly submerged within the near geologic past. (Adapted from Van Riel.)

sea stretching all the way from the Gulf of Siam to the Java Sea. Drowned valleys of the streams that once crossed these lowlands can still be traced on the sea floor. The large animals of Borneo, Sumatra, and Java migrated from Asia across this intervening lowland before it was drowned.

Submergences of North America as great as this have occurred many times in the past, and the outlines of land and sea have thus varied widely from age to age.

Restoring Lost Lands. Layers of sediment spread over a land surface commonly bear evidence of the terrestrial environment in the form of mudcracks, footprints, or skeletons of land animals, remains of land plants, or coal. Fossil animals and plants may even indicate the approximate altitude and the climate under which the beds were deposited. A striking example occurs in Kashmir Valley high in the Himalayas, where at an elevation of 10,600 feet, plant-bearing deposits have yielded a tropical fig and a species of laurel that still inhabits India but lives only below an altitude of 6000 feet and in localities of subtropical climate [1]. It is clear that the region was much lower and warmer when these plant beds were formed.

Since a landmass is constantly exposed to erosion, the sediments that accumulate on its surface are eventually reworked and carried into the sea, except where they have been downwarped or downfaulted below sealevel. Along lowland coasts sediments may accumulate on the flood plains and deltas of rivers emptying into the sea as well as in the shallow coastal waters and the deeper water offshore. Where a body of strata includes such a change from nonmarine to marine the different environments are commonly revealed by characteristic sedimentary structures or by different kinds of fossils. For example, a change in fossil content of the strata from land plants and river clams at one place to seashells at another would indicate that the shoreline of the time lay between these two localities.

Many groups of animals live exclusively in the sea. Corals, brachiopods, crinoids, sea urchins, and cephalopods are but a few examples. Their occurrence as fossils in a rock implies the presence of a sea at some former time, even though the fossils are now far inland and at great elevations in the mountains, as are certain Eocene marine fossils at 20,000 feet in the Himalayas. The coral reefs so common in the Silurian limestone of northern Indiana leave no doubt that a great bay or inland sea like Hudson Bay or the Baltic Sea covered Indiana during Silurian time. If we plot the distribution of Silurian rocks that bear marine fossils, we can determine at least the minimum extent of this ancient seaway. If in other regions we find fossil land plants with stumps or roots in place, or if we find abundant bones or shells of land animals, it is evident that the enclosing rocks were formed above sealevel. Fossils thus indicate the past distribution of land and sea.

Although fossils may afford the more obvious and more reliable clues to the position of land and sea in ancient times, they are not always preserved in strata deposited in the rigorous marine environments along the margins of land areas. But the sediments themselves may throw much light on the position and character of the land from which they were derived. During transport, the sediment is size-graded, the coarsest material always coming to rest first while progressively finer detritus is carried farther and farther. Hence if a mass of strata grades from fine in one direction to coarser in another, it is evident that the ultimate source lay in the direction of increasing coarseness. Plotting of such data for the beds of a given age may indicate the approximate position and extent of the lands. In some circumstances the derivation of sediment from a particular area may be confirmed by demonstrating that the composition of the sediment matches that of the rock at the suspected source.

In most of the late geologic formations the application of these criteria is obvious, for they point to sources in landmasses that are still supplying sediment. When applied to many of the ancient formations, however, they indicate highlands that no longer exist, and in some instances the results are surprising. The San Onofre Conglomerate of California, for example, increases in coarseness toward the west, indicating a source of its boulders in highlands beyond the present shoreline. This is confirmed by the presence of boulders of peculiar types of metamorphic rock that are unknown on the mainland but crop out on Catalina Island, some 40 miles offshore. The inference is that the islands west of southern California are remnants of a larger landmass that stood high while the San Onofre Conglomerate was forming and has since foundered.

Of course the record is not read this easily everywhere. The greater bulk of the sediment that reaches the sea in any age has gone through several cycles of erosion and deposition since its origin from a nonsedimentary parent source. For plotting the distribution of land and sea, the most direct evidence will obviously be those features of the strata that were inherited from the environment at the site of deposition.

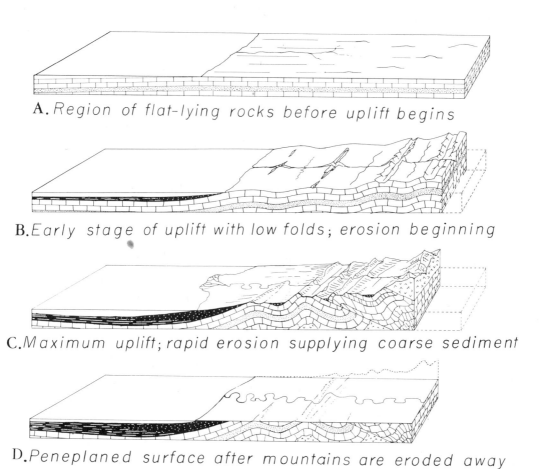

A. Region of flat-lying rocks before uplift begins

B. Early stage of uplift with low folds; erosion beginning

C. Maximum uplift; rapid erosion supplying coarse sediment

D. Peneplaned surface after mountains are eroded away

E. Later submergence buries the roots of the mountains

Figure 1-4 Five stages in the rise and decay of a mountain range. Sediments eroded from the rising range and deposited in the adjacent geosyncline are shown in black. Length of section, some tens of miles.

Vestiges of Mountains. The growth and decay of a mountain range are recorded in the rocks long after the mountains themselves have disappeared. The nature of the evidence is indicated in Fig. 1-4. Block A represents the region before disturbance, formed of flat-lying bedded rocks with a lowland at the right and shallow sea at the left. Uplift is under way in block B, and the strata at the right are being buckled into low folds. Streams are beginning to intrench themselves and the old mantle of fine and well-decayed sediment is being stripped off and transported to the sea, to be spread again as layers of marine mud. Thus far there is but little sand and gravel because the streams still have a low gradient. Block C represents a later stage after strong folding and faulting have produced rugged relief in the mountain region, and rapid erosion is loading the stream with sand and gravel as well as mud. Upon reaching the shore zone the overloaded streams now tend to build an aggraded coastal plain extended locally into deltas. Meanwhile, the uplift in the mountain region is counterbalanced by sinking of a nearby geosyncline (left), and here the sediment derived from the mountains comes to rest, the gravel and coarse sand near shore and the finer sand and mud farther out. Block D shows a later stage after the region of uplift has been peneplaned. The mountains have now vanished, and sluggish streams meander across a lowland, carrying only the finest sediment. The cycle is complete in block E as the region is again submerged, and a shallow sea creeps in over the site of the vanished highlands to bury the roots of the mountains with layers of fine mud.

At the surface all traces of the mountains are now gone; but underground there are records of two sorts, one where the mountains stood and another where their debris accumulated. If the region is uplifted and dissected at some later date, these records will be brought to light. In the area of disturbance a conspicuous unconformity will separate the post-orogenic strata from the buried roots of the mountains, whose truncated folds and faults may indicate the nature of the disturbance and something of the size of individual folds and fault blocks. In block E of our hypothetical case, for example, it is evident that the folds were open to the left of a major thrust and that to the right of it they were more severely mashed and deformed. It would be quite simple in such a case to count the major folds, restore the missing parts, and determine their approximate dimensions. The date of the disturbance is clearly later than the youngest strata involved in the deformation and earlier than the beds immediately above the unconformity.

A complementary record is found in the area of deposition at the left (black). Insofar as the debris of the destroyed mountain range accumulated here, the volume of the detrital sediment bears a definite relation to the size and height of the range. Fossils in these deposits will record the geologic date at which uplift and erosion were taking place, and obviously the coarsest part of the sedimentary record will correspond in time with the most rapid erosion and the greatest relief in the uplifted area. In the region of deposition the growth of the range is reflected in increasing coarseness and ever wider spread of the sands and gravels, and its decline is betrayed by a return to deposits of finer and finer grain. Hence if the region of deposition were available for study, it would still be possible to infer much about the position and size of the uplift and to date it in geologic time even if the site of the ancient mountains remained covered or was repeatedly disturbed until the early record was obscured.

To apply this reasoning to a specific case, consider the sedimentary deposits of Late Cretaceous date which are plotted in Fig. 1-5. They clearly record a vast interior sea that extended from the Gulf of Mexico to the Arctic Ocean. In the eastern part these formations are of fine-grained shale with some interbedded chalky limestone; but toward the west they thicken and coarsen, including along the western margin vast deposits of sandstone with local conglomerates. Clearly the source of this material was chiefly to the west, and it must have been a highland of considerable ruggedness. Furthermore, as we approach the western margin, the rocks contain land plants

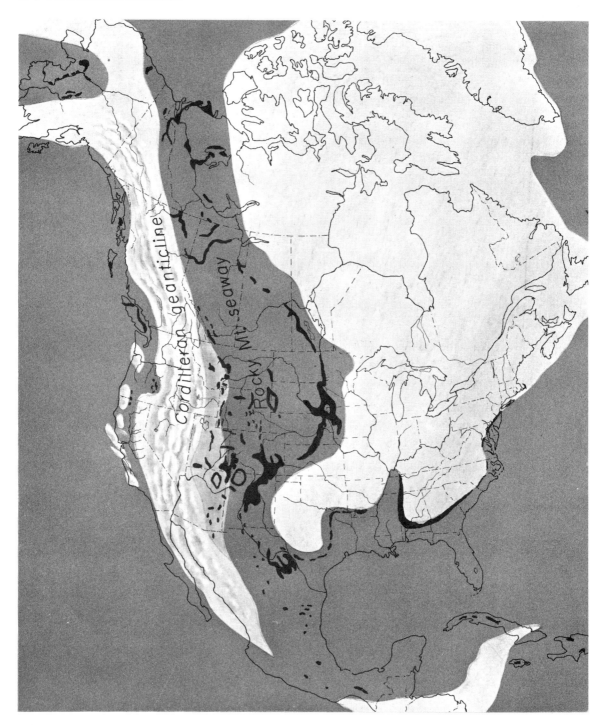

Figure 1-5 Paleogeographic map of North America as it was in Late Cretaceous time. Present outcrops of Cretaceous rocks of this date are in solid black. Coastal plain along the west side of the Rocky Mountain sea lightly shaded.

and dinosaur bones, and locally have much interbedded coal. This must represent a swampy coastal lowland between the mountains and the sea. To supply the enormous volume of these detrital deposits to the Rocky Mountain seaway (estimated at more than 850,000,000 cubic miles) required the erosion of an average thickness of nearly 5 miles of rock from the source area. It is evident, therefore, that this western landmass was either continuously or intermittently rising, and remained mountainous and rugged during the long span of Late Cretaceous time.

By the application of such principles we can restore the major geographic features of North America as they existed a hundred million years ago—a continent separated by a wide interior seaway into two landmasses, the eastern broad and low, the western narrow and mountainous. The reconstruction of ancient lands and seas is called **paleogeography** [Gr. *palaios*, ancient + geography], and the inferred restorations (for example, Fig. 1-5) are **paleogeographic** maps.

Paleogeographic maps may vary greatly in accuracy depending on several circumstances. In Cretaceous and younger deposits the approximate shorelines can in many areas be located and the limits of the seas can be accurately portrayed, but many of the older formations have been extensively eroded away or partly hidden by younger deposits. In this case the approximate limits of the sea may be inferred with varying degrees of confidence; commonly deeply buried formations can be identified in well cores. The uncertainty is increased where the rocks have been intensely deformed, as in parts of the Rocky Mountain region. In large areas of western Canada and in the Arctic Islands the Paleozoic formations are still very incompletely studied and mapped. The accuracy of a paleogeographic map also depends on both the geographic and stratigraphic scale used. In a limited and well studied area the distribution of a minor stratigraphic unit may be accurately delineated. But since the seas ebbed and flowed almost endlessly during an extended period of time no single map can show their limits during a whole period or even an epoch of time. Yet it may be useful to show the maximum extent reached by the seas during a period or epoch. This is the nature of the maps in the paleogeographic folio following Chapter 6; their shorelines are fairly accurately located in some places but in other places they are highly generalized. Such maps are always subject to change with the growth of geologic information.

The Geologic Time Scale

History is a narrative of events that succeeded one another in time, with an interpretation of their causes and interrelations. In this respect geologic and human history have much in common.

The story of civilization, for example, is based upon records of human cultures from many parts of the world that have been built into a composite whole by archeologists and historians. No single country, nor even a continent, could supply all the necessary data, because the locus of important events that make history has shifted from time to time as one great civilization succeeded another. From ancient settlements of the Euphrates Valley, the dominant centers of culture shifted, appearing in Egypt, Greece, Rome, China and, in recent times, Europe and the Americas. Basic to the story is **chronology**—a time scale divided into periods and ages in which events can be assigned their proper dates, permitting an analysis of causes and effects. Thus we recognize the three major divisions of ancient, medieval, and modern history, each further subdivided into ages on the basis of dynasties or cultural changes.

The vastly longer story of earth history, reaching back in time to a beginning more than 4 billion years ago, is based on the rocks of the earth's crust, in particular the sedimentary rocks. The scientific study and interpretation of these strata is the province of **stratigraphy** and it has been the endless job of the stratigrapher to compile a history of the earth from the fragmentary rock record. As different parts of this record were discovered, first in western Europe, then elsewhere, they were pieced together in the form of a vertical sequence, or composite **geologic column,** in which the different local sequences of strata were arranged according to their relative age, each superposed on the next older. From this composite record a **geologic time scale** has been derived. It is not yet a uniformly complete time scale, for just as the path of human history becomes more difficult to discern as it is traced back through the ages, so, to a much greater degree, does the immensely longer path of earth history. Only for the last 600,000,000 years, a time set apart by its abundant record of ancient life, are the events of earth history well enough known to be arranged in a detailed chronologic sequence of periods or ages. Geologists the world over employ a common geologic time scale (Fig. 1-8) for this important part of earth history, which has been named the Phanerozoic Eon [Gr. *phaneros,* visible + *zoo,* life]. The more obscure record of the preceding 3½ billion years, the Cryptozoic Eon [Gr. *kryptos,* hidden + *zoo,* life], does not at present permit construction of a comparable universal scale for all of geologic time. Present success at dating events in this great complex of largely unfossiliferous rocks by means of radioactive isotopes of minerals promises to make a detailed Cryptozoic time scale possible in the future. The nature of this highly important tool for deciphering earth history is described in Chapter 2.

FIRST PRINCIPLES

To understand how the geologic column was built up and the record of past events translated into earth history it is important to know the principles involved. These are discussed in the following paragraphs.

Superposition. Since sedimentary rocks originate as loose sediment spread layer upon layer, in any pile of undisturbed strata the oldest bed is at the bottom, and each in turn is younger than the one on which it rests. Therefore, in a general study of a region of relatively simple structure, the sequence of beds and of formations can readily be ascer-

tained. This simple observation is the principle of superposition.

In disturbed areas, of course, the normal succession may be locally inverted, as in the lower limb of an overturned fold, or it may be interrupted, or duplicated by faults; but such abnormalities will betray themselves in evidences of disturbance and in an unnatural sequence of fossils.

Organic Succession. Most sedimentary rocks include fossil remains of the animals and plants that lived while they were accumulating. The assemblage of animal species living together at a given time and place constitutes a **fauna;** the corresponding assemblage of plants is a **flora.** Thus the animals that now inhabit a region constitute a modern fauna, and the assemblage recorded by fossils in a bed of stratified rock constitutes an older fauna that lived while the rock was forming.

Just before 1800 William Smith, studying the Jurassic rocks of England, discovered that each formation has a distinct fauna, unlike those above or below. He also saw that the characteristic fauna of any formation can be found in outcrop after outcrop as it is traced across the countryside, and so the characteristic fossils served as a **guide** or **index** to distinct formations which he could recognize in any outcrop, without the necessity of careful tracing. A wealth of experience accumulated since the days of Smith has shown that this is a principle of general application—the faunas and floras of each age in earth's history are unique, and permit us to recognize contemporaneous deposits, even in widely separated regions, and so to piece together scattered fragments of the record and place them in proper sequence.

William Smith did not know why each formation possesses a distinct fauna; his inference was based solely on observations in a region of richly fossiliferous rocks where the strata dip gently and the order of superposition is self-evident. We now know that different kinds of animals and plants succeeded one another in time because **life has continuously evolved;** and inasmuch as organic evolution is worldwide in its operation, only rocks formed during the same age could bear identical faunas. The appearance of trilobites and dinosaurs and three-toed horses is not fortuitous and irregular; each lived only at a certain time in geologic history, and each is found fossil only in a certain part of the geologic column. The relative time of existence of a vast number of kinds of ancient animals and plants has now been established, and their place in the geologic column has been confirmed by the research of geologists the world over. This is not a theory derived a priori, but a discovery tediously substantiated by hundreds of systematic studies of the faunas of rock formations carefully located in the geologic column. It is an important natural principle that **fossil faunas and floras succeed one another in a definite and determinable order.**

WAYS AND MEANS IN STRATIGRAPHY

Formations of Rocks. In any attempt to bring order to the immense piles of strata that constitute the geologic record the first need is a practical classification of the rocks themselves. From his earliest attempts to study rocks, man has recognized that they are varied in kind and occur in conspicuously different bodies, or groups of strata. These discrete natural divisions, called **formations,** have become the fundamental units for describing and mapping the geology of local areas.

Each formation is a body of rock distinguished from the rocks above and below by its lithology and appearance. It may consist of one kind of rock, a sandstone, a limestone, a shale, or it may be characterized by interbedded rocks of several kinds. The locality at which it is first described becomes its **type locality** and since the description normally is presented in the form of a measured section of the formation at this locality, this is called the **type section.** Formations are named for some geographic feature near the type locality: the St. Louis Limestone, for example, was named for St. Louis, Missouri; the Kaibab Limestone for the Kaibab Plateau in Arizona; and the Sawtooth Formation for the Sawtooth Range in Montana. As these examples indicate, either

the word formation or the dominant lithology can be used in composing a formation name.

Formations have been called the building blocks of stratigraphy but they should not be thought of as simple uniform building units, for they differ in size and thickness as well as in kind of rock. Many are internally complex and can in some places be further divided into subunits, or **members.** Frequently it is convenient to recognize distinctive characteristics or lithologic peculiarities shared by two or more successive formations and this is done by linking them as a **group.** Both members and groups are also named for geographic features of the areas where typical exposures occur. Formations, however, are the essential units in the classification of local stratigraphic sequences and in the geologic mapping which plots their areal distribution and reveals the geologic fabric of a region. Each is the product of a particular set of depositional events and as such constitutes a fragment of earth history.

Relating Fragments of the Record. The matching of formations from outcrop to outcrop (or from well to well underground) to determine their mutual relations and their degree of equivalence in age is a basic step in working out geologic history. No single area on earth contains the rock record of all earth history, but we need only relate enough of the scattered fragments of the record to build a composite representing all of geologic time. For more than 150 years stratigraphers in all parts of the world have been cooperating in this endeavor and a prodigious literature documents their efforts. If all the local sequences of strata that presently go to make up the composite geologic column were directly superposed, the total thickness would exceed 500,000 feet (95 miles)!

Many criteria are now available for relating strata and some of the more useful ones are illustrated in Fig. 1-6. In this diagram the block at the left represents the formations exposed near the mouth of the Grand Canyon and that at the right represents the section at Bright Angel Trail about 100 miles farther east. A photograph of the wall of the Grand Canyon may be seen in Fig. 1-1.

At Bright Angel Trail ten formations of

stratified rock are recognized, and in the western section there are also ten. When we attempt to compare one with the other, it appears that No. 10 at Bright Angel Trail is the same as No. 10 at the mouth of the canyon, because (1) it presents the same lithologic appearance, being a buff-weathering cherty limestone, (2) it agrees in thickness, and (3) it yields the same kinds of marine fossils, which in both sections are limited to this unit. If more proof were needed, (4) we could follow the rim of the canyon and trace the forma-

tion from one section into the other. This is the Kaibab Limestone that rims the Grand Canyon from end to end.

In the same way, No. 5 of the eastern section can be related to No. 5 of the west. It is the cliff-forming Redwall Limestone (Fig. 13-19) that maintains the same lithologic character and approximately the same thickness for a distance of more than 100 miles. Furthermore, it has distinctive fossils and it crops out in a bold cliff, forming a bench that can be followed continuously along the canyon wall.

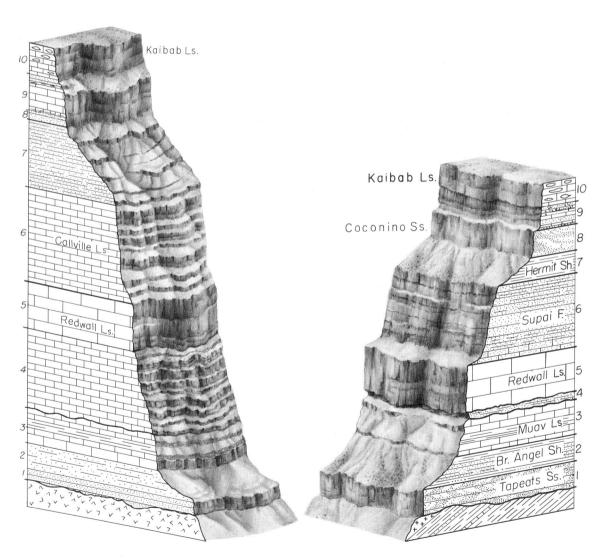

Figure 1-6 Block diagrams showing at the left the section near the mouth of Grand Canyon and at the right the section at Bright Angel Trail about 100 miles farther east.

This gives us two tie-points or key beds for relating the distant sections. The intervening formations, however, present more difficulty. In the eastern section No. 6 is a bright red sandstone and sandy siltstone with fossil plants and footprints of land animals; No. 6 of the western section is a light gray limestone bearing marine fossils. The two seem to hold the same relative position in the sections, just above the Redwall Limestone, but they are totally different in lithology, are unequal in thickness, and have no fossils in common, the one bearing only marine shells and the other only remains of land life. It is evident that both are limited in age to some part of the time between the deposition of the Redwall and the Kaibab limestones, but several alternatives must be considered:

1. The Callville Limestone may be the older, thinning eastward to disappear short of Bright Angel Trail, either because it was never deposited there, or because it was eroded away before deposition of the Supai Formation.
2. The Supai Formation may be the older, thinning westward for either of the reasons suggested in (1).
3. They may be equivalent in age, having formed under different environments.
4. They may be equivalent in part, with one representing more time than the other.

Most of the criteria used to relate the Kaibab or the Redwall Limestone cannot be applied here, for the Callville and Supai formations are dissimilar in lithology, are unequal in thickness, and have no fossils in common. But the Callville can be traced eastward along the walls of the Grand Canyon, and the upper part of it is found to change laterally as the limestone beds first become sandy, then reddish, and finally alternate with layers of red siltstone and sandstone (Fig. 13-20). In short, the upper part of the Callville grades laterally into the Supai Formation and the two formations **intertongue;** they must therefore have been laid down at the same time. The intertonguing of these formations, one marine and the other nonmarine, indicates that the shore fluctuated back and forth across a wide belt during dep-

osition. Fossils in the lower part of the Callville, however, indicate that these beds are older than any part of the Supai; hence this part of the Callville Formation probably never extended as far east as Bright Angel Trail. More field work actually is needed to confirm this belief.

Still another basis for relating the Callville Limestone and the Supai redbeds might have been used, even if the Grand Canyon had not revealed the lateral gradation and intertonguing of one into the other. The fossil plants of the Supai might indicate the same position in the geologic column as the marine fossils of the upper part of the Callville. This could be confirmed if in some other region—say, New Mexico or Kansas—there were an alternation of marine and nonmarine beds carrying the Supai flora and the Callville fauna. Indeed, the **relation of two sections with reference to a third** is common practice.

Similar problems are presented by formation No. 4. This formation holds the same relative position in each section between the Muav and the Redwall limestones, but it differs in lithology and thickness and in faunal content in the two sections.

To recapitulate, formations in separate outcrops may be related because of (1) lithologic similarity, (2) similar thickness, (3) similar position in a sequence of formations some of which are known to be equivalent, (4) continuous tracing between outcrops, (5) lateral gradation and intertonguing of one into the other when the formations are lithologically dissimilar, (6) identical faunas, or (7) fossils that are dissimilar (for example, marine versus nonmarine) but are known to occur together elsewhere and therefore to indicate the same age. These are some but not all of the means by which strata may be related from place to place. All of the means of relating strata may not be applicable in a given situation.

The time relations of rocks are of particular importance in earth history, for they provide the essential framework of chronology. Stratigraphers apply the term **correlation** specifically to the practice of relating rocks in terms of their age, and where strata at two different localities can be shown to be equivalent in age they are said to be correlative. Not all of the means of relating strata that we have applied to the Grand Canyon rocks are reliable criteria for indicating that the strata being related are actually of the same age. For example, lithologic similarity, lateral traceability, and similar position in a sequence permit us to identify a particular formation over a wide area and to deduce that it formed everywhere under similar conditions of deposition, but these criteria alone do not tell us that it is the same age throughout its extent.

Units Nos. 1 and 2 of the Grand Canyon sections present an interesting relationship that illustrates this problem. Number 1, the Tapeats Sandstone, is unfossiliferous; it can be followed in continuous outcrop from one section to the other and is clearly a persistent lithologic unit (Fig. 1-7). The overlying Bright Angel Shale, No. 2, is also a traceable lithologic unit; in the western section it bears two zones of distinctive marine fossils, a lower one of Early Cambrian age and an upper one of Middle Cambrian age. When traced eastward along the canyon walls, however, these fossil zones are found at progressively lower levels in the Bright Angel Shale; they do not parallel the lithologic boundaries. At about 30 miles east of the western section the lower zone disappears into the unfossiliferous beds of the underlying Tapeats Sandstone; only the upper

zone extends all the way to the eastern section and here it is found in the lower beds of the Bright Angel Shale. The fossils clearly show that in spite of their physical similarity throughout the Grand Canyon area, the Tapeats Sandstone and Bright Angel Shale are mostly Lower Cambrian in the western section and become progressively younger to the east, so that at Bright Angel Trail nearly all of the Bright Angel Shale is Middle Cambrian. To put this another way, we can say that the Tapeats Sandstone at Bright Angel Trail is correlative with much of the middle part of the Bright Angel Shale near the mouth of the Grand Canyon.

The explanation of this relationship is that the sea gradually encroached on the region from the west, sand (Tapeats) being spread to form coastal deposits associated with the advancing shoreline while muds (Bright Angel), carried into the sea, were deposited in the shallow marine waters out from shore. The sea did not reach the vicinity of the eastern section at Bright Angel Trail until early in Middle Cambrian time. Such transgressive relations are by no means exceptional; formations commonly vary in age from place to place, especially those that were deposited in shallow water where fluctuations of sea level caused the shore to shift landward or seaward. On the other hand, formations that were deposited off-

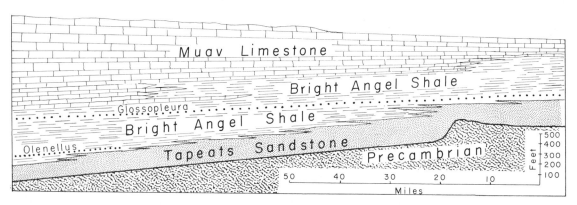

Figure 1-7 Stratigraphic section of the Cambrian formations between the sections represented in Fig. 1-6. Adapted from a figure by Edwin D. McKee, 1945.

shore on the continental shelf or in broad epeiric seas may be of the same age over great areas.

Correlation is one of the most difficult tasks of the stratigrapher who must select from careful study of local rock sequences and their fossils the criteria that will afford the most reliable and most precise indication of age relationships. Fossils, of course, provide the chief method of correlating strata from place to place and they are the only common means of long-range correlation such as that between continents. A body of strata distinguished from strata above and below by the presence of a certain fossil, or assemblage of fossils, is called a **zone,** and is identified by the name of its most characteristic fossil. For example, the *Olenellus* zone of the Bright Angel Shale includes those beds in the formation that contain, among other fossils, the trilobite *Olenellus.* Lower Cambrian strata in many parts of the world contain *Olenellus,* or closely related contemporaries, and its zone provides an excellent marker for this part of Cambrian time. However, the use of fossils in correlation, a practice commonly called **biostratigraphy,** is not as simple as this one, briefly stated example may suggest. More than one kind of zone and a variety of biostratigraphic methods are employed in correlation, as will be seen in later chapters.

In local areas some of the physical relations of strata are useful in correlation. Bedding planes that bound individual rock layers generally approximate time planes and the bed itself is usually the result of a single episode of deposition, brief in terms of geologic time. However, individual layers are difficult to trace beyond a single exposure and even where broadly exposed tend to vanish laterally or grade into other layers. Single layers deposited as the result of one particular widespread event, such as a fall of volcanic ash from an eruption, serve as useful key beds in correlation and may cover many thousands of square miles (p. 372). Intertonguing relationships between different kinds of strata, like that between the Supai Formation and the upper part of the Callville Limestone, are also

very useful indications that strata are equivalent in age. These examples illustrate but do not exhaust the criteria that can be used in correlation.

SUBDIVISIONS OF THE PHANEROZOIC RECORD

The Phanerozoic Eon embraces no more than one-eighth of earth history, but, as we have seen, it includes that part of the geologic record containing abundant evidence of plant and animal life. In the rock record itself the change from generally unfossiliferous Cryptozoic strata to fossiliferous Phanerozoic strata is, with rare exception (p. 163), strikingly abrupt. Yet the change from the primitive life of Cryptozoic time to the more diverse and complex life of Phanerozoic time was undoubtedly transitional. Many theories have been proposed to account for this seeming anomaly; one of the most convincing recent ideas holds that a great acceleration in the rate of organic evolution was brought about by a change in the composition of the earth's atmosphere (p. 167). Whatever its explanation, the abrupt appearance of abundant fossils marks a highly significant event in the history of life and affords a useful subdivision of the geologic record.

The hierarchy of Phanerozoic rock and time units widely accepted today is as follows [2]:

Rank	Units of Strata	Units of Geologic Time
1st order	Erathem	Era
2nd order	System	Period
3rd order	Series	Epoch
4th order	Stage	Age

In the Phanerozoic Time Scale (Fig. 1-8) only the eras and periods are customarily listed because the named units for these ranks are employed universally, whereas named units for the two lower ranks are of more local use and may differ from continent to continent. A common exception to this is the use of worldwide epochs for the Cenozoic era, an explanation for which may be found in Chapter 17.

Relative time scale			Radiometric dates
Phanerozoic Eon	**Cenozoic Era**	**Tertiary Period** — Pleistocene Epoch	2,000,000 years B.P.
		Pliocene Epoch	12,000,000 years B.P.
		Miocene Epoch	25,000,000 years B.P.
		Oligocene Epoch	38,000,000 years B.P.
		Eocene Epoch	55,000,000 years B.P.
		Paleocene Epoch	65,000,000 years B.P.
	Mesozoic	Cretaceous Period	135,000,000 years B.P.
		Jurassic Period	180,000,000 years B.P.
		Triassic Period	225,000,000 years B.P.
	Paleozoic Era	Permian Period	275,000,000 years B.P.
		Carboniferous Period	350,000,000 years B.P.
		Devonian Period	413,000,000 years B.P.
		Silurian Period	430,000,000 years B.P.
		Ordovican Period	500,000,000 years B.P.
		Cambrian Period	600,000,000 years B.P.

Figure 1-8 The geologic column of the Phanerozoic Eon. The time scale is shown on the left page and the corresponding range of the chief vertebrate and plant groups on the right page.

Vertebrate and plant record

Eras of Phanerozoic Time. Three eras of Phanerozoic time based on conspicuous changes in the record of plant and animal life have long been recognized. These are:

1. The **Cenozoic** Era [Gr. *kainos,* recent + *zoo,* life], characterized by such advanced organisms as mammals and flowering plants.
2. The **Mesozoic** Era [Gr. *mesos,* middle + *zoo,* life], characterized by dinosaurs, marine reptiles, and ammonoids.
3. The **Paleozoic** Era [Gr. *palaeos,* ancient + *zoo,* life], characterized by invertebrate animals such as trilobites, graptolites, and tetracorals and by spore-bearing plants.

Systems of Rocks and Periods of Time. Each system of rocks is based upon a region where it was first defined and where it is a natural rock unit separated from rocks above and below by major structural breaks or by important changes in the life record. All synchronous deposits in other regions are then assigned to the same system and the period is based on the time span during which these rocks were formed.

All the generally recognized systems were established in Europe and it is a significant commentary on the importance of the fossil record that all but one were defined within a span of only 23 years following the discovery that fossils provided the key to correlation. The record is shown in Table 1-1.

Table 1-1

System	Author	Date	Type Region
Carboniferous	Conybeare	1822	England
Cretaceous	D'Halloy	1822	Belgium
Triassic	Alberti	1834	Germany
Jurassic	Humboldt	1834	Switzerland
Cambrian	Sedgwick	1835	Wales
Silurian	Murchison	1835	Wales
Devonian	Murchison and Sedgwick	1839	England
Permian	Murchison	1845	Russia
Ordovician	Lapworth	1879	Wales

Since geology was still in its infancy and these units were proposed by pioneers working in widely separated regions, it is not surprising that mutual boundaries between the systems subsequently had to be shifted in many cases, either to prevent overlap or to achieve a more complete or natural subdivision of the record. Whether the present subdivisions, which serve well in Europe and North America, reflect worldwide changes is still a moot question, but in any event it is essential to have a widely accepted scale of reference.

Systems of rocks were named for their type regions, or less commonly for some conspicuous feature of their stratigraphy, and the corresponding periods of time bear the same names. Thus the Cambrian System was so called for **Cambria,** the Roman name for Wales, and the Permian System was so named for the province of **Perm** in Russia. In a few cases the systemic name was based on some conspicuous feature of the strata in the type region. For example, the Carboniferous System was so named for the abundance of coal [L. *carbo,* coal + *ferere,* to bear], and the Triassic [L. *trias,* in threes] for its conspicuous threefold nature in central Germany, its type region.

Series of Rocks and Epochs of Time. Lesser and more local breaks in the stratigraphic record, or changes in the fossil record, have been used to subdivide the systems into series of rocks and the periods into epochs of time. Units of series rank commonly are recognizable only in a single continent, or part of a continent, consequently a particular system may have a different number of series in one region than in another. Systems may be divided into two, three, four, or even more series. These commonly are given geographic names; for example, the **Cincinnatian Series** is based on a sequence of Ordovician strata exposed around Cincinnati, Ohio. On many continents systems are divided into two or three series distinguished simply as **Lower** and **Upper,** or **Lower, Middle,** and **Upper;** the corresponding time units are then **Early, Middle,** and **Late.** In this technical usage the names are capitalized and the terms series and epoch usually dropped; thus **Upper Triassic** rocks were formed during **Late Triassic** time.

Stages of Rock and Ages of Time. The finest formal subdivisions of the Phanerozoic record commonly employed are the stages. A stage is a sequence of rock that is delimited by its fossil content; essentially it is a group of zones. Stages and their equivalent ages bear the same name, usually a geographic one derived from the type locality of the stage. A few systems, principally those of the Mesozoic, have a standard sequence of stages that can be recognized in most parts of the world and for these systems the ages have become the medium of correlation between continents.

REFERENCES

1. Puri, G. S., 1947, *The occurrence of a tropical fig in the Karewa beds at Liddarmarg, Pir Panjal range, Kashmire, with remarks on the sub-tropical forests of the Kashmir valley during the Pleistocene.* Indian Bot. Soc., Jour., v. 26, no. 3, pp. 131–135.

    ———, 1948, *A preliminary note on the Pleistocene flora of the Karewa formations of Kashmir.* Quart. Jour. Geol., Min., and Metal. Soc. of India, v. 20, no. 2, pp. 61–66.

2. International Subcommission on Stratigraphy and Terminology, 1961, *Stratigraphic classification and terminology, Statement of principles.* Intern. Geol. Congr., 21st, Copenhagen, 1960, Rept. Session, Norden, pt. 25, pp. 12–13.

Figure 2-1 Evidence of radioactivity in the black mineral samarskite from Mitchell County, North Carolina. Left, normal photograph of a slice of the mineral; right, autophoto from a negative enclosed in its black wrapper while the specimen lay upon it. It was in comparable manner that Becquerel, in 1898, discovered radioactivity. (Yale Peabody Museum.)

 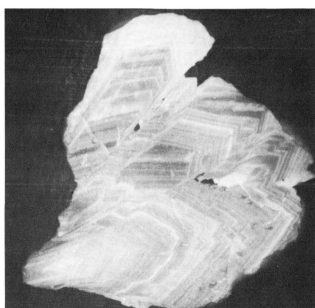

CHAPTER **2** *The Span of Geologic Time*

The poor world is almost six thousand years old.
AS YOU LIKE IT, ACT IV, SCENE 1

EARLY SPECULATION

Philosophers have speculated on the age of the earth since time immemorial, and their beliefs have ranged from that of the Brahmins of India who regarded the world as eternal, to the conviction, widely held for a time in the Judeo-Christian world, that creation occurred only about 6000 years ago.

The Christian belief, which was current in western Europe during the Middle Ages, sprang from the assumption that the Old Testament Scriptures constituted a complete and literal history of the world. It was given added authority about the middle of the seventeenth century by certain Anglican churchmen. Perhaps the first of these was John Lightfoot [9], a distinguished Greek scholar and Vice Chancellor of Cambridge University, who, in 1642, deduced that the moment of creation was "9.00 o'clock in the morning on September seventeenth." He did not at this time indicate the year of creation, but in a later chapter of the same work in 1644, he stated that it was 3928 B.C. On the basis of similar reasoning, Archbishop Ussher, Primate of Ireland, in 1658, wrote [14]:

In the beginning God created Heaven and Earth, Gen. I, V. 1, which beginning of time, according to our Chronologie, fell upon the entrance of night preceding the twentythird day of Octob. in the year of the Julian Calendar, 710 [i.e., 4004 B.C.].

And in the Great Edition of the English Bible, published in 1701, Bishop Lloyd inserted the date 4004 B.C. as a marginal commentary [11]. This was repeated in many later editions of the Authorized English Version of the Bible and thus was incorporated into the dogma of the Christian Church. For more than a century thereafter it was considered heretical to assume more than 6000 years for the formation of the earth and all its features. When, for example, in 1749, the distinguished French scholar Comte de Buffon advanced the theory that the six days of creation may have been six long epochs of time, and that the earth's surface has been shaped and reshaped by processes now going on, he was forced by the Church to recant and to declare that he accepted the Old Testament as a complete and literal history of the world.

Geology grew up under this influence and during its early years supernatural explanations were invoked for many natural phenomena. The glacial drift so widely spread over western Europe, for example, was attributed to the Biblical Flood; deep river gorges were believed to be clefts in the rocks produced by earthquakes, and mountains were believed to have arisen with tumultuous violence.

The uprooting of such fantastic beliefs began with the Scottish geologist James Hutton, whose *Theory of the Earth*, published in 1785, maintained that **the present is the key to the past** and that, given sufficient time, processes now at work could account for all the geologic features of the globe. This philosophy, which came to be known as the **doctrine of uniformitarianism,** demands an immensity of time.

SCIENTIFIC APPROACH

Early attempts to determine the length of geologic time on a scientific basis also fell wide of the mark, not because of poor logic but because not enough critical information was yet available. It was as though scientists had attempted to launch an earth satellite when gunpowder was the most powerful propellant known.

One of the first attempts was based on the rate at which sediment is transported to basins of deposition. Indeed this method had been foreseen by the Greek historian Herodotus about 450 B.C., when he observed the annual overflow of the Nile spreading a thin layer of sediment over its valley. He realized that the Nile Delta had grown by such annual increments of riverborne muds, and concluded that its building must have required many thousands of years. This was confirmed in 1854, when the foundation of the colossal statue of Rameses II at Memphis was discovered beneath 9 feet of river-laid deposits (Fig. 2-2). Since the statue is known to be about 3200 years old, the rate of deposition at this place has averaged about $3\frac{1}{2}$ inches per century. At this rate, burnt brick found some 40 feet below

Figure 2-2 The colossal statue of Rameses II at Memphis. This illustration by Bonomini in 1847 shows the statue as it lay face down, having fallen from its buried pedestal. About 5 years later excavations were made to discover the thickness of river-laid sediment about its base. The pyramids of Gizeh appear in the distance at the left and the Nile River is near the horizon at the right. Two observers, one standing and one sitting, indicate the size of the statue.

the surface at Memphis indicates that humans inhabited the region about 13,500 years ago. Yet these deposits are only a surface veneer of the great delta built by the Nile.

There are other local deposits for which the rate of deposition can be determined. The postglacial varved clays, for example, show summer and winter layers, which, like the growth rings in trees, are seasonal additions. Some of the older sedimentary formations likewise appear to be seasonally layered, notably the banded anhydrites of the Permian in West Texas and the Green River Lake beds of the Eocene in Wyoming. On this basis Bradley estimated that 2600 feet of the Green River Shale required 6,500,000 years to accumulate.

But it was also hoped that by measuring the mass of all the sedimentary rocks of the earth's crust and the increment of sediment transported annually to the sea, it could be calculated how long erosion has been going on. Unfortunately, we have no assurance that the present rate of sedimentation is an average for all geologic time; moreover, the mass of the sedimentary rocks formed since the beginning cannot be determined even approximately since in large part the deposits formed

during one geologic age have been destroyed by erosion and redeposited in another. To an unknown degree, also, these rocks have been altered beyond recognition by metamorphism. The rate of sedimentation therefore can provide no valid measure of the length of geologic time, even though it clearly indicates that the earth is very ancient.

An early attempt by Joly [5] to estimate the age of the earth was based on the amount of salt in the ocean. It assumed that (1) the primitive oceans were not salty, (2) the NaCl produced by weathering of the earth's crust is carried to the sea where nearly all of it remains in solution, and (3) the present rate of addition of salt is an average for all geologic time. Since the salinity is nearly uniform throughout the oceans, the total amount of salt now held in solution can be estimated with only a small percentage of error, and, by setting gauges at the mouths of major streams, an approximate measure of the annual increment of salt now being added could be estimated. When these calculations were first attempted, in 1899, a figure of approximately 100,000,000 years was indicated, and for a time this appeared to be a reasonable estimate of the length of geologic time; but further consideration showed that drastic revisions would be required. This is because, first, we do not know that the primeval oceans were not salty. Second, much of the salt now being carried to the sea is not derived directly from the weathering of the primary rocks but is leached from sedimentary rocks where it was stored during previous erosion cycles. Moreover, some 14,000,000 tons of it is mined annually and thus artificially returned to the streams. All such salt has been previously removed from the sea and makes no permanent addition. Furthermore, we have no assurance that the present rate of erosion is an average for geologic time. In short, there are so many uncertain quantities involved that oceanic salt offers no promise of a reliable age determination.

In a series of papers between 1862 and 1897 the distinguished physicist Lord Kelvin attempted to determine the length of geologic time on physical principles and arrived at the final conclusion that the earth is between 20 and 40 million years old. His reasoning was based on the assumption, however, that the earth has been gradually cooling down from an original molten condition from which all its interior heat is inherited. The discovery of radioactivity in 1895 swept away the basis of his calculations.

DISCOVERY OF RADIOACTIVE CHRONOMETERS

A valid means of measuring geologic time was achieved only after the discovery of radioactivity about the turn of the present century. This now enables us to determine the age, **in years,** of many individual rock masses, and thus will make it possible to transform the **geologic** time scale (Fig. 1-8) into an **absolute** time scale.

The accidental discovery of the radioactivity of uranium by Becquerel in 1895 (Fig. 2-1) was followed by intensive studies of this newly disclosed property of matter, and the understanding that has since emerged may be briefly summarized as follows.

The atom consists of a small nucleus sur-

rounded by a much larger swarm or cloud of electrons. The nucleus consists of protons, each of which carries a positive charge, and neutrons, which are without charge. Each electron carries a negative charge. All the elements are built of these three basic units—protons, neutrons, and electrons—and the differences among the elements are determined only by their number and arrangement. All the atoms of an element have a constant and unique number of protons. This is the **atomic number** of the element and determines its position in the periodic table.

The mass of the proton is the same in all the elements and is considered as unity in the scale of mass. A neutron has almost the same mass as a proton, but that of the electron is negligible. The mass, or **atomic weight,** of an atom is then the sum of the masses of all its protons and neutrons. The simplest of all the elements is hydrogen, which has only one proton (Fig. 2-3). Its atomic number is therefore 1 and its mass 1. Common carbon, on the other hand, has 6 protons and 6 neutrons; its atomic number is 6 and its atomic weight is 12. At the other extreme stand the heavy elements such as common uranium, which has 92 protons and 146 neutrons. Its atomic number is 92 and its atomic weight 238.

It has been found, however, that many of the elements occur in two or more varieties distinguished by differences in the number of neutrons. One variety of carbon, for instance, has 6 neutrons, another 7, and a third 8. The atomic weights are, respectively, 12, 13, and 14, and they are distinguished as carbon-12, carbon-13, and carbon-14 (C^{12}, C^{13}, C^{14}). Such varieties, distinguished by differences in the number of neutrons, are **isotopes.** As another example, uranium, with 92 protons, has one isotope with 143 neutrons (U^{235}) and another with 146 neutrons (U^{238}). In spite of such differences in atomic weight, the several isotopes having the same number of protons belong to the same element since their chemical characteristics are determined not by the neutrons but by the nature of the electron cloud, which in turn is a function of the number of protons in the nucleus.

Certain isotopes of some of the elements

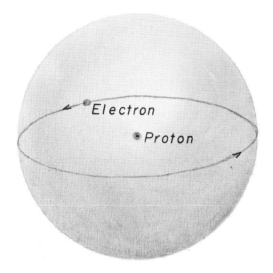

Figure 2-3 The structure of two simple atoms, hydrogen and carbon. The nucleus of hydrogen consists of a single proton with a single electron circling about it at a constant radius but in ever-changing planes, defining a spherical boundary for the atom (shaded). The nucleus of carbon-12 is composed of six protons (white) and six neutrons (black), and the electron cloud consists of six electrons. Two of the electrons revolve at a larger radius, two at an intermediate radius, and two at a lesser radius so that the atom has three electron shells.

are not quite stable, and, in these, transformations occur from time to time within the atoms that eventually reduce them to a stable state. During this process particles are discharged from the nucleus of the atom with high energy. These emitted particles produce the phenomenon known as **radioactivity.**

Uranium is one of the unstable elements, and in its radioactive disintegration to lead, three main types of radiation may be distinguished. These are: (1) **gamma rays,** similar to X-rays but generally more energetic; (2) **alpha particles,** each consisting of 2 protons and 2 neutrons; and (3) **beta particles,** high energy electrons.

Upon its emission an alpha particle, being positively charged, immediately seizes a pair of electrons and thus becomes an atom of helium. The loss of an alpha particle reduces both the atomic number and the atomic weight of the parent atom and transforms it into a different element. Thus, for example, discharge of one alpha particle from an atom of uranium transforms it into an atom of "ionium" (an isotope of thorium), and the loss of another alpha particle transforms it into an atom of radium. Thus, by a series of steps of emission of alpha particles, beta particles, and electrons, an atom of uranium is eventually transformed into an atom of lead, which is stable. Here the alchemist's dream of transforming one element into another is actually realized!

One other type of alteration may take place within the atom:

1. Capture of an electron (which is negatively charged) by a proton (which is positively charged) will neutralize its charge and transform it into a neutron. This decreases the atomic number by one without changing the atomic weight.
2. Discharge of a beta particle converts a neutron into a proton. This increases the atomic number.

The rate and mode of radioactive disintegration are definite and unique for each unstable isotope. The rate is commonly expressed by the term **half-life,** which is simply the length of time that is required for any given quantity of the material to be diminished by half. Table 2-1 indicates the type of basic information now used in geochronometry. In this table α indi-

Table 2-1 Radioactive Isotopes

Parent (element-isotope)	Daughter (element-isotope)	Disintegration Mode	Half-Life (billion years)
uranium-U^{238}	lead-Pb^{206}	$8\alpha + 6\beta$	4.51
uranium-U^{235}	lead-Pb^{207}	$7\alpha + 4\beta$	0.71
thorium-Th^{232}	lead-Pb^{208}	$6\alpha + 4\beta$	13.9
rubidium-Rb^{87}	strontium-Sr^{87}	β	50.0
potassium-K^{40}	argon-A^{40}	electron capture	12.4

After Tilton and Davis [13].

cates an alpha particle and β a beta particle, and the half-life is expressed in billions of years.

Dating by the Uranium/Lead Ratio

In 1907 the American chemist Boltwood observed that uranium and thorium minerals such as uraninite, pitchblende, and samarskite (Fig. 2-1) invariably contain both lead and helium. He therefore concluded that these are the stable end products of the radioactive disintegration of uranium and thorium, and suggested that once the rate of disintegration was known the length of time since the mineral crystallized could be calculated from the lead/uranium and lead/thorium ratios. For instance, using a simplified version of the radioactive disintegration data given in Table 2-1 (and recognizing that U^{238} is presently about 140 times more abundant than its isotope U^{235}), we calculate that 1 gram of uranium will give annually 1/7,600,000,000 gram of lead. At this rate U grams of uranium will yield U/7,600,-000,000 grams of lead in one year and in t years it will yield $t \times$ U/7,600,000,000 grams of lead. Then, the lead (Pb) produced by a given mass of uranium (U) in a certain number of years may be expressed as

$$Pb = \frac{tU}{7,600,000,000}$$

whence

$$t = \frac{Pb}{U} \times 7,600,000,000$$

For example, crystals of uraninite (an oxide of

uranium) from Branchville, Connecticut (Fig. 2-4) show a lead/uranium ratio of 0.050. Solving the equation, $t = 0.050 \times 7,600,000,000 = 380,000,000$ years.

This result is acceptable insofar as the assumptions, explicit and implicit, are valid. It soon became evident, however, that such simple calculations gave only approximately correct results, since when several dates from a single rock mass were determined it frequently turned out that they were not consistent. One of the difficulties lay in the assumption that at the time of crystallization no lead had been incorporated in the mineral. If lead had been so incorporated the calculated age would be too great. Another source of error lay in the assumption that no lead had been added from underground solutions after crystallization.

As a result of pioneer work by A. O. Nier on precision mass spectrometry, it is now possible to determine the actual amounts of each of the lead isotopes. Then independent age determinations can be calculated from the three ratios, U^{235}/Pb^{207}, U^{238}/Pb^{206}, and Th^{232}/Pb^{208}. If there had been loss of lead by solution, the different isotopes would have gone into solution in equal amounts (since they behave alike chemically) and the calculated ages would disagree, since the three forms of lead were generated at different rates; but if they agree there can have been no loss and the calculated age has a very high degree of reliability.

It is even possible to determine whether any lead was incorporated at the time of crystallization or has been added since by ground water. It is known, for example, that common lead, which might be carried in solution, is a mixture of 4 isotopes in approximately the following proportions: 1 percent Pb^{204} (which is not radiogenic), 26 percent Pb^{206}, 21 percent Pb^{207}, and 52 percent Pb^{208}. Once the amount of Pb^{204} has been determined by mass spectrometry, the proper amounts of each of the other isotopes can be deducted before age calculations are made.

Independent age determinations may also be made from the lead/uranium ratios in the common accessory mineral zircon ($ZrSiO_4$). The advantages of this method are twofold:

Figure 2-4 Granite from Branchville, Connecticut, including small black crystals (white × s) of uranite. The lighter minerals are feldspar. The lead/uranium ratio shows that this rock crystallized about 350,000,000 years ago. (Yale Peabody Museum.)

(1) zircons are ubiquitous in igneous rocks, whereas uraninite is generally restricted to pegmatite dikes; (2) zircons are less subject than common uranium minerals to chemical changes during geological vicissitudes.

Dating by the Lead/Alpha Particle Ratio

A simple dating method, first proposed in 1952 [7], is based on the lead/alpha particle ratio in zircon. Since the transformation of an atom of uranium (or thorium) into an atom of lead involves the discharge of eight alpha particles, the rate of discharge of alpha particles from a given mass of zircon gives a measure of the rate at which lead is being formed, since we know experimentally the rate at which the radioactive parents (uranium and thorium) disintegrate.

In practice, an age determination is made as follows. A fresh sample of zircon-bearing igneous rock is selected and crushed so that the grains separate and can be segregated and freed of matrix and fragments of other minerals. A small sample (about 60 milligrams) of pure zircon is thus secured for investigation.

The lead content of the sample, in parts per million, is then determined by mass spectroscopy, and the number of alpha particles emitted per hour is determined by a special type of Geiger counter. The age of the rock is then calculated from the formula

$$A = \frac{Pb \times k}{a}$$

where A is the age in millions of years, Pb is the amount of lead in parts per million, k is a constant representing the rate at which alpha particles are emitted from the parent radioactive substances (uranium and thorium), and a is the observed count of alpha particles per milligrams of the sample per hours. If all the radiation is due to uranium, $k = 2632$; if it is all due to thorium, $k = 2013$; and if both uranium and thorium are present in equal amounts, $k = 2485$. The proportions of uranium and thorium are determined spectroscopically and, whatever the ratio may be, the value of k is then readily calculated.

Because of its relative simplicity and the speed with which these measurements can be made, this method is currently being used extensively. Its proponents believe that there is little likelihood of lead having been incorporated in the zircon crystals when they formed, or of lead being added from underground solutions, because the lead atom, being about 50 percent larger than the zirconium atom, could not fit readily into the tightly packed space lattice, and because lead, being bivalent whereas zirconium is tetravalent, would require a complex and improbable double bonding. These claims are now denied by some of the specialists who express considerable doubt as to the reliability of dates based on the lead/alpha particle ratio.

Ages Based on the Potassium/Argon Ratio

Potassium is one of the common elements found in rocks. One of its isotopes, K^{40}, is radioactive and can be used in geochronometry. There are two paths by which it can disintegrate: (1) by emission of a beta particle it is transformed into calcium-40; and (2) by electron capture it is transformed into argon-40. The second mode is of greater importance in geology not only because of its use in geochronometry but also because it explains the source of the great abundance of argon-40 in the atmosphere (1 percent by volume).

Since argon is a gas, it tends to leak out of most minerals, especially the feldspars. Therefore when independent age determinations are made from the potassium/argon ratios in feldspars and in micas from the same igneous rock, the feldspars give ages far younger than do the micas. On the other hand, where this method has been applied to rock masses that have been independently dated by lead/uranium ratios, the age derived from the micas agrees well with that from the lead/uranium ratios. For this reason it appears that argon cannot escape from the space lattice of the micas, and only they are now used in measuring the K/Ar ratio.

Dating Based on the Rubidium/Strontium Ratio

The last common method for dating rocks is based on the ratio of rubidium-87 to strontium-87. Since this isotope of strontium tends to follow the rather abundant common strontium in its chemical associations, caution must be used to select minerals with a low content of common strontium or to make proper corrections for common strontium in a manner similar to that used making corrections for lead.

The minerals commonly used are those rich in potassium since rubidium is also an alkali metal and thus will tend to be associated with potassium in crystallization, whereas strontium is discriminated against. Two micas, lepidolite and biotite, are the minerals commonly used.

Radiocarbon Dating

The radioactive elements previously discussed, having immensely long half-life periods, serve well to date the ancient rocks, but their disintegration is so slow that changes within a few million years are subject to the limits of experimental error. For rocks less than 10 or 15 million years old we need radioactive isotopes of much shorter half-life periods, and, fortunately, a few are now known, but for the last 50,000 years or so, a radioactive isotope of carbon, C^{14}, is almost ideal. It was first discovered in nature by Willard Libby [8], former member of the United States Atomic Energy Commission, who by 1949 had developed the techniques of radiocarbon dating. Its contributions to late Pleistocene history and to archeology are truly spectacular.

Radiocarbon is formed in the outer atmosphere by the bombardment of cosmic rays which transform some of the atoms of nitrogen (N^{14}) into a radioactive isotope of carbon (C^{14}). This carbon readily unites with oxygen to form CO_2 and is then rapidly mixed with the normal CO_2 in the atmosphere. The radiocarbon is slightly unstable, however, and is gradually transformed back into nitrogen. This takes place at such a rate that half of any given quantity of C^{14} will disappear in 5730 years, and half the remainder will be gone in another 5730 years, and so on. Since radioactive carbon is constantly being formed in the upper atmosphere and constantly disappearing by transformation back into nitrogen, the ratio of radioactive CO_2 to normal CO_2 in the atmosphere must long since have reached a steady state, so that the ratio of radiocarbon (C^{14}) to normal carbon (C^{12}) is constant.

When used by plants, the two forms of carbon are built into their tissues in exactly this ratio, and when assimilated into animal tissue the ratio is also unchanged. Once locked in, however, the radiocarbon slowly reverts to nitrogen and the ratio of C^{14} to C^{12} gradually decreases. This ratio is then a measure of the time that has elapsed since the organism lived. If, for example, the ratio is one-half that in the atmosphere, then the organism lived about 5730 years ago, and if it is one-fourth, the organism lived about 11,460 years ago. After about 30,000 years the amount of C^{14} remaining is so small it is difficult to measure, and the probable error of age determination rises appreciably.

The reliability of radiocarbon dating can be objectively tested by applying it to the great trees in which growth rings record the last

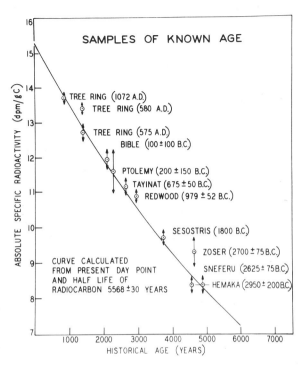

Figure 2-5 Radiometric dates of numerous articles whose dates are known by other means. The solid line indicates the predicted dates based on carbon-14 decay during the last 6000 years. The dots locate the age of historically dated objects. Where the dots fall on or near the line the agreement confirms the radiocarbon dates. (Reproduced from *Radiocarbon Dating* with permission of the author, Dr. W. F. Libby.)

3000 to 4000 years, and to archeologic objects that are independently dated. The high degree of correlation is indicated by Fig. 2-5.

Calibrating the Geologic Time Scale

The geologic time scale, based on the sequence of beds of stratified rocks (Fig. 1-8), shows the **relative** antiquity of geologic events but not their absolute ages (in years). However, if we can establish the ages of rock masses in many places in the geologic column by radiometric criteria, we can transform the geologic time scale into an absolute scale of geologic time.

Within recent decades hundreds of rock masses have been dated by one or another of the methods described, and for many of these masses two or more methods have been used

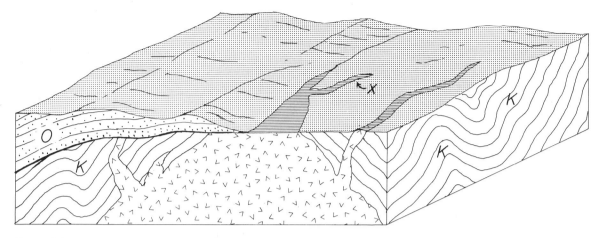

Figure 2-6 Idealized block diagram to illustrate the problem of correlating the radiometric dates with stratigraphic record.

with accordant results, so that the date can be accepted with considerable confidence. But several problems are involved in fitting these dates into the geologic time scale.

Uranium and thorium minerals are most commonly found in igneous rocks that were intruded into the stratigraphic sequence. This creates the problem of determining just when the intrusion occurred. Figure 2-6 illustrates the difficulties. In the northern Rockies the Late Cretaceous formations were deformed by the intrusion of a large granite batholith. A long period of erosion then followed during which the region was peneplaned and the granite and the deformed Cretaceous rocks were exposed. Then in Oligocene time evenly bedded sedimentary formations were spread over the region. Later erosion has exposed the relations as indicated diagrammatically in Fig. 2-6. Now the granite is dated as about 60,000,000 years old. Obviously it is younger than the Cretaceous rocks into which it was intruded, but are these the very youngest beds in the Cretaceous System? Furthermore, a very long period of erosion followed before the Oligocene deposits were laid down. The Paleocene and Eocene series preserved in other parts of the Rocky Mountain region must belong in this great gap. Thus we have a fixed geologic date, younger than the Cretaceous and older than the Oligocene, but this leaves an enormous gap in the stratigraphic record in which to place the date. Since a very long

period of erosion was required to lay bare the granite, we can infer that the intrusion had occurred long before Oligocene time and probably near the close of the Cretaceous, but we can go no further in this area. Fortunately there are many places where the lost interval between intrusion and overlap is much smaller.

More precise dates can be determined where lava has flowed out into a basin of deposition or where volcanic ash has fallen, to be included in a stratigraphic sequence. Here fossils in the adjacent sediments will prove the geologic age and radiometric data will give the corresponding absolute age. Such dates have now been established in many parts of the world. For example, nine specimens of lava imbedded in Lower Devonian strata near Presque Isle, Maine give accordant dates of 413,000,000 ± 5 years [1].

Finally, certain types of sedimentary rocks, notably black shale, have a concentration of radioactive material that give dates tied exactly into the stratigraphic section. A good example is the Chattanooga black shale that lies at the Devonian-Mississippian boundary over a large area in Tennessee and Kentucky and gives consistent dates of approximately 340,000,000 years [2].

A recent summary on the Phanerozoic time scale gives detailed accounts of radioactive dating in 337 areas [12]. Comprehensive evaluation of the data were made by Henry Faul and Arthur Holmes in 1960 [3, 4] and by J. Lawrence Kulp in 1961 [6]; from these sources we have selected the dates inserted in column 4 of Fig. 1-8.

The Precambrian rocks have been studied in many places and give vastly older dates ranging back to more than 3,500,000,000 years.

In summary, the beginning of the Pleistocene Epoch, when humans appeared on earth, falls about 2 million years ago; the beginning of the Cenozoic Era was about 65 million years ago; a good record of life, beginning with the Cambrian Period, takes us back some 600 million years; and the oldest Precambrian rocks are more than 3.5 billion years old. From its origin, the earth is now estimated to be at least 4.5 billion years old. These dates were put in perspective by the Australian geologist Mahoney [10] when he wrote:

To grasp what these figures mean, we may imagine ourselves walking down the avenue of time into the past and covering a thousand years at each pace. The first step takes us back to William the Conqueror, the second to the beginning of the Christian Era, the third to Helen of Troy, the fourth to Abraham, and the seventh to the earliest traditional history of Babylon and Egypt. . . . 130 paces takes us to Heidelberg man, and about $\frac{1}{4}$ mile to the oldest undoubted stone implements of Europe. And should we decide to continue our journey until we meet the most ancient fossil organisms, the journey would exceed 250 miles.

REFERENCES

1. Bottino, M. L., and Fullagar, P. D., 1966, *Whole-rock rubidium-strontium age of Silurian-Devonian boundary in northeastern America.* Geol. Soc. America Bull., v. 77, pp. 1167–1176.
2. Cobb, J. G., and Kulp, J. L., 1960, *U-Pb age of the Chattanooga shale.* Geol. Soc. America Bull., v. 71, pp. 223–224.
3. Faul, Henry, 1960, *Geologic time scale.* Geol. Soc. America Bull., v. 71, pp. 637–644.
4. Holmes, Arthur, 1960, *A revised geologic time scale.* Edinburgh Geol. Soc., Trans., v. 17, pt. 3, pp. 183–216.
5. Joly, John, 1901, *An estimate of the geologic age of the earth.* Smithsonian Inst. Ann. Rept., 1899, pp. 247–288.
6. Kulp, J. L., 1961, *Geologic time scale.* Science, v. 133, pp. 1105–1114.
7. Larsen, E. S., Jr., Keevil, N. B., and Harrison, H. C., 1952, *Method for determining the age of igneous rocks using the accessory minerals.* Geol. Soc. America Bull., v. 63, pp. 1045–1052.
8. Libby, W. F., 1952, *Radiocarbon dating.* Univ. of Chicago Press.
9. Lightfoot, John, 1642, *A few and new observations on the Booke of Genesis, Most of them certaine, the rest probable, all harmless, strange, and rarely heard of before.* Printed by T. Badger in London.
10. Mahoney, D. J., 1943, *The problem of the antiquity of man in Australia.* Nat. Mus. of Australia, Mem. no. 13, p. 7.
11. *Oxford Cyclopedic Concordance.* London, Oxford University Press, p. 54.
12. *The Phanerozic time-scale: A Symposium dedicated to Professor Arthur Holmes.* Edited by W. B. Harland, A. G. Smith and B. Wilcock. Quart. Jour. Geol. Soc. London, v. 120s, 1964, 458 pp.
13. Tilton, G. R., and Davis, G. L., 1959, *Geochronology.* In *Researches in Geochemistry,* Philip Abelson, Ed., New York, John Wiley & Sons, pp. 190–216.
14. Ussher, James, 1658, *Annals of the world.* London at the Sign of the Ship in St. Paul's Churchyard, Fleet Street.

Figure 3-1 The Beresovka mammoth as it lay mortally wounded by a fall into a crevasse in the Siberian ice some 38,000 years ago. It was found frozen in this position (Fig. 3-2). (Etching by Rudolf F. Zallinger.)

CHAPTER **3** *The Living Record of the Dead*

Race after race resigned their fleeting breath — The rocks alone their curious annals save.
T. A. CONRAD

Earth's Cold-Storage Locker. In August, 1900, a Russian hunter, following a wounded deer along the valley of Beresovka River in eastern Siberia, came upon the head of an elephant sticking out of the frozen ground (Fig. 3-2). He chopped off a tusk, which he later sold to a Cossack through whom the news reached St. Petersburg that another frozen mammoth had been found. An expedition organized by the National Academy of Sciences to collect the specimen reached the locality in September, 1901, after a journey of some 3000 miles, and after the back of the animal had been exposed to the warmth of two summers and part of the flesh had been gnawed away by wild animals. Excavation soon revealed, however, that the buried portion was still intact, and on one of the limbs the flesh was "dark red in color and looked as fresh as well-frozen beef or horse meat." Indeed, scraps of it were eagerly devoured by the collectors' dog team.

The animal had died in the position shown in Fig. 3-1. Broken bones (several ribs, a hip, and a shoulder blade), as well as much clotted blood in the chest, and unswallowed food in its jaws, suggest a sudden and violent death; and the wall of ice beside it indicated that the animal had fallen into a crevasse and died struggling to extricate itself.

The presence of this elephant 60 miles within the Arctic Circle, and more than 2000 miles north of the present range of living elephants, was not more surprising than its anatomical peculiarities. Unlike all living elephants, it bore a thick coat of reddish-brown wool interspersed with long black contour hair. Its head was narrow and had a distinctive high-crowned profile. It was, in short, a woolly

Figure 3-2 Frozen carcass of the Beresovka mammoth after partial excavation. (After O. Herz in Proc. U. S. Nat. Museum.)

mammoth belonging to an extinct species that ranged widely over the northern parts of Eurasia and North America during the last great Ice Age, and it was as well adapted as the reindeer to the frigid climate of the time.

Most of the Beresovka specimen is on display in the Zoological Museum at the University in Leningrad, but pieces of the dried flesh, clotted blood, and woolly skin were presented to the United States Government and put on display in the U. S. National Museum.

The woolly mammoth has been extinct so long that no allusions to it appear in the legends of any living people, yet an unmistakable picture of it was left by prehistoric man on the walls of a cavern at Combarelles, France (Fig. 3-3). Beside the nearly complete frozen carcass just described, some 50 others, less complete, have been found in Siberia and in Alaska, preserved since prehistoric times in nature's cold-storage locker. Radiocarbon dates for several of these run as follows [14]:

Specimen locality	Radiocarbon age in years
Moskovaya, Yenesei Valley	32,500+
Sanga Jurjak, Yenesei Valley	39,000+
Lena River Delta	33,000+
Taimyr Peninsula	11,450 ± 250
Beresovka Valley	39,000+
Gyda, Kakut	33,500 ± 1,000

A similarly preserved extinct bison in Alaska is known from radiocarbon dating to have died 28,000 years ago [12].

Images in Stone. Less spectacular but no less real are the bones and shells of animals and the remains of plants entombed in solid rock in many parts of the world. They are the remains of creatures that lived and died and were buried in sediment that later solidified into rock. The ancients observed such "images in stone" and speculated on their meaning. The Romans called them **fossils,** because they were objects dug up [L. *fossilis*, from *fodere*, to dig], and we have borrowed that name. The study of fossils is the science of **paleontology** [Gr. *palaios*, ancient + *onta*, existing things].

The Nature of Fossils. The term fossil was for a time applied to a wide range of "curios"

found in the rocks, both organic and mineral; but it is now correctly used only for the remains of once-living things naturally preserved from the past.

Extinction is not a criterion, since many existing species were already present millions of years ago and are preserved as fossils in Cenozoic rocks while, on the other hand, some species have become extinct within historic time. A good example is the dodo, a giant flightless pigeon, which lived on Mauritius Island in the Indian Ocean until 1681 when the last survivors were killed by white settlers. Remains of this bird now preserved in the British Museum are not fossils because they were artificially preserved. Nevertheless, if remains of the dodo were found in the Cenozoic rocks of the Islands they would be considered fossils.

Figure 3-3 Drawing of the woolly mammoth by prehistoric man on the wall of a cave at Combarelles, France, and modern restoration of the same by Charles R. Knight, based on this and the Beresovka specimen.

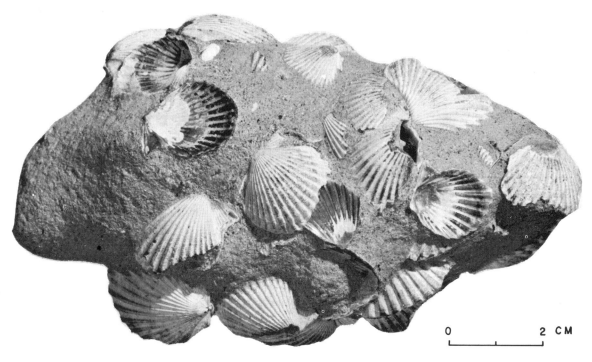

Figure 3-4 Shells of the bay scallop in limestone nodule from postglacial marine sediments in Long Island Sound. The shells are only 4500 years old. (Yale Peabody Museum.)

Remains do not have to be petrified to qualify as fossils. The frozen carcasses of mammoths are fossil, as are the shells in marine deposits of Miocene age in Maryland where the sediment has not been hardened into rock nor the shells in it mineralized. Today these shells wash from headlands into Chesapeake Bay where they accumulate with recent shells in the beach litter; unless one is familiar with the Miocene species it is difficult to distinguish fossil from modern specimens. On the other hand, hard limestone nodules dredged from Long Island Sound contain shells of the present-day fauna (Fig. 3-4) and have been dated by radiocarbon as only 4500 years old [17]. Even the common saying that "if the remains stink they belong to biology, if not, to paleontology" fails as a criterion in the presence of a thawing piece of mammoth! Some degree of antiquity is implied in the definition of a fossil, but since the present grades insensibly into the past no sharp line of distinction need be drawn.

Natural derivatives of organisms such as tar and petroleum or altered masses of organic matter such as coal are not considered to be fossils. Not uncommonly, however, spores, bits of wood, and cuticles or imprints of leaves are recognized as fossils **within** the coal. In general, a fossil is the remains or trace of an organism that can, if well enough preserved, be identified by its form and structure as a particular kind of animal or plant.

Although the word fossil, as a noun, is correctly applied only to the remains or traces of once-living things, it is commonly used as an adjective in a figurative sense for the products of both inorganic and organic phenomena preserved in rocks. Thus, for example, we speak of fossil mudcracks, fossil rain prints, fossil soils, fossil dunes, and fossil fuels.

With similar poetic license we sometimes speak of a species of existing animal or plant as a "living fossil," implying that it belongs to the remote geologic past but has somehow managed to survive. For example, when the coelacanth fish *Latimeria* was caught off the central east coast of Africa in 1939 it was hailed as a living fossil since it belongs to a tribe of lobe-finned fish that had its heyday in the Devonian Period and is not known in post-Mesozoic rocks.

TYPES OF FOSSILIZATION

Actual Preservation. The preservation of flesh and other soft tissues is possible only where bacterial action and decay have been almost miraculously inhibited. Cold storage in the Arctic ice is one means of such preservation. Oil seeps have also, in a few instances, afforded sufficiently antiseptic conditions. A remarkably complete carcass of the woolly rhinoceros was excavated from the muck of an oil seep in the district of Sarunia, Poland, in 1930, and a similar specimen from another oil seep is preserved in the museum at Lemberg, Poland, where the skeleton and the mounted skin stand side by side. The woolly rhinoceros was a contemporary of the woolly mammoth and was similarly adapted to arctic conditions; it has also been found preserved in ice.

Although soft tissues are rarely preserved and occur almost exclusively in late Pleistocene deposits, the hard parts such as bone, shells, or woody tissue have commonly been preserved with little change since Cretaceous time. Logs embedded in Cretaceous clays or in Eocene lignites, for example, have suffered slight discoloration but still retain the woody structure and burn readily. Marine shells in some of the Late Cretaceous formations likewise show little alteration, retaining the original microstructure and, rarely, even the color pattern in spite of nearly 100 million years of antiquity.

Petrifaction. In the older rocks, the hard parts alone are usually preserved, and these commonly appear to be turned to stone, as indeed they are. Such fossils are said to be **petrified** [L. *petra* stone + *facere*, to make]. The change from the original condition is accomplished in one of the following ways.

1. Permineralization. If the original structure is porous, as are bones and many kinds of shells, mineral matter may be added from the

Figure 3-5 A skeleton in the rock. This complete skeleton of a ground sloth, about 8 feet in length, was discovered in a stream bank in Argentina. (Chicago Museum of Natural History.)

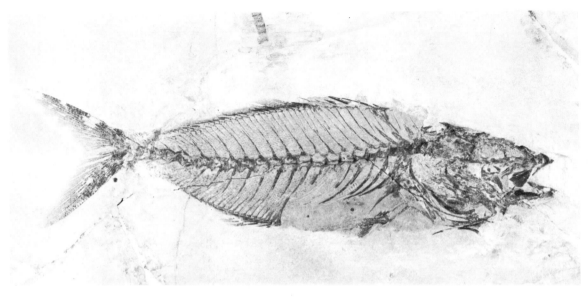

Figure 3-6 A fossil fish related to the pompano, from Eocene beds at Monte Bolca, Italy. (Yale Peabody Museum.)

Figure 3-7 Petrified wood from the Fossil Forest of the Yellowstone National Park. Left, a slightly enlarged piece showing growth rings; right, a thin section cut from this piece and enlarged about 12 times to show the cells in parts of 3 growth rings. (Yale Peabody Museum.)

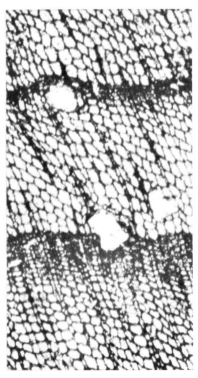

underground water to fill up all the voids without altering the original substances. This makes the object heavier, more compact, and more stonelike, at the same time protecting it from the air or solutions that would dissolve and destroy it. Such fossils are said to be **permineralized.** Petrified bones are commonly of this category (Figs. 3-5 and 3-6).

Wood and other plant structures may be spectacularly preserved by a very delicate permineralization in which cell cavities and the spaces between cells are filled with the replacing material, completely enclosing and protecting the tissues of the cell walls. At one time it was thought that the cell walls had been replaced so perfectly as to preserve their microstructure. But paleobotanists have developed methods of etching away the secondary mineral substances, such as silica, and recovering the cell walls intact; in this way they have demonstrated that the highly resistant plant tissue has been preserved. The method has made possible great advances in the knowledge of microstructures and, consequently, in the relationships of ancient plants [4]. Wood of the petrified forests of the Yellowstone (Fig. 3-7) illustrates this remarkable type of preservation.

2. Replacement. The original substance may, however, be dissolved and **replaced** by mineral matter of a different sort. The substitution may result in the loss of all internal structure while preserving the gross form of the organic object, as, for example, a coral or shell replaced by quartz, calcium carbonate, or dolomite. Such a fossil is a false replica or **pseudomorph,** showing only the external form (Fig. 3-8).

Where calcareous shells are embedded in limestone, some subtle chemical difference between the shells and the matrix commonly causes the shells to be replaced by silica while the surrounding stone is unaltered. In the weathering of such rocks the fossils are freed

Figure 3-8 Silicified shells etched out of Ordovician limestone from Lake St. John, Quebec. Left, a block of the limestone with fossils partially etched by weather; right, snail shells etched free by treatment with HCl. (Yale Peabody Museum.)

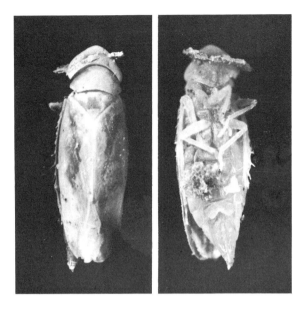

Figure 3-9 A leaf-hopper preserved as a siliceous pseudomorph, etched free by formic acid from a Miocene limestone nodule from Yermo, California. The specimen is 3.3 mm long. (A. R. Palmer, U. S. Geological Survey.)

Figure 3-10 Fossil fern leaves preserved as a carbonaceous film in shales in the Coal Measures of Germany. (Yale Peabody Museum.)

and may be collected as siliceous pseudo-morphs on the surface. It is also possible to free them artificially by dissolving pieces of the limestone in hydrochloric, acetic, or formic acid (Fig. 3-8).

Calcareous nodules from a Miocene lake deposit near Yermo, California have thus yielded amazingly preserved insects and spiders in which all chitinous structures have been replaced by silica [15]. Freed by the use of formic acid, they reveal such delicate structures as the internal trachaea, the muscles, the heart, the alimentary canal, and genitalia (Fig. 3-9).

3. *Distillation.* The volatile elements of organic material may be distilled away, leaving a residue of carbon to record the form of the object. Leaves are generally preserved in this way, and the beautiful "carbon copies" of fern leaves in the shale above some of the coal beds give a vivid picture of the ancient plants (Fig. 3-10). The original waxy cuticle of the leaf may also be preserved in addition to the carbonaceous film. Fossil cuticles, removed and affixed to thin sheets of cellulose acetate and studied under the microscope, reveal details of the cellular structure of the leaf (Fig. 3-11). Animal tissue is recorded, although

Figure 3-11 Left, leaf of *Sapindus eoligniticus,* about natural size, removed intact from Eocene clay, Henry County, Tennessee. Right, piece of cuticle from lower epidermis of *Sapindus,* magnified 400 times, showing epidermal cells, guard cells, and stomata. (Courtesy of David Dilcher.)

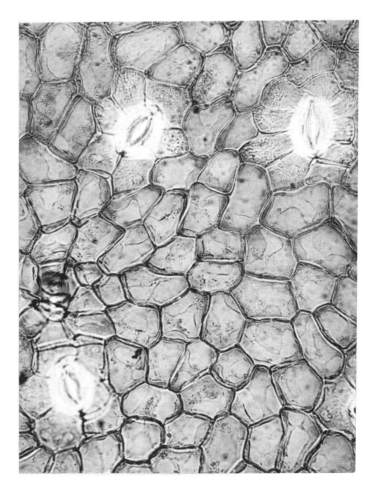

O ____ I CM

Figure 3-12 Fossil ichthyosaur, a marine reptile from the Lower Jurassic shales of Holzmaden, Germany. The skeleton is permineralized, the flesh reduced to a film of carbon. The species ranges from 8 to 10 feet long. (American Museum of Natural History.)

Figure 3-13 Left, natural mold of a brachiopod shell in sandstone (natural size); right, artificial cast from the same made with a molding compound. (Yale Peabody Museum.)

rarely, in carbonaceous residue (Fig. 3-12).

Molds, Casts, and Imprints. Shells or other organic structures embedded in rock may later be dissolved by percolating ground water, leaving an open space that preserves the form of the object. This hole is a **natural mold.** By pressing into it a plastic substance such as dental wax, we may obtain an artificial cast or replica of the original (Fig. 3-13). Percolating subsurface water has in many instances filled such holes with some other mineral substance, usually quartz, thus producing **natural casts.**

The terminology applied to fossils follows that of foundry practice, the **cast** being the replica of the original, its counterpart being the **mold.** Hollow objects may have, besides the external mold, an internal mold or **core.** The molds of thin objects like leaves are commonly spoken of as **impressions** (Fig. 3-14). The pattern of the scales in the skin of some dinosaurs is thus well shown by impressions in the matrix, although no other trace of the skin is preserved (Fig. 3-15).

Among the most remarkable natural molds

Figure 3-14 Right, imprint of a fern leaf in a concretion of Pennsylvanian age from Mazon Creek, Illinois. (Yale Peabody Museum.)

Figure 3-15 Below, portion of the tail of a herbivorous dinosaur, *Corythosaurus,* with part of the skin preserved. Late Cretaceous, Red Deer Valley, Alberta. (C. W. Gilmore, U. S. National Museum.)

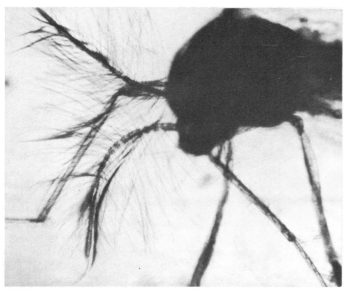

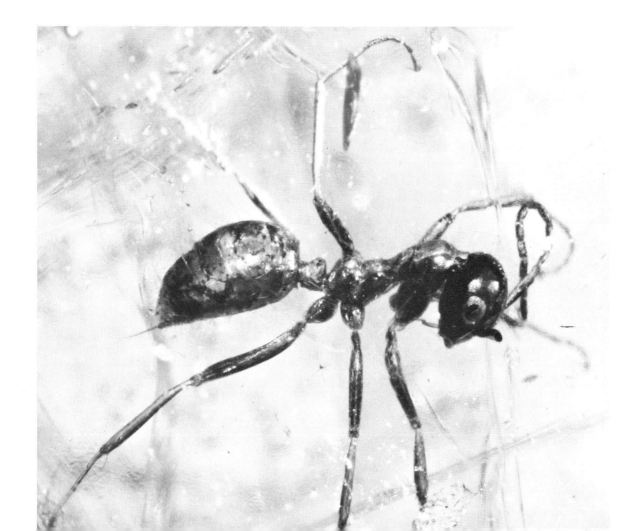

Figure 3-16 Insects preserved in amber. Right, two tiny specimens enlarged 7 times; left, the upper specimen enlarged about 70 times, showing the microscopic filaments on its antennae. From Oligocene beds on the Baltic Coast of Germany. (Yale Peabody Museum.)

Figure 3-17 Worker ant preserved in Cretaceous amber. This oldest known fossil ant marks the first appearance of social insects in the geologic record and constitutes an evolutionary link between the ant family and tiphiid wasps. From the Upper Cretaceous Magothy Formation, Cliffwood, New Jersey. (Courtesy of F. M. Carpenter.)

are those of insects and spiders preserved in amber in the Oligocene beds along the Baltic coast of Germany (Fig. 3-16). Amber is the dried and hardened residue of the resins exuded by evergreen trees such as the white fir. The insects and spiders were entangled while it was still soft and have since been entombed for more than 20 million years. For the most part the organic tissues have dried and almost disappeared, though occasionally dried muscles and even the viscera can be recognized; but the sharp hollow molds in the transparent amber retain the shape of the insects with extraordinary faithfulness.

Although the Baltic deposits are the best known, fossiliferous amber has been discovered in a number of places in beds of Cenozoic and Cretaceous age. Amber preservation is particularly important in working out the history of insects (Fig. 3-17), which are rarely well preserved in other media.

Trace Fossils. Structures such as burrows, tracks, and trails left in the sediment by living organisms are considered fossils. They supplement other fossil remains in interesting ways. From the tracks of a land animal we can tell whether it was bipedal or quadripedal, whether it moved by running, leaping, or sprawling, and whether it was agile or ponderous.

Thus we know that although the bipedal dinosaurs were shaped much like the kangaroos, they nevertheless ran like ostriches. Tracks alone give us some of the earliest records of land vertebrates, and in some formations tracks alone prove that animals were once abundant where no skeletal remains are preserved (Fig. 3-18).

The greatest abundance and diversity of trace fossils are found in strata deposited in shallow-water marine environments. This fact in itself is often helpful in determining whether a particular sequence of strata is marine or nonmarine. Marine sediment is commonly mixed and its bedding disrupted by organisms that burrow or plow through it while it is still a soft, water-saturated deposit on the sea floor. Reworked sediments like those in Fig. 3-19 tell us that burrowing organisms were plentiful in the environment

Figure 3-18 Dinosaur tracks exposed on a bedding plane of shaly sandstone of Triassic Age. At Dinosaur State Park, Rocky Hill, Connecticut. All the tracks on this layer are of the same species; tracks of this and other species occur on other layers. Over 6000 square feet of rock showing tracks has been exposed at this unusual site. (John Howard for Yale Peabody Museum.)

even though no parts of them may be preserved. More informative than the mixed sediment are burrows and trails of well-defined structure that occur repeatedly in strata (Fig. 3-20). The major problem with these, as with all trace fossils, is their identification with a particular animal. When an identification can be made trace fossils commonly prove to be

of great value in the interpretation of ancient environments. They may also help establish the geologic record of soft-bodied creatures that have left no actual remains.

Fossil excrement constitutes another class of trace fossils, known as **coprolites.** These often contain undigested hard parts of animals or plants that were devoured. Wherever asso-

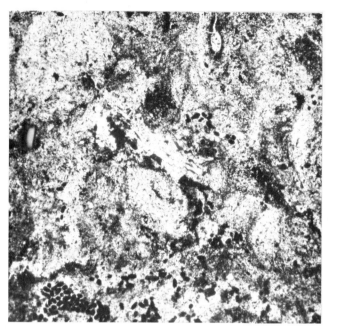
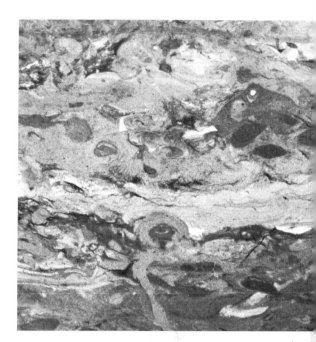

Figure 3-19 Sediment disturbed by organisms. Left, a vertical section of about 2 inches of bottom sediment from the intertidal zone in Buzzards Bay, Cape Cod. The small black objects are fecal pellets. Right, a vertical thin section from Silurian marine strata in Nova Scotia. Both the same scale, slightly enlarged. The trails and swirled structure are characteristic. (Courtesy of Donald C. Rhoads.)

Figure 3-20 Feeding tracks of a bottom-dwelling invertebrate that derived its nourishment from the surface sediment of the sea floor. From deep water deposits of Early Eocene beds in Switzerland. Almost identical traces, made by acorn worms, have been recorded in photographs of present day deep-sea bottoms. (Yale Peabody Museum.)

ciated with skeletal remains in such a way that their source can be recognized, coprolites are of special significance for the light they throw on the food and feeding habits of the animal in question. For example, the dried dung found with fossil ground-sloths gives the only proof we have of the type of vegetation preferred by that race of extinct giants (Fig. 3-21).

Figure 3-21 Floor of Rampart Cave in Lake Mead National Recreation Area, Nevada, covered by the dung of an extinct ground sloth. (R. K. Grater, courtesy of National Park Service.)

Conditions Favoring Preservation

The buffalo carcasses strewn over the plains in uncounted numbers two generations ago have left hardly a trace. The flesh was devoured soon after death by vultures, predators, and wolves and even the bones and teeth have crumbled to dust. On the other hand, excavations in downtown New Haven, Connecticut, recently uncovered remains of a wharf dating from colonial times. The black mud around the old pilings, once the bottom of the harbor, contained fresh-looking clam and oyster shells and a variety of trash discarded by the colonists, including delicate chicken bones and a boot of thin leather, intact and pliable enough to wear. These examples suggest that special conditions are required for preservation. Two such conditions are paramount.

Possession of Hard Parts. Animals or plants with hard parts have an overwhelming advantage in the chance of preservation. Leaves are commonly preserved as imprints or as carbonized films, but flesh decays so readily that preservation is rarely possible. Shells, wood, bones, and teeth, on the contrary, form most of the fossil record. It is not surprising that many groups of soft-bodied animals such as worms and jellyfish have left almost no geologic record; the rare instances in which they were preserved, as in the Mid-Cambrian black shale of Mount Wapta (Chapter 8) serve only to emphasize what a wealth of animal life existed in the ancient seas that is virtually unrecorded.

Quick Burial. A carcass left exposed after death is almost sure to be torn apart or devoured by carnivores or other scavengers, and if it escapes these larger enemies, bacteria insure the decay of all but the hard parts, and even they disintegrate after a few years if exposed to the weather. If buried under moist sediment, however, weathering is prevented, decay is greatly reduced, and scavengers cannot disturb the remains. For these reasons burial soon after death is the most important condition favoring preservation.

An obvious corollary to the condition of quick burial is that the buried organisms remain undisturbed over a great length of time. Deeper and more permanent burial takes place where sediment is accumulating continuously, or relatively rapidly, and where it is not interrupted by episodes of erosion. The sea bottom affords such optimum conditions and the shallow seas of the continental shelves, abounding in organisms, are today the greatest region for the preservation of organic remains. The ancient shallow marginal seas and those that spread over the continents many times in the past were likewise optimum areas for preservation and **by far the greater part of the fossil record is preserved in marine strata** formed from their accumulated sediments.

Water-laid sediment is also the more common agent of rapid burial on the land and most of the record of ancient terrestrial life is found in the strata formed from deposits on flood plains of aggrading rivers, in swamps, and in lakes. But here continuous sedimentation is usually the exception rather than rule. Floods, which accomplish most of the sediment transport and deposition on land, are also powerful agents of erosion, and what one flood deposits another may take away.

Whereas relatively stable conditions favoring preservation exist in the sea, the conditions above sea level favor erosion. Terrestrial deposits preserved in the geologic record are dominantly those that formed at or near sea-level or those that owe their preservation to great sediment traps like the fault-bordered basins of the Triassic (p. 329). The fossil record is heavily biased toward the environments of shallow seas and the edges of the lands, but these apparently have been the areas of greatest organic diversity, so our re-construction of life history is probably not too seriously distorted. Yet even in the environments that favor preservation, relatively few of the organisms that die are buried. Still fewer remain buried long enough, or survive the rigors of change after burial, to become a part of the fossil record. As scanty and unrepresentative of the total pageant of life as it may be, this record is a complex and fascinating one that deserves closer study.

Burial in the Sea. Marine sedimentary rocks of the Phanerozoic are rarely without fossils of some kind. The reason for this is apparent when we consider the immense productivity of the sea, with its swarming microscopic plankton, the hordes of organisms in the water column, and the vast communities of organisms living on and in the sediments of the sea floor. Bottom-dwelling animals with hard parts are naturally the most likely to be preserved, but the constant rain of dead organisms and discarded shells and husks from the floating and swimming populations of the water column also contributes to the accumulation of organic remains on the bottom, and some of it is commonly preserved.

The sea, like the land, has its scavengers and its bacteria that immediately consume and decompose the soft parts and scatter the hard parts unless the remains are quickly buried. Hard parts may also be ground to unrecognizable grains in the mill of wave action along the shores. But vast areas of the shallow sea lie below effective wave base and are undisturbed except by the more intense waves generated by storms and earthquakes, or by shifts in currents. Here it is common to find the sea bottom heavily populated by animals living both on the bottom (epifauna) and in the sediment just beneath the bottom (infauna). Very few of these animals can survive under conditions of rapid sedimentation; thus parts of the sea floor that are heavily and rapidly blanketed by sediment are largely devoid of bottom life. Rapid spread of sediment into areas where bottom-dwelling organisms are living may bury them alive and preserve them in the environment in which they lived. Other drastic changes in the environment may locally kill great quanti-

Figure 3-22 Clams washed ashore at Ocean City, New Jersey, in the winter of 1961. The mass killing was due to the coincidence of low tides and extreme cold. (Courtesy of World Wide Photos.)

ties of organisms whose remains are then concentrated in sea floor depressions or spread about in windrows on the sea floor.

In February, 1961, the beaches of Ocean City, New Jersey, were buried under tons of clams, chiefly surf clams and quahogs (Fig. 3-22), which normally live in the muddy sand bottom just off shore beyond the area of breaking waves. A number of other beaches along the coast were similarly affected. This mass killing of millions of clams was caused by the coincidence of a period of exceptionally cold weather with a period of lower than normal tides. The clams, either killed or rendered moribund by the cold, could not burrow to escape the usual scour and shifting of the bottom due to changes in intensity of the incoming waves. As a result, wave action excavated

and winnowed them out of the bottom sediment and threw them on the beach, where they formed a clam chowder that cost the taxpayers of Ocean City thousands of dollars to remove.

Local mass killings of marine animals from the cold are yearly occurrences in many parts of the world; hurricanes and the blooms of microorganisms that cause the destructive "red tides" are other common agents of mass killing in the sea. Although we think of these extreme conditions as both local and unusual, in terms of geologic time they are probably a major factor in the formation of the large concentrations of marine fossils frequently found in the Phanerozoic record. The scale of some of these concentrations may be illustrated by an occurrence of the extinct, epifaunal bivalve

Gervillia in the Upper Cretaceous rocks of South Dakota. Great masses of this small, elongate shell are found concentrated in a layer that extends over an area of some 400 square miles. Local counts of specimens, counting two valves per specimen, show that the density over 1 square mile of the layer in the heart of the area amounts to over 100 million specimens. Even if the entire area has only half this concentration, the number of specimens must at least be 20 billion!

Some marine animals and plants have tended to live together in great colonies and their lime-secreting bodies have collectively built lasting monuments to themselves in the form of organic reefs. Although popularly called coral reefs, the reef-building habit was adopted by many kinds of animals and a variety of algae, not just corals, hence the term organic reef is preferable. In every period of earth history from the Cambrian to the present day, organic reefs have formed in the warmer shallow seas of the world; even in the Cryptozoic (Chapter 7) we find similar bodies composed of concentrically laminated sedimentary structures, called stromatolites, which may owe their peculiar form to some organic agent. Organic reefs range in size from small patch reefs a few feet across to great continuous masses like our modern atolls and barrier reefs that extend for tens of miles. All are easily preserved and can be readily identified by their form, structure, and peculiar associations of animals and algae. The discovery that fossil reefs are a common reservoir rock for the accumulation of petroleum has led to extensive exploration for them and to detailed studies of their structure and paleoecology; thus they are probably better known and understood than any other marine environment.

The sea is also the site of some unusual examples of preservation. Perhaps best known is the incredible throng of delicate soft-bodied creatures recorded in thin films of carbon on black Middle Cambrian shale from Mt. Wapta in British Columbia, described in Chapter 8. Similar thin films of carbon were discovered surrounding the bones of some of the ichthyosaurs (Fig. 3-12) found in Lower Jurassic slate at Holzmaden, Germany. Here hundreds of these marine reptiles have been excavated in an excellent state of preservation, some even containing skeletons of unborn young and others the remains of their meals of squidlike belemnites. The carbonized film surrounding some of the ichthyosaur skeletons is so close in appearance to the black shale matrix of the fossils that it went undetected for years. It is said to have been discovered by accident in the 1890s when a glass of water overturned on a slab during its preparation and the differential drying of the water revealed the less porous area of carbonization [13]. Until this discovery the fishlike outline of the ichthyosaur was unknown and these reptiles had been reconstructed with lizardlike bodies.

The black muck of the sea bottom on which the ichthyosaur carcasses accumulated was obviously toxic. No scavengers were present to tear apart the bodies and scatter the bones; indeed the Holzmaden fauna lacks any kind of indigenous bottom-dwelling animals. Swimmers are dominant and include plesiosaurs, marine crocodiles, a few fish, squidlike cephalopods, and a few others in addition to the great numbers of ichthyosaurs. A small flying reptile and some sea lilies attached to a piece of wood are obviously introduced. Sulfur compounds in the shale indicate that the bottom was made toxic by hydrogen sulfide. Stagnant and toxic areas of black mud are known in areas of present-day seas where the bottom is in a depression or otherwise cut off from circulation of oxygen-bearing water. What attracted the ichthyosaurs and other animals into the stagnant bay that apparently existed at the Holzmaden site is not known, but their death was most likely due to toxic or poorly oxygenated water and their preservation was assured by the antiseptic mud that buried them.

Burial on the Land. Water-borne sediments are so much more widely distributed than all other kinds, that they include the great majority of all nonmarine fossils. Flooding streams drown and bury their victims in shifting channel sands or in the silts and muds of the flood plains (Fig. 3-23). Whether the organisms are driftlogs or leaves, shells of fresh-water mollusks, or bodies of animals, the remains are.

Figure 3-23 A modern forest killed and in process of being buried by alluvium on the delta of the Yahtse River, Alaska. (E. S. Dana.)

Figure 3-24 Bone bed in Lower Miocene deposits of Agate, Nebraska, showing a remarkable concentration of bones of a small twin-horned rhinoceros (Diceratherium). Slab preserved in the museum at Scotts Bluff National Monument, Gehring, Nebraska. (Courtesy of John W. Henneberger.)

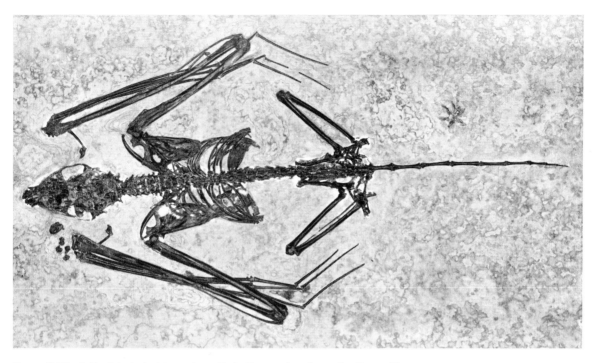

Figure 3-25 Articulated skeleton of an Early Eocene bat from the Green River Formation in southwest Wyoming. (×1.5) (Glenn L. Jepsen.)

likely to be concentrated and buried in low places and covered with sediment or standing water as the flood recedes. This probably accounts for the extraordinary concentration of fossil skeletons in some ancient fluvial deposits (Fig. 3-24).

For many years the vast exposures of continental Cenozoic beds in the American Western Interior, which have yielded magnificent vertebrate faunas (Chapter 19), were thought to be lake deposits. But as both the fossils and the sediments enclosing them became better known, it became evident that most of the varicolored deposits making up the many western badlands were alluvial deposits characterized by irregularly bedded clays and silts interspersed with discontinuous, lenticular bodies of sand representing channel deposits (Figs. 17-15 and 17-16). Bones and teeth of large and small quadrupeds and wood, the principal fossils, are found chiefly in the channel sands.

Here and there among the Cenozoic alluvial deposits are **lake deposits**, some covering hundreds of square miles. The thin, regular bedding characteristically formed in bodies of standing water contrasts sharply with the less regular stratification of fluvial deposits, and so too do the kinds of fossils and their type of preservation. Lake deposits typically preserve the more delicate, smaller members of the local aquatic and terrestrial flora and fauna. Leaves, insects, fish, freshwater mollusks, and aquatic arthropods are commonly found and, infrequently, birds and bats are preserved (Fig. 3-25).

In southwestern Wyoming and adjacent parts of Colorado and Utah the extensive alluvial deposits of the Eocene pass laterally into a vast deposit of lake beds known as the Green River Formation (p. 417). Where they can be found together, alluvial and lake deposits commonly afford the most complete documentation possible of the land life of the time. But

Figure 3-26 Asphalt trap at Rancho La Brea in Los Angeles. Above, reconstruction of a scene at the asphalt seep during Late Pleistocene time by Charles R. Knight. In left foreground a saber-toothed tiger snarls at a group of giant ground sloths while vultures wait overhead, and in the middle distance a herd of Imperial elephants is in view. Lower left, a modern jack rabbit recently trapped in the asphalt; lower right, a mass of fossil bones preserved in the asphalt. (Upper picture courtesy of the American Museum of Natural History; lower pictures courtesy of the Los Angeles Museum.)

much less common types of burial have also enriched our knowledge of land animals; some of these are described in the following paragraphs.

Bogs and deposits of quicksand or of asphalt form natural death traps in which animals commonly mire and are quickly covered. The peat bogs that formed in the northern states after the last Pleistocene ice sheet melted away must have been a particular hazard to the elephants that still roamed the United States, for isolated skeletons of mastodons and mammoths have been found in the peat at many places. Perhaps the most remarkable deposit of Pleistocene vertebrate fossils is that of Big Bone "Lick," a bog deposit about 20 miles south of Cincinnati from which more than 100 mastodons have been recovered along with skeletons of bison, moose, reindeer, and the wild horse. This locality was known to Thomas Jefferson, who sponsored extensive excavation and, during his presidency, reserved a room in the Presidential Mansion as a museum for the fossils he had secured.

In 1943 an artesian spring deposit of Pliocene date was discovered in Meade County, Kansas, containing a large fauna of mammals that had either been trapped by quicksand or were mired in the bog around the spring [9]. The fauna includes elephants, giant camels, wild horses, wolves, and great cats.

In 1942 a locality was discovered in the Pleistocene beds of San Pedro Valley, Arizona, where mastodons had mired in the marsh about a salt lake, their limb bones still remaining upright in the deposit [6].

In several parts of the world natural **petroleum seeps** have formed traps as the more volatile constituents evaporated away leaving pools of sticky asphalt. The way in which they work is illustrated by an incident that occurred at Mount Pleasant, Michigan in 1939, when four small boys ventured onto a pool of such asphalt discharged as waste from a coke plant [11]. After a few steps the first boy began to sink slowly. When two companions rushed to aid they, too, were soon trapped like flies on sticky paper. The fourth lad, fortunately, ran for help and the others were rescued, but by the time help arrived the first boy was submerged up to his neck and the other two were waist-deep. The whole incident had happened within about half an hour!

An amazing deposit of extinct fossil mammals trapped in this way is found at Rancho La Brea in Los Angeles (Fig. 3-26). It was the scene of extensive oil seeps since at least early Pleistocene time, and animals coming to the

Figure 3-27 Volcanic ash fall from the eruption of Mt. Katmai, Alaska, in 1912. The white line indicates the profile of the mountain before eruption. Timber in the foreground was killed and largely buried by the ash fall. It is estimated that 5 cubic miles of ash and pumice were spread over an area some 200 miles across. (National Geographic Magazine.)

seeps for water or attempting to cross patches of asphalt were trapped, like the Michigan boys. Their struggles and death cries attracted carnivores and scavengers, which in turn became trapped. Bones of these victims have been recovered by the hundreds of thousands as the asphalt was excavated for use. No flesh has been found here, but in similar oil seeps in Poland almost entire carcasses of the extinct woolly rhinoceros have been recovered.

Caves and underground caverns are another source of fine fossils. They were often used as lairs by carnivores which dragged their prey in to devour the flesh and leave the bones. Here the remains are largely protected from the weather and, not uncommonly, they are soon covered by a limy deposit precipitated from dripping water. Some sink holes have served as traps into which animals have fallen. One such, near Cumberland, Maryland, has yielded the bones of 46 species of vertebrate animals, including wolves, bears, mastodons, tapirs, wild horses, deer, and antelope [7].

The cave shown in Fig. 3-21 is floored with deposits more than 20 feet deep in which skeletons of ground sloths occur along with

Figure 3-28 Imprints of fern leaves on the under surface of a thin lava flow, formed during the eruption of Kilauea in 1868. (Harold S. Palmer.)

bones of the mountain lion, an extinct horse, mountain sheep, and numerous other creatures [8]. This cave is so dry that decay is inhibited, and considerable portions of the skin and hair of the extinct ground sloths have been recovered.

During the Ice Age primitive man took shelter in caves of this sort in various parts of Eurasia, tossing the refuse outside the entrance to accumulate, layer upon layer, through the centuries of occupation. Such deposits include the bones of animals he used for food along with discarded or broken implements and ornaments. At the same time primitive man commonly buried his dead in the caves where they would be safe from scavenging wild beasts. Such caves provide the most important records of prehistoric peoples and their culture (Fig. 20-5).

Falls of volcanic ash (Fig. 3-27) commonly kill and bury, as the tragic fate of Pompeii reminds us. Many fine fossil deposits in the Cenozoic rocks of western United States are in volcanic ash, like those of the John Day Basin in Oregon and Lake Florissant in Colorado. Flows of lava sometimes overwhelm timber, charring the tree trunks and then solidifying before actually burning the wood; tuff and volcanic breccia also overwhelm and cover trees. In this way the magnificent fossil forests of Yellowstone were buried. Figure 3-28 shows a remarkable instance where fern leaves left an imprint on a thin flow of lava. In this unusual case the leaves were probably green and wet and the flow was very thin, so that the lava hardened before the leaves were burned. A cavity observed in the base of a lava flow in the Grand Coulee, Washington, in 1935 proved to be a natural mold of the body of an extinct twin-horned rhinoceros [3]. The animal evidently had been killed and enveloped in the lava, which chilled and hardened before burning the flesh.

In deserts and along the seashore, wind-blown sands may overwhelm the living or bury the dead. Dry sands are not a good medium for fossilization, since oxygen can penetrate to great depths and solution after rainfalls is very active. Hence fossils are rare in

a

0   2 CM

b

Figure 3-29 The Cretaceous ammonite *Hoploscaphites* preserved and uncrushed in a limestone concretion (*a*), and crushed in shale (*b*). From the Pierre Shale and Fox Hills Formation in South Dakota. (Yale Peabody Museum.)

desert deposits. Nevertheless, the nests of dinosaur eggs found in Mongolia (Fig. 16-18) were apparently preserved in this way by drifting sands.

Occurrence of Fossils. It is hardly necessary to state that fossils never occur in plutonic rocks and are not to be found in any igneous rocks except where ash falls or nearly cooled lavas have overwhelmed plants and animals. They occur in all types of sedimentary rocks but generally are least common in pure sandstones and most abundant in calcareous shales and limestones. Red sandstones and shales usually have few fossils, except footprints, because the red color is due to complete oxidation, which destroys organic compounds. Even where abundant tracks prove

the existence of plentiful life, as in the Triassic redbeds of Connecticut, skeletal remains are rare.

Once buried, organic remains are subject to the changes that transform the enclosing sediment into rock. One of the first of these changes is the compaction that takes place from the growing weight of sediment as burial becomes deeper and deeper. In mud, which compacts considerably more than sand, even hard parts such as shell and bone tend to be squeezed flat (Fig. 3-29*b*). Commonly, however, chemical action around buried organisms begins very soon after burial and carbonates of calcium and iron may form solid nodules or **concretions,** enclosing the remains and so protecting them from compaction (Fig. 3-29*a*). Even relatively soft structures may be preserved by the early formation of concretions (Fig. 3-30).

After preservation, fossils may be distorted or completely destroyed during the deformation and recrystallization of rocks involved in strong folding or thrusting (Fig. 3-31). The flow of weak rocks during folding commonly distorts fossils into grotesque forms. Metamorphic rocks frequently show blurred traces of fossils (as in many marbles), but where recrystallization has been complete, all traces of organic remains are commonly destroyed. Nevertheless, poorly preserved and generally distorted fossils are found locally even in highly metamorphosed rocks. Here they are of importance for three reasons:

1. They give clear evidence that the rocks were originally sedimentary.

Figure 3-30 Remarkable preservation of the tentacles of a Middle Pennsylvanian squid in an ironstone concretion from the famous Mazon Creek collecting area of northeastern Illinois. Tiny hooks occur on the arms and X-rays have revealed a very fragile, flat shell. (Courtesy of Ralph G. Johnson and Eugene S. Richardson.)

2. Their shapes may give evidence as to the nature and amount of distortion the rocks have undergone.
3. They may serve to date the original rocks, at least approximately [2].

INTERPRETATION OF FOSSILS

Early Interpretations

History does not record the first observation of fossils. A Jurassic brachiopod has been found among the amulets of Neanderthal man at Saint Léon (Dordogne), France, and there are other evidences of aboriginal fossil hunters. In the writings of the Greeks, there are occasional references to such objects long before the beginning of the Christian Era. Some of these early observers correctly interpreted the organic remains and drew significant inferences from their presence in the rocks. Herodotus, for example, during his African travels about 450 B.C. observed fossil seashells in Egypt and the Libyan desert, from which he correctly inferred that the Mediterranean had once spread across northern Africa. On the other hand, much fancy and mysticism was associated with fossils even by great thinkers like Aristotle who, while believing them to be organic remains, thought they had grown in the rocks. His ideas of how this came about were obscure, but his pupil, Theophrastus, clearly expressed the idea that eggs or seeds buried with accumulating sediment had germinated and grown after burial.

The natural interpretation of fossils as organic remains of some sort seems to have been general until the beginning of the Dark Ages, when the Christian church came to insist upon a belief in a special Creation accomplished in six days, and a beginning of the earth at a time only a few thousands of years ago. This belief left no place for extinct creatures or great changes in the position of land and sea. Under this influence, interest in fossils was so subdued that few references to them appear in the literature of the Dark Ages. With the Renaissance, however, and the growth of natural science, fossils again claimed attention and

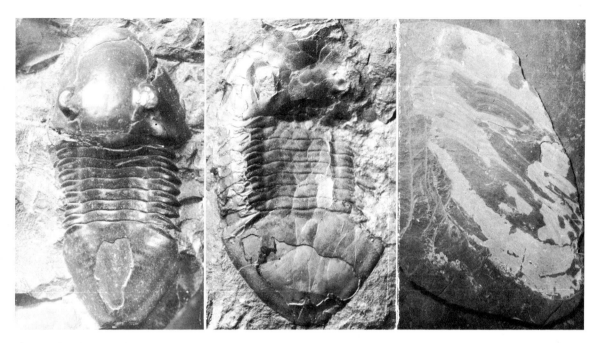

Figure 3-31 Deformation and destruction of fossils. Left, an uncrushed specimen of the trilobite *Isotelus;* center, the same species badly crushed by compaction of the sediment after burial; right, a similar trilobite reduced to a "ghost" when the enclosing shale was crushed and recrystallized into slate. (Yale Peabody Museum.)

soon became the subject of a great controversy.

The "Fossil Controversy." The controversy really began in Italy about A.D. 1500, when the digging of canals through Cenozoic marine formations brought to attention abundant shells so obviously like those of the present seacoast that their significance could hardly be doubted. Leonardo da Vinci, who besides being an artist was trained as an engineer, took a great interest in these fossils, and argued clearly that they were shells of animals once living in the places where they were found. There were many who flatly denied this, and for two centuries the controversy raged, advocates of the Organic Theory being frequently subjected to persecution. The most fantastic explanations were invented to avoid the biologic interpretation of fossils. Some were entirely mystical, attributing them to a "plastic force" at work in the rocks, while a few simply declared them "devices of the Devil," placed in the rocks to delude men.

One of the most remarkable illustrations of such fanaticism occurred in Germany as late as 1696, when parts of the skeleton of a mammoth were dug from the Pleistocene deposits near Gotha. These fell into the hands of Ernst Tentzel, a local teacher who declared them to be the bones of some prehistoric monster. This heretical idea raised a furor, as a result of which the bones were examined by the medical faculty of the school and dismissed as but a "freak of nature."

Eventually when even devout Christians could no longer doubt the organic nature of fossils certain churchmen attributed them to creatures that had been buried by the Flood described in the Scriptures. The extreme to which this fallacy was carried is immortalized in a small Latin volume by Johann Jacob Scheuchzer published in 1726, entitled *Homo diluvii testis* (The man who is proof of the flood). It contained illustrations and description of articulated skeletons from the Oligo-

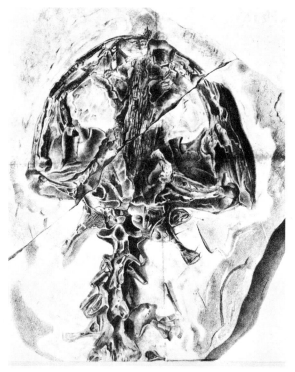

Figure 3-32 *Homo deluvii testis*. Skull and part of backbone of a large salamander from the Oligocene lakebeds at Oeningen, Switzerland, mistaken by Scheuchzer for the remains of a human drowned during the Biblical Flood. About ½ natural size.

cene lakebeds near Oeningen, Switzerland (Fig. 3-32). When the great paleontologist Cuvier later restudied one of these skeletons and found it to be only that of a giant salamander, he described and named it *Andrias scheuchzeri!* Two vertebrae were likewise figured by Scheuchzer as "relics of that accursed race that perished with the flood" and his figures were reproduced in the "Copper Bible" of 1731. Cuvier found these to be vertebrae of Mesozoic reptiles.

The first attempts at the reconstruction of extinct animals were made before anyone had a very general knowledge of anatomy or comparative morphology. It is not surprising, therefore, that these early efforts were fantastic. The oldest known attempt is the reconstruction of the unicorn by Otto von Güricke, burgomaster of Magdeburg, in 1663 (Fig. 3-33). It was composed of various Pleistocene elephant bones, the "horn" being, in reality, a tusk. Such early attempts were based more on legends and myths than on knowledge of animal life.

In 1706 a mastodon tooth discovered in a peat bog near Albany, New York, was sent to Governor Dudley of Massachusetts, who wrote to Cotton Mather about it as follows.

July 10, 1706

I suppose all the surgeons in town have seen it, and I am perfectly of the opinion it was a human tooth. I measured it, and as it stood upright it was

Figure 3-33 Early reconstructions based on bones of fossil elephants. Right, as a "Unicorn," by Otto von Güricke in 1663; left, as "The Missourium" by James Pedder in 1841. (The latter is after George Gaylord Simpson from *Natural History*.)

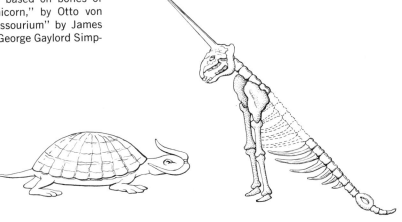

six inches high lacking one eight, and round 13 inches, lacking one eight, and its weight in the scale was 2 pounds and four ounces, Troy weight.

I am perfectly of the opinion that the tooth will agree only to a human body, for whom the flood only could prepare a funeral; and without doubt he waded as long as he could keep his head above the clouds, but must at length be confounded with all other creatures and the new sediment after the flood gave him the depth we now find.

It is difficult to realize that these words concerning a mastodon tooth were written in pious seriousness and not as a jest! Yet as late as 1784 Thomas Jefferson wrote to Ezra Stiles, President of Yale, for his opinion about similar remains, to which, after a lengthy discussion, President Stiles concluded "Perhaps the sensible rational and anatomical Virtuosi will judge those dug up at North-Holston, at Claverack and elsewhere . . . (of this enormous Description) and the Mammoths of Siberia all truly belong to an Animal Race in the shape of Men, called Giants in the Scriptures," etc. [1].

A famous paleontological hoax perpetrated on Professor Johannes Beringer of the University of Würzberg, serves as another illustration of the uncertainty about the significance of fossils that marked the early eighteenth century. Much of the legend that has grown up around this hoax, including the judgment that Beringer's students were the perpetrators, has recently been stripped away [10]. Beringer, a Professor of Medicine much interested in the local rocks, hired several boys to collect "figured stones" for him. The University librarian, Eckhart, and a mathematics professor, Roderick, apparently driven by jealousy of Beringer, bribed one of the boys to "salt" the outcrops with bits of limestone on which Roderick had carved various objects. These included flowers, insects, frogs, a seahorse, astronomical signs, and even characters from the Hebrew alphabet. In 1726, Beringer published *Lithographiae Wirceburgensis* with plates of the figured stones and a scholarly treatment of their probable origin. Deceived by the hoax, he believed "the figured stones" to be natural, although he realized that they were not true organic

remains and ascribed them to some "unique" manifestation of nature. Eckhart and Roderick, worried that they had gone too far, informed him of the hoax shortly before publication; but Beringer would not believe them and included in the work a denunciation of the men for attempting to keep his contribution from the world! Very soon after publication Beringer realized that he had been deceived and that same year instituted legal proceedings against Eckhart and Roderick. Although the guilt of these men was established, the harm had been done and Beringer attained lasting fame in paleontology as a dupe.

The controversy over fossils served only to kindle interest in these objects, and by the beginning of the nineteenth century fossil collecting had become a hobby with many devotees in the church and out, one of the first large collections being made at the Vatican. By the year 1800 the organic nature of fossils was almost universally recognized, and learned men were generally agreed that they represented the life of the geologic past.

Modern Interpretations

Reconstructions. The majority of fossils represent only the hard parts of organisms, and even these are commonly incomplete. Scientists and laymen alike are interested in what ancient animals, particularly vertebrate animals, looked like when alive. Modern museums, therefore, display fossil skeletons with missing parts restored and even attempt to clothe the bones in flesh.

Restoration is not a mere feat of the imagination. Simple and natural principles together with a sound knowledge of comparative anatomy are employed in restoring a fragmentary specimen. Assembling a heap of disarticulated fossil bones is no more difficult than solving a jigsaw puzzle. The anatomist knows the basic pattern of skeletal parts common to all vertebrates and can fit the bones together in their proper places. Especially helpful to him in this work are the articulated specimens of similar living vertebrates for comparison. Commonly bones or parts of bones are missing and these are replaced with plaster replicas. For example, if some of the bones in a right limb are lacking but those of the corresponding left limb are preserved, the whole can be reconstructed with absolute fidelity because vertebrates are bilaterally symmetrical. Missing parts of one specimen may also be reconstructed from another specimen of the same species. Thus, if it so happens that the front end of one has been destroyed by erosion and the tail end of another is missing, we can still restore an entire, but composite, skeleton faithfully representing the species. The restoration of fossil skeletons therefore leaves little to guesswork.

However, the clothing of the bones with flesh in lifelike reconstructions does involve some powers of the imagination guided by profound knowledge of comparative anatomy. The ridges, pits, and roughened areas on bones were the seats of muscle attachment in life, and the muscles that manipulate the skeleton are those that are most apparent in its external form. By modeling the necessary muscles on the articulated skeleton one by one, the muscle sheath of the fleshy body is restored, revealing the basic soft-part outline of the animal. Beyond this step the external restoration of fat, skin, and hair is largely an educated guess, but it too is guided by comparison, either with closely related living forms or, if there are none, a living form of similar habits. The potential for error may be illustrated by the woolly rhinoceros mentioned earlier. Had only bones of this animal been found it probably would have been restored with a thick, relatively hairless hide like its present-day relatives in Africa and India. On the other hand, some specialist just might have reasoned that if contemporary mammoths had hair at that latitude during the Ice Age, the rhinoceros might also have had it.

The restoration of an extinct vertebrate helps to visualize what animals of the past looked like, an interesting but hardly scientific objective. What the paleontologist is looking for in restoring a skeleton are clues to how the animal functioned, and what it did to make a living. The kind of teeth and the mechanics of chewing indicated by bone structure and restored musculature of the jaws and adjacent

parts of the skull will commonly reveal both what an animal ate and how it ate it—perhaps even how it killed it. This study of form to deduce function, called **functional morphology,** is an important method by which fossils can be studied as living things.

Fossils as Living Organisms. A single tooth or bone fragment is often enough evidence for the identification of an animal but rarely enough to tell much about its life habits. The great canine teeth of saber-tooth cats may at first suggest that their function was stabbing, but this is only a part of the story. Adaptations to a particular way of life affect more than one part of an organism. *Smilodon*, the lion-sized, sturdily built saber-tooth cat of the Pleistocene was the last of a long and successful lineage of stabbing cats that extended at least 40 million years back into the Cenozoic. The articulation of its lower jaw was so constructed that the mouth could be opened to an angle of 90°, clearing the great upper canines for action. The skull was further modified for attachment of unusually powerful muscles which pulled down the head in the act of stabbing. The heavy, stocky body of *Smilodon* contrasted markedly with the lighter, faster, and more agile biting cats that were its contemporaries. It was admirably adapted for feeding on large, slow-moving, thick-skinned beasts like mastodons, mammoths, and the giant ground-sloths that were common in its time (Fig. 3-26). With the disappearance of these great Pleistocene herbivores, *Smilodon,* and with it the saber-tooth lineage of cats, became extinct. It was ill-adapted to compete with the biting cats for the smaller, swifter herbivores that dominated the postglacial scene.

Use of the technique of functional morphology to reveal living habits is not restricted to vertebrate animals. Many kinds of invertebrate animals have hard parts whose structures indicate particular activities. Because the hard parts of most invertebrates are external and relatively simple, they cannot be expected to reveal as much as an internal skeleton, which is more directly involved with the life activities of its owner. Nevertheless, much can be deduced from simple shells; take, for example, *Tancredia americana,* shown in

Figure 3-34 Shells of the Cretaceous bivalve *Tancredia americana,* left, a cluster in sandstone with tapering anterior ends pointing downward; right, shells freed from the matrix, the upper a view of the interior of a valve showing strong hinge articulation and muscle scars, the lower a view of joined valves from above showing the pronounced posterior gape and anterior taper. Note the thickness of shell of this sturdy burrower in the broken specimen on the left. (Yale Peabody Museum.)

Fig. 3-34, the last surviving species of an extinct family of Mesozoic bivalves. The shell itself is thick and strong with a great posterior gape indicating large siphons and an infaunal habit. Viewed from any position, the valves of *Tancredia* taper anteriorly, presenting a streamlined shape for rapid burrowing, a habit that is also suggested by internal shell structures. In all bivalves muscles that hold the valves in place and work the digging organ, or foot, are attached to the inner surface of the valves, leaving traces or scars. These may be faint if the muscles are relatively weak or prominent if they are strong. Active bivalves today typically have strong musculature; judging from its large and well-defined scars, so did *Tancredia*. The features of the distinctive shell of *Tancredia* indicate that it led an active life as a shallow burrower in areas of high cur-

rent energy, where the ability to dig rapidly in or out of shifting bottom sediment was a necessity. The mode of occurrence of fossil *Tancredia* shells supports this interpretation. They occur locally in great numbers in conspicuously current-bedded bodies of clean, fine sand whose close proximity to ancient shorelines suggests that they were beaches and shallow sand bars. Very few other shells are found with *Tancredia*, which in itself is suggestive of a rigorous environment. Most commonly it is associated with a branching burrow, having a cemented wall made up of rounded sandy pellets. Identical burrows are known today [19] along the coast of the southeastern United States in the current-washed sands of the shore zone, in and just below the lower part of tidal range. A peculiar shrimp-like crab, *Callianassa*, makes the burrows by plastering their sides with the pellets. Three lines of evidence thus agree in reconstructing the mode of life and habitat of the extinct *Tancredia;* its own morphology, the current-bedded nearshore sand, and its association with burrows linked by similarity with those of a living creature of the present-day shore zone.

The example of *Tancredia* indicates some of the principles and methods, other than functional morphology, that are used to interpret fossils as living animals. **Analogy with living organisms and their environments** is a principle underlying all methods of the interpretation of fossils as living animals. Like other aspects of the uniformitarian doctrine, however, it cannot be applied too rigidly with success. For example, a number of relatively rare kinds of marine invertebrates living today in deep water are virtually identical to related forms that literally crowded the waters of shallow seas of the past. The constant evolutionary change of organisms has all the while imposed some measure of constant change on the environments of which they have been a part. Interpretations based on analogy must always take this into consideration.

The nature of the enclosing sediment may reveal much about the life habitat of a fossil provided that it was buried where it lived. So also can the other associated fossils, provided that they too were part of the living community and were not transported from a number of different habitats to a common burial ground. Naturally, the fossil record contains all kinds of organic accumulations ranging from the rare occurrences of fossils in their actual living position to the most diverse assortments of unrelated remains imaginable.

The importance of determining to what degree fossil assemblages reflect ancient associations of organisms in their natural habitat can hardly be overemphasized. Not only may it provide a key to the life habits of particular fossils, but on a broader level it leads to the reconstruction of natural communities and their environments. Studies of this kind lie in the field of **paleoecology,** one of the most fascinating and active areas of modern paleontologic research. Like ecology, its sister field

in the biological sciences, paleoecology seeks to understand the relationships of organisms to their environment. Unlike ecology, however, it must first resurrect the life and environments of the past from long dead and much altered organisms and sediments before it can attempt to interpret relationships.

Fossil Thermometers. Most kinds of animals and plants are now restricted to definite climatic environments. Palms and crocodiles characterize the tropics and subtropics, as the reindeer and musk-ox do the Arctic. The presence of the former as common fossils in the Oligocene rocks of the Dakotas bears a very strong implication that the winters at that time were much milder than now in the Great Plains region. The presence of fossil musk-ox in New York and Arkansas and of reindeer in France in Pleistocene sediments similarly accords with unmistakable evidence of glacial climates at that time.

In reasoning thus from the known distribution of living types, we are fairly secure for late geologic time, but less and less so as we go back to older rocks where the species and even the genera are different and may have had different habits. Caution must be exercised even for relatively recent geologic time, since animals and plants are highly adaptive and certain extinct species may have been adjusted to different climatic extremes than related living species are. Of course the implication of a **group** of species is more trustworthy than that of one.

Lines of evidence other than direct relationships make it possible to interpret climatic implications of older flora and fauna. For example, flora of the Carboniferous coal fields of the world have commonly been considered tropical because of their obvious great abundance. These flora have few plants in common with the flora of modern lush tropical lowlands. Structurally, however, they possess features pointing to growth in an exceedingly humid environment and the rarity of any cyclical growth patterns, like tree rings, indicates an equitable climate without well-defined seasons. Thus it is a fair assumption that Carboniferous coal-measure floras indicate either tropical or subtropical climate. Again,

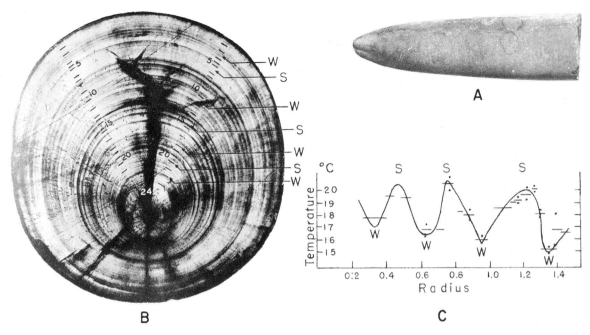

Figure 3-35 Evidence of paleotemperature. A, lateral view of a belemnite shell; B, enlarged cross section of a similar shell from the Upper Jurassic rocks of Scotland showing concentric growth layers; C, temperature curve recorded by the O^{18}/O^{16} ratios in successive layers of the shell. (B and C after Urey, Lowenstam, Epstein, and McKinney.)

the great Paleozoic fossil reefs had a markedly different composition of organisms than do the organic reefs of today, but it is reasonable to suppose that the ancient reefs were similarly restricted to the warmer seas. It is the reef structure itself, a product of intense organic activity, and the concentration of diverse forms of life in and about it that suggest tropical to subtropical seas. Increase in organic activity and diversity toward the tropics has long been observed and is a principle of animal and plant distribution useful in delineating the broader climatic zones of the past.

In 1950 Harold Urey and a group of colleagues at Chicago developed a technique for determining the actual temperature of the sea water in which certain marine animals lived, even millions of years ago, by measuring the ratio of common oxygen, O^{16}, to its rare isotope, O^{18}, in the calcium carbonate ($CaCO_3$) built into their shells. Complicated techniques are required for making measurements of sufficient accuracy and only exceptionally well-

preserved and unaltered fossil material can be used. Nevertheless, paleotemperatures derived from oxygen isotope studies have been used with apparent success with fossils as old as Permian, and the technique has had some outstanding success in Cretaceous and younger rocks (Fig. 3-35).

The method is based on the fact that the ratio of the two isotopes incorporated in the shell as carbonate varies with the temperature at time of shell formation since the ratio of the carbonate dissolved in sea water is a function of the temperature. Thus if the shell has not been altered by recrystallization or by the diffusion of oxygen at a later date, it retains a built-in record of the temperature of the water in which it was formed. Of all the kinds of shells tested, the belemnites and foraminifera appear to be the most reliable.

The belemnite, a Mesozoic cousin of the squid, developed an internal shell with a cigar-shaped piece built up by successive additions during growth of the animal, so that it has con-

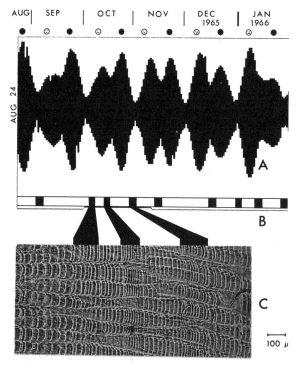

Figure 3-36 Relationship between tides and growth increments in the shell of a quahog, *Mercenaria mercenaria*. Tidal cycles are reflected by alternations of groups of thick and thin daily growth increments. A peel from a section of shell (C) enlarged 200 times, shows the alternating clusters of thick and thin increments. When the clusters of thin increments are plotted (B) against the tidal ranges (A) of the days when they were deposited, they are seen to correspond to periods of low tidal range, the neap tides. Adjacent thick and thin clusters represent a single 14-day tidal cycle. (Courtesy of G. Pannella and C. MacClintock.)

centric layers similar to the growth rings of trees. Figure 3-35 shows the application of this method to a belemnite from the Jurassic rocks of Scotland [18]. The O^{18}/O^{16} ratios determined for successive concentric zones in this shell indicate rhythmic fluctuations of temperature from a winter low of about 15°C to a summer high of about 22°C. From these data it can be deduced that this animal was born in spring, lived through three summers, and died in the fourth winter at the age of about $4\frac{1}{2}$ years, and that the temperature range in the shallow sea in which it lived some 140,000,-

000 years ago was about that of the modern shallow sea off Gibraltar.

This method has an important application in the study of deep sea cores which penetrate the Pleistocene sediments of the ocean floors [5]. Since the shells of pelagic foraminifera (family Globigerinidae) have a built-in record of the temperature of the surface water in which they lived, climatic fluctuations will be recorded by variations in the O^{18}/O^{16} ratio of these shells at different depths in the cores. Of course such fluctuations at a single locality could be due to shifts in the position of warm, or cold, surface currents; but if and when such fluctuations at widely distributed places in the oceans can be synchronized by radioactive dating, we shall have evidence of worldwide temperature fluctuation that will permit correlation of the sediments of the ocean floor in low latitudes with the record of glacial and interglacial ages on the land. However, isotopic dates associated with Pleistocene deep sea deposits do not at present extend back much beyond 30,000 years, and substantiation of widespread paleotemperature fluctuations may be well in the future. Meanwhile other methods of correlation, such as the paleomagnetic technique employing reversals of the earth's magnetic field (see p. 75), are being applied to the deep sea deposits and may well supplant stratigraphy based on paleotemperature changes. But the great importance of paleotemperature studies is not in correlation but in their potential for quantitative expression of ancient marine climates.

Shell Growth and Fossil Calendars. Many invertebrates with limy shells or skeletons provide for growth by successive additions of small increments of calcium carbonate to these structures. Commonly such increments appear as fine growth lines on the surface of a shell, and it has long been observed that on some shells annual growth periods are clearly marked by larger undulations. Not until 1963, however, was the potential significance of growth increments realized. At that time John Wells [20], using corals as an illustration, presented the idea that the fine growth lines represented daily increments. He noted that astronomers believe the period of rotation of the earth around its polar axis has been gradually slowing down at a rate of about 2 seconds every 100,000 years, due to the slight damping effect of tidal forces on rotational energy, and that in the course of geologic time the length of the day has been increasing and the number of days in the year decreasing. Reasoning that if this were so, fossil corals should have more daily increments in their annual fluctuations than modern corals, Wells examined several Devonian corals and found that all have more than 365 growth lines, and that in general they average about 400 growth lines to the annual increment. Astronomical data indicate that the Cambrian year should have about 420 days, so that 400 days for the Devonian year is in line with the astronomical calendar based on the theory of decelerating rotation. Wells emphasized that if more detailed studies supported his idea, fossils with patterns of additive growth were potentially useful chronometers of absolute time because the length of the year indicated by growth lines could be correlated directly with the astronomical scale of deceleration.

From this beginning a whole new frontier of paleontological investigation is rapidly unfolding. Working experimentally with the living quahog *Mercenaria*, Pannella and Mac-Clintock [16] established that the basic growth units of the shell were, indeed, daily increments. Moreover, they found a number of distinct cyclical patterns of increments, the most obvious being cycles reflecting seasonal changes in series of thicker (summer) and thinner (winter) increments. Recognizable throughout the seasonal cycles is a more delicate but no less distinct cyclical pattern of alternating thicker and thinner clusters of increments reflecting 14-day tidal cycles, that part of a cycle with thinner increments being deposited during neap tides, the part with thicker during spring tides. In addition to these universal cycles, events in the life of the animals, particularly spawning periods, were found to be clearly expressed in the daily increment record. Once these interpretations of the daily increment patterns were substan-

tiated by observations in quahogs and other living bivalves, the investigators turned to the fossil record. As might be expected, the same patterns of daily increments were found in a variety of fossil bivalves. Growth increments, as is evident in Fig. 3-36, are delicate, essentially microscopic structures and their preservation in the fossil record becomes rarer the farther back in time the sample is taken. Moreover, even in the shells of living bivalves it may be difficult to distinguish clearly each daily increment. All increments are not defined with equal clarity owing to the variety of physiological and environmental events that can affect their deposition. In using the bivalves as calendars Pannella and MacClintock employed, because of the greater accuracy, the number of daily increments in a lunar month, the latter being clearly delimited by the tidal increment cycles. Their preliminary results have shown a decreasing trend in the mean number of days in a lunar month from the Pennsylvanian to the Recent. This and other recent work employing the technique of measuring daily growth increments in fossils bear out the idea presented by Wells and lay the foundation for a new kind of chronology based on fossils.

REFERENCES

1. Anonymous, 1951, *Early letters concerning the Mammoth.* Soc. Vert. Paleontologists News Letter, n. 33, pp. 27–31.
2. Bucher, W. H., 1953, *Fossils in metamorphic rocks; a review.* Geol. Soc. America Bull., v. 64, pp. 275–300.
3. Chappell, W. M., Durham, J. W., and Savage, D. E., 1951, *Mold of a rhinoceros in basalt, Lower Grand Coulee, Washington.* Geol. Soc. America Bull., v. 62, pp. 907–918.
4. Delevoryas, Theodore, 1962, *Morphology and evolution of fossil plants.* New York, Holt, Rinehart and Winston.
5. Emiliani, Cesare, 1966, *Isotopic paleotemperatures.* Science, v. 154, pp. 851–857.
6. Gazin, C. L., 1942, *The Late Cenozoic vertebrate faunas from the San Pedro Valley, Arizona.* U. S. Nat. Museum Proc., v. 92, pp. 475–518.

7. Gidley, J. W., and Gazin, C. L., 1933, *New mammalia in the Pleistocene faunas from Cumberland Cave.* Jour. Mammalogy, v. 14, pp. 343–357.

8. Grater, R. K., 1958, *Last stand of the ground sloth.* Ariz. Highways, v. 34, no. 7, pp. 30–33.

9. Hibbard, C. W., and Riggs, E. S., 1949, *Upper Pliocene vertebrates from Keefe Canyon, Meade County, Kansas.* Geol. Soc. America Bull., v. 60, pp. 829–860.

10. Jahn, M. E., and Woolf, D. J., 1963, *The lying stones of Dr. Johann Bartholomew Adam Beringer; Being his Lithographiae Wirceburgensis.* Berkeley, Univ. California Press, 221 pp.

11. Kelly, W. A., 1940, *Tar as a trap for unwary humans.* Am. Jour. Sci., v. 238, pp. 451–452.

12. Kulp, J. L., Tryon, L. E., Eckelmann, W. R., and Snell, W. A., 1952, *Lamont natural radiocarbon measurement, II.* Science, v. 116, p. 411.

13. Ley, Willy, 1951, *Dragons in amber.* New York, Viking Press.

14. Nydal, Reidar, 1962, *Trondheim natural radiocarbon measurements III.* Radiocarbon, v. 4, pp. 160–181.

15. Palmer, A. R., 1957, *Miocene arthropods from the Mojave Desert, California.* U. S. Geol. Survey Prof. Paper 294 G, pp. 237–280.

16. Pannella, Giorgio, and MacClintock, Copeland, 1968, *Biological and environmental rhythms reflected in molluscan shell growth.* Paleont. Soc. Mem. 2, pp. 64–80.

17. Stuiver, Minze, 1969, *Yale natural radiocarbon measurements, IX.* Radiocarbon, v. 11, no. 2 (in press).

18. Urey, H. C., Lowenstam, H. A., Epstein, S., and McKinney, C. R., 1951, *Measurement of paleotemperatures and temperatures of the Upper Cretaceous of England, Denmark, and the southeastern United States.* Geol. Soc. America Bull., v. 62, pp. 399–416.

19. Weimer, R. J., and Hoyt, J. H., 1964, *Burrows of* Callianassa major *Say, geologic indications of littoral and shallow neritic environments.* Jour. Paleontology, v. 38, no. 4, pp. 761–767.

20. Wells, J. W., 1963, *Coral growth and geochronometry.* Nature, v. 197, pp. 948–950.

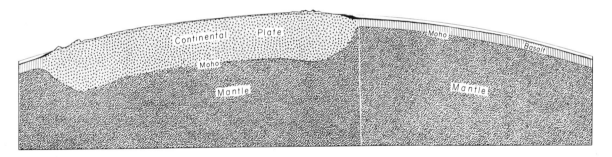

Figure 4-1 Idealized section of a sector of the crust and its relation to the Moho and the mantle. Bottom sediments on the ocean floor in black. Vertical scale greatly exaggerated.

CHAPTER 4 *The Restless Crust*

There rolls the deep where grew the tree,
O Earth what changes hast thou seen!
There where the long street roars hath been
The stillness of the central sea.
TENNYSON

Throughout its long history the earth's surface has been endlessly deformed. Time and again vast shallow seas have spread across parts of each continent and then slowly retreated again. Mountain systems have risen to impressive height only to be eroded away as new ones came up out of the sea. The towering white summits of the Carnic Alps, like those of the Canadian Rockies, are made of marine limestones, and fossil seashells are embedded in the flanks of the high Himalayas at elevations as great as 20,000 feet. The continents have drifted as the vast tabular icebergs do in the polar sea.

The energy required to deform the earth's crust in such manner staggers the imagination and its source has long been one of the great enigmas in geology. But the scientific ad-

Figure 4-2 Diagrammatic section of the earth showing the course of selected seismic vibrations radiating through the deep interior from an earthquake focus.

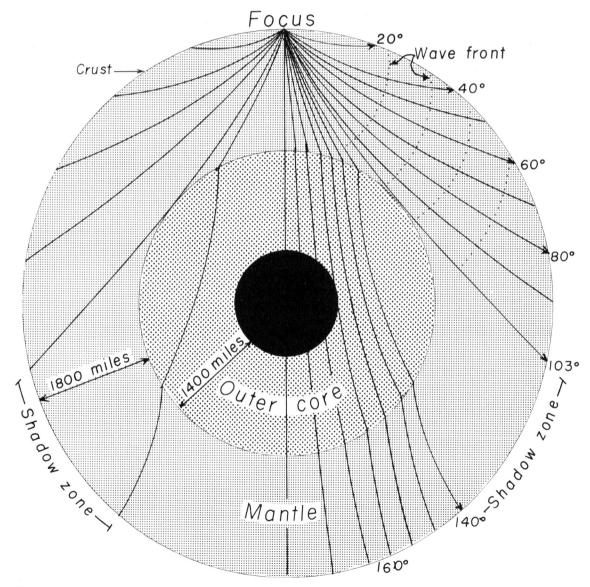

vances of recent decades that have enabled us to explore outer space and even to visit the moon have given us new insights into the forces that shape the earth, and it will be our purpose in this chapter to examine several of these revelations.

THE NATURE OF EARTH'S CRUST

As indicated in Fig. 4-1, the crust is a relatively thin veneer of the earth consisting of two dissimilar elements. The continents are vast plates of granitoid rocks each about 25 miles thick (except in mountainous regions where they range up to some 40 miles in thickness). The oceanic segments of the crust are only about 5 miles thick and consist largely of basaltic rocks with a relatively thin veneer of oceanic sediments. Both elements of the crust (continents and ocean floors) rest on the **mantle** (discussed later), which can yield under slow pressure so that the crust is literally floating on it in isostatic equilibrium; and because granite is lighter than basalt and the continental plates are about 5 times as thick as the basaltic segments, the continents ride high, as icebergs do in a sea of flow ice. This much is not new; the principle of isostasy has been known for more than half a century; but recent developments in geophysics have produced a vast amount of data about the thickness of the crust.

If we could remove the water from the oceans and view the naked earth from a high-flying satellite, two major features would stand out. The continents would appear as vast plateaus standing about 3 miles higher than the ocean floors. From this view mountain ranges would appear almost trivial, like the decorations on a cake.

It may be noted in passing that the crustal rocks consist of materials that have come to the surface from deep within the earth, having risen to the surface as slag rises above the metal in a furnace or as cream rises to the surface of a vessel of milk. This assumption leads to the question of why the granitic material is concentrated in the continental plates, which occupy less than one-fourth of the earth's surface, instead of forming a continuous layer

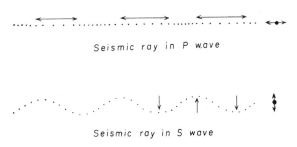

Seismic ray in P wave

Seismic ray in S wave

Figure 4-3 Diagram to contrast the movement of particles in a seismic ray of P waves and of S waves.

over the entire planet. We shall return to this problem.

THE INTERIOR OF THE EARTH

The model illustrated in Fig. 4-2 indicates what we know of the nature of the interior of the earth. On this scale the crust would be embraced within the thin surface line of the diagram. Beneath the crust lies the **mantle,** which extends to a depth of about 1800 miles, consists of ultrabasic silicate rock, and comprises about 80 percent of the entire mass of the earth. It surrounds the metallic **core,** which consists of iron with a minor admixture of nickel. The **outer core** is molten and within it lies an **inner core** of solid iron and nickel.

Since the deepest wells do not reach even the bottom of the crust, this interpretation rests on indirect evidence. The evidence is derived partly from the behavior of seismic vibrations [Gr. *seismos,* earthquake], which, after passing deep into the earth, reach the surface at points far from their source and are recorded by seismographs. Waves of such vibrations are of two distinct sorts: (1) **compression waves,** commonly referred to as P waves, and (2) **shear waves** called S waves (Fig. 4-3).

Compression Waves. From the source of disturbance alternate waves of compression and rarefaction spread into the earth causing particles of the rock to oscillate back and forth in the line of propagation (as they do in a coiled spring struck end on). Such waves will travel in either solids, liquids, or gases. In water, for example, they are known as sonar waves, and in the atmosphere they are sound waves. The velocity at which compression waves travel depends on the rigidity of the material through which they are passing and increases with the rigidity. In the atmosphere, for example, sound waves travel at a velocity of about 1087 feet per second, but in granite the velocity of seismic waves is between 6.1 and 6.7 km (about 3.5 and 4.0 miles) per second.

To determine the velocity within rocks of any given type an outcrop several miles wide is selected and a bomb is exploded at a predetermined distance from a portable seismograph. Thus the time lapse between the explosion and the arrival of the P waves at the seismograph station gives the velocity of travel. In this way it is known that in loose sediment the velocity of P waves ranges from 1.5 to 2.5 km per second (depending on the degree of compaction); in sedimentary rocks the velocity ranges from 3.0 to 3.5 km per second (depending on the degree of consolidation); and in granite or basalt the velocity ranges from about 6.2 to 6.7 km per second (depending on the depth). If the material were more rigid than granite, the velocity would of course be greater. Furthermore, since the rigidity of any solid increases with the confining pressure, the velocity must gradually increase as the vibrations pass deeper into the earth even if the material were homogeneous.

Shear Waves. From the focus of an earthquake (or a massive explosion) waves of a different sort are generated in which particles in the rock vibrate at right angles to the line of propagation (as they do in a telegraph line struck from the side). The waves pass outward from their source but the disturbed particles do not. Such are the shear waves. They follow the same course as the P waves, but travel only about half as fast. As a result the S waves arrive at a distant seismograph later than the P waves and the time lag is proportional to the distance. In this way a seismologist can determine the distance to an earthquake center as soon as the record is received. Unlike the P waves, **S waves can travel only in solids.**

The Mohorovicic Discontinuity. The lower boundary of the crust was first discovered about 1909 when the Hungarian geophysicist Mohorovicic completed his study of the records of a great number of earthquakes scattered over central Europe where the basement rocks are granitic. He discovered that seismic vibrations that had penetrated not more than 25 miles deep had traveled at the velocity appropriate for granite, whereas those that went deeper were abruptly accelerated to 8.1 or 8.2 km per second. They had evidently encountered material more rigid than granite or any other kind of surface rock. This level marked by an abrupt change in the velocity of seismic vibrations soon came to be known as the **Mohorovicic discontinuity** or in modern parlance the **Moho.** Subsequent investigations in many parts of the world have shown that all the continents have a similar lower boundary generally lying about 25 miles below the surface but in mountainous regions ranging to as much as 40 miles deep. Study of the ocean floor has revealed the Moho to be equally well defined but to lie only about 5 miles beneath the ocean bottom. On this basis the crust is now defined to include the rocks above the Moho and the denser material below it is the **mantle.**

The Mantle. The only crustal rock known experimentally to transmit vibrations at the velocity they have in the mantle is **dunite,** a rather rare ultrabasic silicate that is found chiefly in the form of chunks or blocks embedded in the lavas of some of the oceanic islands. These may be actual pieces of the mantle rock brought up from great depth by the rising magma. It is therefore believed that the mantle consists of dunite or some closely allied ultrabasic silicate rock.

Special interest attaches to a weak zone in the lower part of the upper mantle indicated by a decline in the velocity of seismic waves. The reason for this zone of weakness has been a great puzzle, but H. H. Hess recently suggested a plausible explanation [10]. It is well known that pressure increases the rigidity and that heat reduces it. Hence the rigidity at any depth depends on the ratio of these opposing influences. Radiogenic heat generated deep within the mantle would build up progressively if it were not carried to the surface to be dissipated into space. This can be done in only two ways, by conduction or by convection. Rock is a very poor conductor of heat at low temperatures, but recent studies at Harvard have shown that at temperatures over 1100°C dunite will transmit infrared light (and presumably heat) very effectively. Hess therefore postulates that in the lower mantle where the temperature exceeds this critical limit the heat escapes by conduction, but in the upper mantle, where the temperature is lower, the heat builds up and convection, carrying the hot rock upward by plastic flow, is the only effective mode of heat transfer. Thus the lower part of the upper mantle between a depth of about 60 and 750 km is a mobile zone where the forces arise that deform the overlying crust.

The Molten Outer Core. When seismic vibrations reach the base of the mantle two highly significant changes occur: (1) the P waves are deflected deeper, indicating that they have encountered materials less rigid than mantle rock, and (2) the S waves are completely damped out. This can only mean that the outer core is molten. As a result of the deflection the P waves return to the surface at a distance of more than 140° from the locus of the disturbance, leaving a "shadow zone" between 103° and 140° from the disturbance, in which no direct waves appear at the surface.

The Solid Inner Core. P waves that pass within about 800 miles of the center of the earth are accelerated, indicating that they have encountered material more rigid than the outer core. The inner core is therefore inferred to be solid.

The chemical nature of the core is indicated by other lines of evidence. For example, the specific gravity of the earth as a whole is about 5.5 and is almost twice as great as any of the crustal rocks. Even granting that pressure deep within the earth would increase the density of any material, it still appears that the core must be heavier than rock. And, since it is molten in spite of the pressure, it must be chemically different from the mantle. If it is iron, this would explain both considerations.

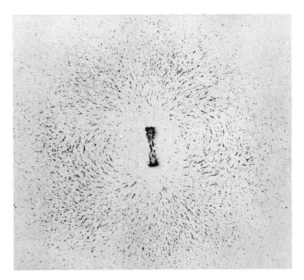

Figure 4-4 Iron filings oriented by the invisible lines of force about a bar magnet.

Meteorites afford a third line of evidence. They are believed to be fragments of a disrupted minor planet and should include representatives of both the outer and the deep interior regions. They are of two sorts—**stony** meteorites made of silicate minerals and **metallic** meteorites made of iron with minor amounts of nickel. Study of all the known meteorites collected reveals that the ratio of metallic to stony ones compares favorably with the ratio of the core to the mantle.

EARTH'S MAGNETIC FIELD

Ever since the invention of the compass it has been evident that the earth is a colossal magnet with a dipole axis running through it from the north to the south magnetic pole. The reason for its magnetism remained a mystery, however, until a few years ago.

If we should sprinkle iron filings over a simple bar magnet, they are immediately drawn to it; but if we shield it by laying a sheet of paper over it, the filings leap into a festoon of rows arranged as shown in Fig. 4-4. These rows reveal the invisible lines of force in the magnetic field that surrounds it. The force flows from the north to the south pole where it arches around in a series of subconcentric ellipses to the north pole. The compass proves that comparable lines of force form a vast magnetic field about the earth (Fig. 4-5). In the equatorial region the compass needle lies parallel to the earth's surface but when carried northward, it dips down toward the earth, at first gently and then more steeply as it is carried to higher latitudes, and finally at the north magnetic pole it dips vertically downward. On the contrary it points vertically upward at the south magnetic pole and when carried northward its inclination decreases progressively until it lies horizontally at the equator. Rocket probes show that lines of force far out in space are arranged like those about our bar magnet; a few of these are indicated in Fig. 4-5. These account for the inclination of the compass needle.

The reason for the earth's magnetism was first convincingly explained by W. M. Elsasser in 1968 [5]. If a bar of steel is brought into a strong magnetic field, it becomes a hard or permanent magnet. This is the way compass needles are magnetized. It also explains why a watch must never be worn near magnetic equipment. Now obviously the earth is not made of steel. But there is another kind of magnet. If a core of soft iron is wrapped within a cover of insulated electric wire, magnetism is induced when an electric current flows about it but disappears immediately when the current is turned off. This is the nature of the powerful induction magnets with which scrap iron is handled.

Elsasser reasoned that weak electric fields exist in the mantle and, if so, the spin of the earth on its axis must carry them around the iron core to make it an induction magnet. Furthermore, he has calculated that even though the electric fields in the mantle are very weak, on the scale of the earth they would be self-sustaining.

MAGNETISM IN THE CRUST

As a magma crystallizes it shows no magnetism until it has cooled below a point known as the **Curie temperature.*** Then each crystal of ferric and titanium oxide is magnetized with its dipole axis oriented like a compass needle aligned with the ambient lines of force in the earth's magnetic field. (Since the crystals are already frozen in at random, their magnetic axes are independent of the shape and the crystallographic axes but are readily recognized by means of a magnetometer.) As they cool further these crystals become hard or permanent magnets and their dipole axes will then persist even if the rock mass is later rotated by crustal movements.

If the rock mass should be so moved, a secondary magnetism will be superposed with its dipole axis oriented with the then current lines of force but, fortunately, this secondary magnetism can be destroyed by suitable treatment in the laboratory. Two chief methods are used for removing the secondary magnetism. Heating to a suitable temperature will destroy it, as will spinning a specimen of the rock in a suitable reversing electric field. Such "clean-

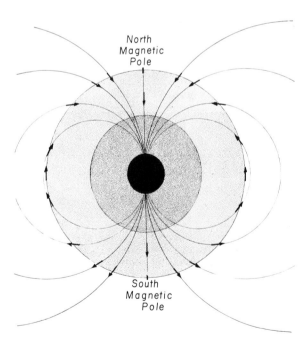

Figure 4-5 Diagram showing representative lines of force in the earth's magnetic field. The black arrows indicate the inclination of the compass needle according to latitude.

* This temperature varies from mineral to mineral, depending on the composition, but ranges between 200 and 680°C [4].

ing" then leaves the original "hard" magnetism as a permanent record of the orientation of the rock with respect to the earth's magnetic field at the time of its formation. This is the **remanent magnetism, or paleomagnetism.**

Two aspects of the magnetic orientation are significant. The **declination** is a measure of the horizontal divergence of the dipole axis from true north. The **inclination** is a measure of the dip of the dipole axis into the earth and will indicate the latitude at which the rock was formed (Fig. 4-5).

Thus far we have considered remanent magnetism only in igneous rocks, but the methods of paleomagnetic study can also be applied to sedimentary rocks. When magnetized particles are carried as sediment to settle slowly in standing water they come to rest with their dipole axis aligned with the lines of force in the earth's magnetic field just as the compass needle does.

When applied to either igneous or sedimentary rocks younger than the Oligocene Epoch the magnetic declination agrees with the modern world geography. But when applied to many of the older rocks the paleomagnetism shows spectacular departures from the modern orientation.

REVERSALS OF THE EARTH'S POLARITY

One of the most unexpected and spectacular discoveries of recent years is that from time to time in the past the polarity of the earth has been reversed, so that the magnetic force was reversed by 180°. The reasons for this are at present unknown but the evidence is quite convincing [3]. It was first extensively studied in a sequence of lavas in Hawaii where flows could be dated by included radioactive isotopes and were found to range from about 4,000,000 years old to the present. In some of these the magnetic pull was found to be directed toward the earth's north magnetic pole and in others it was reversed 180°. The dating established the sequence of reversals as indicated in column 1 of Fig. 4-6. Since the periods of normal and reversed polarity differed

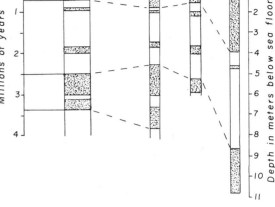

Figure 4-6 Diagram showing a time scale based on periods of normal and reversed polarity of the earth in lavas (left) and in oceanic sediments. Normal polarity shaded and reversed polarity blank. (From Cox et al. [3].)

greatly in length, this sequence establishes a chronology against which data from other regions may be compared. The sequence in widely separated parts of the world has been found to agree, proving that the reversals were a worldwide phenomenon. The significance of this will be shown in following sections.

THE DRIVING FORCE WITHIN THE EARTH

Theory of a Shrinking Earth. During the nineteenth century, geologists sought an explanation for diastrophism in a shrinking earth. It was then widely believed that the internal heat is a residue from a primordial molten stage and that as it slowly cooled the earth began to solidify at the surface, thus developing a rigid crust.[*] Then, as the interior continued to lose heat and to shrink, the crust, like the skin of a shriveled apple, had to accommodate itself to a smaller radius. At the time this seemed to be a plausible explanation of mountain folds and thrust faults.

So deeply entrenched was the belief in a

[*] This is the origin of the term crust but, as explained earlier, it now has a quite different connotation.

cooling earth that by the middle of the century the distinguished physicist Lord Kelvin calculated, on the basis of present heat flow to the surface, that the earth was still completely molten as late as 100,000,000 years ago and probably less than 40,000,000 years ago. Geologists and biologists were reluctant to accept such limitations of geologic time but lacked convincing evidence to oppose the physical "proof" until the discovery of radioactivity around 1900. Then it became evident that the internal heat is probably radiogenic and that it may have been increasing throughout geologic time.

Subsequent geochemical studies by Urey [25] have provided convincing evidence that during its formative stages the earth was relatively cool throughout. Furthermore, we now know from radioactive isotopes that crustal rocks were already in existence more than 4 billion years ago and that life was present more than 3 billion years ago. All this, however, left the cause of diastrophism as much of an enigma as ever.

THEORY OF THERMAL CONVECTION IN THE MANTLE

A breakthrough to new thinking occurred in 1934 when the Dutch geophysicist Vening-Meinesz advanced the theory of thermal convection in the mantle [27]. It is well known that if a flame is placed under a shallow vessel of fluid, the part directly above the flame ex-

pands and is forced to the surface where it flows laterally until it cools and condenses and settles back to the bottom to be reheated. Thus a thermal convection cell is maintained (Fig. 4-7).

Vening-Meinesz knew that the mantle is not fluid but is dense and rigid. But he also knew that under suitable conditions of heat or pressure, rocks can yield slowly by plastic flow. Ice, for example, is a rock, and it is well known that a glacier will flow down its valley as a river of solid ice which remains rigid and brittle throughout. And in the eroded stumps of old mountains we see great masses of granitic rock that have flowed like the dough of a marble cake (Fig. 4-8).

Vening-Meinesz reasoned that the heat in the mantle is radiogenic and may not be uniformly distributed. Then wherever it reaches a critical level the mantle rock will expand and move upward by plastic flow, lifting the crust in a wide bulge from which it spreads laterally until it gradually cools and becomes dense enough to sink slowly back into the depths, only to be reheated and rise again. Thus a convection cell is maintained as long as excess heat persists in depth. The flow must be slow like that of a glacier but vastly more irresistible, and as it moves laterally it must exert a drag on the overlying crust and will place the bulge in the crust under tension. Meanwhile, where the current descends again it must tend to pull the crust down as it is known to sag in the geosynclines. As a corollary it may be supposed that the descent of cooled material will in the course of millions of years restore the heat balance in depth so that convection will cease here. Meanwhile the temperature may build up in other regions to generate new convection cells.

This spectacular theory almost immediately gained the support of a few geophysicists, notably Harry Hess [9], David Griggs [8] and C. L. Pekeris [19], and as sensational supporting evidence has come to light during the last two decades it has become the dominant working hypothesis of nearly all geologists and geophysicists. We must therefore examine the evidence.

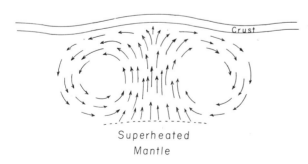

Figure 4-7 Diagrammatic vertical section of a thermal convection cell.

Figure 4-8 Flow structure in granite gneiss near Stone Mountain, Georgia. (Carl O. Dunbar.)

EVIDENCE OF CONVECTION IN THE MANTLE

The Mid-Ocean Ridges. Within the last two decades oceanographers have discovered several enormous linear bulges in the ocean floors. One in the Pacific west of South America has been under intensive study by the staff of Scripps Institute of Oceanography; in 1960 Menard [18] described and named it the East Pacific Rise (see Map I, paleogeographic map section, p. 120). It runs generally north-south through the South Pacific but passes under the edge of the North American continent in Mexico and continues under the mountainous highlands as far north as Alaska. In the Southern Pacific it is several hundred miles wide and its crest rises as much as 3 miles above the abyssal plains on either side. Temperature probes at more than 100 places along the summit revealed the heat flow to range from 3 to 8 times as great as is normal for the ocean floor, and seismic studies indicate that the crust is thinner than normal over the Rise. These discoveries seemed to indicate that the Rise is underlain by abnormally hot rock and that it is under tension.

Further spectacular discovery was made on the sea floor west of California in 1952 [16, 21]. When Ronald Mason was trailing a magnetometer behind a research vessel sailing west he observed a succession of alternating positive and negative magnetic anomalies. To discover the reason for this unexpected relation a regional survey was made by airborne magnetometers flown in east-west traverses a few miles apart. These revealed that Mason had crossed a series of linear bands of magnetic anomaly running generally north-south. It was immediately suspected that belts of high magnetism were due to the intrusion of dense magma along subparallel rifts in the crust and that they were intruded while the crust was under tension. Later discoveries led to an important modification of this interpretation (see following paragraphs).

Meanwhile, the Mid-Atlantic Ridge (Fig. 4-9) was being studied by the Lamont Oceanographic Laboratory and many traverses were made across it with sonic depth recorders. These proved that it is as large as the East Pacific Rise. It follows the middle of the Atlantic Ocean from the Antarctic Sea to the Arctic, it is several hundred miles wide, and its crest rises as much as 3 miles above the level of the abyssal plains on each side. In 1960, Heezen announced the sensational discovery that throughout its length the crest of

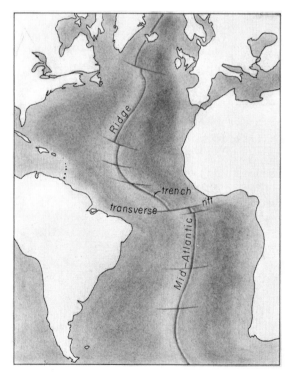

Figure 4-9 Map to show the relations of the Mid-Atlantic Ridge. Adapted and greatly simplified from a map by Heezen et al. [11]. The ridge is crossed by numerous great transverse rifts only a few of which are shown. Along some of these rifts the ridge is offset by many miles.

the Ridge is divided by an enormous trench several miles wide and as much as a mile deep [12]. This is obviously not an erosional valley but a great fault-bordered rift, indicating that the ridge is under tension and here the sea floor is being pulled apart.

The discovery of reversals of the earth's polarity added further insight into the spreading of the ocean floor [3]. As the crust is pulled apart along the crest of a mid-ocean ridge, magma rises along the rift to form new crust. As it cools its ferric minerals are magnetized and in middle or high latitudes of the northern hemisphere they add to the normal magnetic pull while the polarity is normal (page 75 and Fig. 4-6). This forms a belt of positive magnetic anomaly. Then at a later time of reversed polarity the magnetized crystals oppose the normal magnetic pull and, even though there is no change in the composition, this forms a belt of negative anomaly along the middle of the previous band of positive anomaly. As spreading continues the center of the rise is bordered by symmetrical bands of alternating positive and negative anomalies as indicated in Fig. 4-10.

Since the width of each band of anomaly is directly determined by the length of the period of polarity, it is possible to compare them with the chronology established in the Hawaiian basalts, as indicated in Fig. 4-6. This provides the means of measuring the rate of spreading of the ocean floor. Detailed survey of two areas on the Mid-Atlantic Ridge [3, 20], one near latitude 30°N and another near latitude 70°N, show the anticipated arrangement of paired bands of magnetic anomaly (Fig. 4-11) and they indicate that for the last few million years the spreading of the ocean floor has been at the rate of about 1.5 cm per year [20]. Of course the bands of anomaly are more irregular than those in our idealized diagram because they represent tears in the crust.

When the bands of magnetic anomaly were first discovered off the California coast this explanation could hardly have been foreseen since only the western flank of the East Pacific Rise is exposed there. But similar studies have since been made in the South

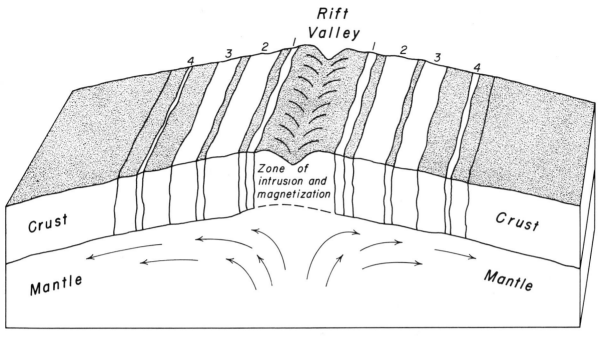

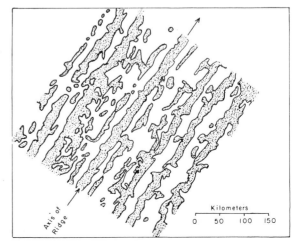

Figure 4-10 Diagram to illustrate how spreading of the crust along a mid-ocean ridge during alternating periods of normal and reversed polarity will produce symmetrically paired bands of positive and negative magnetic anomaly, here numbered in sequence. (Adapted from Cox et al. [3].)

Figure 4-11 Map of actual bands of positive and negative magnetic anomaly in a large area straddling the Mid-Atlantic Ridge near 70°N latitude. Positive bands are shaded. (After Cox et al. [3].)

Pacific and in the South Atlantic where the rate of spreading has ranged from 2 to 4 cm per year [28].

The distribution of bottom sediments across the Mid-Atlantic Ridge gives quite independent evidence of spreading and shows that it has not been continuous. In 1967 Ewing and Ewing [6] reported that there is a belt 100 to 150 miles wide along the Mid-Atlantic Ridge in which bottom sediment is very thin so that coring tools frequently reach the basement rock. These deposits thicken slowly toward the flanks of the Ridge. Farther out on the flanks, however, the sediment is much thicker, and for a considerable distance the thickness is fairly uniform. They infer that the thicker deposits were formed during a long period of quiet before about 10,000,000 years ago and that since then the floor has been spreading. On the contrary, they find that on the East Pacific Rise an early cycle of spreading ended in Late Mesozoic or Early Cenozoic time, but was now renewed. It may be noted in passing that this coincided with the dying out of the great diastrophism in the Rocky Mountain region. The East Pacific Rise does not possess a great rift valley along its crest comparable with that on the Mid-Atlantic Ridge but in-

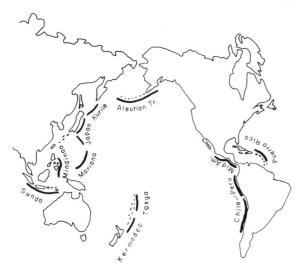

Figure 4-12 Island arcs and ocean trenches about the Pacific Ocean basin.

stead wears a long slender belt of basalt rock recently dated as about a million years old. Apparently the breach is being filled with new crust [1].

UNDERTHRUSTING OF THE CONTINENTAL MARGINS

Study of the margins of the Pacific Ocean reveal another striking fact. This vast ocean basin is rimmed by great chains of mountains on the continental margins; with one exception, these are closely paralleled by deep ocean trenches (Fig. 4-12). Frequent volcanic eruptions and great seismic activity have earned this circumpacific belt the name "ring of fire."

The eastern margin of the Pacific is particularly interesting. The lofty Andes stand close to the coast of South America all the way from Panama to southern Chile, and they are closely paralleled by a chain of great ocean trenches. Deep focus earthquakes are common along the Andes and the depth of the focus of each has been determined by seismographs [30]. They increase in depth inland under the

mountains indicating a zone of ruptures in the crust that dips inland at about 45°. Roughly paralleling this zone of disturbance but far out in the Pacific lies the East Pacific Rise from which the ocean floor is spreading eastward. These relations indicate that the spreading ocean floor is being thrust under the continent of South America (Fig. 4-13).

The contrast with relations of the west coast of North America is striking and significant. Here the Rise is under the continental margin and the spread of the ocean floor is westward away from the continent. This is the only part of the Pacific margin that lacks ocean trenches (Fig. 4-12).

In contrast, the western side of the Pacific is rimmed by a series of great island arcs, each closely paralleled on the seaward side by profound ocean trenches. Each of the island arcs is crowned by volcanoes and is the locus of intense earthquake activity. And here, as along the coast of South America, the focus of the earthquakes increases in depth toward the land, indicating a zone of underthrusting plunging down under the island arcs [7, 30].

Recent discovery that the sedimentary deposits within the trenches along the coast of South America are undeformed [24] might seem to throw doubt on the underthrusting, but Elsasser has shown that if the crust is rigid enough to carry the thrust, then the trenches are the result of local tension and have been simply downfaulted [5].

THE MISSING SEDIMENTS OF THE OCEAN FLOOR

Another far-reaching discovery of recent years concerns the general absence of ancient sedimentary rocks on the ocean floor. Estimates of the current rate of deposition, based on several lines of evidence, have been made by many different oceanographers. Most of these estimates have been based on bottom cores many feet in length and the rate obviously differs from region to region and with distance from land, but most of the estimates for the deep ocean range between 250 and 500 years per centimeter for the uppermost deposits. It is obvious that the more deeply buried sediments must have suffered compaction and may have been reduced to perhaps half or less than half of their original thickness. One particularly interesting core, studied by Hough [13], was taken by the U. S. Navy about 1200 miles northwest of Peru where the water is 11,880 feet deep. The core is 194 cm long and consists of alternating layers of red clay and of globigerina ooze. Radiometric dates were determined at 28 depth levels in the core and the oldest indicates that the bottom of the core is about 800,000 years old. The estimated rate of deposition for the upper layers was 1 cm in 440 years but for the older part between 338,000 and 700,000 years old the rate was only 1 cm in 6033 years. Now if we accept an average rate of 1 cm per 4000 years, then in one million years the thickness should amount to 8.2 feet; thus since Early Cambrian time 4920 feet of sediment should have accumulated and since Early Precambrian time about 26,220 feet should have accumulated.

A more significant record was secured from 3 cored wells drilled by the Mohole Project in the Pacific floor about 150 miles west of the coast of Mexico where the water is almost 12,000 feet deep [22]. These cores went to a depth of 560 feet, ending in basalt that has been dated by radioactive isotopes as 32,000,000 years old. This date is in good agreement with fossils found near the bottom of the cores which prove a Miocene age. The average rate of accumulation proves to be about 1 cm in 1880 years, and it is obvious that much of this material has suffered about as much compaction as possible. At this rate the deposits formed since Early Cambrian time should be about 10,640 feet thick and since Early Precambrian time it should be 56,000 feet. This is nearly twice the thickness of the entire suboceanic crust.

Seismic reflection studies have now been made at many localities over all the ocean basins and by many oceanographers and they have shown that the sedimentary rocks on the ocean floor are surprisingly thin, and no Paleozoic deposits have been recognized anywhere.

Figure 4-13 Diagram to illustrate thrusting of the crust under the western margin of South America. As the basaltic crust of the ocean floor reaches a depth of 25,000 to 30,000 feet, the rising temperature causes partial melting of some elements in the basaltic crust and this generates magma that rises to feed the great chain of volcanoes that crowns the Andes. (Based on a diagram by J. Tuzo Wilson)

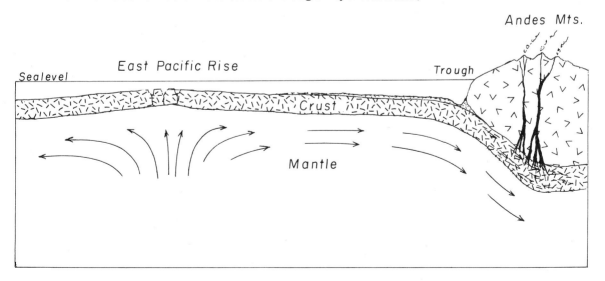

The continental shelf along the eastern side of North America is formed of Cretaceous and younger rocks [11]. In cross section they have the relations indicated in Fig. 4-14. Only the outer part of the shelf deposits are shown here; they thin to a feathered edge in the coastal plain about 150 miles farther west.

This vast deposit of Cretaceous and younger sediments built seaward from the land forms a wedge about 200 miles wide and at least 2 miles thick at the edge of the shelf that rests directly on basement metamorphics. If the Atlantic Ocean bordered the continent during the long Paleozoic Era and the Early Mesozoic, a vastly greater mass of sediment should have built out into the ocean. But no such sediments are present under the shelf and seismic studies show that they are not buried beneath the ocean floor. This absence of sedimentary record combined with the evidence of spreading of the ocean floor led Dietz to the daring and attractive hypothesis that all the Paleozoic and Early Mesozoic deposits of the ocean floor and the continental shelf were pushed under the continental margin by spreading of the ocean floor. There they were metamorphosed beyond recognition, thickening the edge of the continental plate and thus contributing to the rise of the Appalachian region [4]. Can this

be true, or did the Atlantic Ocean not appear until the continents drifted apart early in Mesozoic time? Or did both processes play a part?

CONTINENTAL DRIFT

For almost a hundred years geologists have been intrigued by the distribution of Late Paleozoic glaciation when vast ice sheets like that of modern Antarctica covered much of central and southern Africa and of Australia, New Zealand, and parts of South America and peninsular India. Equally impressive was the association of the "tongue ferns," *Glossopteris* and *Gangamopteris*, which in each of these regions are found with the glacial deposits.

In 1883 and 1885, two distinguished Viennese geologists advanced the hypothesis that the glaciated regions are parts of a great continent which they called **Gondwanaland** and that in the intervening oceanic regions the crust had foundered. It is now known that such foundering of parts of Gondwanaland never occurred because if such masses of crustal rocks were buried in the ocean floor, they would be easily revealed by gravimetric and seismic investigation.

In 1912 Alfred Wegener [29] advanced the

Figure 4-14 Cross section of the outer part of the Atlantic continental shelf at Cape Hatteras. (Simplified from a figure by Heezen et al. [11].) The section is based on logs of deep wells, on the consistent dip in numerous wells and on benches along the continental rise. The Hatteras No. 1 well reached the basement crystalline rocks at a depth of 9840 feet.

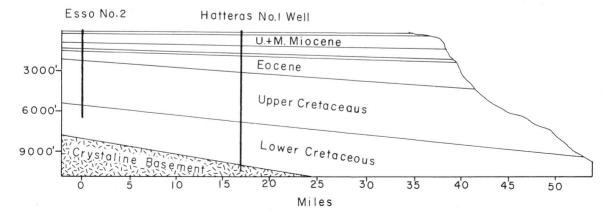

new theory of continental drift. He proposed that the southern continents and India are fragments of a great southern landmass centered about the south pole in Paleozoic time — a landmass which he called **Pangea** (Fig. 13-25). Thus he accounted for the glaciation and the associated *Glossopteris* flora. Then, he reasoned, this great landmass broke up into the present continental units which slowly drifted apart to their present positions. As supporting evidence, he pointed to the remarkable parallelism of the opposite margins of the Atlantic Ocean.

Wegener's concept immediately challenged the interest of both geologists and biologists, for if the southern continents had once been united, the fossil record should reflect the fact. But to physical geologists there was one fundamental obstacle. The force required to push a continent thousands of miles would be colossal. No such force could then be imagined and it was argued that if a continent had been so moved it would act like a great bulldozer pushing mountains of rock ahead of it and leaving a wide belt of tension in its wake. Thus continental drift became the subject of one of the great controversies in geology with an enormous and contradictory literature, much of which now seems irrelevant. If convection currents in the mantle are as real as now believed, they supply the force and the continents merely float and drift like great tabular icebergs in the sea, carried along without friction on the flowing mantle. The fossil record can tell us much about the time of the breakup of Pangea and it proves conclusively that the continents have been separated at least since Cretaceous time.

The Evidence from Paleomagnetism. The remarkable discoveries of geophysics on the nature of the earth's crust and mantle during the past two decades have stimulated new evaluations of the theory of continental drift [2, 14, 17]. Some of the most recent and most convincing support for the theory has come from the study of paleomagnetism [15, 23]. When the dipole of remanent magnetism in a rock mass is determined, it will indicate the direction of the magnetic pole, and the inclination will indicate the latitude with respect

to the magnetic pole when the rock was formed (page 75). If this does not agree with present world geography, we must conclude either (1) that the magnetic pole has migrated or (2) that the land mass has moved. Wherever tested in rocks of Oligocene or younger age the paleomagnetism fits modern world geography, but when applied to Mesozoic and Paleozoic rocks, the results are surprisingly different, showing that the magnetic poles have wandered or the continents have moved. For example, a large number of measurements in the Permian rocks of North America agree well in indicating the north magnetic pole to lie near Peking in eastern China (Fig. 4-15). But comparable measurements in the Permian rocks of Europe indicate that the pole then lay in the Pacific Ocean east of Japan, while measurements in the Permian rocks of Africa place the pole in central Canada. Such discrepancies could not possibly be accounted for by polar wandering.

Van Hilten [26] has shown that another significant relation can be inferred. Once the inclination is determined, we know the distance, in degrees, to the magnetic pole, and a circle about the pole running through the mass of rock concerned must form an isocline along which the inclination will be the same. Then, by using the knowledge of how the inclination varies with the latitude (Fig. 4-5), other isoclines can be drawn at any desired latitudes and the equator will be a great circle about the magnetic axis.

Paleomagnetism does not indicate how far the continents have moved longitudinally. But the isoclines suggest rotation. In Fig. 4-15, for example, it is evident that if North America were rotated counterclockwise and pulled somewhat to the south, its isoclines could be brought into line with those of Greenland and Europe. At the same time, the polar position for these three landmasses would be brought into close agreement. Asia, east of the dashed line, would need to be moved slightly to the north. Africa could be brought into agreement if it were rotated clockwise and moved far to the west. In short, one can hardly avoid the conclusion that since Permian time, the continents have drifted **and** rotated from their

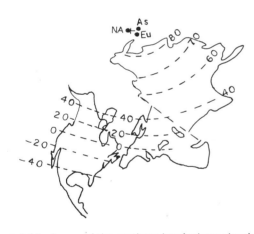

Figure 4-15 A map of the northern hemisphere showing orientation of the paleomagnetism in Permian rocks. Cross marked NP is present north pole; other crosses mark estimated Permian north poles based on paleomagnetic data from Africa (Af), Asia (As), Europe (Eu), Greenland (Gr), and North America (NA). (Upper map after Van Hilten [26].)

original positions. It is highly significant also that when similar maps are prepared for Triassic, Jurassic, Cretaceous, and Eocene times, the divergences among the continents de-

crease progressively and in Oligocene and younger rocks all the continents are oriented in harmony with modern world geography.

SUMMARY

Most of the concepts reviewed in this chapter have grown out of recent discoveries, still the subject of intensive study, and some of them are controversial. In such brief treatment it is impossible to do justice to the enormous literature dealing with these problems. Our aim has been to give a simple introduction to modern thinking on some of the exciting, current problems that profoundly affect the interpretation of the earth's history.

If our reasoning is valid thus far, we can understand why the rises produced by convection generally lie close to the middle of the ocean basins—the lateral flow of the mantle has simply floated the continents into the intervening more stable regions. We can also understand why the granitoid crust is concentrated into continental units instead of forming a universal sheet over the planet. Such relatively light material, floating on the mantle, has been carried away from convection cells and added to the intervening continents. Indeed there is clear evidence that each of the continents has grown by such lateral accretion.

By the same token, if a new convection cell should arise under a continent, it would produce a rift from which opposite sides would drift apart. Such a movement is actually in progress now where the East Pacific Rise passes under northwestern Mexico and a slice of the mainland has been pulled away to form the peninsula of Baja California, leaving the Gulf of California in its wake. This break extends northwest across California as the great San Andreas Rift. The land west of the rift is still moving to the northwest and Crowell has found evidence that during Cenozoic time it has traveled some 300 miles. Recently discovered bands of gravity anomaly in the floor of the southern end of the Gulf of California indicate that it has been widening at the rate of 1 to 2 cm per year.

One remarkable feature of the ocean floors remains and is not yet understood. The mid-ocean ridges are crossed in many places by enormous slip faults. Several of these are indicated in Map I. In many of these, differential movement on opposite sides of a fault is to be measured in tens or even hundreds of miles and they have resulted in offsets in the axis of the rises.

REFERENCES

1. Bonatti, Enrico, 1968, *Fissure basalts and ocean-floor spreading on the East Pacific Rise.* Science, v. 161, pp. 886–888.
2. Bullard, E. C., 1964, *Continental drift.* Quart. Jour. Geol. Soc., London, v. 120, pp. 1–33.
3. Cox, Alan, Dalrymple, G. B., and Doell, R. R., 1967, *Reversals of the Earth's magnetic field.* Scientific American, v. 216, no. 2, pp. 44–54.
4. Dietz, R. S., 1962, *Ocean-basin evolution by sea-floor spreading.* Natl. Research Council Publ. 1035, pp. 11–12.
5. Elsasser, W. M., 1968, *Submarine trenches and deformation.* Science, v. 160, p. 102A.
6. Ewing, John, and Ewing, Maurice, 1967, *Sediment distribution on the Mid-Atlantic Ridges with respect to spreading of the sea floor.* Science, v. 156, pp. 1590–1592.
7. Gutenberg, B., 1959, *Physics of the earth's interior.* New York and London, Academic Press.
8. Griggs, David, 1939, *A theory of mountain building.* Am. Jour. Sci., v. 237, pp. 611–650.
9. Hess, H. H., 1938, *Gravity anomalies and island arc structures.* Am. Philos. Soc. Proc., v. 79, pp. 71–95.
10. Hess, H. H., 1968, *Drifting continents and spreading sea floor.* Rept. 11th Ann. Mtg., Natl. Research Council, pp. 61–67.
11. Heezen, B. C., Tharp, Marie, and Ewing, Maurice, 1959, *The floors of the oceans.* Geol. Soc. America Spec. Paper 65, pp. 1–122.
12. Heezen, B. C., 1960, *The rift in the ocean floor.* Scientific American, v. 203, no. 4, pp. 98–114.
13. Hough, J. L., 1953, *Pleistocene climatic records in a Pacific Ocean core sample.* Jour. Geology, v. 61, pp. 252–262.
14. Hurley, P. M., 1968, *The confirmation of continental drift.* Scientific American, v. 218, no. 4, pp. 52–64.
15. Irving, E., 1964, *Paleomagnetism and its application to geological and geophysical problems.* New York, John Wiley & Sons.
16. Mason, R. G., and Raff, A. D., 1961, *Magnetic survey of the west coast of North America. 32°N latitude to 42°N latitude.* Geol. Soc. America Bull., v. 72, pp. 1259–1266.
17. Maxwell, J. C., 1968, *Continental drift and a dynamic earth.* Am. Scientist, v. 56, no. 1, pp. 35–51.
18. Menard, H. W., 1961, *The East Pacific Rise.* Scientific American, v. 205, no. 6, pp. 52–61.
19. Pekeris, C. L., 1935, *Thermal convection in the interior of the Earth.* Mon. Not. R. Astron. Soc. Geophys. Sup., v. 3, pp. 343–368.
20. Phillips, J. D., 1967, *Magnetic anomalies over the Mid-Atlantic Ridge near 27°N.* Science, v. 157, pp. 920–922.
21. Raff, A. D., and Mason, R. G., 1961, *Magnetic survey off the west coast of North America, 40°N latitude to 52°N latitude.* Science, pp. 1267–1270.
22. Reidel, W. R., et al., 1961, *Preliminary drilling phase of the Mohole project 2, Summary of coring operations.* Am. Assoc. Petroleum Geologists Bull., v. 45, pp. 1793–1798.
23. Runcorn, S. K., 1963, *Paleomagnetic methods of investigating polar wandering and continental drift.* Soc. Econ. Paleontologists and Mineralogists Spec. Pub. 10, pp. 47–54.
24. Scholl, D. W., von Huene, R., and Ridlon, J. B., 1968, *Spreading of the ocean floor: Undeformed sediments in the Peru-Chile Trench.* Science, v. 159, pp. 869–871.
25. Urey, H. C., 1952, *The planets, their origin and development.* New Haven, Conn., Yale Univ. Press.
26. Van Hilten, D., 1964, *Evaluation of some geotectonic hypotheses by paleomagnetism.* Tectonophysics, v. 1, pp. 3–71.
27. Vening-Meinesz, F. A., 1964, *The Earth's crust and mantle.* Amsterdam, London, and New York, Elsevier.
28. Vine, F. J., 1966, *Spreading of the ocean floor: new evidence.* Science, v. 154, pp. 1405–1415.
29. Wegener, A., 1966, *The origin of continents and oceans.* New York, Dover Publications.
30. Wilson, J. T., 1950, *An analysis of the pattern and possible cause of young mountains and island arcs.* Geol. Assoc. Canada Proc., v. 3, pp. 141–166.

Figure 5-1 Remote galaxies. *A*, Ursa major, seen from a polar view; *B*, Coma Berenices, seen from its equatorial plane. (Mount Wilson and Palomar observatories.)

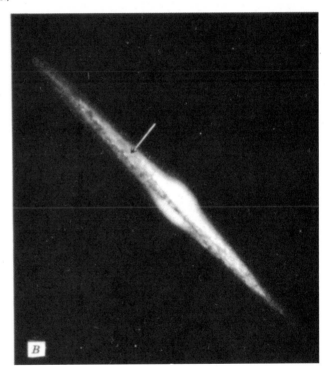

CHAPTER **5** *Cosmic History of the Earth*

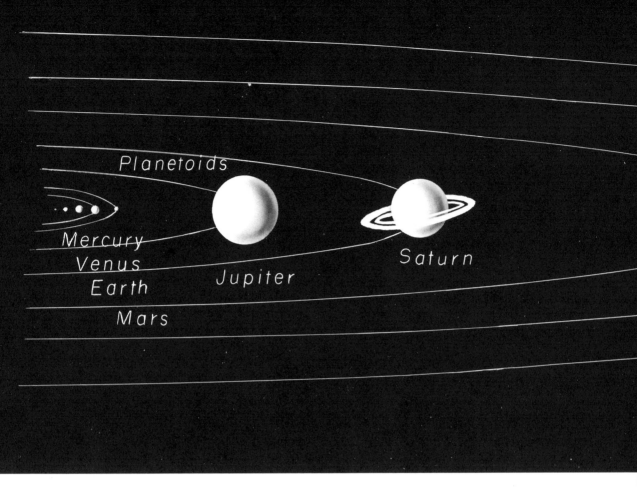

Figure 5-2 Diagram of the Solar System seen from an oblique view above the plane of rotation. If viewed from a polar direction the orbits would be seen to be almost circular.

Beginnings. The origin of the world has intrigued man since time immemorial. Few cultures have been without legends of the Creation, but until the Renaissance such beliefs were generally fanciful because men had little real knowledge of the physical universe or of the natural laws that govern it. Even in this scientific age there are questions which we cannot hope to answer—the origin of matter, the beginning of time, and the limits of space lie far beyond the scope of modern scientific inquiry and of human understanding. Nevertheless, it is now certain that the earth, as such, is not eternal—it was "molded out of star dust" and set wheeling about the sun a few billions of years ago. Any attempt to understand this cosmic event must therefore begin with a brief

survey of the universe about us.

Our Galaxy. All the stars visible to our eyes belong to a well organized system. Since we are deep in its midst its shape is not readily perceived, but there are many similar galaxies far out in space that can be seen through the great telescopes. Two of these are illustrated in Fig. 5-1, one seen from the polar view and the other from the equatorial plane. The shape of Ursa Major strongly suggests that it is rotating so that centrifugal force prevents it from collapsing into a single great mass. This can be confirmed in Coma Berenices, for when the spectroscope is trained on one limb it is seen to be receding from us at a very high velocity while the opposite limb is approaching at equal velocity.

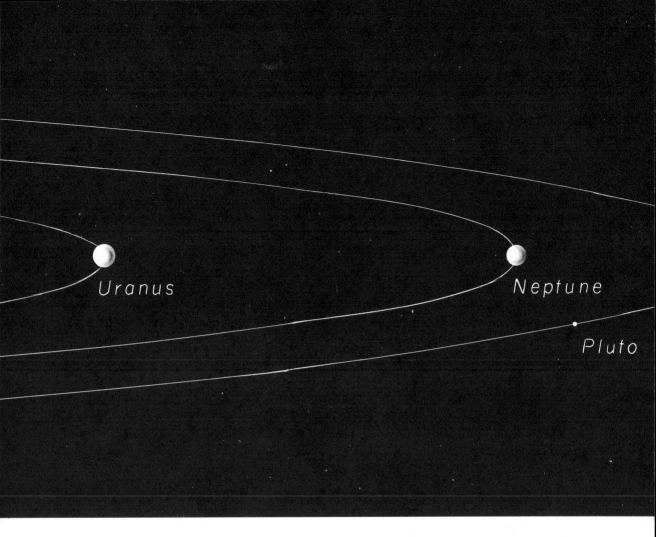

Our galaxy is similarly shaped, its equatorial diameter being more than 100,000 light years * and its polar diameter about one-tenth as great. We are situated in about the position indicated in the white arrow in Fig. 5-1B.

As we look out in the polar direction the stars appear bright and widely spaced, but as we turn our gaze toward the equatorial plane more and more stars crowd into view, most of them appearing small and far away, and finally they blend into a belt of faint light that girdles the night sky. The ancients observed this belt of light and called it the **Milky Way** and from this a universe of stars has come to be termed a galaxy [Gr. *laxa*, milk]. As in Ursa Major, the stars in our galaxy are not evenly distributed but are in places grouped in clusters.

THE SOLAR SYSTEM

The sun is relatively isolated in our galaxy, most of the near stars being tens or hundreds of light years away, with the nearest about 4.2 light years away.

Revolving about the sun are nine planets and their moons, more than 1500 asteroids, and an unknown number of comets and meteors (Fig. 5-2). These constitute the Solar System. It is

* A light year is not a unit of time but of distance—the distance light will reach in a year's time traveling at a velocity of 186,000 miles per second. This amounts to 5,865,696,000,000 miles.

Figure 5-3 A portion of the sun's surface photographed in the red light of the hydrogen α line revealing swirling clouds of hot gases. (Courtesy of Mount Wilson Observatory.)

completely dominated by the sun, which includes more than 98 percent of its total mass and holds the other members in its gravitational control while supplying them with light and warmth. These celestial bodies form a close-knit family and evidently have shared in a common origin.

The Sun. The sun is a true star some 860,000 miles in diameter and so hot that it is self-luminous and entirely gaseous. Recent research by means of rockets [15] has shown that the outer reaches of its atmosphere have a temperature of about 4727°C, increasing to about 100,000°C at a greater depth; it is inferred that near its center the temperature may be as great as 20,000,000°C. As a result of this great heat the sun constantly radiates energy and a great flux of atomic particles into space.

Although we intercept only about one-two-billionth of this radiant energy, it is sufficient to provide the life-giving warmth to the earth, to enable plants to build organic compounds by photosynthesis, to lift the vapors that return as rain, to keep our atmosphere in motion, and thus to motivate all the forces of erosion on the earth. Furthermore, geologic evidence now proves that this prodigious flow of energy has never flagged nor greatly varied for more than 4000 million years. Obviously the history of the earth could not be understood without considerable knowledge of the sun.

The sun is made of the familiar chemical elements that exist on earth, but in proportions quite different. About 99 percent of the sun's mass consists of hydrogen and helium. The source of its heat was a mystery until atomic energy was understood. Now it seems evident that the interior of the sun is a furnace in which hydrogen is being transformed into helium by atomic fusion as in a hydrogen bomb and perhaps other light elements are being built into heavier ones. As the specter of hydrogen bombs hangs over a troubled world we are inclined to regard atomic energy with fear, little realizing that without it our world would be cold and lifeless!

The visible disc of the sun, the **photosphere,** appears smooth to the naked eye, but seen through a great telescope it presents a billowy surface of seething white-hot clouds (Fig. 5-3). And with the aid of the spectroscope we see an outer layer of crimson gas from 5000 to 10,000 miles thick, which because of its color is known as the **chromosphere.** This outer atmosphere is clearly visible at total solar eclipse, when the moon blocks out the white light from the photosphere, and then we see solar prominences which from time to time leap up from the chromosphere like great tongues of crimson flame reaching heights of many thousands of miles, bearing evidence of the enormous explosive energy in the sun. At time of total eclipse we also see the chromosphere surrounded by the **halo** of faint light that reaches far out into space (Fig. 5-4), at times even touching the earth. This is the flux of atomic particles that give rise to the **aurora borealis** and cause the electric storms that interfere with radio and television transmission.

Although the sun is the center of our solar system, it is not stationary in space; with its retinue of satellites it is plunging through space toward the bright star Vega at a velocity of about 11 miles per second. Meanwhile, it is rotating on its own axis in a period of about 25 days. There is no prospect of a collision since Vega is extremely remote and is moving across our path so that we will probably never be nearer to her than we are now.

The Planet Earth. Nine planets revolve about the sun in nearly circular and concentric orbits. In Fig. 5-2 they are represented as if the solar system were viewed from an oblique view, the orbits appearing elliptical instead of circular. The sizes of the planets are represented on a uniform scale and the orbits on a uniform but much smaller scale. The reason for the different scales needs explanation. If the scale of distances were made to agree with the sizes of the planets, then in our diagram Mercury would be placed 24.8 feet from the sun, Earth 64 feet, Jupiter 332.8 feet, and Neptune 1923.3 feet from the sun!

The planets fall into two dissimilar groups. The inner planets, Mercury, Venus, Earth, and Mars, are relatively small and are solid, whereas the outer planets, Jupiter, Saturn, Uranus, Neptune, and Pluto, are vastly larger and are almost completely gaseous.

Earth has a diameter of about 8000 miles and is about 93,000,000 miles from the sun. It travels at a speed of about 18 miles per second in its yearly journey about the sun and rotates daily on its polar axis which is inclined 23° to the plane of the ecliptic (the common plane of its orbit). The nature and structure of the body of the earth was discussed in Chapter 3. Its atmosphere is relatively thin, consisting of nitrogen (78 percent), oxygen (21 percent), the inert gas argon (0.93 percent), and carbon dioxide (0.003 percent). It also holds a highly variable amount of water vapor.

The composition of the earth's atmosphere differs greatly from that of the other planets and is uniquely fitted to support life. Carbon dioxide is used by plants to build organic compounds and these in turn provide food for all

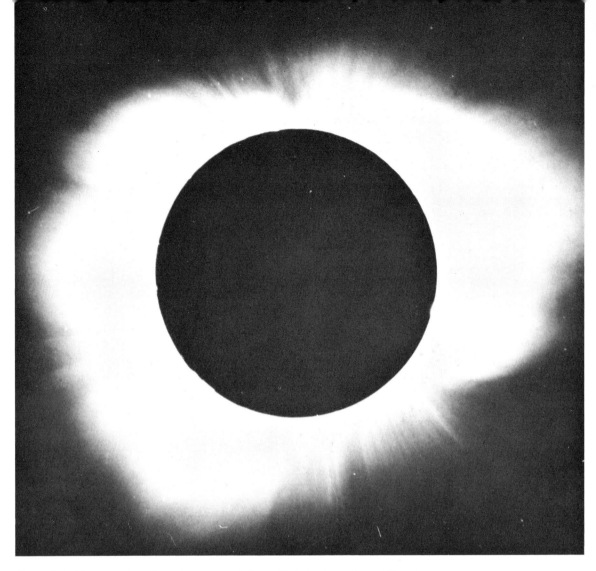

Figure 5-4 Photograph of the solar corona at time of total solar eclipse. (Mount Wilson and Palomar observatories.)

animal life. But if CO_2 were more abundant— say as much as 3 percent of the atmosphere— it would be lethal to all kinds of animal life. Oxygen is absolutely essential for the metabolism of animal life, but oxygen has such great affinity for carbon and hydrogen that if the atmosphere were pure oxygen all organic matter would be consumed in a sudden flash of fire. Nitrogen is relatively inert and serves to dilute the oxygen to a safe level.

Until the present century the atmosphere could be explored only by climbing to mountain tops, and it was then supposed that its density decreased upward with little change in composition and that its temperature de-

clined gradually with elevation. With the development of high-flying planes, and especially of space rockets, it was discovered that the atmosphere is layered as indicated in Fig. 5-5.

The troposphere is composed of the zone of winds and of weather. Clouds seldom rise above this zone. The stratosphere is more tenuous and is relatively still, except for a jet stream in each hemisphere that flows sinuously at high velocity in a west to east direction. Near its summit is a zone rich in ozone (O_3). Here the impact of incoming ultraviolet light breaks some of the oxygen molecules apart $(O_2 = 2O^+)$ and the isolated atoms then join other molecules to form ozone ($O_2 + O = O_3$). This zone reflects radio and television signals back to earth and makes broadcasting possible. This relatively thin zone also absorbs most of the ultraviolet light, thus shielding the earth from these lethal rays. The temperature steadily rises through the stratosphere, then drops to far below 0°C at the top of the mesosphere and then rises rapidly through the ionosphere and eventually reaches a point of as much as 1700°C where much of the energy from the sun is absorbed and transformed into heat.

Small meteorites are heated to incandescence by friction and burn up at a height of around 50 miles above the earth's surface (Fig. 5-5).

The Van Allen radiation belts (Fig. 5-6) discovered by means of high-flying rockets [17]

Figure 5-5 Right, layers of the atmosphere. (Adapted from Nicolet.)

Figure 5-6 Below, diagram to illustrate the relation of the Van Allen radiation belts. It is designed to represent a median axial section of the system.

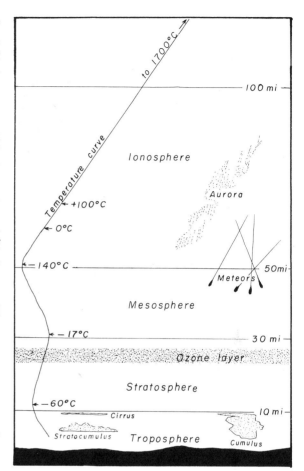

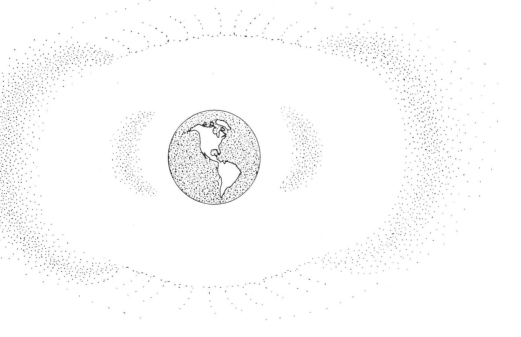

lie far above the atmosphere where lines of force in the earth's magnetic field trap incoming ionized atomic particles from the sun. Such particles then spiral back and forth along the lines of force until enough accumulate to spill out of the ends of the radiation belts and plunge earthward where their impact on atoms in the ionosphere produces the auroras. This explains why the aurora appears only in high latitudes.

It is evident, then, that the atmosphere plays a vital role in making the earth a safe abode for life.

The Moon as a Satellite. As the nearest to us of all the celestial bodies, 238,860 miles distant, our moon is in some respects the most interesting, and it may be the most significant in its bearing on the early history of the earth. Having a diameter of only 2100 miles and a mean density of 3.46, its mass is only about one-eightieth that of the earth and the pull of gravity at its surface is only one-sixth of that at the earth's surface. A 150-pound astronaut on the moon will weigh only 25 pounds; we can imagine his exhilaration as he leaps about!

Because the moon is not large enough to hold any atmosphere, its stark, barren surface is never obscured by haze or clouds and its major features were well known even before the recent soft landing of lunar satellites sent back such marvelous photographs at close range. The two major features are the great maria and the cratered uplands. The maria were once thought to be oceans [L. *mare*, sea] and they would be if water were present on the moon, for they are relatively level plains lying well below the rest of the lunar surface (Fig. 5-7). Several lines of evidence suggest that they are covered by layers of basaltic lavas [4]. Figure 5-8 shows isolated craters in the Sea of Tranquility, and Figure 5-9 shows an upland area on the back side of the moon completely covered with meteor pits.

The uplands appear lighter because their rough surface reflects more of the sunlight. The craters are enormous circular pits with relatively flat floors and lofty rims. They range in size from a fraction of a mile in diameter to giants such as Clavius, which is

146 miles across with a rim that towers about 20,000 feet above its floor. They were once thought to be volcanic craters, and possibly some of the small ones are volcanic, but most of them almost certainly were produced by the impact of great meteors that plunged below the surface and then exploded like atom bombs. The energy carried by a falling meteor is measured by the product of its mass and the square of its velocity [2], and since they travel at velocities of many miles per second a large meteor carries an enormous charge of energy. And when it is suddenly stopped this energy is transformed into heat. According to Beals [2], a meteor traveling at 20 miles a second possesses energy equivalent to that of 65 times its weight in nitroglycerine, and Baldwin has estimated that it would blast out a crater about 60,000 times its own volume [1]. Moreover, since the explosion occurs after burial, the crater is symmetrical and in the absence of an atmosphere the debris falls symmetrically back to form a circular rim. Since the debris in the rim is about equal in volume to that of the pit, it provides further evidence that the craters are not volcanic [2].

Of course meteors must have fallen at random, and since there are only scattered craters in the maria we must infer that the surface of the maria was formed relatively late in lunar history. This is also confirmed in places around the margins of the maria where they extend into great semicircular bays in the highlands as though part of the rim of a great crater had foundered and the lava of the floor of the maria had flowed into the crater.

The reason for the absence of a lunar atmosphere is well known and will have an important bearing on our interpretation of the history of the earth. The atoms or molecules in a gas behave like a swarm of Ping-Pong balls in violent motion. When two collide they rebound as perfectly elastic bodies and are

Figure 5-7 The surface of the moon at first quarter. (Lick Observatory.)

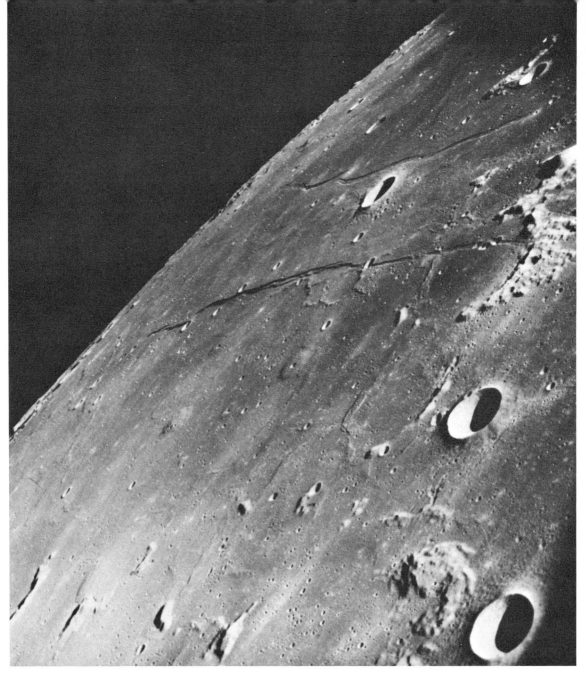

Figure 5-8 Picture of the Sea of Tranquility showing the flat surface with only a few large craters. (Courtesy of NASA.)

thus relatively far apart. In the lower part of our atmosphere they are so abundant that none travels far before colliding with others and rebounding. But in the rare upper reaches of the atmosphere they are so far apart that some fly away without collision. If the pull of gravity is strong enough, these atoms eventually fall back like spent missiles and rejoin the atmosphere. But if they leave with sufficient

velocity, or if the pull of gravity is weak enough, they will drift off into space and never return. The atoms of the lightest gases such as hydrogen and helium travel much faster than heavy atoms and thus for each gas there is a critical **escape velocity** that depends on the pull of gravity with which it is held. The earth is big enough to hold most gases, but hydrogen and helium have leaked away. But

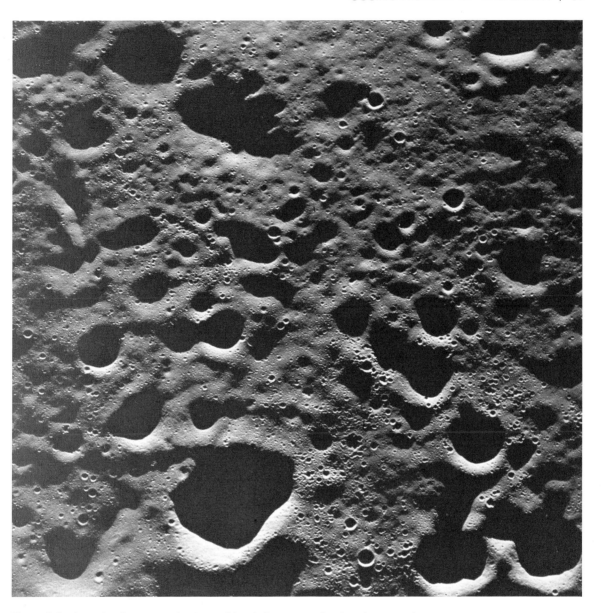

Figure 5-9 An upland area on the back side of the moon, showing the rugged surface almost covered by meteor craters. (Courtesy of NASA.)

the weak gravity of the moon would allow all the gases to escape and even if they were sweated out of the interior they would be lost as fast as they appeared at the surface. Lacking an atmosphere, the moon's surface experiences great extremes of temperature. Recently Surveyor 1 satellite found the temperature to be 235°F at lunar midday and −250°F during the lunar night [8].

Other Planets. Mercury, the nearest to the sun, is only 3000 miles in diameter and is too small to hold an atmosphere. Its surface probably resembles that of the moon but it is too remote from us and too near the sun for us to see any surface details.

Venus is slightly smaller than the earth and its mass is about four-fifths as great; but conditions on Venus are vastly different from those on the earth. It has a thick atmosphere which perpetually hides its rocky surface. The Russian satellite Venus 4 made a soft landing on the planet in October, 1967, and as it slowly descended through the atmosphere the temperature rose from 104°F to 536°F at the surface of the planet. This extreme temperature is believed due to the "greenhouse effect" of its opaque atmosphere, which lets the sun's heat penetrate to the rocky surface but prevents it from radiating back into space. Venus' atmosphere is about 15 times as dense as ours, exerting a pressure of about 323 pounds per square inch at the planet's surface. It was found to be 90 to 95 percent carbon dioxide with no more than 1.5 percent of O_2 plus H_2O, and 7 percent of nitrogen. The U. S. satellite Mariner 5 bypassed Venus on October 19, 1967, and generally confirmed the finding of the Russian satellite [11]. The satellites revealed that Venus keeps one face toward the earth and thus has very slow retrograde rotation [12] and that it lacks a magnetic field. These facts tend to confirm the explanation that the earth's magnetism is due to its spin about its iron core, as discussed earlier.

Mars has only about half the diameter of the earth and is able to hold only a very tenuous atmosphere. The remarkable photographs sent back by the U. S. satellite Mariner 4 in 1965 shows a surface like that of the moon scarred by great craters, one of which is 75 miles across [6].

Jupiter, the largest of all the planets, has a diameter of 88,600 miles and a volume more than a thousand times that of the earth, but its low specific gravity (1.36) proves that it is almost completely gaseous. Spectroscopic analysis of the light reflected from its upper layers reveals ammonia as the dominant gas and methane as second, but there is no water, and free oxygen certainly cannot exist in the presence of the other gases. This planet is so far from the sun that its surface temperature ranges between 100 and 140°C below zero and the clouds that mantle it are believed to be frozen crystals of ammonia, even as the high cirrus clouds on earth are formed of ice crystals.

Saturn has a diameter of 74,000 miles but its mean density of 0.72 indicates that it is entirely gaseous. Its upper layers, like those of Jupiter, consist of ammonia and methane.

The more remote planets, Uranus, Neptune, and Pluto, are also gaseous but they are so far away that we have little knowledge of their composition. We can surmise that at such immense distances from the sun their surface temperature must be very low.

From this brief survey it is evident that the earth is a unique planet—the only one having abundant water and an atmosphere of nitrogen and free oxygen.

Asteroids. With a single exception the orbits of the planets are so spaced that each is a little more than twice the distance of the next nearest to the sun. The exception is that, according to this scheme, another planet should exist between Mars and Jupiter. This is the belt of the asteroids—small solid bodies ranging from less than 5 miles in diameter to about 485 miles. Eros, one of the largest of these, varies in brightness in a way that suggests it is a solid chunk of rock of unequal dimensions tumbling end over end as it travels. The asteroids travel in more elliptical orbits than the planets do. In June, 1968, one came within 5 million miles of the earth. From all these relations it seems probable that the asteroids are fragments of a small planet that was disrupted early in the history of the solar system by tidal stresses as it passed and repassed planet Jupiter, and that some of the more eccentric orbits brought fragments near enough to the earth to be captured as meteors and comets.

Meteors. The "shooting stars" that streak across the night sky are meteors. Each is a solid body heated to incandescence by friction as it plunges into the atmosphere at a velocity of some miles per second. It is estimated that millions of them fall daily, but most of these are no larger than sand grains and are completely vaporized before reaching the earth. Large ones occasionally drive through the atmosphere so quickly that only a surface shell is vaporized while the interior is still cool. If their velocity is great enough, they plunge into the earth and then explode to form a crater like

Figure 5-10 Aerial view of Meteor Crater in Arizona, looking westward toward San Francisco mountain in the distance. The crater is approximately 4000 feet across and 550 feet deep. (Spence Air Photos.)

those on the moon. Meteor Crater, shown in Fig. 5-10, was formed in this way [2]. It is three quarters of a mile across and 550 feet deep and was blasted out of massive sandstone. Deep wells drilled in its floor penetrated hundreds of feet of broken and shattered rock and thousands of fragments of meteoric iron have been found on the surrounding surface out to a distance of 5 miles from the crater. In the rim of the crater also some of the quartz sand has been altered into coesite, a mineral that is formed only at extremely high temperature. A number of similar craters are known. Chubb Crater in the Ungava peninsula of Canada was blasted out of solid granite and is almost 2 miles across; it is now occupied by a lake about 800 feet deep [7].

Undoubtedly the earth has been scarred by

meteors as often as the moon has been, but here all the ancient craters have been destroyed by erosion, whereas those on the moon are the accumulation of time.

Occasionally one of the large meteors approaches slowly enough to be cushioned by the atmosphere and lands on the surface. These are then termed **meteorites**. The Weston meteorite (Fig. 5-11), which fell about 25 miles west of New Haven, Connecticut, in 1807, is especially interesting as the first to be seen to fall and it created such a sensation at the time that President Jefferson is said to have exclaimed, "It is easier to believe that two Yankee Professors lied than to believe that stones fall out of the sky!" Since then many meteorites have been picked up on the surface and a few have been seen to fall. In

Figure 5-11 Weston meteorite. This stony meteorite was one of several seen to fall, near the town of Weston, Connecticut, on the morning of December 14, 1807. The meteor was first seen over Rutland, Vermont, where it appeared as a "fire ball" about one fourth the diameter of the moon. To observers in Connecticut it appeared somewhat larger and was estimated to be traveling at about 3 miles per second while still some 18 miles above the earth's surface. A loud explosion was heard as it burst into fragments that rained to the ground. This specimen, weighing about 33 pounds, plunged through the sod in a pasture field and buried itself a foot or so below the surface. It was the first meteorite seen to fall, and the first accounts of it were received with skepticism. Thomas Jefferson, for example, who beside being President of the United States was also president of the American Philosophical Society, wrote to Daniel Salmon on February 15, 1808, "It may be difficult to explain how the stone you possess came into the position in which it was found. But is it easier than to explain how it got into the clouds from whence it is supposed to have fallen." (Yale Peabody Museum.)

1954, for example, an $8\frac{1}{2}$-pound meteorite crashed through the roof of a house in Sylacauga, Alabama, slightly injuring a woman[13].

Two distinct types of meteorites are recognized. **Stony meteorites** are made of silicate minerals and **metallic meteorites** consist of iron with small amounts of nickel. If meteorites are fragments of a disrupted planet the stony ones must represent the outer part and the metallic ones the core. Meteorites thus bear on the nature of earth's interior. In another respect they tell us something very interesting about the early history of our world. Radioactive minerals have enabled us to determine the age of many of the meteorites and it is significant that the dates all fall about 4,500,000,000 years. This is the best evidence we have of the date when the solar system came into being.

Comets. Comets appear from time to time as faint luminous objects in the sky and then fade again in a period of months as they recede from us in highly elliptical orbits (Fig. 5-12). Each displays a dense nucleus or "head" surrounded by a gaseous envelope that streams out into a long "tail." In its journey about the sun the comet grows brighter and its tail longer during its approach; and the tail is constantly directed away from the sun, trailing behind as it approaches, then swinging through a wide arc at perihelion, and rushing ahead like the flame of a blowtorch as the comet recedes. This behavior indicates that the tail is made of atomic particles driven away from the head by the radiant energy of the sun. Spectroscopic analysis of the tail shows that it consists of ionized gases, chiefly carbon, carbon monoxide, and carbon nitride; from this it is inferred that the head is a mass or masses of ices of H_2O, methane (CH_4), and ammonia (NH_3). It may be noted that the latter two are the predominant gases in the upper atmosphere of the major planets [3].

Origin of the Solar System

With this background we are prepared to contemplate the origin and early history of the earth. Three different hypotheses of the origin of the solar system were advanced during the last hundred years and each for a time seemed promising, but when critically examined each in turn proved unacceptable. These were (1) the Laplacian hypothesis, (2) the Planetesimal hypothesis of Chamberlin and Moulton, and (3) the Gaseous Tidal hypothesis of Jeans and Jeffreys. In the light of modern discoveries recounted in earlier chapters, these no longer are even considered.

The Dust Cloud Hypothesis. What happens

to the material expelled by the radiant energy of the sun and the myriad of other bright stars in the cosmos? It seems probable that it accumulates in the voids of space between the stars. If so, it eventually forms diffuse nebulae —a flux of atomic particles of many sorts. Then when enough such debris has accumulated it will begin to be drawn together because of mutual gravitational attraction and will form a gaseous nebula or "dust cloud." The modern great telescopes reveal many such diffuse nebulae among the stars (Fig. 5-13).

The Dust Cloud hypothesis has grown out of the work of several astromoners, notably Whipple [19], Spitzer [14], Kuiper [5], and von Weizsäcker [18]. It postulates that the solar system evolved out of such a nebula. For some

Figure 5-12 Brooks comet. (Yerkes Observatory.)

Figure 5-13 Nu Scorpii, a diffuse Nebula or dust cloud. (Yerkes Observatory.)

unknown reason the nebula began to rotate. Mutual attraction of all its particles caused it to shrink and become densest at the center. Kuiper showed that as it shrank the angular momentum must have increased so that the centrifugal force in the equatorial plane gradually transformed the nebula into a thin lens. And he thinks centrifugal force and internal gravity came into balance at all depths so that by the time it had shrunk to the diameter of the orbit of Pluto it had become stable and rotated as a unit, each particle revolving in a free, circular orbit. But in accordance to Kepler's third law, the period of revolution varied with the distance from the center, the inner part of the nebula turning much faster than the outer.

Von Weizsäcker continues the argument to the effect that this must have caused the nebula to subdivide into several concentric zones (which he called shells) where the fric-tion between levels traveling at different velocities generated great cells of turbulence as indicated in Fig. 5-14. This, however, creates a dilemma, for all such cells must have rotated in a retrograde sense, whereas all the planets are prograde. To avoid this difficulty von Weizsäcker explains that the friction between adjacent cells in any zone would generate an intermediate cell rotating in a prograde direction and he argues that condensation would be most rapid here and the prograde cells would develop gravitational fields capable of draining material from the retrograde cells until they eventually disappeared. Thereafter each zone in the nebula would have only prograde cells. Centrifugal force would keep each cell in its proper zone, but within a zone mutual attraction would be free to draw the cells into one big mass which would thus become a protoplanet. One or more cells near the top or bottom of the zone might remain free but would be captured to form a satellite or satellites revolving about the protoplanet. We must then assume that the nebula had become subdivided into ten concentric zones, one for each of the planets plus one for the belt of asteroids.

In the nebula the elements were presumably present in the proportions indicated

Figure 5-14 Idealized diagram to illustrate the evolution of a dust-cloud nebula into the solar system. (Adapted from Von Weizsäcker.) The three sectors represent three distinct stages in the evolution of the system. Sector 1 represents an early stage when the nebula had become lenticular and the outer part was tenuous, consisting chiefly of hydrogen and helium and it increased in density toward the center where more and more of the heavier atoms were concentrated. Sector 2 represents a later stage when, because the angular velocity increased toward the center, the nebula was separating into concentric shells in each of which large, turbulent eddies developed retrograde rotation. The friction between each pair of such eddies caused the development and growth of intervening countereddies (dark) with prograde revolution. Sector 3 shows a later stage after the prograde eddies have grown at the expense of the retrograde eddies until the latter have disappeared. At a later stage the several eddies in a shell will be drawn together to form a protoplanet whose orbit will be within the parent shell of the nebula. This evolution is indicated only for the two outer shells.

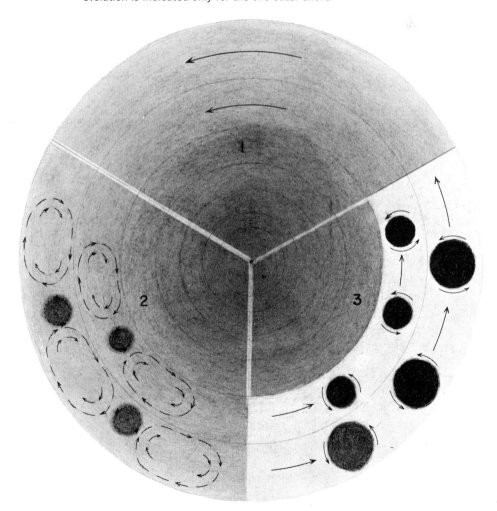

Table 5-1. Relative Abundance of the Commonest Elements in the Present Solar System

Element	Abundance
Hydrogen	5100.
Helium	1000.
Oxygen	28.
Iron	2.7
Nitrogen	2.1
Magnesium	1.7
Carbon	1.0
Silica	1.0
Sulfur	0.43
Aluminum	0.10
Sodium	0.10

Adapted from Mason after Unsöld

in Table 5-1, in which silica is chosen as unity. It will be noted that hydrogen and helium vastly exceed all the other elements. It appears certain, however, that the heavier elements were more concentrated deep within the nebula while the outer parts consisted chiefly of hydrogen and helium. Thus, from the start, the inner planets were made of heavier matter than the outer ones.

As the system evolved, moreover, an important new factor came into play. As the sun condensed to smaller radius its temperature rose to a point at which atomic fusion began and hydrogen was transformed into helium (as in a hydrogen bomb). It then began to pour out a flood of radiant energy as it does now. This raised the temperature at the surface of the near planets to a point where their gases began to leak away into space. Thus the minor planets lost all, or nearly all of their atmospheres, which were literally blown away. Meanwhile the major planets were so far from the sun that they were scarcely affected. They may actually have grown by capturing some of the gases blown away from the minor planets.

The theory thus envisioned offers a logical explanation of many features of the solar system; notably the spacing of planets, their nearly circular and concentric orbits, their revolution in a common plane and in the same direction, the difference in size and composition between the minor and major planets, and the high temperature of the sun. Never-

theless, a number of special features remain that are not fully accounted for and will require supplementary explanations.

COSMIC HISTORY OF THE EARTH

If our reasoning thus far is valid, the protoplanet that was to become the earth separated from the primeval nebula as a rotating mass of gases and dust particles—a mass greatly exceeding the present dimensions of the earth not only because it was still diffuse but also because it included a large volume of gases that would subsequently be lost. It is now our purpose to explore the steps by which this nebulous globe was transformed into the solid earth. For this we are indebted to Urey [16] for a penetrating study of the geochemical development. Among his findings the following are vital to our problem.

1. Oxygen combined actively with silica, aluminum, magnesium, iron, calcium, and potassium to form complex silicates. It also combined with hydrogen to form water. But judging from the cosmic abundances of the elements involved (Table 5-1), hydrogen should have been in excess abundance and in its presence **no oxygen could remain free.**

2. Hydrogen united with nitrogen to form ammonia (NH_3) and with carbon to form methane (CH_4), and except for free hydrogen and helium these were the chief gases of the primeval atmosphere.

3. The fate of iron (Fe) depended on the temperature. Below 25°C iron oxides (FeO_3 and Fe_3O_4) could form and would be stable. But if the temperature rose above 227°C, iron oxides would be unstable and above 327°C metallic iron was the stable form.

4. Molecules of the silicates and of iron probably appeared as dustlike particles and Urey suggests that water vapor and liquid ammonia may have moistened the dust particles so that they stuck together like wet snowflakes as they were drawn together to form the solid body of the earth.

5. The solid body of the growing earth must have consisted largely of silicates and iron, and its atmosphere consisted of hydrogen, helium, ammonia, and methane. (The major planets still consist of such gases.)

6. In view of the cosmic abundances of the chemical elements in the original nebula, the earth during its formative stage was vastly larger than it is now. Kuiper [5] has estimated that its mass may have been 1200 times that of the present earth.

7. Having accumulated in the void of space, the parent nebula must have been cold. But as the protoplanet earth gradually contracted under its internal gravity its temperature inevitably increased and may eventually have reached 2000 to 3000°C at the center. Meanwhile, for the same reason, the immense sun had reached a temperature at which it became an atomic furnace discharging radiant energy as it does now. The effect on the protoplanets near the sun was profound. None of them was large enough to hold hydrogen or helium at a temperature of 2000°C and the radiant energy of the sun conspired to blow away the atmospheric gases just as it now forms the tails of comets. Thus the minor planets were stripped of their primeval atmosphere and were reduced to solid, naked bodies.

Quite aside from these deductions, which are based on well established principles of thermodynamics, there is convincing evidence that the earth, at an early stage, lost nearly all of its original atmosphere. This is based on the relative abundance of argon in the present atmosphere, which far exceeds the cosmic abundance that it must have had in the original nebula. Since it is inert it could not have combined with other elements to be buried in the solid earth and to be sweated out at a later date — it must have been present in the primeval atmosphere. Also, its escape velocity is much higher than that of hydrogen and helium and thus while they leaked away into space most of the argon remained. The cosmic ratio of argon to these lighter gases thus gives us some measure of the amount of the primeval atmosphere that was lost. It is evident, then, that our present atmosphere, so different from that of all the other planets, was secondarily derived. How this came about we can now understand.

After the solid earth had condensed, radioactivity in the interior caused the temperature to rise, possibly to the point of melting (the outer core is still molten). The rising temperature caused some of the minerals to break down, releasing oxygen and water and ammonia which, being volatile, were sweated out to the surface through volcanic activity. The residue of ammonia from the primeval atmosphere as well as that sweated out of the interior was attacked by free oxygen to form water and free nitrogen ($4NH_3 + 3O_2 = 4N + 6H_2O$). Residual methane reacted with the emerging oxygen to form water and carbon dioxide ($CH_4 + 2O_2 = CO_2 + 2H_2O$). Metallic nitrides were broken down into metallic oxides and free nitrogen and the nitrogen escaped to the surface, where it is virtually inert and continued to accumulate in the atmosphere.

Thus far no free oxygen existed in the atmosphere because it has such strong affinity for other elements. Carbon dioxide would have been abundant save that it reacts so readily with silicates to form carbonates, such as $Ca(CO_3)_2$, $NaCO_3$, and $Mg(CO_3)_2$. In this connection water has played an important role in continually eroding the rocky surface and exposing fresh silicates to be attacked by CO_2. Water is one of the chief volatiles brought to the surface by volcanic activity and has been added much more rapidly than it was dissociated. Rubey has shown [10] that if CO_2 were not continually supplied to attack the silicates, all the carbonates would cease to form, the present supply of CO_2 would be exhausted in a few million years, and instead of carbonate sediments such as limestone and dolostone, brucite ($Mg(OH)_2$) would spread over the shallow ocean floor. Since no significant deposits of brucite are known it is evident that CO_2 has been continually added to the atmosphere.

The appearance of plant life led to a profound change in the composition of the atmosphere. Plants feed on CO_2 and water and produce oxygen throughout their life. They are thus the source of the oxygen in the earth's atmosphere. And since animals cannot live without free oxygen, we can understand why the evolution of animals followed by many millions of years the appearance of plants (see page 163, Chapter 7).

Rubey [10] has reasoned that both the atmosphere and the oceans have been largely derived from the interior of the earth, since the loss of the primeval atmosphere, by continuous volcanic activity. Another theory agrees in deriving the atmosphere and oceans by degassing of the earth but holds that this occurred during a single great episode of expulsion of volatiles during a very early stage in the earth's formation. As yet we lack the evidence to choose which if either theory is more likely to be correct. Gradual degassing implies that the oceans have grown throughout geologic time, but the critical evidence for this, if it exists, has yet to be deciphered from the complex geology of the Cryptozoic (Chapter 7).

REFERENCES

1. Baldwin, R. B., 1949, *The craters of the moon.* Scientific American, v. 181, no. 1, pp. 20–24.
2. Beals, C. S., 1958, *Fossil meteor craters.* Scientific American, v. 199, no. 1, pp. 33–39.
3. Biermann, L. F., and Lüst, Rhea, 1958, *The tails of comets.* Scientific American, v. 199, no. 4, pp. 44–50.
4. Hapke, Bruce, 1968, *Lunar surface: composition inferred from optical properties.* Science, v. 159, pp. 76–79.
5. Kuiper, G. P., 1951, *On the origin of the Solar System,* in *Astrophysics,* J. A. Hynek, Ed., Ch. 8, pp. 327–427. New York, McGraw-Hill.
6. Leighton, R. B., 1965, *Photographs from Mariner IV.* Scientific American, v. 214, no. 4, pp. 54–71.
7. Meen, V. B., 1951, *The Canadian meteor craters.* Scientific American, v. 184, no. 5, pp. 64–68.
8. Newell, H. E., 1966, *Surveyor: Candid Camera on the Moon.* National Geographic, v. 130, pp. 578–592.
9. Nicolet, M., 1960, *The properties and constitution of the upper atmosphere,* in *Physics of the upper atmosphere.* Ch. 2, pp. 17–71. New York and London, Academic Press.
10. Rubey, W. W., 1951, *Geologic history of sea water.* Geol. Soc. America Bull., v. 62, pp. 1111–1148.
11. Shapiro, I. I., 1967, *Resonance rotation of Venus.* Science, v. 157, pp. 423–425.
12. Jastrow, Robert, 1968, *The Planet Venus.* Science, v. 160, pp. 1403–1410.
13. Swindel, G. W., Jr., and Jones, W. B., 1955, *Meteoritics,* v. 1, pp. 125–132.
14. Spitzer, Lyman, 1939, *The dissipation of planetary filaments.* Jour. Astrophysics, v. 90, p. 675.
15. Tousey, R., 1961, *Solar Research from Rockets.* Science, v. 134, pp. 441–448.
16. Urey, H. C., 1952, *The Planets: their origin and development.* New Haven, Conn., Yale Univ. Press.
17. Van Allen, J. A., 1959, *Radiation belts around the earth.* Scientific American, v. 200, no. 3, pp. 39–47.
18. Von Weizsäcker, C. F., 1944, *Über die Entstehung des Planetensystems.* Zeit. für Astrophysics, v. 22, p. 319.
19. Whipple, F. L., 1964, *The history of the solar system.* Proc. Natl. Acad. Sci. (USA), v. 52, pp. 565–593.

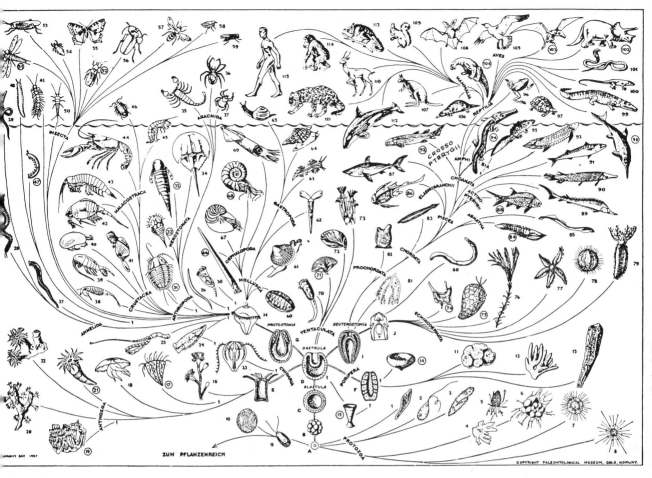

Figure 6-1 A family tree of the animal kingdom, after a colored wall chart by Heintz and Störmer. (Courtesy of the authors.)

CHAPTER **6** *The Constant Change of Living Things*

THE ORIGIN OF LIFE

One cannot contemplate the spectrum of life stretching down through all time without wondering how it started. Indeed, philosophers of all ages have speculated on the origin of life, and their beliefs have fallen generally into one of two categories: either life began by a supernatural act of creation, or it developed spontaneously from inorganic matter.

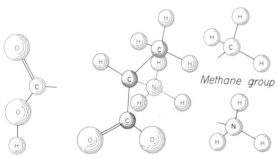

Carboxyl group   Alanine, an amino acid   Amine group

Glycine + Glycine = Glycylglycine + Water

Portion of a polypeptide chain — a protein

Figure 6-2 Building units of organic compounds. Alanine, one of the amino acids, is a fundamental unit of proteins. The center figure above shows the space relations and the relative sizes of the atoms of carbon, hydrogen, nitrogen, and oxygen in the alanine molecule which is built up of the simpler units shown at either side. The central figures show how two glycine unite to form a bigger molecule. In the proteins simple units are combined to form long chains of complex configurations in which constituent smaller units are held together by carbon atoms. The lower figure represents part of a single relatively simple protein molecule. (Adapted from figures in *General Chemistry* by Linus Pauling.)

Spontaneous generation was a common explanation during the Middle Ages, when worms were thought to develop from mud, maggots from decaying flesh, and mold from refuse. But the experiments of Pasteur, around 1860, convincingly proved that maggots develop from eggs laid by flies, and even germs of disease and decay do not appear in sterilized organic matter unless the matter is contaminated by living organisms from outside sources. Since then, abundant experience in medicine and in food preservation has made it clear that life is not spontaneously generated in the modern world. For this there are two chief reasons:

1. In the presence of free oxygen, organic matter is readily destroyed by oxidation and is reduced largely to CO_2 and H_2O. Thus, for example, in the forests where organic tissue is rapidly built up, the fallen leaves and trees disappear about as fast as they are formed and, whether they are consumed by fire or by the slower process of decay, the end products are the same.

2. The modern world is populated by an enormous complex of living creatures that feed upon and destroy organic matter—especially the ever-present molds and bacteria that cause putrefaction and decay.

But if the earth developed as outlined in Chapter 5, it passed through an early stage when neither of these conditions existed, for until life had appeared, it was an utterly sterile world. Furthermore, contrary to previous conception, the primeval atmosphere probably contained no free oxygen but was composed largely of methane (CH_4), ammonia (NH_3), carbon dioxide (CO_2), and water vapor.

Of all the elements, carbon is unique in its ability to unite with a great variety of other elements to form metastable compounds, and to join relatively simple units into long chains or rings (Fig. 6-2) and thus build up large and complex molecules such as the proteins of which living matter is made. In industry catalyzers now are used to speed up such reactions, but they do not cause chemical reactions—they merely speed up reactions that can and do take place naturally.

Now, on our primitive planet, bathed in an atmosphere of carbon dioxide, a great variety of organic compounds must have formed; and in a sterile and oxygen-free world they may have persisted, accumulating in the seas and lakes until some of the shallow waters were virtual soups of organic compounds. This then was the environment, so different from our modern world, under which life may have begun naturally [4, 6].

The problem is not simple, for a living organism not only includes a great variety of such compounds but has them organized in very definite and almost infinitely complex systems, so that some serve as catalyzers while others react in complex ways to store, and again to free, the energy that is manifest in living things.

The first great source of such energy was sunlight, which is still the means by which plants build up organic matter through photosynthesis; and when free oxygen began to appear in the atmosphere a still more effective source of energy was available in the oxidation of proteins. This is the present source of energy in animal metabolism.

Thus far, by deductive reasoning, we have pictured conditions under which proteins may have formed and the simplest types of life may have developed in nature. Experiments have gone further. In the geochemical laboratory at Chicago Dr. S. L. Miller [3] circulated a mixture of the gases methane (CH_4), ammonia (NH_3), water vapor (H_2O), and hydrogen (H_2) over an electric discharge (to supply energy), and after a time found that the water in the bottom of the apparatus contained certain amino acids (both alanine and glycine), which are the basic units proteins are made of. On the early earth lightning may have supplied the energy to form the complex proteins. No living tissue has yet been synthetically produced, but many of the complex chemical building blocks have. However, the organization of the proteins in even the simplest forms of life is amazingly complex. The chemical reactions leading to life left no fossil record, but geochemistry can establish limits of speculation and indicate probable pathways, and we are already beginning to understand how biochemistry, through protein synthesis, can reconstruct the steps leading to living matter.

Although the fossil record cannot reveal how life began, it does afford some evidence as to when it began and what its early forms may have been like. Probable evidence of life in the form of minute, bacterium-like bodies and substances which appear to be made of organic chemical compounds have been found in some of the oldest known sedimentary rocks. It is therefore reasonable to suppose that life was present on earth more than 3 billion years ago. Throughout most of Cryptozoic time, which amounts to nearly seven-eighths of earth history, life apparently existed in primitive form, for it left only rare traces of such simple organisms as blue-green algae and bacteria. Recent discoveries and advances in methods of investigation have attracted increased attention to the Cryptozoic record and led to some interesting interpretations of the early environment and evolution of life on earth. We will explore these in more detail in Chapter 7 on the Cryptozoic Eon.

THE DOCTRINE OF ORGANIC EVOLUTION

To early men who knew at most a few hundred kinds of animals and plants and had only superstition to guide them, it seemed reasonable to believe that the particular god they feared and worshiped "molded each species out of the dust of the earth and breathed into it the breath of life." The story is much the same among primitive peoples throughout the world. It seemed as plausible as the belief that the earth was flat and was the center about which sun and moon and stars revolve. Handed on through the ancient scriptures, the concept of **special creation** became deeply imbedded in religious dogma and, in the Western world, came to dominate thought until after the Renaissance.

But by late in the middle ages explorations

in Africa and other parts of the world had led to the discovery of thousands of new kinds of animals and plants (now recognized as more than a million). Moreover, the domestication of animals and plants had opened new vistas. By careful selection of the best animals in a herd for reproduction, the stock could be improved. If stock breeding could change the wild asiatic pony into breeds as distinct as the ponderous Percheron, the rugged little Shetland pony, and the sleek Arabian stallions, and if the wild dog could be bred into varieties as distinct as the mastif, the bulldog, and the poodle — possibly nature in some fashion has acted as a stock breeder developing new species from old ones.

During the seventeenth century several scientists began to speculate along this line, and by the mid-nineteenth century it was openly debated. But it was not until 1859, when Darwin published his classic work, *The Origin of Species by Means of Natural Selection,* that the idea of organic evolution was placed on a solid, well-documented, scientific basis. In essence, this idea is that from very primitive ancestors all the diverse forms of life have gradually been derived through natural selection, so that all living things are related, however remotely (Fig. 6-1).

Darwin's monumental work shook the intellectual world as no other work had, for it challenged not only earlier biological beliefs but the teachings of the Christian Church as well. Its dramatic impact may be illustrated by the following incident. When the book appeared it was enthusiastically received by some of the leading scientists, but many of them were skeptic or even hostile. It was therefore made the subject of a symposium before the British Association at Oxford on June 30, 1860. At this meeting Bishop Wilberforce of Oxford appeared, determined to squelch this rank heresy. He spoke in a scoffing, insolent manner and in closing turned to Thomas Huxley, who was Darwin's chief protagonist, with the question: "Mr. Huxley, was it through your grandfather or your grandmother that you claim descent from a monkey?" When the cheers of the Bishop's followers died down Mr. Huxley rose to reply

as follows: "If there were an ancestor whom I should feel shame in recalling, it would be not an ape but a man—a man of restless and versatile intellect who, not content with success in his own sphere of activity, plunges into scientific questions with which he has no real acquaintance, only to obscure them by aimless rhetoric and distract the attention of his hearers from the real point at issue by eloquent digressions and skilled appeals to religious prejudice" [1]. The effect on the audience was electric. The exchange enhanced Huxley's stature and with it his influence among the intelligentsia of his time.

The debates gradually waned and as evidence accumulated to support it, organic evolution passed beyond the stage of theory and, by early in the present century, had taken its place as the most important guiding principle in biology. A leading student of evolution in our time, the paleobiologist George Gaylord Simpson, sums up this exciting chapter in the history of human thought in the following words:

Here we have started with the premise that life has evolved and has had a history. This premise has been so conclusively established by generations of study and the resultant accumulation of literally millions of concordant facts that it has become almost self-evident and requires no further proof to anyone reasonably free of old illusions and prejudices [5].

EVIDENCE OF EVOLUTION

The evidence of evolution is so varied and so extensive that volumes would be required to review it all, and much of it is too technical for simple presentation. Therefore we can only suggest here the nature of the evidence. Our illustrations are chosen from three basic lines of investigation, the first two biologic and the third paleobiologic.

Comparative Anatomy. It is a striking fact that in related groups of animals each organ or anatomical structure is built on a common plan. This is illustrated repeatedly in Fig. 6-3. In spite of the impressive differences between reptile, lemur, ape, and man, their skeletons are basically similar. The forelimb in each in-

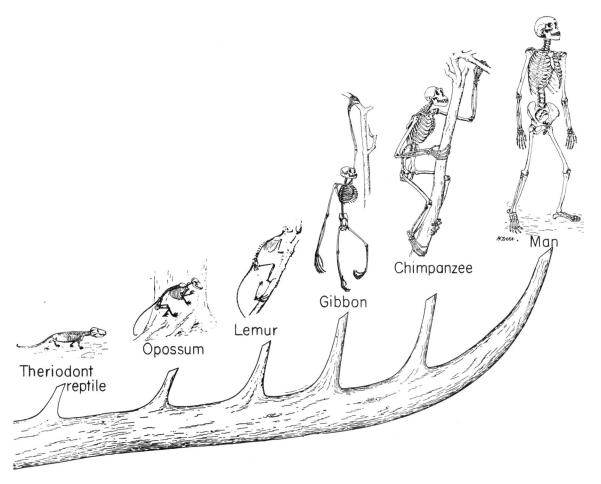

Figure 6-3 A series of skeletons from reptile to man showing homologies in skeletal structures. (From the American Museum of Natural History.)

cludes a single upper arm bone (humerus), a pair of lower arm bones (ulna and radius), a series of wrist bones, and five digits. We can extend the comparison to distantly related mammals that are highly specialized for different modes of life, for example, man, bat, seal, and dog. All have the same complement of bones, which differ only in size and proportions (Fig. 6-4). Organs that agree thus in fundamental structure are said to be homologous [Gr. *homologos*, agreeing]. Comparison of the soft parts such as muscles, nerves, and viscera also shows an amazing number of homologies. They admit of no other rational explanation than descent from a common ancestor that possessed all these structures.

Equally telling evidence of kinship may be seen in many vestigial structures—structures that have lost their original function. In most humans the muscles attached to the ear are not under voluntary control and serve no useful purpose, but they correspond to the muscles that in most other mammals do move the ears. The human appendix is not only useless but is commonly a source of infection, yet in many of the lower animals it is an important part of the digestive system. About 180 such structures are found in the human body. One of the striking misfits is the way in which our viscera are supported in the body cavity. All the lower mammals walk on all fours with the body in a horizontal position and the delicate mesenteries which attach the visceral organs to the body wall give effective support.

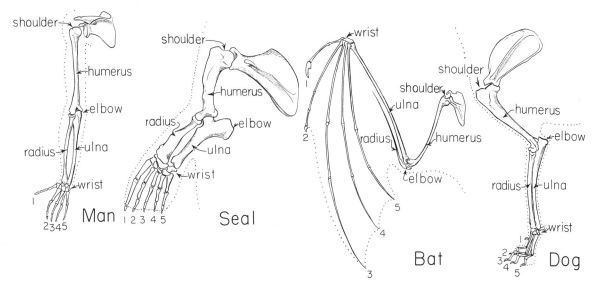

Figure 6-4 Bones of the forelimbs of man, seal, bat, and dog.

But in learning to walk upright man and the apes have rotated the body into a vertical position in which the visceral support is very ineffective. As a result we humans suffer such afflictions as paunchiness, hernia, fallen stomach, and prolapse of the womb. Surely if the human body had been created *de novo* it would not have been endowed with useless and dangerous structures, aping those that function well in the lower animals. They admit of no other rational explanation than that man has descended from remote ancestors that walked on all fours.

Embryology. In its development from the egg to maturity an individual animal undergoes a progressive series of changes that constitute its individual history, or **ontogeny.** Other genera and species derived from the first commonly repeat many of the same ontogenetic stages at least up to the point where the descendant form begins to specialize away from its ancestors. Ernst Haeckel recognized this in the latter part of the nineteenth century and saw in it an aid in determining the ancestors in a racial history, or **phylogeny.** He concluded that "ontogeny recapitulates phylogeny." It is now well known, however, that this idea of recapitulation is a great oversimplification of fact. Many stages of the racial history are commonly omitted from ontogony (the embryology of the modern horse, for example, shows no five-toed stages) and special larval adaptations may cause new stages to be inserted that have no bearing on racial history. Nevertheless, it is commonly agreed that animals having closely similar embryos are probably related and that those with distinctly different embryos are at the most remotely related.

For example, the early embryologic stages of the starfish and the sea urchin indicate that they are related and that although both are radially symmetrical, they descended from a remote ancestor that was bilaterally symmetrical.

Although the toad is a terrestrial animal breathing by means of lungs and lacking a tail, its young are tadpoles, confined to the water where they breathe only by means of gills, and have a long tail for swimming. Such early life stages strongly suggest that it, like other amphibians, evolved from fishes. The paleontological record completely supports this inference. The early embryos of man and the apes are so nearly identical that they are difficult to distinguish. Only descent from a common ancestor could account for such similarities.

Documentation in the Fossil Record

Inferences from comparative anatomy and embryology enable us to build a family tree of the animal kingdom like that in Fig. 6-1. If it is correct, the myriad of fossils entombed in successive beds of sedimentary rock should provide much documentary evidence, as they certainly do. Many illustrations will be found in later chapters of this book, but we may anticipate by citing a few examples here.

The keyhole shell, *Pygope*, will illustrate both the development of a physical feature through time and the record of it in the early stages of a single descendant species (Fig. 6-5). The adult shell is shaped almost like a doughnut, but a graded series of young shells (A to C) show how the hole was developed. In a very young shell the front edge was merely notched. But in successive stages the sides grew forward and finally coalesced. The growth lines on the adult shell show the same

history of development. From this we might infer that *Pygope diphyoides* descended from a lineage of species whose mature forms showed successively the progression from a notched front edge to the stage before complete enclosure of the "keyhole." This is confirmed by a sequence of distinct species (E to H) found at successive levels in the Jurassic and Early Cretaceous formations of Europe.

The skeleton of the modern horse's leg (Fig. 6-6) shows vestigial structures that throw light on horse ancestry. In the hind leg, for example, if we begin with the hip joint it is easy to identify the thigh bone, the knee cap, and below these a pair of lower leg bones (tibia and fibula). Thus far comparison with the leg of man is obvious. Furthermore, the hock joint is obviously the heel, the short bones beside it are the ankle bones, and the long bones below are the metacarpals. From this examination we may conclude that the horse walks on the end of one toe and that the lower part of

Figure 6-5 The Jurassic keyhole shell, *Pygope.* In the upper row several growth stages of *P. diphyoides* represented by young shells. In the lower row adult shells of four distinct species found in successive horizons in Jurassic and Early Cretaceous rocks.

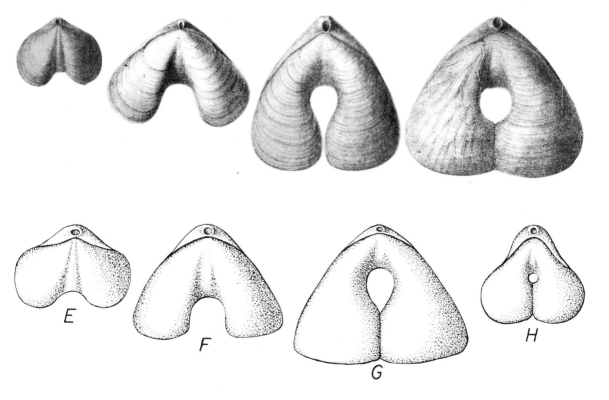

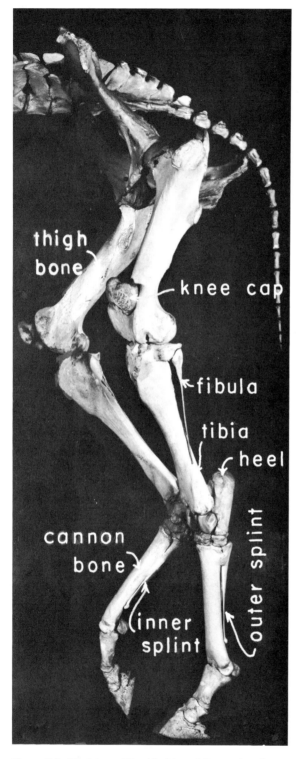

Figure 6-6 Skeleton of the hind leg of the modern horse showing vestigial structures. (After Chubb, American Museum of Natural History.)

the leg has been greatly lengthened as the heel came up off the ground.

In primitive mammals, and even in the specialized ones that use their hands and feet for grasping or digging, the paired lower limb bones are subequal in size and provide for a rotary motion of the hands and feet. But in the horse, which can bend its leg only fore and aft at the heel, the fibula is reduced to a mere sliver, and its lower end does not even make contact with the foot bones. The presence of this vestigial fibula implies, however, that the remote ancestor of the horse had a less specialized limb with freer movement of the foot. And on each side of the main toe bone (the cannon bone) there is a slender bone known as the splint, which can be nothing but the vestige of another toe. The splint bones clearly imply that the ancestor of the modern horse had three toes, as do the rhinoceros and the tapir.

The series of fossil horses recovered from successive Cenozoic formations in the western United States proves this inference to be true beyond any doubt. For example (Fig. 19-11), all the fossil horses from the Pleistocene, and most of those from the Pliocene, are like the modern horse, but all those of the Miocene have three toes on each foot. In these, however, the side toes are slender and in most species apparently were functionless, probably dangling like the dew-claws of cattle. In the underlying Oligocene beds, on the contrary, the three toes are subequal in size on each foot, and all shared in bearing the animal's weight. Finally, in the still older Eocene strata, vestiges of a fourth toe have been found in the front feet. No fully five-toed ancestor has yet been identified, but there can be no doubt that such a one existed. Illustrations of the paleontological evidence of evolution could be multiplied to any length, for they form the essence of the geological history of life.

WAYS AND MEANS OF EVOLUTION

Ever-increasing knowledge of the evolutionary process, built on a colossal amount of documentation and experimentation, has advanced and refined the Darwinian concept. In particular, the study of **genetics,** which deals with the complex mechanisms of heredity, has added greatly to our understanding of how evolutionary change takes place.

One fundamental contribution of genetics has been the knowledge that the agents of heredity are exceedingly minute chemical entities, the **genes,** which are grouped in larger bodies called **chromosomes** and contained within the nucleus of the cells of organisms. In reproduction it is the genes that carry the complex genetic pattern from one generation to the next, and with it the changes which, through many generations, mark the path of evolution.

Variation. Heredity is perhaps most commonly linked with the idea that "like begets like," and in a general way this is certainly true. Mice beget mice and sparrows beget sparrows. But we need only add that dogs beget dogs to make the point that there may be considerable variation among individuals of the same species. In dogs, of course, selective breeding by man has exploited to a spectacular degree the potential of variation in the species. But such selective breeding would not have been possible had there not been some differences among individuals to begin with. No offspring is exactly like one of its parents, nor exactly like any of its brothers and sisters.

Only those variations that affect the agents of heredity, the genes, are significant in evolution, for it is only through inheritance that minute differences can be transmitted and eventually lead to change. Environmental factors and disease may cause conspicuous differences between parents and offspring, but these are not hereditary. A population of fish in a lake with a very low supply of nutrients is atypical of its species in that the body is dwarfed in size. Eggs from the dwarfed population when introduced to waters with abundant food supply develop into fish of normal size for the species, showing that the effects of environmental pressure are not inherited.

The principal source of hereditary differences between parents and offspring is the genetic change brought about by **sexual reproduction,** a process found in all but the most primitive organisms. The genetic pattern of the offspring is a **combination** of the genetic patterns of its parents. The chromosomes from each parent normally have differences in their suites of genes; consequently the resulting combination of genes found in the chromosomes of the offspring will be different from that of either parent. In this way the genetic patterns of individuals in a freely interbreeding population are constantly being shuffled and the potential number of possible variations is incredibly great.

Variations provided by the mechanism of sexual reproduction bring about relatively small changes that may lead to new varieties or breeds. A second and much less common kind of variation is brought about by **mutation:** a change, usually abrupt, in the nature of a gene. Geneticists do not yet know what causes gene mutation or how the chemistry of the substance of the gene is altered. Experiments over a long period of time have shown that some mutations work little change whereas others effect radical changes and are commonly lethal. Potentially, mutations can cause great changes in organisms and most evolutionists believe they are involved in major evolutionary steps.

Natural Selection. Recognition that natural selection was the key to evolution was Darwin's greatest contribution. It is now common knowledge that in the struggle of the individual organism for survival chance plays an important role, but, by and large, individuals that are best adapted to the particular environment into which they are born have the best chance to live to sexual maturity. Nature serves, like a breeder of fine stock, to further the survival of the fittest individuals, who will then pass on their genetic patterns to the next generation.

It was not until well into the twentieth century that it was realized that evolutionary change was accomplished by natural selection not as it affected individual organisms, but as it affected entire interbreeding populations. Figuratively, the interbreeding population of a species pits its spectrum of ever-changing individual variations against the array of selective pressures in the environment.

In this way life has responded to the randomness of environmental change—the broader the spectrum of variability within a population, the better the chance of quick adaptation to changes in the environment. Great range of variability may also lead to new adaptations permitting a spread to new niches in the environment. Natural selection is thus a great deal more than a negative process eliminating the unfit; it is creative in a very real sense.

A single illustration may serve to show how natural selection may lead to the development of new species from a parent stock (Fig. 6-7). The crossbill, which ranges over Europe from the Alps to Siberia, feeds on the seeds of evergreen trees. In the Alps the dominant evergreen is the pine, whose cones are hard and tough. Only birds with a stout beak can successfully break them apart to secure the nuts. In Siberia, however, the dominant evergreen is the cedar, whose cones are softer but seeds lie deeper and here a long, slender beak is a decided advantage. In turn the Himalayan crossbill lives on small, soft cones of the larch.

When the crossbill appeared in Europe, whether by migration or evolution from some less specialized stock, those that settled in the Alps faced a special environmental situation. Here those members of the population that by chance variation had the stoutest beaks were better adapted to the available food supply and had the best chance of survival. Among those that settled in Siberia, the birds with longer, slender beaks tended to survive. Thus distinct species gradually evolved.

Isolation may also lead to speciation. If a species (an interbreeding population) spreads over a large region of varied environments, as the crossbill has done, more or less isolated communities may specialize in different directions as they become adapted to new types of food or new habits of life. Thus new geographic races or subspecies arise. Until the specialization has reached a critical limit, however, a subspecies is still capable of interbreeding with related subspecies. Where the geographic habitats of two subspecies are contiguous or overlap somewhat, interbreeding does take place and the characters of the two subspecies blend and lose their distinctiveness. Similarly, if a few individuals of one subspecies should migrate into a province occupied by a related subspecies, interbreeding with the larger population will so dilute their contribution to the common genetic system, or gene pool, that within a few generations their peculiarities will disappear. For the same reason, it is improbable that a new subspecies could arise in direct competition with a parent subspecies. However, where subspecies are geographically isolated for a sufficient length of time, the specializations of each may become so great that they are incapable of interbreeding. Each thus becomes a distinct species. Isolation is thus an important factor in speciation. This was perceived by Darwin more than a century ago when he visited the Galapagos Islands and found finches, presumably derived from the mainland of South America, to be widespread but with a distinct species inhabiting each island.

Two species with identical habits could not long survive in direct competition because one would be more successful and would soon drive out or exterminate the other; but even closely related species adapted to different feeding habits or to different niches in the environment may successfully survive together.

Figure 6-7 Adaptive speciation in the beak of the crossbill, *Loxia*. (Data from David Lock.)

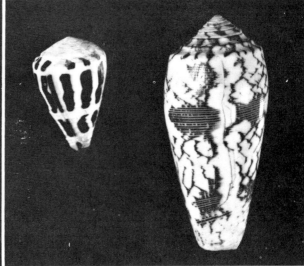

Figure 6-8 Four species of the genus *Conus* that live together on the subtide flats in Hawaii, but feed on different organisms. From the left, *Conus eboraceus; C. abbreviatus; C. sponsalis; C. striatus.* (Yale Peabody Museum.)

Figure 6-9 Three sympatric species of the marine snail *Crepidula.* From the left, *C. fornicata; C. plana; C. arcuata.* (Yale Peabody Museum.)

In Hawaii several species of *Conus*, a genus of marine snails (Fig. 6-8), live together on the floor of the subtidal flats, but Alan Kohn [2] found that no two feed on the same prey. *C. eboraceus* and *C. abbreviatus* feed on different species of worms, while *C. striatus* feeds on fish which they are able to harpoon with poisoned darts. On the Atlantic and Gulf beaches of the United States three quite different species of the gastropod genus *Crepidula* may be collected (Fig. 6-9). They feed on microscopic organisms which they sieve out of the water. Their food may be the same, but each inhabits a distinct niche in the environment. *C. fornicata* lives in clusters on solid objects on the open sea floor; *C. plana* inhabits the empty shells of large gastropods; and *C. convexa* clings to seaweed.

Thus isolation may be geographic or it may be by habits of feeding or of reproduction.

Survival and Extinction. For many groups of organisms the ever-changing pattern of variation in species has been adequate to insure their survival throughout long periods of geologic time. Other groups were not so fortunate. Many limited exterminations in the geologic record can be attributed to local extreme changes in environment but widespread extinctions are more difficult to explain. The pattern of evolution is certainly one of gradual perfection and elaboration of adaptations and a number of extinctions of once dominant groups in the fossil record can be attributed to replacement, usually gradual, by better-adapted competitors. But the great changes in the history of life that mark the divisions between the eras remain an enigma (pages 312 and 399). These mass extinctions together with the great radiations of organisms at certain times in the past are but two of many aspects of the intricate patterns of organic evolution that remain to be explained. All approaches to evolution — the genetic, biochemical, biogeographic, paleontologic and others — are active areas of intensive research, for much has yet to be learned. But study of the patterns of evolution in the fossil record is among the most neglected.

REFERENCES

1. Huxley, Leonard, 1903, *Life and letters of Thomas Huxley.* V. 1, pp. 259–274.
2. Kohn, A. J., 1959, *The ecology of Conus in Hawaii.* Ecological Monographs, v. 29, pp. 47–90.
3. Miller, S. L., 1953, *A production of amino acids under possible primitive earth conditions.* Science, v. 117, pp. 528–529.
4. Oparin, A. I., 1953, *Origin of Life,* 2nd Ed., New York, Dover Publishing Co., 270 pp.
5. Simpson, G. G., 1949, *The Meaning of evolution.* New Haven, Conn., Yale Univ. Press, p. 338.
6. Wald, George, 1954, *The origin of life.* Scientific American, v. 191, no. 2, pp. 44–53.

PALEOGEOGRAPHIC MAPS

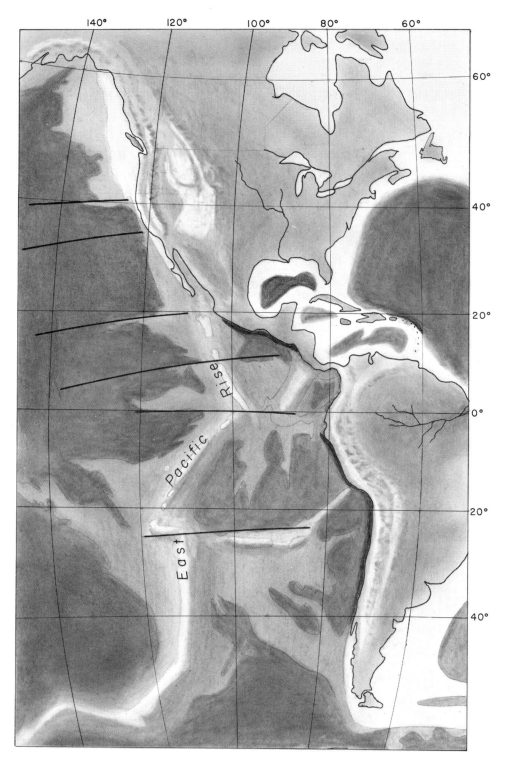

Map I The East Pacific Rise (adapted from Menard). In the seas the lightest shade indicates depth less than 3 km, intermediate shading 3 to 4 km, and darker shading more than 4 km. The narrow black band along the shore of South and Central America indicates oceanic trenches. The irregular belt of intermediate depth running northwest from near the tip of South America toward Hawaii is the residue of a much older convection cell not related to the East Pacific Rise. Heavy black lines mark the location of great rifts in the ocean floor.

Map II Cambrian lands and seas.

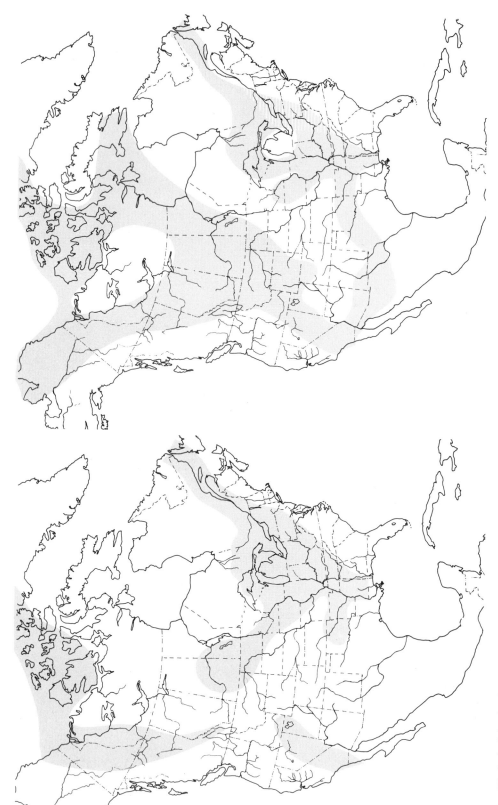

Map III Ordovician lands and seas. Left, maximum submergence of Mid-Ordovician time; right, the same for Late Ordovician. The subaerial surface of the Queenston Delta is marked by an overlay of white lines. The boundaries of the seas in the Arctic islands are very tentative; it is known that the Ordovician rocks are widespread in this region but only reconnaisance surveys have been made.

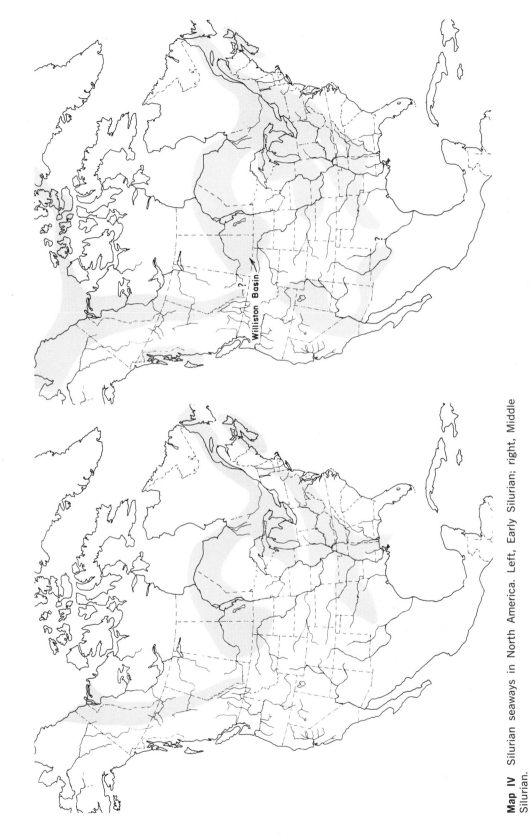

Map IV Silurian seaways in North America. Left, Early Silurian; right, Middle Silurian.

123

Map V Left, maximum spread of early Devonian seas; right, maximum spread of middle Devonian seas.

Map VI Three stages in the Mississippian cycle of submergence. Bottom, Kinderhookian Stage; middle, Osagian Stage; top, Mid Chesterian.

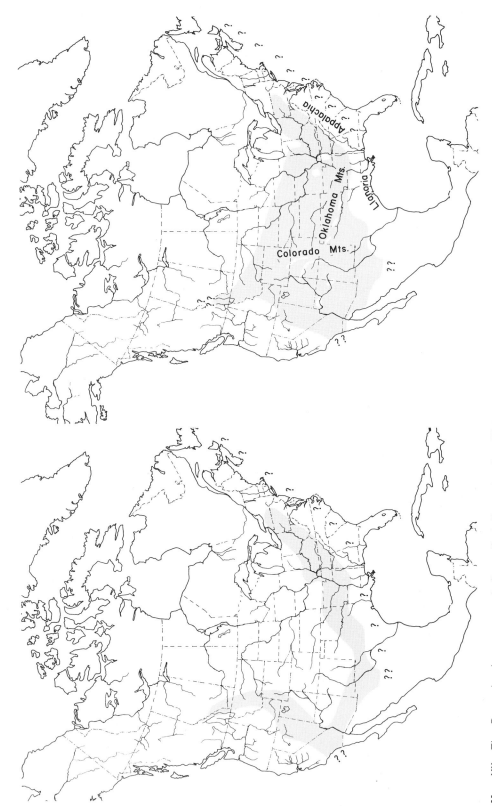

Map VII The Pennsylvanian seaways. Left, earliest Pennsylvanian (Morrowan Stage); right, maximum submergence during the Desmoinesian Stage.

Map VIII Paleogeographic maps showing the distribution of Permian deposits. The light shading crossed by horizontal white lines denotes essentially nonmarine deposits, largely redbeds. Left, generalized map of Early Permian (Wolfcampian); right, map of Upper Permian (Guadalupian) deposits. The Oklahoma and Colorado mountains were undergoing erosion. The three deep Permian basins of West Texas are deeply shaded.

Map IX Triassic paleogeographic maps of North America. Left, Early Triassic time; right, Late Triassic time. Darker shading marks seas; lighter shading marks basins of nonmarine deposition.

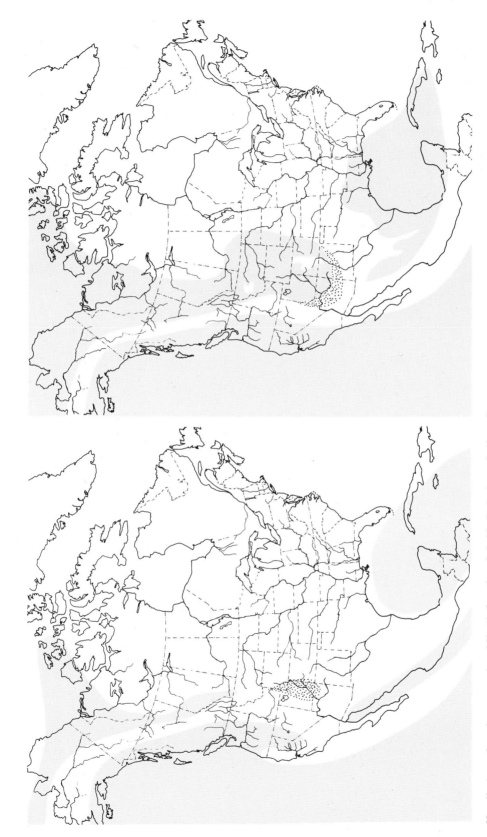

Map X Jurassic paleogeography of North America. A (left), Early Jurassic time; B (right), middle Late Jurassic (Oxfordian) time. Stippled areas were dune fields of windblown sand.

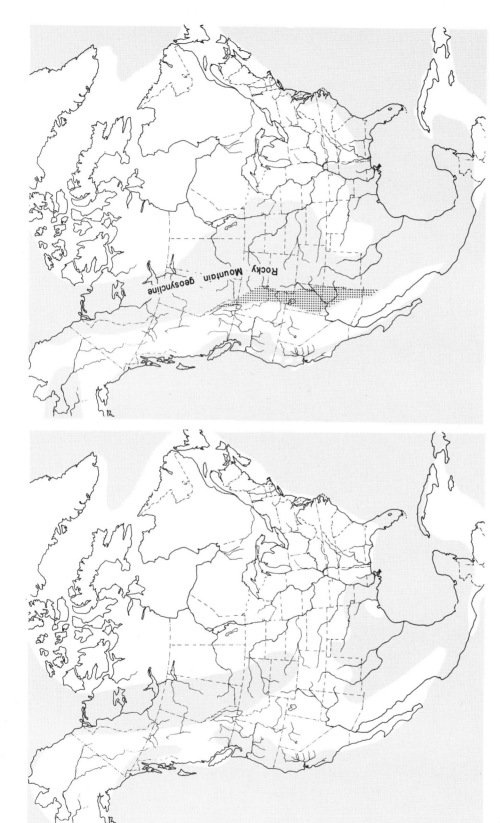

Map XI Cretaceous paleogeography of North America. Left, late Early Cretaceous (Albian). Right, a composite for the Late Cretaceous showing the maximum submergence.

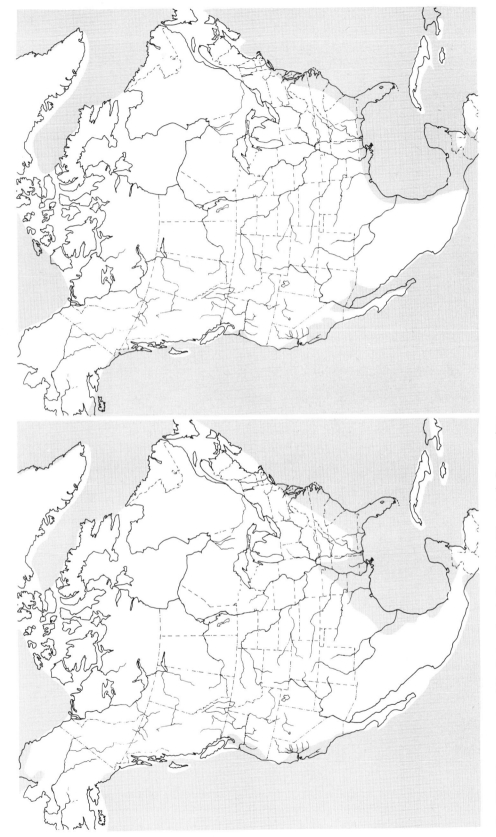

Map XII Paleogeography of North America. Left, extent of the seas in Eocene time; right, Miocene time.

131

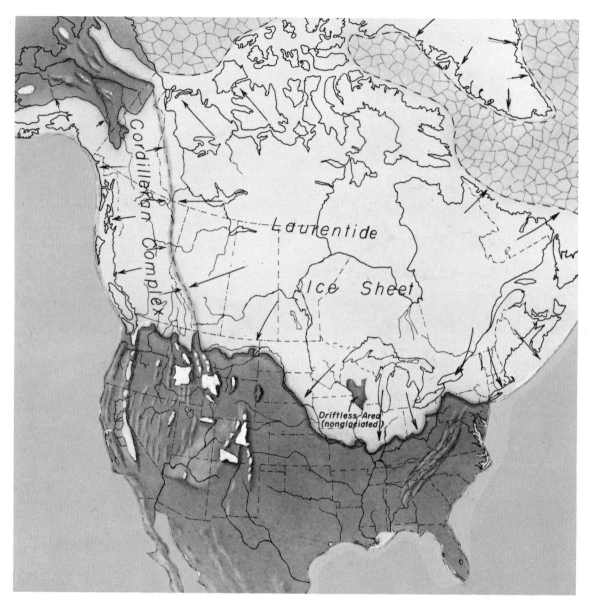

Map XIII Pleistocene ice fields of North America somewhat generalized to show the maximum extent of the glaciation. Arrows indicate the direction of ice movement. (Adapted from R. F. Flint.)

Figure 7-1 Precambrian structure of folded strata in the Canadian Shield, Belcher Islands, Hudson Bay. National Air Photo Library, Department of Energy, Mines and Resources, Number A3914-5-8.

CHAPTER **7** *The Cryptozoic Eon*

*What seest thou else
In the dark backward and abysm of time?*
THE TEMPEST, ACT 6, SCENE 2

The Ruins of Time. Earth's beginning was followed by long eras that are veiled in the shadows of antiquity. More than three-quarters of its history is hidden in enormous groups of ancient rocks that lie in tangled confusion below the Paleozoic. Without fossils to date them, these incredibly complex rock groups have lain unsorted like scattered sheets of an unpaged manuscript. Most have been intensely deformed and have suffered metamorphism and intrusion during one or more orogenic episodes. Collectively this impressive but enigmatic rock record is most commonly called the **Precambrian**; that it has as yet no generally accepted worldwide subdivision of any kind is a measure of the extreme complexity of its geology.

On a much smaller scale, the early part of human history presents similar problems. The millennia that preceded written records are known to us only from scattered ruins, repeatedly looted and not uncommonly rebuilt

and ruined again. The buried cities of Mesopotamia, the stone implements of Neanderthal man, and the skulls of *Pithecanthropus* record chapters of human history no less real because they are but vaguely known. With this viewpoint we must approach the Cryptozoic history of the earth. We are dealing with isolated fragments of the geologic record that have escaped complete destruction, but not the ravages of time.

Distribution of Precambrian Rocks. Although they presumably are present under younger rocks throughout the continental parts of the crust, Precambrian rocks are exposed over less than a fifth of the present land surface of the earth. The principal areas of exposure are the so-called **shields,** which are stable, gently arched, relatively low-lying areas that have not been deeply covered with younger rocks since the Precambrian. Each continent displays at least one shield (Fig. 7-2). Other Precambrian exposures are limited

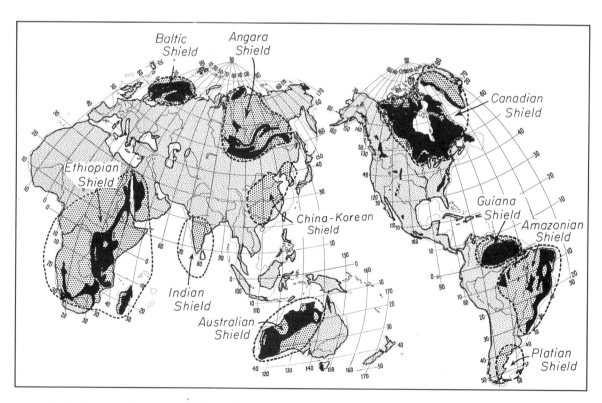

Figure 7-2 Principal shield areas of the world. Precambrian outcrop is shown in black. Antarctica, not shown, also has a shield. (Base map from American Museum of Natural History.)

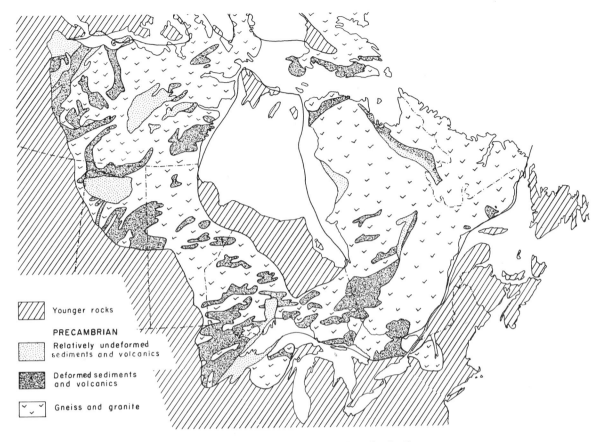

Younger rocks

PRECAMBRIAN

Relatively undeformed
sediments and volcanics

Deformed sediments
and volcanics

Gneiss and granite

Figure 7-3 Generalized geologic map of the Canadian Shield showing distribution
of major bodies of different rock types.

to the cores of mountain ranges where great
uplift and erosion have laid them bare, and to
a few gorges, such as the Grand Canyon, cut
deeply into high plateaus.

Precambrian shields have invited extensive exploration because of the richness of
their mineral resources and detailed studies
have been made of many mineralized areas. In
addition, the mysteries of the Precambrian
began to attract geologists to its problems in
the mid-nineteenth century and the more accessible areas of outcrop, in particular the
Scottish Highlands, the Baltic Shield, and the
southern part of the Canadian Shield, became
classical areas of study. Other shield areas are
remote and large parts of many are accessible
only with extreme difficulty; even today some
areas have received only cursory study. However, the principal Precambrian shields of the
earth are at least superficially known and it is
apparent that they possess a general similarity

of structure and of rock types. In the architecture of continents the Precambrian shields
have long been recognized as the cores against
which the later Phanerozoic sedimentary
troughs, or geosynclines, formed and were
thrust. The Cryptozoic sequence of events
may have differed from one shield area to the
next, but the overall nature of the Precambrian
record and the geologic problems it poses are
global. These broader aspects of the Precambrian are well exemplified in the Canadian
Shield.

PRECAMBRIAN ROCKS OF THE CANADIAN SHIELD

The Canadian Shield includes the largest
exposure of Precambrian rocks in the world
(Fig. 7-3). The larger North American portion
occupies more than 1,800,000 square miles,
consisting largely of a flat-lying glaciated plain

with Hudson's Bay in the center. Davis Strait, a fairly deep arm of the sea, separates the Greenlandic portion of the shield from North America; the Precambrian rocks of Greenland crop out around the periphery of the inland ice sheet and presumably underlie the greater part of it. The Canadian Shield is today one of the great focal points of the recently intensified study of the Precambrian. Its overall geology, which has recently been summarized [36, 3], is relatively little known in spite of the great amount of work that has been done on its more accessible parts. For a long time it was thought to be one of the younger shield areas of the world but studies of lead isotopes [31] indicate a history extending back beyond 3200 million years.

The Older Precambrian Rocks

Granite and Gneiss. At least three-quarters of the Canadian Shield is formed of granite and granite gneiss (Figs. 7-3 and 7-4). The remainder consists of irregular lenses and patches of sedimentary and volcanic rocks that are in most places intensely deformed and metamorphosed. At first it was believed that the granite and gneiss were part of the original crust of the once molten earth upon which the

sedimentary rocks and interbedded lava flows had accumulated. But in 1885 A. C. Lawson, then a young geologist working in the Rainy Lake area northwest of Lake Superior, found on examining the contacts of the granite that it clearly intruded the overlying sediments and volcanics; it was therefore younger than these overlying rocks and could not have been the original crust. Subsequent work in both the Canadian and Baltic shields supported Lawson's observation; the original floor on which these ancient sedimentary and volcanic rocks was laid down is unknown; it has been engulfed by granite magmas or altered beyond recognition.

A second fact learned about the granites and gneisses is that their apparent occurrence as a single great homogeneous mass is misleading; they are in fact complex, consisting of numerous bodies of granite and granite gneiss that formed at different times during a great number of orogenic episodes. Lawson was one of the first to recognize this multiphase nature of the shield granites, deducing it from the stratigraphic relations of his rock units in the Rainy Lake area. His work can serve as an example of how such relationships are interpreted; Fig. 7-5 illustrates the interpretation diagrammatically. Piecing together the se-

Figure 7-4 Gneiss, Grenville Township, Quebec. The dark layers are rich in pyroxenes. The foliation dips steeply to the right. (M. E. Wilson, Geological Survey of Canada.)

quence of rocks from different parts of the area revealed that the oldest layers were a sequence of sediments and volcanics (Keewatin Series) over 20,000 feet thick which had been intensely deformed, metamorphosed, and intruded by granite during an ancient orogeny. Lawson used the name Laurentian for the granites of this episode. In another part of the area a second thick series of sedimentary rocks (Seine Series) was found unconformably overlying the Laurentian granite and the Keewatin Series. Boulders of Laurentian granite occur in the basal part of the Seine Series (Fig. 7-6), indicating that erosion had stripped to their core the mountains formed during intrusion of the Laurentian granites. Lawson also discovered that the Seine Series had in turn been intruded by granite which must therefore be younger than his Laurentian; he named the younger granite the Algoman.

The local geology of the Rainy Lake area illustrates that the mass of granite and gneiss in this particular part of the Canadian Shield is not of one age. The example can be duplicated many times throughout this and other shield areas. In the late nineteenth century, before isotopic dating was available, the only means to bring order out of the many studies of individual areas, like that of Rainy Lake, was to attempt to relate the different successions of Precambrian rocks on the basis of their general appearance, using the sequence of metamorphosed sedimentary and igneous rocks, the number and position of unconformities, and the sequence of intrusions of granitic rocks. From the immense complexity of the Precambrian record certain general relationships emerged. Many of the elongate, podlike bodies of highly metamorphosed sediments and volcanics like the Keewatin had very similar suites of rocks, made up chiefly of greenstones and graywackes. On the other hand, many of the presumably younger and less metamorphosed sedimentary terranes of the shield were characterized by quartzose sandstones, limestone and dolostone, shale and some arkosic sandstone and shale.

Greenstones and Graywackes. Metamorphosed lavas take on a greenish color from cer-

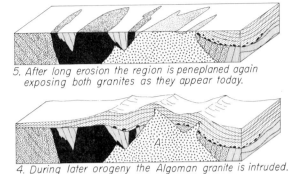

5. After long erosion the region is peneplaned again exposing both granites as they appear today.

4. During later orogeny the Algoman granite is intruded.

3. The Seine Series accumulates over the region, with boulders of the granite in its base.

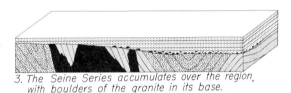

2. After long erosion the granite is exposed and the region is peneplaned.

1. Laurentian granite (L) is intruded into folded Keewatin sedimentaries (Ks) and volcanics (Kv).

Figure 7-5 Block diagrams to show the relation of the two granites in the Rainy Lake region.

tain minerals produced in their alteration and are commonly called greenstones. Alteration, however, has not completely obliterated the structure that proves they were surface flows. Pillow structure, formed by sudden chilling where viscous lava flows into standing water, is a common feature; in these areas of great structural complexity it is often a useful key in determining which side of a bed is the top. The frequency with which pillow structure occurs in the greenstones of the shield indicates that these enormous masses of lava were extruded into water, presumably into the seas of the time. Associated and locally interbedded with the greenstones are thick sequences of

Figure 7-6 Basal conglomerate of the Seine Series in which the lighter colored boulders were derived from the "Laurentian" granite that is intrusive into the Keewatin Series. West of Mathieu. (A. C. Lawson, Geological Survey of Canada.)

graywacke, a well-lithified sediment that originated as a muddy, poorly sorted sand or grit made up of a wide variety of rock fragments largely derived from rocks with abundant dark minerals. In these distinctive rocks an original sedimentary feature, graded bedding, has in many places survived the metamorphism. Graded beds form when currents heavily charged with sediment flow down the slopes of lake or ocean basins; the load settles out, heaviest grains first, as it spreads into the still water of the basin floor, leaving a layer that grades from coarse at the bottom to fine at the top.

Water-laid volcanics and graywackes with graded bedding are a combination of rocks not restricted to the Precambrian. In the more easily read Phanerozoic record these rocks characterize regions where tectonically active islands shed great quantities of sediments and volcanics into deep adjacent basins. As far as is known, the site of these mobile regions is limited to the outer, deeper parts of those geosynclines that border continents; they lie between the ocean basins and the shallower, more stable parts of the geosynclines which adjoin continents and receive their sediments from them. The Precambrian greenstone and graywacke terranes are many, and mobile belts of volcanic islands must have been a common feature of the earth's crust in these ancient times. If these early sites of deep geosynclinal deposition also had shallower parts accumulating sediment from adjacent stable areas, they are not now recognizable in the sea of granite gneiss that surrounds them.

Younger Terranes with Well-Sorted Sediments

The better sorted, less metamorphosed sediments of the shield consist chiefly of quartzites, slates, limestones, dolostones, and banded iron formations. Arkoses, graywackes, and lavas also occur but are not widespread. A pronounced unconformity generally separates these less deformed rocks from the underlying granite gneiss and greenstone-graywacke sequences.

Great thicknesses of these dominantly sedimentary formations are concentrated in several parts of the Canadian Shield where they are the locus of valuable mineral deposits. At some places the rocks are strongly folded in with the underlying granite gneiss, but in many places they are little deformed. Every-

where they are complex, recording a number of orogenic episodes and subject to abrupt lateral changes in rock type.

The best known terranes of younger Precambrian rocks on the shield are in the area lying north of Lake Huron in southeastern Ontario, and the larger region around the western part of Lake Superior in Wisconsin, Minnesota, and Ontario. The fact that the rocks of these two closely adjacent areas have been studied in detail for over half a century and still remain to be satisfactorily correlated is a measure of their great complexity. A few highlights of the geology in these two areas will serve to indicate the nature of the younger Precambrian rocks.

Huronian Rocks. North of Lake Huron a laterally variable sequence of quartzites, mud-

Figure 7-7 Gowganda Tillite near the base of the Cobalt Series at Drummond mine, near Cobalt, Ontario. The large boulder at the right above is 30 inches across. (C. W. Knight, Ontario Bureau of Mines.)

stones, conglomerates, and limestone with an aggregate thickness of more than 10,000 feet has long been called the Huronian and commonly separated into the Brucé Series below and Cobalt Series above. The Bruce rests with conspicuous angular unconformity on the granite gneiss of the older Precambrian and has a basal conglomeratic graywacke layer with a variety of cobbles and boulders of greenstone, granite, and other rocks. Above this the sequence appears to contain several cycles consisting of basal boulder conglom-

erates overlain by fine-grained sedimentary rocks, which in turn grade upward into coarse-grained sediments [26]. The upper of three similar boulder conglomerates, a part of the Gowganda Formation, lies at the base of the Cobalt Series and is commonly known as the **Gowganda tillite.** This is a poorly sorted, unstratified conglomeratic graywacke that includes faceted and striated boulders as much as 10 feet in diameter (Figs. 7-7 and 7-8). Long heralded as evidence of Precambrian glaciation, the origin of the Gowganda as a

Figure 7-8 Striated boulder from the Gowganda Tillite at Cobalt, Ontario. Slightly enlarged. (Royal Ontario Museum.)

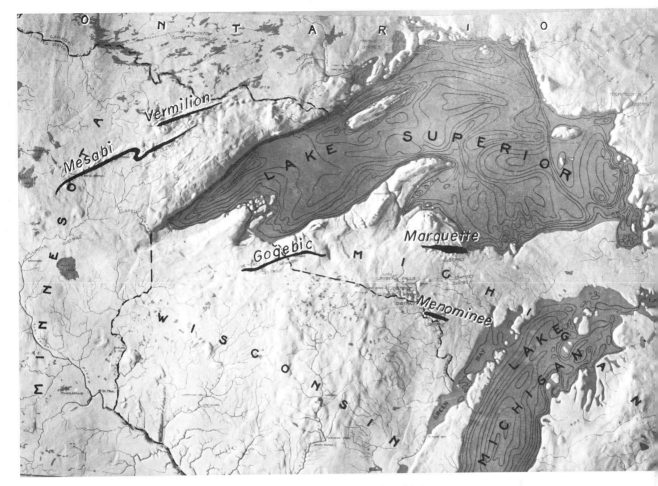

Figure 7-9 Model of the Lake Superior region showing distribution of the chief "iron ranges." (Photographs of the model used by permission of the Director, U. S. Geological Survey.)

fossil tillite has been questioned by many. That it is indeed a tillite in some areas has been unquestionably demonstrated by the discovery that at the south end of Lake Timigami in Ontario it lies on a polished and striated pavement; here the evidence indicates that the ice sheet advanced three times from the northeast during deposition of the Gowganda [29]. The Gowganda is a widespread formation, suggesting a large ice sheet rather than montane glaciation. If the two similar boulder conglomerates in the Huronian sequence beneath the Gowganda should also prove to be tillites, an extended period of Precambrian glaciation would be indicated.

Animikie Rocks. South of Lake Superior in the Menominee district of northwestern Michigan and Wisconsin (Fig. 7-9) a different sequence of younger Precambrian rocks is exposed. Resting unconformably on granite gneiss of some part of the older Precambrian are nearly 50,000 feet of younger Precambrian rocks. The lower 12,000 feet or more of this sequence consist of quartzites overlain by dolostones. The great remaining thickness is made up of slates, graywackes, greenstones with pillow structure, and cherty banded iron formations. The entire sequence is presently called the Animikie Series [22]. In one part of the area the basal formation contains tillite,

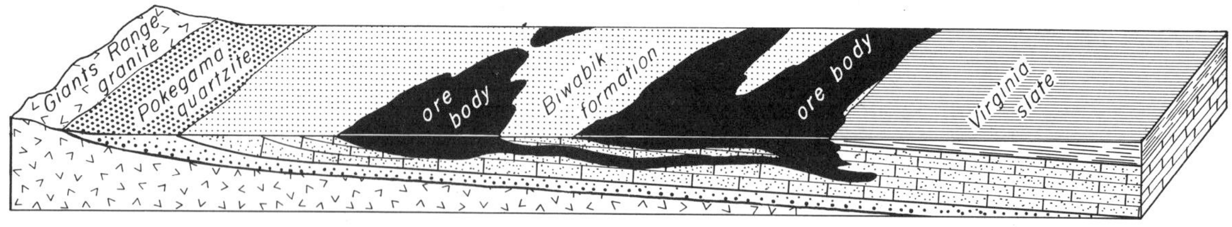

Figure 7-10 Block diagram showing a section across the Mesabi Range at Hibbing, Minnesota. Length of section about 4½ miles. (Adapted from Van Hise and Leith, U. S. Geological Survey.)

but this is the only known occurrence of this rock type, so distinctive of the Huronian Series in Ontario, in the Animikie rocks of the Lake Superior region. West and northwest of the Menominee area the Animikie Series thins and in the Mesabi and Vermillion iron ranges at the west end of Lake Superior only its middle part with a basal quartzite, an iron formation, and overlying slates (Fig. 7-10) are present.

Banded iron formations are unique concentrations of iron and silica known only in Precambrian rocks. They are of great interest not only as the backbone of American economic development, for they have been its chief source of iron, but in addition they contain some of the earliest known forms of life on earth and suggest environments radically different in some ways from those about us today. Characteristically, iron formations are well-bedded, locally laminated bodies of rock consisting of chert and siliceous shale. Iron-rich layers alternate with iron-poor layers to give the formations a banded appearance. Layers of slate and carbonate rocks are also locally interbedded with the cherts. Stromatolites (p. 159) of chert are a fairly common occurrence and some of them at least are of algal origin (p. 160). The iron content of the cherts varies appreciably but in general is about 25 to 30 percent.

Banded iron formations are widespread in the Precambrian of other parts of the world. Wherever they occur, it has been observed that they were deposited during an immense span of Cryptozoic time extending back from about 1700 to 3000 million years ago, but none are known from younger rocks. Lepp and Goldich [24] point out that this distribution in

time conforms to the widely held theory that the primitive earth atmosphere was deficient in oxygen. They postulate that iron and silica derived by thorough weathering were transported without oxidation to the sea. Here they were deposited together, precipitated either directly or indirectly by biologic processes, to form the cherty iron formations, the liberation of oxygen by photosynthesis in primitive marine algae probably playing a critical role.

Keweenawan Rocks. A second great sequence of younger Precambrian overlies the thinning Animikie Series in the central and western parts of the Lake Superior region. This is the Keweenawan Series of redbeds and lava flows. The redbeds include sandstones, siltstones, and shales with lesser amounts of conglomerate and have an aggregate thickness of some 15,000 feet. They are interbedded with and succeeded by lava flows totaling nearly 35,000 feet in thickness. The surfaces of successive flows are still rough and slaggy and much of the lava, originally vesicular, is amygdaloidal. In contrast to the subaqueous origin of greenstones with pillow structure, the vesicular lavas flowed out over the land. This enormous outpouring of molten rock allowed the crustal rocks to sag, forming the great structural basin now occupied by Lake Superior. Not all of the basic magmas came to the surface; great bodies of it were intruded into the lower part of the Keweenawan Series and cooled there to form sheets of gabbro while lavas were still building the upper part of the series. The largest such mass known is the Duluth Gabbro, which forms a body over 100 miles long and in places more than 40,000 feet thick. With upwelling of these lavas came copper-bearing solutions, which in places filled the vesicles of the lava with metallic copper and in other places cemented the pebbles of interbedded conglomerates.

Most Keweenawan rocks are only slightly metamorphosed. In Minnesota they are overlain unconformably by fossiliferous strata of Upper Cambrian age. The Duluth Gabbro, which lies approximately between the middle and upper parts of the Keweenawan Series, was intruded approximately 1100 million years ago, according to isotopic dates [16]. As young as the Keweenawan rocks may seem in terms of other Precambrian rocks, the length of time that elapsed between their deposition and the beginning of the Cambrian is nearly equivalent to that of the entire Phanerozoic Eon!

Grenville Rocks. A deformed and strongly metamorphosed series of sedimentary and volcanic rocks called the Grenville Series crops out in southern Quebec and southeastern Ontario and in the Adirondack Mountains of New York. Metamorphism of the Grenville rocks has altered the limestone to marble, the sandstone to quartzite, and the volcanic rocks to amphibolite. The whole has been intruded by granite. In the more intensely metamorphosed parts of the Grenville, intimate mixture of igneous and metamorphic rocks in alternating layers has produced conspicuously banded rocks (Fig. 7-11). Because of their high grade of metamorphism and great deformation the Grenville Series was once thought to be among the oldest of the Precambrian rocks, but dating by radioactive isotopes has shown their deformation and intrusion to be only about 1000 million years ago.

Grenville rocks clearly point out the fallacy of attempting to relate Precambrian rocks on the basis of degree of metamorphism or deformation. Apparently they were deposited at about the same time as the very slightly metamorphosed Keweenawan and Belt (p. 150) Series.

Figure 7-11 An early stage in the formation of migmatite. Bodies of pegmatite have intruded along the axes of folds and have locally begun to spread laterally into the layers of metamorphosed sediments. From a Precambrian outcrop near Helsinki, Finland. (Courtesy of P. M. Orville.)

Mineral Wealth

The Precambrian rocks of the Canadian Shield have yielded iron, copper, nickel, zinc, gold, silver, and platinum beyond the dreams of Midas. The iron of the banded iron formations is the sole sedimentary deposit, the other metals occurring in association with igneous rocks. The exploitation of these metallic riches, and other shield minerals, has played no small part in the industrial development of both the United States and Canada.

Iron. Since they were first mined in 1845, the iron ranges of the Lake Superior District have accounted for the bulk of the iron produced by the United States. Here the banded iron formations of the Precambrian, exposed to weathering over millions of years, had been deeply leached and oxidized, the silica being carried away and the iron concentrated. Whereas the fresh rock of the cherty iron formations, called **taconite**, contains only 25 to 35 percent iron and is not an ore, the weathered product was enriched to 42 to 56 percent iron and reduced in hardness so that it could be easily mined. The iron ranges are more or less linear belts of outcrop of the folded banded iron formations (Fig. 7-9). That of Mesabi, which has produced twice as much as all the other ranges together, will serve for illustration (Fig. 7-10). Here the banded iron formation is from 400 to 750 feet thick and dips gently to the southeast. The outcrop averages over a mile in width and extends northeastward for 110 miles; ore bodies occur along it in pockets up to 1.5 square miles in surface area. These rich bodies are mined by earth-moving equipment in enormous open cuts (Fig. 7-12). In the Gogebic Range of Michigan, where similar iron formations are more strongly folded, the enrichment of the taconite reached to depths of about 3000 feet below the present surface so that underground mining has been necessary.

Production of iron from the Lake Superior District rose to a wartime high of 91,000,000 tons in 1943, declining in the early postwar years then rising to an all-time high of 110,096,199 tons in 1953. The rich deposits of weathered taconite were seriously depleted by the end of World War II, but technological

Figure 7-12 Open-cut mine in the Mesabi Range, Minnesota. (L. P. Gallagher.)

advances, in particular jet-piercing drills and methods of concentrating the iron in low grade taconite by fine crushing and separation, then forming pellets from the concentrate, brought phenomenal rebirth to the waning Lake Superior District. In the 1960s concentration and pelletizing of taconite has become a major new industry with well over a billion dollars invested in new processing plants and equipment. In the Mesabi Range alone it is estimated that at least 5 billion tons of taconite is available to surface mining and will yield 1.7 billion tons of pellets averaging about 60 percent iron. An additional 31 billion tons of Mesabi taconite is accessible by underground mining [17].

Technical advances have turned a colossal amount of previously worthless Precambrian taconite the world over into valuable ore. Along with new discoveries of high grade ore, most of which is in Precambrian terranes, economic geologists believe the world has an adequate supply of iron in sight for the next 250 years [35]. This represents an incredibly large reserve supply considering that world production of iron more than doubled between 1950 and 1962.

Great deposits of rich oxide ore similar to that of the Mesabi Range were discovered in the vicinity of Knob Lake in Labrador near the Quebec boundary in 1937. This ore, lying in the midst of subarctic wastes, could not be immediately exploited, but extensive prospecting indicated the ore bodies were great enough to justify building a railroad north from the port of Seven Islands on the St. Lawrence River and building a town, Schefferville, to house the mining operations. More than half a billion dollars has been invested in these operations, which started shipping ore in 1954 and are now producing over 7 million tons of high grade ore annually. This important resource development on the Canadian Shield was one of the principal reasons for building the St. Lawrence Seaway, which permits cheap transportation of the ore to the steelmaking centers of the Great Lakes region.

Copper. Native copper occurs in the lavas and conglomerates of Keweenaw Peninsula and was known and worked by the Indians

long before the advent of white men. It forms amygdules in the scoriaceous Keweenawan lavas and not only serves as a cement between the pebbles in the conglomerate, but also has partly replaced the pebbles themselves. As the copper is concentrated in certain beds, and the Keweenawan Series dips steeply to the northwest, the producing area is a belt only 3 to 6 miles wide, running for about 70 miles along the axis of the peninsula.

The famous Calumet and Hecla mines are located on this belt. At its peak of production in 1916 this region alone produced 135,000 tons of copper, valued at $66,300,000. The yield had declined to 30,400 tons in 1945, but the grand total for the hundred years 1845–1945 was over 4,800,000 tons, slightly more than one-seventh of the total production of the entire United States. Production in the 1960s has averaged over 70,000 tons of copper a year. In 1966, 7,990,770 tons of ore were mined in Michigan, from which 73,449 tons of copper were recovered.

The Canadian Shield in Canada yielded 443,500 tons of copper in 1966. Nearly half of Canada's production of copper comes from the nickel-copper sulfide ores of the Sudbury District. The remainder comes from sulfide deposits elsewhere in Ontario, Quebec, Manitoba, Labrador, and Saskatchewan; these same deposits also supply much of Canada's zinc and important amounts of gold and silver.

Nickel. Over 50 percent of the world's nickel is now secured from the Canadian Shield, most of it from the mining district of Sudbury, Ontario, where a great sill of gabbro has a border of nickel-copper ore. Production from this small area began in 1889 and through 1966 it had totaled well over 5 million tons of nickel, an equivalent amount of copper, and significant amounts of gold, silver, and platinum. Ore resources of the Sudbury area are estimated to be more than 350 million tons containing approximately 10,500,-000 tons of combined nickel and copper.

Silver. Silver is found in many parts of the Canadian Shield. Up to the beginning of World War II a number of great mines were operated primarily for their silver. Chief among these were the rich ore bodies at Cobalt, Ontario, where native silver and other minerals occur in veins associated with a great sill of diabase. Between 1904 and 1937 the Cobalt mines alone produced over 380 million ounces of silver. Other by-products of these mines included cobalt, arsenic, bismuth, nickel, lead, and copper. Since World War II most of the silver produced from the Shield has been the by-product of mines worked primarily for cobalt, uranium, nickel, and base metals. Silver production in Canada during the 1960s has averaged more than 30 million ounces a year, more than 80 percent of it from by-product recovery. Roughly two-thirds of this production has been from the Canadian Shield.

Gold. Most parts of the Canadian Shield have yielded gold but the bulk of Canada's production for many years has come from the great vein and lode deposits in the provinces of Ontario and Quebec. Most spectacular has been the gold production of the area around Porcupine, a settlement in Ontario nearly 100 miles northwest of Lake Timiskaming. From its discovery in 1909 until 1955 the Porcupine District produced more than $1,200,000,000 worth of gold and it still accounts for more than a fifth of Canada's total gold production. The Hollinger mine has been the major mine in the area and one of the richest gold mines known. Gold in the Porcupine District occurs chiefly in the metallic form in veins associated with granitic porphyry intrusions in altered Keewatin lavas and sedimentary rocks.

For many years the Porcupine and Kirkland Lake districts of Ontario were the only major gold-producing areas in Canada, but together they assured the country of second or third rank among the gold-producing nations of the world. Highest production since World War II was in 1960 when more than 4.5 million ounces of gold valued at more than $157,000,000 was recovered. Although gold mining on the Shield has been declining, Canada still ranked third in 1966 with 8 percent of the world total. The leading producer is the Republic of South Africa with an amazing 64 percent of the world's gold in 1966; this is also from Precambrian rocks of shield areas.

More than one-third of the gold produced in the United States comes from a small area in the Precambrian core of the Black Hills at Lead, South Dakota, where the great Homestake Mine continued its long dominance with the recovery, in 1966, of more than 600,000 ounces of gold and 100,000 ounces of silver having a combined value of some $21,000,000.

All these valuable metals and other important mineral resources are in veins associated with intrusives, and so the cumulative igneous activity over the immense span of Precambrian time has had a very direct effect on humans and their economy. In the year 1966, for example, mines in the Shield accounted for over 75 percent of the total metals production of Canada and these products were valued at more than a billion and a half dollars [5]. Other shield areas of the world are also producers of great mineral wealth and it would not be an exaggeration to say that these great areas of Precambrian rocks have supplied man with the sinews of his modern civilization, and the power either to extend or destroy it.

PRECAMBRIAN ROCKS OUTSIDE THE SHIELD

Grand Canyon Precambrian

Lying with conspicuous unconformity beneath the horizontal Cambrian strata in the inner walls of the Grand Canyon are two markedly distinct series of Precambrian rocks (Fig. 7-13). The older of these is the Vishnu Schist, a complex body of quartz mica schists, amphibolite schists, quartzites, and gneisses intruded by granitic stocks and pegmatite dikes. Relic bedding structures are preserved in the quartzites and mica schists, revealing a sedimentary origin. The amphibolite schists, interpreted as altered volcanic rocks, occur as separate masses within the other rocks. Both the relic stratification and the foliation imposed during intense folding and metamorphism are generally parallel and are vertically oriented in most places in the inner gorge of the Canyon. The aggregate thickness of metamorphosed sedimentary and volcanic rocks is great, perhaps as much as 25,000 feet,

Figure 7-13 Two great series of Precambrian rocks lying beneath horizontal Paleozoic strata in the Grand Canyon. At lower left is the older Precambrian Vishnu Schist, whose foliation and relic bedding is vertical. Lying with conspicuous unconformity on its bevelled surface are the gently dipping beds of the younger Precambrian Grand Canyon Series, here only a thin remnant but elsewhere as much as 10,000 feet thick. These tilted beds are in turn overlain, with angular unconformity, by basal Cambrian strata. Note how the latter are draped over irregularities on the surface of unconformity. (U. S. Geological Survey.)

but accurate measurement is not possible because of the duplication of beds by isoclinal folding and the effects of metamorphism and intrusion. The Vishnu rocks represent a thick sequence of sediments and volcanics that was intensely folded and metamorphosed during one or more episodes of mountain building. The intrusive granitic rocks associated with the Vishnu are believed to have formed during a late phase of this orogeny. Isotopic dating of some of these granitic bodies that cut the schists and of related rocks in Arizona and Nevada suggest that they were intruded between 1600 and 1800 million years ago [34].

Truncating the Vishnu Schist with striking unconformity along a nearly flat erosion surface is a great thickness of unmetamorphosed sedimentary and volcanic formations known as the Grand Canyon Series. These once covered the whole region in regular, horizontal beds, but before Cambrian time they were broken into great fault blocks and then eroded to a nearly plane surface. Only the downfaulted portions of the blocks were preserved, tilted at low angles to the horizontal Cambrian strata above. A comformable sequence of these strata more than 2 miles thick is exposed locally in one of the fault blocks. The formations consist chiefly of sandstones and shales, predominantly of reddish hues, and there are some beds of limestone; in appearance and lithology they much resemble the younger rocks that unconformably overlie them. The limestones and associated ripple-marked shales were deposited in shallow water, but the red shales marked by mudcracks and raindrop prints as well as ripple marks indicate deposition above sealevel. The region is inferred to have been a vast, low-lying deltaic plain over which submarine and subaerial deposition alternated. Since the surface remained near sealevel while several thousands of feet of sediment were deposited, the region obviously was slowly subsiding. Unlike the younger strata above, the Grand Canyon Series contains only rare traces of life: the trails of wormlike animals in the ripple-marked shale and some algal deposits in the limestone beds.

The Precambrian history of the Grand Canyon region is summarized in the block diagrams shown in Fig. 7-14. A thick prism of sediments and volcanics was folded and recrystallized deep in the roots of a growing mountain range (Block A) to form the Vishnu Schist. Granitic intrusion followed during a late phase of the deformation, which extended far beyond the Grand Canyon region. It has been named the **Mazatzal orogeny** for an area in south-central Arizona where it produced intense folding and imbricate thrusting, accompanied by granitic intrusion on a larger scale. The surface relief shown for the range of Mazatzal Mountains is purely symbolic, but whatever the nature and height of that ancient highland, it was eventually worn down to a remarkably flat surface (Block B). Brought near to sealevel, the region then began to subside slowly as delta plain sediments accumulated, layer on layer, to a thickness of more than 12,000 feet (Block C). For a short time during deposition lava flows poured out locally.

Eventually the region was uplifted and, with but slight folding, was broken by great faults that gave rise to a system of block mountains (Block D) much like those that now form the Basin and Range Province east of the Sierra Nevada. Of course these blocks were eroded as they rose, so that their height at any particular time cannot be determined. However, remnants of a dozen or so great blocks are still preserved, and their bounding faults are clearly evident, so that the size and orientation of the ranges can be plotted; and if we project the dip of the beds in portions of the fault blocks still extant, the restored edges reach elevations more than 2 miles above their present surface.

In time these mountains were destroyed by erosion and the region was again reduced to a surface of very low relief (Block E) before the Early Cambrian seas spread over it to mark the beginning of a new era (Block F).

The two profound unconformities in the Precambrian sequence of the Grand Canyon (Fig. 7-13) each represent an immense interval of lost record during which the entire region was slowly reduced by erosion from bold mountainous relief to a low-lying, nearly flat surface. It is probable that the erosion surface on the Vishnu Schist represents as much time as the entire Phanerozoic Eon.

Rocky Mountain Region

From Utah and Colorado northward through Wyoming, Montana, and Idaho, and far into British Columbia, the Rocky Mountains have brought extensive areas of Precambrian rocks to the surface. From range to range these rocks vary in character, most are strongly deformed and considerably metamorphosed. Isotopic dating has shown that deformation of these ancient rocks took place at widely different times, spanning an interval of at least a billion and a half years. For example, in a considerable area about the Beartooth Mountains of southern Montana, the isotopic age of granites cutting a sequence of older Precambrian rocks is about 2750 million years and a still greater age of 3100 million years is esti-

Figure 7-14 Six stages in the Precambrian history of the Grand Canyon region. The view is northward, and the sections represent an east-west distance of about 15 miles. The wedge of Grand Canyon formations near the right end of block F is pictured in part in Fig. 7-13.

A. Mazatzal orogeny — metamorphism forms Vishnu Schist

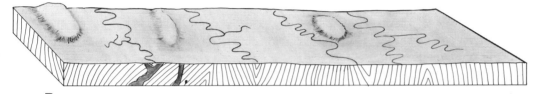

B. Erosion leaves only roots of Mazatzal mountains

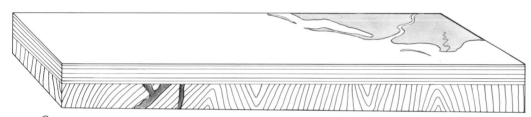

C. Sediments of Grand Canyon Series deposited

D. Block mountains formed by faulting

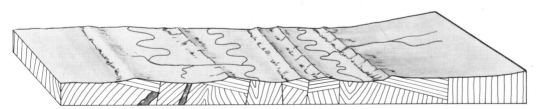

E. Erosion nearly levels region in Late Precambrian

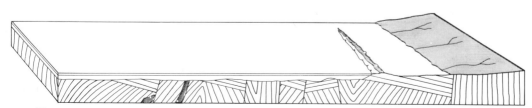

F. Cambrian seas spread beginning a new era

mated from isotopic determinations on associated rocks [6]. These are among the oldest rocks known in North America. Precambrian rocks in the cores of the Colorado Rockies, on the other hand, yield isotopic dates that indicate major deformation between 1200 and 1500 million years ago [14].

Belt Series. In the Rocky Mountains, as in the Grand Canyon region, relatively undeformed Precambrian sedimentary rocks locally overlie the more intensely metamorphosed and deformed Precambrian rocks. A thick sequence of these younger Precambrian rocks, named the Belt Series for exposures in the Big Belt Mountains of southwestern Montana, crops out over an area 200 miles wide that extends through western Montana and adjacent Idaho northward well into Canada. The Belt Series ranges in thickness from more than 40,000 feet in some of its western and central outcrops to less than 10,000 feet on the east. Exceptionally fine exposures occur in Glacier National Park near the northwest corner of Montana, where a thickness of 25,530 feet has been measured (Fig. 7-15) [12].

Many formations comprise the Belt Series and the stratigraphy is complex and not yet well known. Its units are thick and dominantly fine grained, consisting largely of shales or claystones, fine-grained quartzites and limestones. Especially noteworthy are two great limestones, a lower one about 2000 feet thick

Figure 7-15 The "Garden Wall" overlooking Iceberg Lake in Glacier National Park. This towering cliff exposes about 2000 feet of the Siyeh Formation of the Belt Series. The *Collenia* bed is a reef of calcareous algae. (Photograph by Heilman, courtesy of Great Northern Railway.)

Figure 7-16 Reef of calcareous algae in the Siyeh Formation in Hole-in-the-Wall Cirque, Glacier National Park, Montana. (C. L. and M. A. Fenton.)

and a higher one some 4000 feet thick, which are prominent in the Glacier Park area and along most of the eastern half of the Belt terrane. Basic lava flows are interbedded with the sediments at several horizons in the upper part of the series. These are mostly thin, only locally ranging up to 300 feet thick. The entire series appears to have been deposited in very shallow marine environments and adjacent coastal plains. The predominantly gray shales commonly show oscillation ripples suggesting wave action in shallow standing water. Mudcracks and raindrop prints in both red and gray shales indicate periods of emergence, which also attest to generally shallow conditions over much of the basin of deposition.

Perhaps some of the best evidence suggesting shallow seas is afforded by the limestones of the Belt Series, which characteristically contain **algal stromatolites.** These are laminated masses of various shape apparently formed by algae that grew in matlike bodies on the sea floor. Present-day species of green and blue-green algae form similar laminated bodies and it is reasonable to suppose that their ancient counterparts were responsible for the Precambrian stromatolites. Studies of modern stromatolite-forming algae [25] reveal that many are found in very shallow or intertidal marine environments and in saline lakes. The bulk of the carbonate in the Belt Limestones is believed to have been organic in origin [27], presumably the work of the algae. Reeflike deposits of the calcareous algae (Fig. 7-16) are locally prominent in the limestones and in some of the gray shales, and rare trails re-

Figure 7-17 Structures, possibly burrows of organisms packed with sand, in the Siyeh Formation of the Belt Series. Dawson Park, Glacier National Park, Montana, (C. L. and M. A. Fenton.)

sembling those of wormlike animals (Fig. 7-17) have been found. No other undisputed remains of life have been found in Belt rocks.

The regional distribution, great thickness, and abundant evidence of shallow deposition of the Belt Series are the first evidence we have of a broad depositional trough, or geosyncline, lying roughly north-south along the Cordilleran region flanking the west side of the more stable continental platform. Well over a billion years ago this cordilleran trough began receiving sediments, carried into the shallow Belt sea from adjacent highland to the west and southeast. From that time through the Early Triassic it was a persistent feature of continental geography, intermittently occupied by seas and accumulating a vast sedimentary record. There is no direct evidence that the deposits of the Belt Series were once

coextensive with those of the Grand Canyon Series. Too many orogenies have deformed and broken the Cordilleran region since these series were deposited, and we are left with only suggestive bits of information. A possible reconstruction of the extent of the Cordilleran trough [27] during the long time represented by deposition of the Belt and Grand Canyon series is shown in Fig. 7-18.

Except for one small area where it can be seen resting unconformably on older Precambrian granite gneiss, the base of the Belt Series is not exposed. Since over such a large area the Belt formations are but slightly metamorphosed, they commonly appear conformable with the overlying strata, which are Middle Cambrian quartzites. At a number of localities slight angular unconformity is apparent between the Belt and the Cambrian strata but elsewhere there is little evidence of a physical break. For this reason the Belt Series was at one time considered Late Precambrian, but dating by radioactive isotopes shows how deceptive physical appearance can be. In the Coeur d'Alene district of Idaho, Belt rocks are cut by pegmatite veins carrying pitchblende dated at 1190 million years [9]. This indicates a hiatus of perhaps as much as 500 million years between the Belt Series and overlying Cambrian strata in Montana. No evidence of orogeny or block faulting, as in the Grand Canyon, marks this great time gap between the Belt and the Cambrian rocks. Regional study has shown, however, that the Cambrian is transgressive on different parts of the Belt Series and equivalent rocks in adjacent parts of British Columbia. In most of the Canadian portion of the Cordilleran trough, as much as 10,000 feet of Lower Cambrian rocks overlie the Precambrian, but in Montana only Middle Cambrian rocks overlie the Belt Series. In addition, Precambrian rocks younger than those of the Belt Series appear to the north in British Columbia beneath the Lower Cambrian strata. This evidence together with the local examples of slight angular unconformity of the Middle Cambrian on the Belt Series indicate that the great prism of Belt sediments had been uplifted, gently warped, and leveled by erosion before it was transgressed by Middle Cambrian seas. The history has obvious parallels with that of the Grand Canyon Series and the major events may have been approximately contemporaneous.

Interpretation of Precambrian Rocks

To early workers in the Precambrian of the Canadian Shield the widespread relationship of intensely deformed rocks overlain with conspicuous unconformity by less metamorphosed rocks offered a natural basis for subdivision. The name **Archaean** for the older rocks and **Proterozoic** for the younger became common usage not only in this country but in many parts of the world. As new terranes of

Precambrian rocks were studied, subdivisions of the Archaean and Proterozoic were recognized locally. The names of some of these local rock units, for example, the Huronian Series, were then applied in a broader sense

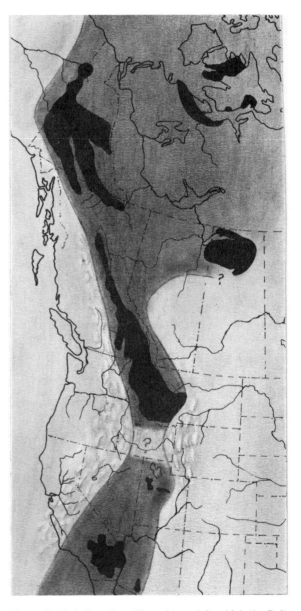

Figure 7-18 Inferred position of trough in which the Belt Series and possibly related Late Precambrian sedimentary rocks were deposited. Black areas indicate distribution of rocks of Beltian age.

to other series of rocks that were believed to be correlative with them. In this way the Huronian and other local names became widely used as major subdivisions of a Precambrian time scale superficially similar to that of the Phanerozoic. The details of this classification of Precambrian rocks need not concern us here, for it has become obsolete, but the reasons that it has not survived are instructive. They provide an introduction to present-day methods of attack on the enormous problem of unraveling Precambrian history.

By the early part of the present century the gradual accumulation of information about Precambrian rocks was beginning to point to a far longer and more complex history than had originally been suspected, but it was the development of isotopic dating that drastically changed prevailing ideas about Precambrian stratigraphy and time. Arthur Holmes, one of the pioneers of modern geochronology, describes the impact of isotopic dating on the study of the Precambrian in these words:

There is no branch of geology in which it has been necessary to discard so many "reasonable" correlations and to abandon so many provisional hypotheses [21].

Isotopic dates showed that many Precambrian sequences which had been correlated with one another on the basis of lithologic similarity and limiting unconformities were different in age, some by as much as hundreds of millions of years. Moreover, the assumption that the close of the Archaean was marked by a great, worldwide orogeny also proved erroneous when isotopic dates revealed that supposed "Archaean" granites differed widely in age and were the product of a number of orogenic episodes. But the most significant effect of the isotopic dating of Precambrian rocks was the revelation of the immense span of Cryptozoic time (Fig. 7-19). At the beginning of the present century the age of the earth was generally accepted to be between 50 and 150 million years, with the Cryptozoic no longer than the Phanerozoic in duration. Today the age of the earth is estimated to be on the order of 4500 million years and the Cryptozoic is judged to be between 7 and 8 times as long as the Phanerozoic.

Figure 7-19 Changes in our concept of geologic time during the last 80 years. The great increase from Joly to Barrell marks the advent of isotopic dating. (Adapted from Harold L. James.)

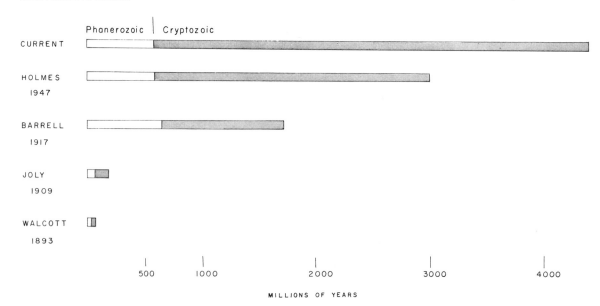

MILLIONS OF YEARS

MOUNTAIN ROOTS AND STRUCTURAL PROVINCES

The intensely deformed granitic areas of Precambrian shields are made of the roots of many ancient mountain ranges that were essentially no different from the younger ranges of Phanerozoic mountains which border them today. The first students of the Precambrian did not comprehend this fact but were impressed by the great extent of granitic rocks, and indeed it does stir the imagination to contemplate the million or more square miles of granite and granite gneiss exposed on the Canadian Shield. But we are accustomed to seeing only the elevated upper parts of mountains that still retain a jacket of thick sedimentary or slightly metamorphosed rocks about their limited granitic cores. Shield mountain systems have been leveled by eras of erosion which have laid bare horizons once several miles below the surface. The granitic core of one great orogenic belt lies next to that of another giving the granitic areas of the shield a superficial uniformity of appearance. Early geologists were misled by this, thinking at first that they were dealing with the original crust of the earth, but later believing in one great, worldwide period of orogeny characterized by the emplacement of enormous amounts of granitic rocks.

The large masses of granite and granite gneiss are almost everywhere closely associated with folded rocks, but the relationship is not a simple one. Bodies of granite in many places obviously have been intruded as a magma into the folded rocks, breaking across the layering and structures. Mixed rocks, or migmatites (Fig. 7-11), on the other hand, show intimate interlayering of granitic rock and metamorphosed sedimentary rock. In addition, many bodies of granite gneiss that retain relict bedding structure were apparently once bodies of sedimentary rock and have been entirely transformed into gneiss. Rocks of this latter type led to the concept of **granitization**, which holds that some granitic rocks originate from intense metamorphism and reconstitution of sediments and volcanics. Opinion is still divided over the origin of certain granites and the relative importance of

"granitization," but it is evident that granitic rocks have not all formed in the same way. Their direct association with orogenic belts suggests that the different kinds and modes of occurrence of granitic rocks are but phases of the great changes taking place in the root zone during mountain-building activity. Here metamorphism, mixing, transformation and, ultimately, melting must all have been operative as the sediment-filled troughs of the ancient geosynclines were deeply infolded and subjected to increasing heat, pressure, and pervading solutions and gasses with depth.

Recognition that the shield areas were composed of the roots of Cryptozoic mountain systems was a contribution of J. J. Sederholm, a renowned Finnish geologist whose classic work on this subject appeared in 1907. The same year a physicist, B. B. Boltwood, having previously recognized that lead was the ultimate product of the radioactive decay of uranium, used this knowledge to measure the geologic age of a number of uranium-bearing minerals [4]. This was the pioneer attempt at isotopic dating. That these two scientific breakthroughs date from the same year is a convenient coincidence, for together they form the basis for modern interpretations of Precambrian geology. The isotopic method for determining the age of rocks, not widely accepted among geologists until the 1930s, has made possible the dating of the Precambrian orogenic belts first recognized by Sederholm.

Today a number of large geographic subdivisions, called **structural provinces**, are recognized in the Canadian Shield (Fig. 7-20). Each is distinguished by the direction in trend of its structural features, such as the axes of folds, and by the age of its last important orogeny, which produced the dominant metamorphism and major intrusions of granitic rocks. Province boundaries are generally well-defined and may be marked by abrupt changes in metamorphism and structural trends or by faults or intrusive contacts. For example, the northwest boundary of the Grenville province, conspicuous enough locally to have earned the name "Grenville front," truncates the east-

Figure 7-20 Structural provinces of the Canadian Shield. The times of last major deformation in these provinces is given in Fig. 7-21.

trending formations and fold axes of the older Superior province. At some places it is a fault and at other places it is the limit of metamorphism imposed during the Grenville orogeny.

Although the structural provinces have each been subjected to more than one orogeny, their structural trends have remained approximately the same throughout their individual histories. In this and most other respects they are comparable to younger major mountain systems, as Sederholm observed. The Appalachian mountain system, for example, was deformed in part during the Taconic orogeny (Ordovician), the Acadian orogeny (Devonian), and the Alleghanian orogeny (Late Paleozoic). In each of these orogenies the deforming forces acted in approximately the same direction and the resultant structures are similarly oriented.

Isotopic dating of minerals formed during the metamorphism and intrusion of rocks in the structural provinces of the Canadian Shield has revealed four principal episodes of orogeny (Fig. 7-21). Three of these have also been detected in shield areas of other continents. Recognition of orogenic episodes affords a new basis for broad subdivision of Precambrian history and eventually a worldwide Cryptozoic time scale based on isotopic dating may be expected. Generally the last major episode of intrusion and deformation in a structural province has erased the isotopic evidence of earlier orogenies by the alteration of minerals during metamorphism. But there are exceptions to this and, in addition, techniques for detecting earlier deformation despite the masking effects of the dominating orogeny have had limited success. Geologic criteria, however, remain the principal source of information about the history of a structural province prior to its last deformation.

GROWTH OF THE CONTINENTAL CRUST

We cannot point to an outcrop of rock anywhere on the earth's surface today and claim it is a part of the "original" crust. Wherever it has been possible isotopically to date a body of ancient rocks it has also been possible to

show by geologic relationships that still older rocks are associated with it. Nevertheless, there is some indication in the complex Precambrian record that the early crust was similar to the present crust of the Pacific Oceanic region and was largely made up of basaltic volcanic rocks and graywackes, their disintegration products. In the Canadian Shield the oldest Precambrian terranes are those of the Superior and Slave structural provinces (Fig. 7-20) where great bodies of granitic rocks invaded an older terrane composed predominantly of greenstones and graywackes (p. 137). Similar relationships in the older Precambrian rocks of other continents also point to early dominance of basaltic volcanics and graywackes in the geologic record. Large areas of granitic rocks first seem to become a common element of the crust approximately 2500 mil-

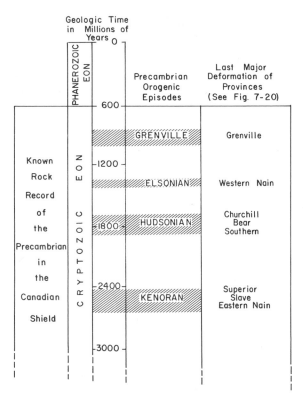

Figure 7-21 Principal orogenic episodes in the Precambrian of the Canadian Shield. The considerable spread in time of an episode reflects inaccuracies in dating as well as the duration of deformation. (Data from C. H. Stockwell [37] and S. S. Goldich [16].)

lion years ago, with the widespread formation of the granites and gneisses of the Superior and Slave provinces and contemporary terranes elsewhere in the world [10]. How widespread granitic rocks were before this time is not evident in the earlier, more obscure part of the Precambrian record; they are present but not abundant in the very oldest terranes yet dated.

The striking architectural symmetry of the North American continent with its central core of Precambrian rocks and younger flanking mountain systems long ago suggested the idea that the continent had grown by accretion. Successive marginal geosynclines formed along the edge of the stable continental plate and were subsequently incorporated in it by deformation and granitic intrusion. The theory persists today, in much less simplified forms, and has been employed to account for the internal structure of the Canadian Shield itself. On the map of the Canadian Shield (Fig. 7-20) it is evident that the older rocks of the Slave and Superior provinces are bordered by the younger rocks of the Churchill and Grenville provinces. The pattern is by no means symmetrical and the structural trends of one province are at an angle to those of another. Studies along the margins of the structural provinces reveal that the younger partially overlap the older. For example, at some places along the Grenville front metamorphism has not completely masked the structural trends of the Superior province and they can be traced well into the Grenville province. Evidently the successive ancient geosynclines formed both marginal to the preexisting block of granitic crust and over its thinner parts. The orogenic episodes which deformed these sedimentary troughs and welded them to the continental nucleus also redeformed portions of earlier orogenic belts.

Different structural trends evident in the provinces of the Canadian Shield indicate either changes in the direction of deforming forces in the earth's crust and mantle, or change in the position of continental nuclei with respect to these great episodic forces. In the Precambrian rocks of the Canadian Shield two such changes are discernible. The first follows the major deformation (Kenoran) of the Superior and Slave provinces, the second follows the major deformation (Hudsonian) of the Churchill and Nain provinces. Since the second change, the orientation of structural trends in all succeeding orogenic episodes in North America have been similar, the fold belts of the Phanerozoic follow the trends of the Grenville orogenic episode. What is known to date of the structural trends and date of deformation in Precambrian rocks of other continents suggests that these two changes were worldwide in distribution [8] and constitute significant events in the physical evolution of the earth.

AN EMERGING PATTERN OF CRUSTAL EVOLUTION

The amazing modern expansion of knowledge about the earth's crust and the forces at work in it (see Chapter 4) finds its greatest challenge in the Precambrian rocks of the world. In this complex record of the vast spread of Cryptozoic history lies the evidence that points to the mechanisms and patterns of crustal evolution. Theories that seek to explain such large-scale phenomena as sea-floor spreading and continental drift, to be validated, must also account for the patterns of Precambrian geology.

The convection current hypothesis (p. 76) is today the favored mechanism for explaining crustal forces. The impressive substantiation it has had from recent investigation of sea-floor spreading effectively removes convection from the realm of hypothesis, but its relative importance as a mechanism through time has yet to be determined. If convection currents in the mantle controlled the development of geosynclinal fold belts, then the marked changes in orientation of structural trends seen in the Precambrian rocks point to changes in the pattern of convection currents. Of several recent theories on convection and crustal evolution those by Sutton [32] and Dearnley [8] most effectively marshal and synthesize the data of Precambrian geology. Both regard the diverse orientations of structural trends as evidence of differences in the pattern of convection currents brought about by change in size or in number of the convection cells in the mantle, but they differ appreciably in detail. Runcorn [28] presents still another convection hypothesis that emphasizes continental drift. Although the three theories differ from one another in many respects, they agree, significantly, in recognizing the same periods of change in structural trends and in attributing these to changes in the structure of the convection system in the mantle.

Innovations in the study of Precambrian geology during the last few decades, particularly the isotopic dating and the structural approach, have converged with lines of evidence from modern geophysical studies to greatly change our concepts of Cryptozoic history. As yet only the broadest outlines of this history have emerged and the details remain to be filled in. At the same time we appear to be on the verge of explaining the mechanism of the physical evolution of the earth, a step comparable to the revelation of organic evolution.

Cryptozoic Life and Environments

The nature of the Precambrian earth and its life, once almost solely the province of the geologist and a few evolutionary biologists, has in the last few decades become a focal point of intense interest for biologists, geochemists, geophysicists, meteorologists, biochemists, and space scientists. The origin of the crust, the oceans, the atmosphere, and life itself are today among the most active areas of research. This cooperative participation of scientists from diverse fields in the mysteries of Cryptozoic history has, among other things, stimulated search for new evidence of life in Precambrian rocks and has led to a reappraisal of the old evidence. Over the years a host of odd-looking and suggestive structures from Precambrian rocks have been attributed to organisms, or to organic activity. Critical restudy has eliminated as organic many of these structures, including such old standbys as the Precambrian jellyfish from the Grand Canyon. There remain a number of structures and sub-

stances that are organic beyond a reasonable doubt and a very small number of forms about which we know too little to be certain. Only the unquestioned Precambrian remains are reviewed here; a more comprehensive treatment of Precambrian fossils and artifacts is given by Preston Cloud [7] as part of a most comprehensive study of early life and its evolution.

PRECAMBRIAN FACTS OF LIFE

Algal Stromatolites. The only abundant Precambrian fossils are the thinly laminated structures called stromatolites (Fig. 7-22).

Figure 7-22 An algal stromatolite from the Precambrian of Mexico. (U. S. National Museum.)

These are widely distributed in Precambrian sedimentary rocks where they occur most commonly in limestones, dolostones, and banded iron formations. The stromatolite form is highly variable in size and shape, ranging from massive clumps (Fig. 7-16) to digitate clusters. The laminae may form gentle arches or steep conical peaks, or they may be irregular in structure.

Present-day colonies of dominantly blue-green algae form similar laminated structures, and it is reasonable to suppose that some of their ancient relatives were responsible for most of the stromatolites in the fossil record. The algae apparently form stromatolites by trapping sediment between the minute protruding filaments of the matlike colony and binding it with a film of gelatinous organic matter, successive increments resulting in the laminated structure. Many stromatolites, recent and ancient, are calcareous but it is not certainly known whether some algal colonies caused precipitation of the carbonate or whether only carbonate sediment was available for trapping. Most Precambrian stromatolites lack any structure within their laminae, but algal filaments have been found in some from cherty iron formations.

Undoubted stromatolites have been found in rocks more than 2000 million years old and some occurrences may be considerably older.

The Bulawayan Series of southern Rhodesia locally contains stromatolites and isotopic dates on intrusive rocks indicate that the series is more than 2700 million years old. However, the isotopic dates are based on rock samples in the Bulawayan Series taken some 250 miles away from the stromatolite locality. In much of the Precambrian under 2000 million years old stromatolites are common enough to be potential guide fossils and in Russia the younger Precambrian sequences are zoned and classified on the basis of different forms of stromatolites. The reliability of correlation based on stromatolite form is open to serious question, and many students believe the stromatolite form to be shaped primarily by environmental factors. In the Pethei Formation of the Great Slave Lake region on the Canadian Shield, Paul Hoffman was able to relate the shape and orientation of stromatolites to paleocurrent directions based on sedimentary structures [19].

Microorganisms. Microscopic filaments, rod-shaped and spherical bodies, are widespread in Precambrian cherts and the preservation at some localities is exceptional enough to confirm their organic nature. Some of the more significant sources of Precambrian microorganisms, with their approximate age in millions of years (My), are:

Figure 7-23 Algae and fungi from chert the Gunflint Formation 4 miles west of Schreiber, Ontario. A, colonies and free filaments of algae (×190); B, colony of fungus (×423); C, fungus mycelium (×700). (Elso S. Barghoorn.)

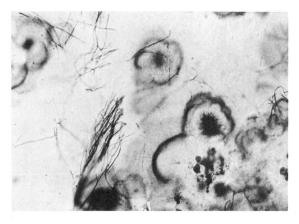
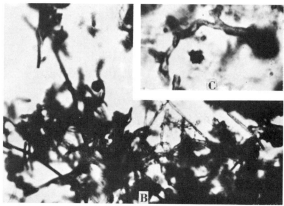

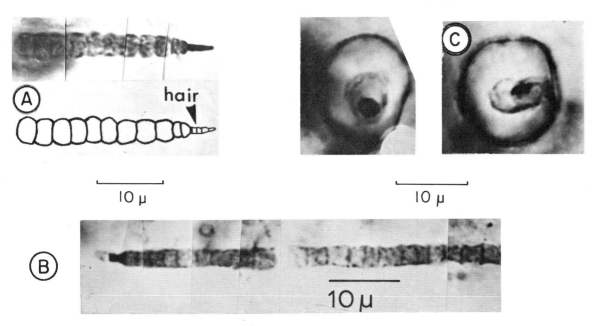

Figure 7-24 Remarkably preserved microorganisms from chert in the Late Precambrian Bitter Springs Formation of central Australia. Filamentous blue-green algae (A and B) and solitary spheroidal algae (C) containing an internal body interpreted as the nucleus. These are the oldest nucleated cells yet discovered. (Courtesy J. William Schopf [30].)

Bitter Springs Formation	Central Australia	1000 My
Belt Series	Montana	1100 My
Gunflint Iron Formation	Southern Ontario	1900 My
Fig Tree Series	South Africa	3100 My
Onverwacht Series	South Africa	3200 My

The first microorganisms well enough preserved to be identifiable were discovered within stromatolites of black chert in the Gunflint Iron Formation, which crops out locally along the north shore of Lake Superior [33]. A number of kinds of blue-green algae (Fig. 7-23), bacteria, and other forms unlike any living microorganisms have been found to date. The Gunflint fossils are significant as the oldest varied assemblage of undoubted microorganisms.

More varied and even better preserved than the Gunflint material are the microorganisms from the much younger Bitter Springs Formation of Australia [30]. This assemblage, also found within black chert stromatolites, numbers over 30 species including green and blue-green algae, bacteria, and funguslike organisms. Perhaps the most noteworthy feature of the exceptional Bitter Springs assemblage (Fig. 7-24) is that it contains spheroidal green algae that preserve an unmistakable nucleus—the earliest record of cells with well-defined nuclei! Presence of such a nucleus indicates cells with a hereditary apparatus like that in all higher organisms. The acquisition of this critical feature was a major step in the development of life, making possible the evolution of the great diversity of plants and animals that have populated every part of the earth. Apparently this step was taken at least a billion years ago.

The most ancient algalike microfossils come from two successive series of Precambrian rocks in South Africa, the Fig Tree Series and underlying Onverwacht Series. Spheroidal forms are found in both series and in addition a rod-shaped, bacteriumlike body is recorded from the Fig Tree Series. Geochemical and carbon isotope studies of the Fig Tree cherts indicate the presence of hydrocarbons and organic carbon, which lends support to the interpretation of the spheroids as algae [1]. Spheroids in the older Onverwacht Series are less well preserved but similar to those in the Fig Tree. If the Onverwacht microstructures are indeed organic, they are the oldest known remains of life on earth [11].

The Ediacara Fauna. In 1947 a small collection of fossils was found in flaggy beds of the Pound Quartzite, a supposed Lower Cambrian formation, in the Ediacara Hills north of Adelaide, South Australia. The complete uniqueness of the fauna was later recognized and because it had nothing in common with any known Lower Cambrian fauna, Martin Glaessner postulated that it was Late Precambrian in age [15]. Undoubted Lower Cambrian fossils occur some 600 to 700 feet above the fauna in the Ediacara stratigraphic sequence.

The fossils consist of external molds in fine-grained sandstone of what apparently were soft-bodied marine animals (Fig. 7-25). Among the hundreds of specimens are a number of

Figure 7-25 Fossils from the Pound Quartzite, Ediacara Hills, Australia. A. *Charnia*, a branching structure superficially similar to a sea fan. B. *Parvancorina*, an enigmatic body apparently not like any living form. C. *Spriggina*, a segmented, worm-like organism. D. *Dickinsonia*, a flat, wormlike form. (Courtesy of M. F. Glaessner.)

different kinds of discoidal impressions attributed to jellyfish, segmented worms, stalked forms resembling sea pens (stalked coelenterate colonies), and several even more perplexing fossils whose identity is unknown. The supposed sea pens, *Rangea* and *Charnia*, also occur on other continents, as do a few other elements of the Ediacara fauna. Nowhere, however, have reliable isotopic dates been secured for beds containing Ediacara fossils.

The Ediacara assemblage is the oldest known fauna of undoubted multicellular animals. Its appearance in the very latest Precambrian—so late that some geologists have suggested redefining the Cambrian to include it, does little to bridge the gap between the very primitive algal-bacterial associations that dominate Precambrian rocks and the advanced faunas of the Lower Cambrian. The discontinuity between primitive and advanced organisms now falls between the Ediacara fauna and the older Precambrian primitive organisms. Nevertheless, the Ediacara fossils are unique in their apparent lack of hard parts and in their total difference from the earliest known Cambrian animals.

Traces of Life. Carbonaceous rocks have long been known in Precambrian rocks and a number of chemical methods have been used to determine the presence of organic carbon and other organic compounds. These methods, still in an active stage of development, have not been uniformly successful, but some can detect a wide range of organic compounds. At present they are useful chiefly in substantiating the presence of organic matter in rocks having forms suspected of being micro-organisms. The organic nature of the Fig Tree microstructures, for example, is bolstered by the fact that analyses of the rocks have yielded hydrocarbons of a kind characteristic of organic matter associated with photosynthesis [1].

Sedimentary structures that may possibly be trace fossils offer some of the most baffling problems in the interpretation of the Precambrian record. Trace fossils in Phanerozoic rocks are made by mobile metazoans (many-celled animals) such as worms and mollusks

and the inference is that similar structures in Cryptozoic rocks would indicate the presence of metazoans. In the absence of any supporting evidence such as the imprints of soft-bodied animals it is not possible to prove unequivocally that a structure was made by an organism rather than some inorganic agency. Consequently the interpretation of supposed trace structures is controversial. Structures such as those from the Belt Series shown in Fig. 7-17 look very much like trails or burrow fillings, but other interpretations have been proposed. It is of considerable significance that supposed trace fossils such as these are the only indication of the existence of metazoans below the level of the Ediacara fauna.

A particularly perplexing form, called *Rhysonetron* (Fig. 7-26), affords an example of the difficulties attending interpretation of Precambrian problematical structures. These are curved, spindle-shaped structures occurring as sand casts in the troughs of a ripple-marked sandstone [20]. The casts have a pattern of corrugations extending at an angle from a central axis. Were it not for this ornamentation they would most likely be passed over as the fillings of arcuate desiccation cracks, for desiccation cracks are present in the same bed, although they are mostly nonarcuate ridges. Arcuate desiccation cracks are known from many sedimentary rocks and are commonly associated with ripple marks. On the other hand, tapering, arcuate, chitinous worm tubes and bodies of worms or wormlike animals are frequently observed to collect in ripple-troughs on present-day tidal flats. The ornament of *Rhysonetron* is hard to account for by inorganic agency; yet it does not resemble that of any known organism. The structure remains a baffling puzzle. It is found north of Lake Huron in the Bar River Formation of the Cobalt Group. The Bar River has been isotopically dated as older than 1390 million years and probably older than 2130 million [20]. Even at the lesser age the interpretation of *Rhysonetron* as a metazoan would extend the known range of these advanced organisms as far back in time beyond the Ediacara fauna as the Ediacara fauna lies behind us today.

SEDIMENTARY EVIDENCE OF ENVIRONMENT

In rocks so generally devoid of fossils what little indication we have of Cryptozoic environments and climates is derived from the sediments. The thick limestones and dolostones with widespread algal colonies and reeflike bodies were probably deposited in warm and relatively shallow parts of seas.

In the younger Precambrian rocks saline deposits and redbeds are locally associated, or coextensive with, the carbonate rocks and indicate evaporite basins that today are characteristic of the coasts of arid lands. Most other formations do not reveal as much, but in regions where exposures are broad and metamorphism slight both the marginal sands of seas and the much more rare terrestrial deposits are recognizable from bedding characteristics and their lateral relationship with marine carbonates. In the U.S.S.R. remarkably complete paleogeographic maps have been put together from such evidence for parts of the last 1500 million years of Precambrian time [23]. In this part of the Precambrian the variety of types of sedimentary formations is very much like that in the Phanerozoic.

Widespread Glaciation. The most direct evidence of Precambrian climate is the glacial deposits that are found in several parts of the Precambrian sequence but are particularly plentiful and amazingly widespread during the very late Precambrian. Before the enormous length of Precambrian time was appreciated, the discovery of these evidences of early glaciation seemed a serious obstacle to the belief that the earth was originally molten. To the older geologists, who conceived of the Precambrian earth as a hot planet clothed in a steaming primal atmosphere, it was, indeed, a shocking discovery. In view of our present knowledge, however, the oldest of the known Precambrian ice ages followed the origin of the earth by 1500 million years or more.

The conclusive evidence for the glacial origin of the Gowganda tillite has been described previously. Its distribution over the southern part of the Canadian Shield shows that a continental ice sheet more than a thousand miles in diameter lay over central Canada sometime between 1600 and 2000 million years ago.

0 _____ 5 CM

Figure 7-26 The problematic structure Rhysonetron, from quartzite in the Bar River Formation of the Cobalt Group, estimated to be between 2000 and 2500 million years old. An inorganic origin is presently favored for these "wormlike" structures. (H. J. Hofmann, Geological Survey of Canada.)

Figure 7-27 Precambrian tillite in Utah. South side of Little Mountain west of Ogden. The compass above the large angular block is 3 inches across. (Eliot Blackwelder.)

Glacial deposits are widespread in northern Utah beneath the Cambrian formations, cropping out in the islands in Great Salt Lake and in the mountain ranges both east and southwest of the lake (Fig. 7-27). Tillite and glaciofluvial deposits occur at different horizons in a series of sandy and shaly formations that exceed 12,000 feet in thickness and are supposedly of Precambrian age, possibly Beltian.

The tillite includes boulders of many kinds of rock, some of them as much as 10 feet in diameter, enclosed in a nonbedded slaty matrix. Some of the boulders are faceted and striated. The tillite reaches the exceptional thickness of 700 feet at one locality and 1100 feet at another, but is generally much thinner. Its age, unfortunately, is not definitely fixed.

The series bearing the tillite rests upon a much older and intensely metamorphosed group of schists, and is overlain with local unconformity by beds high in the Lower Cambrian. The glaciation is believed to have occurred late in Cryptozoic time.

The oldest known tillite is one of several striking glacial deposits in the very thick Precambrian section of southern Africa. The **Chuos tillite** of the state of South-West Africa lies within the "Primitive Systems," the lowest of three major divisions of the Precambrian rocks. It has been recognized over an area of some 20,000 square miles and locally attains a thickness of 1500 feet. Over most of this area it has the character of a moraine, but toward the south it is replaced by marine beds with widely scattered boulders believed to have been rafted to sea by icebergs.

Geologic evidence recently brought together by Harland [18] points to a spectacular ice age very near the close of Precambrian time. The conglomeratic deposits included have not all been satisfactorily demonstrated to be tillites, nor are all of them certainly dated, but the evidence is enough to indicate one of the major, and possibly the greatest, glacial event in earth history. Apparently all the continents except Antarctica were affected!

One of the finest displays of these glacial deposits is in the Flinders Range of south-central Australia where the **Sturtian tillite**, exceeding 600 feet in thickness, forms the backbone of the range and is exposed in great cliffs. The glacial deposits are known to extend for at least 300 miles from south to north. It consists of a nonbedded matrix of ancient "rock flour" and angular chips, in which boulders of many kinds of rock are embedded. These boulders are in part rounded and in part angular, and some of them are faceted and striated. The underlying, well-bedded strata include scattered boulders of glacial origin that are believed to have been dropped from icebergs in a body of water in front of the ice sheet. The Sturtian tillite is about 16,000 feet **below** the base of the Cambrian System in this section, and the intervening beds include a second tillite horizon.

In the Yangtze Valley of China there are also extensive deposits of Late Precambrian glaciers. Scattered exposures of tillite indicate that glacier ice once spread over an area of at least 800 miles from east to west across central China during Precambrian time.

There are other tillites, less certainly dated but probably of Late Precambrian age, in Norway, East Greenland, India, and Africa. Some of these underlie the Cambrian so closely that some geologists believe they belong in the Cambrian rather than the Precambrian.

Life and Atmosphere. As far back in time as a legible sedimentary record is found, some 3200 million years, there is evidence of the existence of a crust with sedimentary environments and what appear to be the remains of primitive life. This enormous span of time

during which the earth seemingly was populated only by the simplest kinds of organisms greatly accentuates the abrupt appearance of abundant and advanced forms of life in the Cambrian. Was the transition really as abrupt as it appears? Did life exist at a primitive level for nearly 2500 million years, then in the short space of 100 million years or so attain the metazoan level and begin its spectacular radiation? Or is a more gradual evolution to the metazoan stage still hidden in the obscure record of the Precambrian? As we have seen, unequivocal evidence favoring either of these, or any other alternatives, is lacking.

Until recently the idea of gradual evolution of Precambrian life was assumed and the sudden appearance of the Cambrian fauna attributed to the abrupt development of hard parts, usually in response to some ecological stimulus. Some of the suggestions put forth to explain the appearances of hard shells and carapaces are reviewed in Chapter 8 (p. 187). In the absence of proof to the contrary, this theory of gradual evolution and sudden acquisition of hard structures is still very much alive. A more recent approach to the problem rejects the idea of gradual evolution and interprets the abrupt appearance of diverse faunas as indicating both the origin and immediate radiation of metazoans. There is no contradictory evidence to eliminate this theory either. Both ideas seek a major environmental event, or change, to trigger the evolutionary burst that produced either hard parts in metazoa or the metazoa themselves. The great ice age of the Late Precambrian has been suggested as such a major event [18], but its connection with organic evolution is obscure and its influence was very brief in terms of the span of life preceding it.

Biochemists and geochemists have for a long time been interested in the early earth and the origin of its seas, its atmosphere, and its life. From this source new insights have recently come to the problems of Precambrian life. It is now widely held that the early atmosphere and hydrosphere of the earth were devoid of free oxygen, and that the atmosphere evolved as oxygen, gradually produced photosynthetically by aquatic algae, accumulated. The oxygen played a double role, permitting the development of the first oxygen-dependent organisms once the waters of the earth were saturated, and eventually forming a protective envelope around the earth that ultraviolet rays, lethal to life, could not penetrate. A theory proposed by Berkner and Marshall [2] holds that when oxygen accumulation reached a point at which free oxygen respiration was possible for organisms, adaptation to this environmental change brought about a major evolutionary radiation. The theory correlates this with the abrupt appearance of metazoan faunas at the beginning of Cambrian time. Since this link between the evolution of the atmosphere and the evolution of life was proposed, a number of elaborations and modifications of the hypothesis have been suggested.

The origin of the unique banded iron formations, widespread in Precambrian rocks between the ages of 1.8 and 3.5 billion years ago, has been attributed [24] to oxidation of iron-laden sediment in the reduced state as it was deposited in Precambrian seas. Here waters were oxygenated by photosynthetic microorganisms. Once the waters were saturated and oxygen began to accumulate in the atmosphere, the iron in sediment was oxidized on the surface and subaqueous iron formations ceased to form. Oxidation of iron with weathering of surface rocks may produce red sediment, and it is significant that Precambrian redbeds of this type make their appearance approximately at the time the banded iron formations disappear from the record [7]. While Berkner and Marshall stressed that the abrupt appearance of abundant fossils reflected the explosive character of metazoan evolution in response to available oxygen, others have pointed out that the theory of gradual evolution can also be adapted to the same mechanism. This viewpoint is well presented by Fischer, who suggests that:

Marine animals developed in "oxygen oases" — marine plant communities — during the latter half of Pre-cambrian time as small, naked forms, essentially incapable of leaving a fossil record. Attainment of the "Pasteur point" [one percent of present oxygen pressure] in the atmosphere emancipated them from their host plants, and allowed them to spread widely over the seas. Oxygen dependence forced many of them to the surface and into the shallowest water, where they developed exoskeletons as radiation protection [13].

The theory that Precambrian history was profoundly influenced by an evolving atmosphere, featuring a slow buildup of oxygen, is a stimulating idea and one that seems to be able to integrate and account for the few scattered facts of Precambrian life. It must be remembered, however, that it is now a theory and no more. Parts or all of it are opposed by many geochemists and geologists, but in the proving or disproving of the theory, the knowledge of Precambrian history, and particularly the history of life, cannot help but be advanced.

REFERENCES

1. Barghoorn, E. S., and Schopf, J. W., 1966, *Microorganisms three billion years old from the Precambrian of South Africa.* Science, v. 152, pp. 758–763.
2. Berkner, L. V., and Marshall, L. C., 1964, in *The origin and evolution of atmospheres and oceans*, Brancazio, P. J., and Cameron, A. G. W., Eds., New York, John Wiley & Sons, pp. 102–126.
3. Berthelsen, Asger, and Noe-Nygaard, Arne, 1965, *The Precambrian of Greenland, ibid.*, pp. 113–262.
4. Boltwood, B. B., 1907, *On the ultimate disintegration products of the radioactive elements, Pt. II. The disintegration products of uranium.* Am. Jour. Sci., v. 23, pp. 77–88.
5. *Canadian Minerals Yearbook 1966.* Canada Dept. Energy, Mines, Resources, Mineral Rept. 15, 582 pp.
6. Catanzaro, E. J., and Kulp, J. L., 1964, *Discordant zircons from the Little Belt (Montana), Beartooth (Montana) and Santa Catalina (Arizona) Mountains.* Geochim. et Cosmochim. Acta, v. 28, pp. 87–124.
7. Cloud, P. E., 1968, *Pre-metazoan evolution and the origins of the Metazoa. In Evolution and Environment*, New Haven, Conn., Yale Univ. Press, pp. 1–72.
8. Dearnly, R., 1966, *Orogenic fold-belts and a hypothesis of earth evolution. In Physics and*

Chemistry of the Earth, V. 7, pp. 1–114. Oxford, Pergamon Press.

9. Eckelmann, W. R., and Kulp, J. L., 1957, *Uranium-Lead method of age determination, Part II: North American localities.* Geol. Soc. America Bull., v. 68, pp. 1117–1140.

10. Engel, A. E. J., 1963, *Geologic evolution of North America.* Science, v. 140, pp. 143–152.

11. Engel, A. E. J., et al., 1968, *Alga-like forms in Onverwacht Series, South Africa: Oldest recognized life-like forms on earth.* Science, v. 161, pp. 1005–1008.

12. Fenton, C. L., and Fenton, M. A., 1937, *Belt Series of the North.* Geol. Soc. America Bull., v. 48, pp. 1873–1970.

13. Fischer, A. G., 1965, *Fossils, early life and atmospheric history.* Nat. Acad. Sciences, Proc., v. 53, no. 6, pp. 1205–1215.

14. Giffen, C. E., and Kulp, J. L., 1960, *Potassium-Argon ages in the Precambrian basement of Colorado.* Geol. Soc. America Bull., v. 71, pp. 219–222.

15. Glaessner, M. F., 1961, *Pre-Cambrian animals.* Sci. American, v. 204, pp. 72–78.

16. Goldich, S. S., 1968, *Geochronology in the Lake Superior region.* Canadian Jour. Earth Sci., v. 5, pp. 715–724.

17. Gruner, J. W., 1954, *A realistic look at taconite estimates.* Mining Engineering, v. 6, p. 287.

18. Harland, W. B., 1963, *Evidence of Late Precambrian glaciation and its significance,* in *Problems in Paleoclimatology,* Nairn, A. E. M. Ed., London, Interscience, pp. 119–149.

19. Hoffman, Paul, 1967, *Algal stromatolites: use in stratigraphic correlation and paleocurrent determination.* Science, v. 157, pp. 1043–1045.

20. Hoffman, H. J., 1967, *Precambrian fossils (?) near Elliot Lake, Ontario.* Science, v. 156, pp. 500–504.

21. Holmes, Arthur, 1963, *Introduction.* In Rankama, Kalervo, Ed., *The Precambrian,* V. 1, pp. xi–xxiv. New York, Interscience Publishers.

22. James, H. L., 1958, *Stratigraphy of pre-Keweenawan rocks in parts of northern Michigan.* U. S. Geol. Survey Prof. Paper 314-C, pp. 27–42.

23. Keller, B. M., Korolev, V. G., Semikhatov, M. A., and Chumakov, N. M., 1968, *The main features of the Late Proterozoic paleogeography of the U.S.S.R.* 23rd Internat. Geol. Cong., Proc., v. 5, pp. 189–202.

24. Lepp, Henry, and Goldich, S. S., 1964, *Origin of Precambrian iron formations.* Econ. Geology, v. 59, pp. 1025–1060.

25. Rezak, Richard, 1957, *Stromatolites of the Belt Series in Glacier National Park and vicinity, Montana.* U. S. Geol. Survey Prof. Paper 294-D.

26. Roscoe, S. M., 1957, *Stratigraphy, Quirke Lake-Elliot Lake Sector, Blind River area, Ontario.* In Gill, J. E., Ed., *The Proterozoic in Canada,* Roy. Soc. Canada, Spec. Publ., no. 2, pp. 54–58.

27. Ross, C. P., 1963, *The Belt Series in Montana.* U. S. Geol. Survey Prof. Paper 346, 119 pp.

28. Runcorn, S. K., 1962, *Towards a theory of continental drift.* Nature, v. 193, pp. 731–735.

29. Schenk, P. E., 1965, *Precambrian glaciated surface beneath the Gowganda Formation, Lake Timagami, Ontario.* Science, v. 149, pp. 176–177.

30. Schopf, J. W., 1968, *Microflora of the Bitter Springs Formation, Late Precambrian, central Australia.* Jour. Paleontology, v. 42, pp. 651–688.

31. Slawson, W. F., Kanasewich, E. R., and Ostic, R. G., 1963, *Age of the North American crust.* Nature, v. 200, pp. 413–414.

32. Sutton, J., 1963, *Long-term cycles in the evolution of the continents.* Nature, v. 198, pp. 731–735.

33. Tyler, S. A., and Barghoorn, E. S., 1954, *Occurrence of structurally preserved plants in Pre-Cambrian rocks of the Canadian Shield.* Science, v. 119, pp. 605–608.

34. Wasserburg, G. J., and Lanphere, M. A., 1965, *Age determinations in the Precambrian of Arizona and Nevada.* Geol. Soc. America Bull., v. 76, pp. 735–755.

35. Wertime, T. A., 1963, *A new era begins in iron,* in *Symposium on iron ores.* Central Treaty Org., pp. 15–22.

36. Wilson, M. E., 1965, *The Precambrian of Canada: The Canadian Shield.* In Rankama, Kalervo, ed., *The Precambrian,* V. 2, pp. 263–416. New York, Interscience Publishers.

PART II

PART II

THE PALEOZOIC WORLD

Figure 8-1 Mount Lefroy in the Canadian Rockies near Banff, Alberta, showing a great thickness of Cambrian beds. Most of the exposure is of Middle Cambrian limestone but the upper part of the peak which reaches an altitude of 11,660 feet is probably basal Upper Cambrian limestone (fide E. J. Mount joy, personal communication.)

CHAPTER **8** *The Cambrian Period*

The race of man shall perish, but the eyes
Of trilobites eternal be in stone,
*   And seem to stare about in mild surprise*
At changes greater than they have yet known.
                                    T. A. CONRAD

A Date of Reckoning. With the basal Cambrian rocks, fossils make their appearance in abundance, supplying at once a record of life and the certain means of correlating the physical record from place to place. Henceforth events recorded all over the world can be pieced together as parts of a continuous story, and it is possible to present the history of a whole continent, or of the whole earth, in systematic fashion, period by period. Thus the beginning of the Cambrian is a date of reckoning in geologic history.

PALEOGEOGRAPHY AND PHYSICAL HISTORY OF NORTH AMERICA

The Cambrian Submergence. At the very outset of the Paleozoic Era, Cambrian seas began to spread over the lowest lands, and before the close of the period more than 30 percent of the present continent was submerged. When the older Cambrian deposits are plotted on a map a striking pattern is seen. The Lower Cambrian formations are not distributed over the present lowlands or along

Figure 8-2 South face of Mt. Bosworth from Kicking Horse Pass, Alberta, showing in simple succession more than a mile of Cambrian strata. (C. D. Walcott photo.)

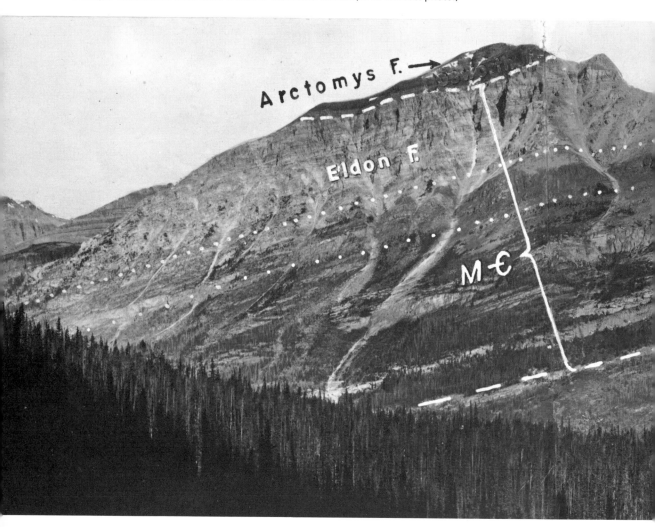

the margins of the present continent but are confined to two relatively narrow belts well inland (see Map II). These were slowly and progressively sinking troughs in the crust, and here the Cambrian formations attain a thickness of several thousand feet and generally abound in marine fossils (Fig. 8-2). But since the sediments and the fossils are of types that form in shallow water, it is clear that these troughs subsided slowly and that fill almost kept pace with subsidence; that is, these were typical **geosynclines.** The eastern trough followed the trend of the present Appalachian folds and is therefore known as the **Appalachian geosyncline.** It was here that the concept of the geosyncline was first perceived, hence this is the **type** geosyncline. The western trough following the axis of the great Cordilleran highlands is known as the **Cordilleran geosyncline.**

Between the geosynclines lay the vast stable interior, or **craton,** of the continent. In each geosyncline flanking the North American craton the Lower Cambrian sedimentary deposits increase in coarseness and in thickness away from the craton, indicating that they were derived largely from marginal lands outside the geosynclines; but the nature and extent of these marginal highlands is problematical since both margins of the continent were intensely deformed by later orogeny. The eastern landmass has been called **Appalachia** and the western **Cascadia.** It has been suspected by some geologists that Appalachia

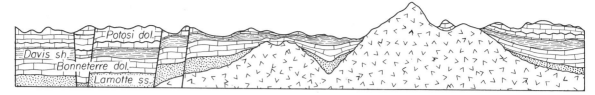

Figure 8-3 Cambrian formations lapping up on the old Ozark Dome. Length of section about 12 miles. Vertical scale exaggerated about 5 times the horizontal. Normal faults on the left flank were formed after Cambrian time. (After C. L. Dake, Missouri Bureau of Geology and Mines.)

arose as a great island arc like those that now fringe the eastern margin of Asia and that its uplift was caused by pressure from the Atlantic Ocean floor. On the other hand, if we accept the idea of continental drift, the Atlantic Ocean may not have existed in early Cambrian time. We now have no convincing evidence of the date when North America and Europe began to drift apart, if, indeed, they ever did so, but neither do we have any evidence of an Atlantic Ocean until Jurassic or Cretaceous time.

One consideration indicates that Appalachia must have been extensive and included old crystalline rocks of granitic nature. This relates to the enormous volume of sandstones and conglomerates that form a large part of the Lower Cambrian deposits all the way from New England to Georgia. This sand could only have come from the quartz grains in granitic source rocks or from quartz veins in metamorphics. Basic lavas such as those derived from volcanoes in the modern island arcs would include no free quartz crystals and, even in granites, quartz crystals form only a minor part of the rock.

Before the end of Lower Cambrian time Appalachia had been eroded down to a low elevation so that little sand was being carried into the geosyncline and fine shale and limestone spread widely over it. In the meantime, during Middle and Late Cambrian time, the interior of the continent slowly subsided and shallow seas crept over a large part of the United States so that during Late Cambrian time approximately 30 percent of the continent was submerged.

By Late Cambrian time two additional geosynclines were taking form, the **Ouachita** trough curving from Trans Pecos Texas, into eastern Oklahoma and Arkansas, and the **Franklinian** trough running through the Arctic Archipelago. They continued to subside until late in the Paleozoic Era. Upper Cambrian formations flanked, and eventually covered, the Old Ozark mountains of southeastern Missouri, a large monadnock inherited from Precambrian time (Fig. 8-3).

Lower, Middle, and Upper Cambrian formations in eastern Massachusetts, southern New Brunswick, Nova Scotia, and southeastern Newfoundland contain trilobite faunas so different from those of the Appalachian trough that they suggest a separate seaway. This has been called the **Acadian trough.** If an actual barrier ever existed between it and the Appalachian trough, later orogenies have completely obscured the original relationships.

In the west the Cordilleran geosyncline was invaded by seas from both the north and the south, but they were isolated by a barrier in the Idaho region until about the beginning of Middle Cambrian time. Relations here are obscured by the much later intrusion of the Idaho batholith, but C. P. Ross has shown [5] that as the Cambrian formations are traced from both north and south into Idaho they become coarser and lap out against the old barrier. Furthermore, there are differences in the faunas between the northern and the southern seaway that indicate a lack of free migration. As in the east, Lower Cambrian rocks are restricted to the geosyncline but the Middle Cambrian seas spread over the craton from the Cordilleran trough, reaching the mid-continent by Late Cambrian time.

Many of the major structural features of our continent present in Cambrian time persisted until late in the Paleozoic Era. The borderlands were from time to time worn low but were repeatedly uplifted into mountainous highlands; the Appalachian, Cordilleran, and Ouachita geosynclines sank progressively and in them we find the most complete record of Paleozoic time in strata that reach an aggregate thickness of from 30,000 to 50,000 feet along the axes of the troughs. Most of the time they were occupied by shallow seas but at times they were drained or filled above sealevel by fluvial sediments. The broad craton, on the contrary, remained stable and was only intermittently covered by shallow seas, and here the sedimentary deposits are relatively thin.

As explained in Chapter 1, the character of a sedimentary deposit indicates the conditions under which it was laid down and at the same time reflects the character of the adjacent lands. The Cambrian strata are in this way the basis for the interpretation of the physical history sketched earlier.

SUBDIVISIONS OF THE CAMBRIAN SYSTEM

Although the Cambrian System of rocks was first studied and named in Wales, a country which the Romans called Cambria, it is more grandly displayed in the mountains of western North America where a great thickness of evenly bedded strata is exposed like pages of a colossal manuscript in stone preserving this chapter of earth history (Fig. 8-2). In different areas the succession of strata is divisible into numerous formations, each differing in lithology from that above and below (Fig. 8-4).

Figure 8-4 Correlation of important Cambrian sections.

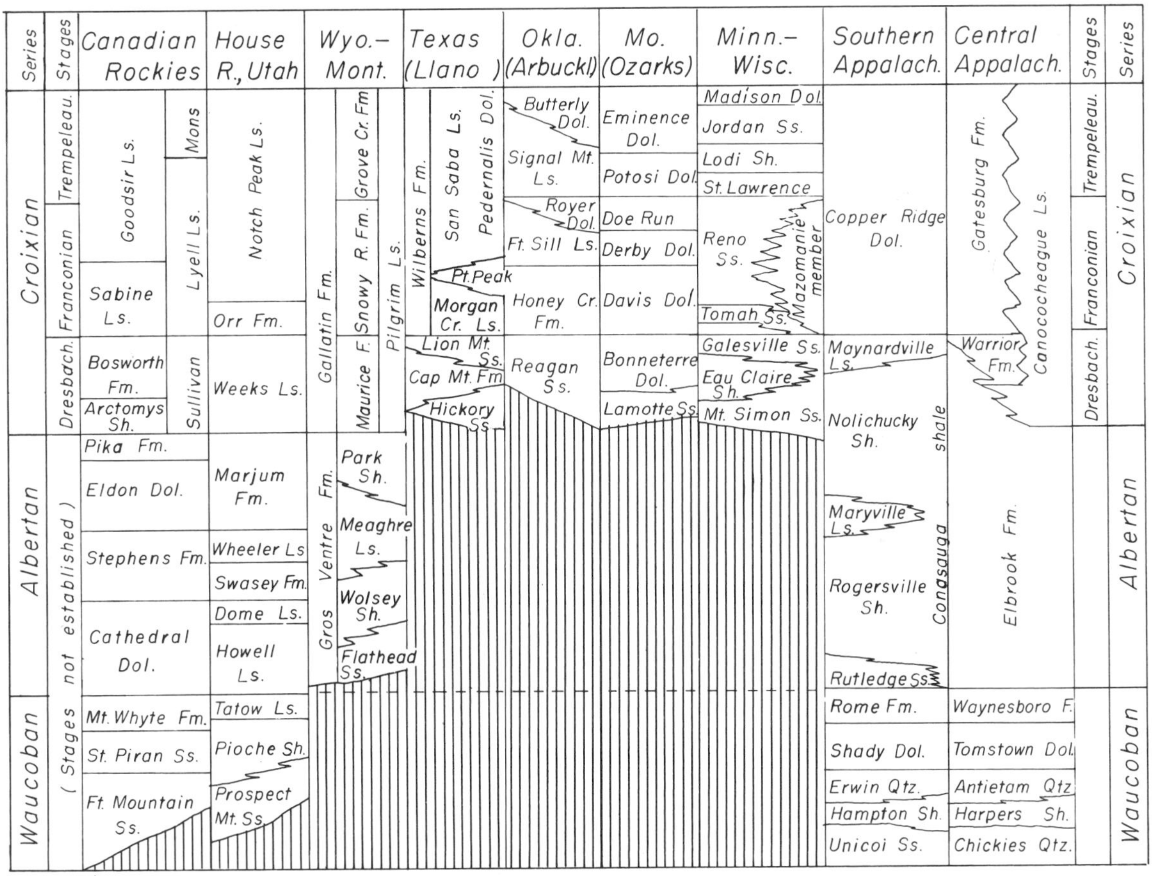

Series	Stages	Canadian Rockies	House R., Utah	Wyo.-Mont.	Texas (Llano)	Okla. (Arbuckl)	Mo. (Ozarks)	Minn.-Wisc.	Southern Appalach.	Central Appalach.	Stages	Series
Croixian	Trempeleau	Goodsir Ls.	Mons / Lyell Ls.	Notch Peak Ls.	Wilberns Fm. / San Saba Ls. / Pedernalis Dol.	Butterly Dol. / Signal Mt. Ls. / Royer Dol.	Eminence Dol. / Potosi Dol. / Doe Run	Madison Dol. / Jordan Ss. / Lodi Sh. / St. Lawrence	Copper Ridge Dol.	Gatesburg Fm.	Trempeleau	Croixian
Croixian	Franconian	Sabine Ls.	Sullivan / Orr Fm.	Grove Cr. Fm / Snowy R. Fm. / Pilgrim Ls.	Pt. Peak / Morgan Cr. Ls.	Ft. Sill Ls. / Honey Cr. Fm.	Derby Dol. / Davis Dol.	Reno Ss. / Mazomanie member / Tomah Ss.	Copper Ridge Dol.	Conococheague Ls.	Franconian	Croixian
Croixian	Dresbach	Bosworth Fm. / Arctomys Sh.	Weeks Ls.	Maurice F. / Gallatin Fm.	Lion Mt. Ss. / Cap Mt. Fm / Hickory Ss.	Reagan Ss.	Bonneterre Dol. / Lamotte Ss.	Galesville Ss. / Eau Claire Sh. / Mt. Simon Ss.	Maynardville Ls. / Nolichucky Sh.	Warrior Fm.	Dresbach	Croixian
Albertan	(Stages not established)	Pika Fm. / Eldon Dol. / Stephens Fm. / Cathedral Dol.	Marjum Fm. / Wheeler Ls. / Swasey Fm. / Dome Ls. / Howell Ls.	Park Sh. / Meaghre Ls. / Wolsey Sh. / Flathead Ss. (Gros Ventre Fm.)					Maryville Ls. / Rogersville Sh. / Rutledge Ss.	Conasauga / Elbrook Fm.		Albertan
Waucoban	(Stages not established)	Mt. Whyte Fm. / St. Piran Ss. / Ft. Mountain Ss.	Tatow Ls. / Pioche Sh. / Prospect Mt. Ss.						Rome Fm. / Shady Dol. / Erwin Qtz. / Hampton Sh. / Unicoi Ss.	Waynesboro F. / Tomstown Dol. / Antietam Qtz. / Harpers Sh. / Chickies Qtz.		Waucoban

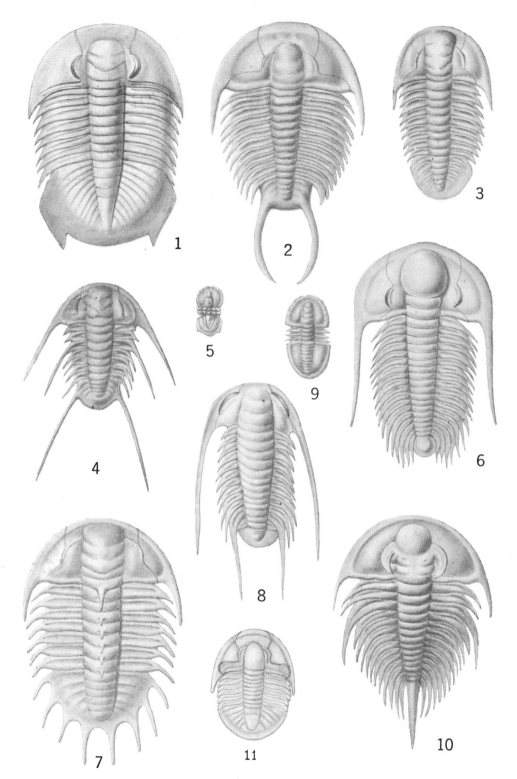

Plate 1 Cambrian Trilobites.

Upper Cambrian: Figure 1, *Dikelocephalus minnesotensis;* 2, *Tricrepicephalus texanus;* 11, *Cedaria.* Middle Cambrian: Figure 3, *Bathyuriscus rotundatus;* 4, *Albertella helenae;* 5, *Agnostus interstrictus;* 6, *Paradoxides harlani;* 7, *Olenoides curtisei.* Lower Cambrian: Figure 8, *Bathynotus holopyga;* 9, *Serrodiscus speciosus;* 10, *Olenellus thompsoni.* All natural size except no. 6, which is about ⅛ natural size, and no. 10, which attains a length of about 8 inches. (Drawings by L. S. Douglass.)

176

Some of these are local but others can be recognized over a considerable area. The chief formations now recognized are represented in Fig. 8-4 where they are arranged according to their time span, which has no close relation to their individual thicknesses. This chart is not to be memorized but is presented only to show the nature of the mappable units of the Cambrian rocks.

The entombed fossils permit the grouping of these varied formations into three major series, each of which represents an epoch of Cambrian time. And in accordance with the custom of using geographic names for geologic units, the Lower Cambrian is called the **Waucoban Series** for the very thick section about Waucoba Springs, California; the Middle Cambrian is named the **Albertan Series** for the fine sections in the Canadian Rockies of Alberta; and the Upper Cambrian is the **Croixan Series** (croy-an), named for the exposures about St. Croix Falls, Minnesota.

The Waucoban Series is readily identified by the common occurrence of the large trilobite *Olenellus* and other genera of the family Olenellidae, none of which range above the Lower Cambrian. The Albertan Series is characterized by several genera of large-tailed trilobites such as *Bathyuriscus,* and, in the Acadian trough, by the genus *Paradoxides* (Fig. 8-7). And the Upper Cambrian is distinguished by several genera among which *Crepicephalus* and *Dikelocephalus* are prominent. These genera are shown in Plate 1. The three Cambrian Series are further subdivided into zones (p. 15) based upon the stratigraphic range of individual species of fossils, chiefly trilobites, since these are the most common Cambrian fossils. Thus, for example, the zone of *Cedaria* is a relatively thin unit marking the base of the Upper Cambrian or Croixan Series [4]. It occurs in both the Appalachian and the Cordilleran geosynclines.

One feature of the Cambrian System deserves further comment. Throughout both the Cordilleran and the Appalachian troughs the initial deposits of the Lower Cambrian are predominately sandstones. This must represent an extensive cover of sandy deposits that had accumulated on the lands during Late Precambrian time which was swept off by rejuvenated streams and carried into the geosynclines in Early Cambrian time. Once this was accomplished and the marginal lands were eroded to low relief, the seas cleared and the ensuing Cambrian deposits were mostly calcareous (either limestones or dolostones). An interesting exception is to be seen on the northern part of the craton, which was not reached by the sea and received no deposits of sediment until Late Cambrian time. Here the predominately sandy Upper Cambrian formations (Fig. 8-5) are relatively thin because the region was stable but was receiving detritus swept off of the low Canadian Shield.

A stratigraphic cross section from deep in the Cordilleran geosyncline up onto the craton (Fig. 8-6) shows the influence of the contributions of sediment from the Canadian Shield. Although relatively thin, the deposits on the craton are nearly all sandstone and sandy shale; these detrital deposits grade westward and intertongue with the thick limestones of the geosyncline. In effect they form a single continuous transgressive sand formation at the base of the Cambrian deposits which is Middle Cambrian in age where it intertongues with the geosynclinal carbonates and Upper Cambrian at its innermost outcrops on the craton. The Flathead sandstone (Fig. 8-6) of western Wyoming is continuous with the lithologically similar Deadwood Formation of the Black Hills, which contains Upper Cambrian fossils.

THE LIFE OF CAMBRIAN TIME

The Curtain Rises. Although fossils are exceedingly rare in Precambrian rocks, they appear suddenly and in great abundance at the base of the Cambrian, revealing a highly varied life as though a curtain had suddenly lifted on a drama already in progress. We have seen that the rare and primitive fossils in Precambrian rocks prove that plant life had been present on earth for probably at least 3 billion years. By the beginning of Cambrian time nearly all of the great branches of the animal

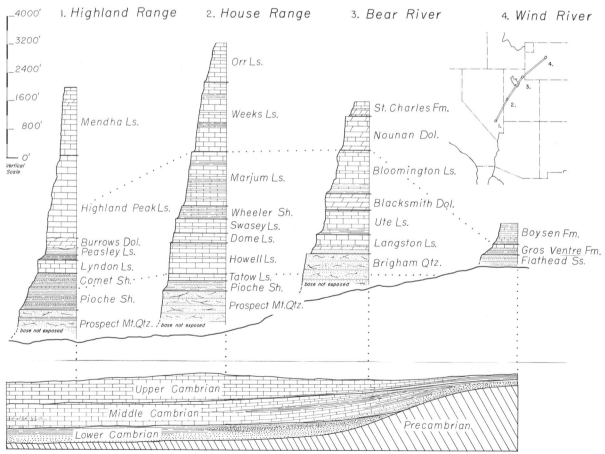

1. *Highland Range* 2. *House Range* 3. *Bear River* 4. *Wind River*

4000'
3200'
2400'
1600'
800'
0'

Vertical Scale

Orr Ls.

Weeks Ls.

St. Charles Fm.

Nounan Dol.

Mendha Ls.

Marjum Ls.

Bloomington Ls.

Highland Peak Ls.

Wheeler Sh.
Swasey Ls.
Dome Ls.

Blacksmith Dol.

Ute Ls.

Boysen Fm.

Burrows Dol.
Peasley Ls.

Howell Ls.

Langston Ls.

Gros Ventre Fm.

Lyndon Ls.

Comet Sh.

Tatow Ls.
Pioche Sh.

Brigham Qtz.

Flathead Ss.

Pioche Sh.

Prospect Mt. Qtz.

base not exposed

Prospect Mt. Qtz.

base not exposed

base not exposed

Upper Cambrian

Middle Cambrian

Precambrian

Lower Cambrian

kingdom were present and complex forms of arthropods such as trilobites held the center of the stage.

But while the shallow seas teemed with life, the lands must have presented scenes of stark desolation, for Cambrian rocks bear no clear evidence of land plants of any sort. It is not improbable that moist lowlands were clothed by primitive soft-tissued plants such as lichens, but, lacking roots and vascular tissue by which to draw moisture from the ground, such lowly plants could not have lived on the uplands. They were too soft and delicate to leave a fossil record.

Stars of the Cast. The dominant fossils of Cambrian times were the trilobites, which occur in bewildering variety (Plate 1). They were bottom-dwellers, groveling over the sea floor or swimming near the bottom (Fig. 8-7) and they make up fully 60 percent of the known fauna. At this time they boasted the greatest size and probably the highest intelligence of any animals on earth. Even so they were small creatures, generally ranging between 1 and 4 inches in length. The giant of them all was *Paradoxides harlani*, found near Boston, with an overall length of about 18 inches. Such giants probably weighed less than 10 pounds. It is a striking commentary on the history of life that such seemingly feeble folk held undisputed sway over the earth throughout Cambrian time—a span exceeding that of human existence by perhaps a hundredfold.

Second in importance were the brachiopods (Fig. 8-8). This is a phylum of "glorified worms" that grew attached to the rocks by a caudal stem and protected the body with a pair of shelly "valves" borne on front and back of the animal and hinged at the posterior end. Brachiopods still exist, chiefly in the East Indies and along the shores of Australia and New Zealand, but they are uncommon in the Atlantic Ocean and are rarely seen on the beaches. They are of two major kinds. Primitive forms, such as *Lingula*, make thin shells of chitin. The more advanced forms make calcareous shells somewhat resembling those of clams except that, being borne fore and aft, each valve is bilaterally symmetrical. Most of the Cambrian brachiopods were small and had

Figure 8-5 Upper Cambrian, Potsdam sandstone. Ausable Chasm on the east flank of the Adirondacks, near Keeseville, New York. (C. O. Dunbar.)

Figure 8-6 Stratigraphic cross section from southern Nevada to central Wyoming showing overstep of the Lower Cambrian by Middle Cambrian along the eastern margin of the geosyncline and change of latter to detrital sediments on the craton. (Data from Charles Deiss and Harry E. Wheeler.)

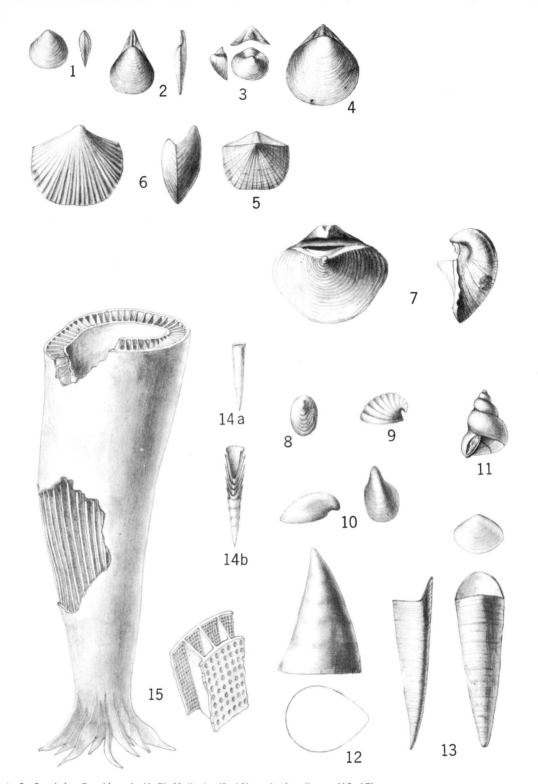

Plate 2 Cambrian Brachiopods (1–7), Mollusks (8–12), and miscellanea (13–15).
Figure 1, *Dicellomus politus* (ventral and lateral views); 2, *Lingulepis pinnaformis*; 3, *Paterina bella;* 4, *Obolus aurora;* 5, *Billingsella coloradoensis;* 6, *Eoorthis texanus;* 7, *Kutorgina cingulata;* 8, 9, *Helcionella rugosa;* 10, *Helcionella* sp.; 11, *Matherella saratogensis;* 12, *Hypseloconus elongatus* (side view and cross section); 13, *Hyolithes princeps* (lateral and side view and oper-culum); 14, *Salterella rugosa;* 15, *Pycnoidocyathus profundus* (lateral view and enlarged detail). All slightly enlarged except Fig. 15 in part. (Drawings by L. S. Douglass.)

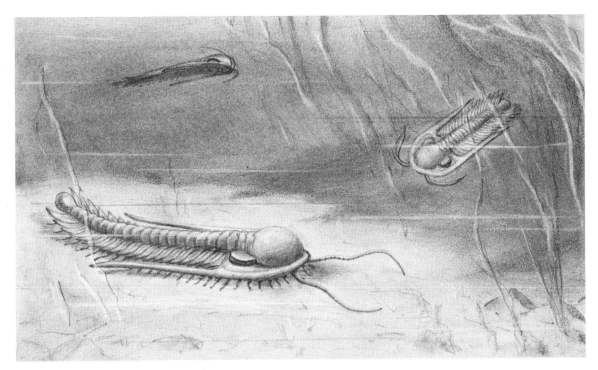

Figure 8-7 Restoration of the Cambrian trilobite *Paradoxides harlani,* one of the giants of its time. About ¼ natural size. (Drawing by Lathrop Douglass.)

Figure 8-8 Living brachiopods, about ½ natural size. At the left, *Lingula,* a persistent primitive type with a thin shell of chitin; at the right, *Laqueus,* having a calcareous shell. (Yale Peabody Museum.)

Figure 8-9 Archaeocyathids in the rock. From an extensive reef in Australia. A millimeter scale at the left. (T. G. Taylor.)

chitinous shells no larger than a fingernail, but in places they are so abundant they almost cover the bedding planes of the rocks. Toward the end of Cambrian time calcareous shells became more common but few were as much as an inch across (Plate 2, Figs. 5 and 6). From the next period to the end of the Paleozoic Era, however, brachiopods were the dominant shellfish of the seas.

A third important group, confined to the Lower Cambrian, was a strange phylum of uncertain biologic relations known as the <u>Archaeocyatha</u>. They made calcareous structures of conical or subcylindrical shape and grew in profusion sufficient to make extensive reefs. The skeleton consisted of an outer and an inner wall separated by radial septa, and both walls were perforated by abundant small pores (Plate 2, Fig. 15). They are unmistakable when seen in the rock (Fig. 8-9) and are one of the best guide fossils to the Lower Cambrian.

Colonies of microscopic blue-green algae also formed limy reefs on the Cambrian sea floor. Like their Precambrian ancestors these **stromatolites** are hemispherical, irregular, or spreading masses of finely laminated calcite. Since these deposits (Figs. 8-10 and 8-11) preserve only the gross form of the colony, the algae cannot be identified biologically and the deposits are simply called stromatolites. The name *Cryptozoon*, once given to all Lower Paleozoic stromatolites, is no longer used.

Figure 8-10 Stromatolites weathering out of the Upper Cambrian Gros Ventre Formation in Grand Teton quadrangle, Wyoming. (Eliot Blackwelder.)

Figure 8-11 Natural exposure of part of a stromatolite reef in the Upper Cambrian near Saratoga, New York. Glacial scouring during the Pleistocene epoch removed the summits of these colonies and exposed them in cross section. (H. P. Cushing. New York State Geol. Survey.)

3 Cm

Figure 8-12 Vertical tubes of burrowing organisms generally called *skolithus*, particularly widespread in Cambrian marine sandstones, although they are also present in other Phanerozoic systems. This specimen is from Silurian sandstone in Virginia. (Yale Peabody Museum.)

Figure 8-13 Dr. Charles Doolittle Walcott.

Minor Characters in the Drama. The remaining groups of fossils play a minor role in the record of Cambrian life. Siliceous spicules are locally common, proving that sponges were present, and snails are represented by several small species, and two groups of uncertain relationships, the salterellas (Plate 2, Fig. 14) and the hyolithids (Plate 2, Fig. 13) are locally abundant. Vertical tubes (Fig. 8-12) in sandstones indicate that burrowing organisms were locally abundant. The nautiloid cephalopods are represented by a few rare and quite small species in the Upper Cambrian. The phylum **Echinodermata** is represented from Lower Cambrian on by an increasing variety of small and rare species of primitive kinds. But many of the classes of animals that would become important in the next period had not yet appeared or did not make shells. For example, we have no record of corals, clams, bryozoans, starfish, echinoids, or crinoids and, of course, no vertebrates.

The Burgess Shale Fauna. To what has just been written there is one amazing exception. In 1910 Dr. C. D. Walcott, then Director of the U. S. Geological Survey, was leading a pack horse up a trail over the Middle Cambrian section on the flank of Mt. Wapta above the town of Field, British Columbia, when the horse stumbled over a fallen chunk of slate. As the specimen fell apart, Dr. Walcott's eye caught the gleam of glossy spots on the bedding plane, which proved to be the carbonaceous imprint of soft-bodied creatures. It was probably the most important fossil discovery ever made, for it was a treasure trove of the remains of delicate, soft-bodied creatures such as have never been seen elsewhere to this day. Dr. Walcott (Fig. 8-13) opened a quarry (Fig. 8-14) and split tons of the slate from which he was able to describe 70 genera and 130 species, almost all of which were delicate, soft-bodied animals. Although they were all reduced to mere carbon films, most of the animals had been whole when buried and they clearly show the appendages, even delicate bristles, and in some specimens traces of the viscera. It was a fauna of worms, sponges, and a great variety of small arthro-

Figure 8-14 Walcott's quarry in the Burgess shale (Middle Cambrian) on the slope of Mt. Wapta above Field, British Columbia. (C. D. Walcott.)

pods as delicate as the living brine shrimp (Plate 3). Among them was a specimen of *Aysheaia* (Fig. 8-15), which is related to the living onychophorans [3], long recognized by zoologists as a persistently primitive group structurally like the expected common ancestor of the annelid worms and arthropods. The Burgess shale fauna is important chiefly because of the perspective it gives. Here we get one clear view of the world of soft-bodied creatures that inhabited the early Paleozoic seas. Without these fossils we could only infer that life was abundant, and guess what it was like; and we might have believed that tri-lobites were more abundant than other animals just because they are the common fossils. This glimpse of the primitive soft-bodied creatures of the Cambrian sea proves the incompleteness of the fossil record; it also presents an exciting challenge. The wonderful discovery came by accident as a pack horse, high up on a mountain trail, turned over a slab of slate that caught the eye of a trained paleontologist. It was a spot where the forces of nature had conspired to preserve an almost incredible record from the past. There must be many others yet undiscovered—perhaps some in the Precambrian slates.

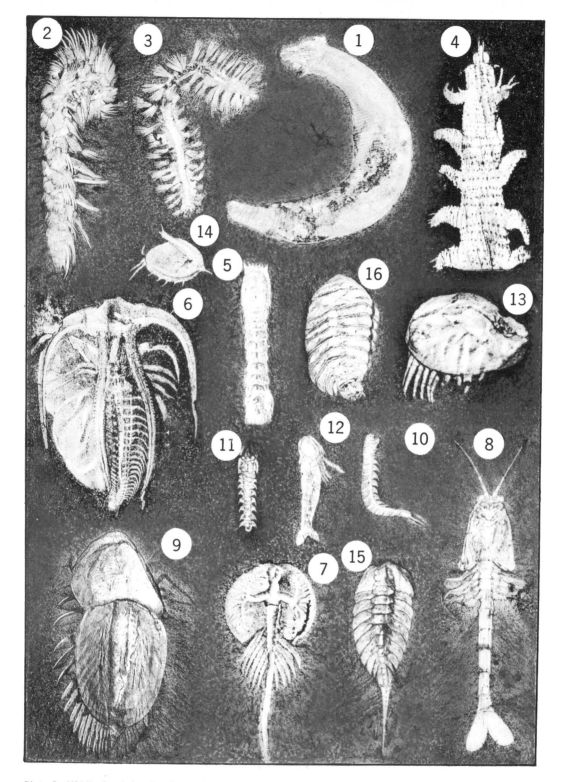

Plate 3 Middle Cambrian Fossils from the Burgess Shale.
Figures 1–3, annelid worms; 4, an onycophoran (see also Figures 8–14); 5–16, trilobitoids.

Figure 1, *Ottoia prolifica;* 2, *Canadia prolifica;* 3, *C. setigera;* 4, *Aysheaia pedunculata;* 5, *Mollisonia gracilis;* 6, *Marella splendens;* 7, *Burgessia bella;* 8, *Waptia fieldensis;* 9, *Naraoia compacta;* 10, 11, *Yohoia tenuis;* 12, *Y. plena;* 13, *Hymenocaris(?) circularis;* 14, *H. parva;* 15, *Molaria spinifera.* Figures enlarged a 1.5 to 2. (After Walcott.)

One may infer the circumstances under which this unusual deposit formed. Muds so fine of grain and so rich in decaying organic matter must have formed a soft, oozy sea floor, deficient in oxygen and reeking in hydrogen sulfide. Occasional storm waves stirring the bottom muds liberated the H_2S and allowed it to rise toward the surface, poisoning such swimming or floating organisms as were present. The locality of Mt. Wapta was probably a depression somewhat deeper than the surrounding sea floor, and when the dying organisms drifted into it they settled below wave base and were never again disturbed. Because of the poisoned water, no scavengers were present to devour them and so they remained whole, with limbs and other delicate appendages intact. Comparable mass destruction of organisms by H_2S gas rising from mud bottoms has been observed in modern seas.

The Acquisition of Hard Parts. Although the vast length of Precambrian time was not suspected when Walcott, in 1890, first brought out the richness and variety of Middle-Cambrian soft-bodied life, the zoologist Brooks [1] exclaimed that these forms, instead of being "the simple, unspecialized ancestors of modern animals, are most intensely modern themselves in the zoological sense and . . . belong to the same order of nature as that which prevails at the present day." Obviously the evidence of the truly primitive stages of life and of their differentiation into the great phyla must be sought earlier than the Cambrian.

From the preceding review of Precambrian fossils it is apparent that we still lack any irrefutable evidence of animal life before the appearance of the Ediacara fauna of Australia, and that even this remarkable fauna is very near the base of the Cambrian. Indeed its position serves to emphasize the change that marks the beginning of Cambrian time, for the Ediacara fauna consists only of soft-bodied creatures, yet the Cambrian rocks abound in shell-bearing animals; those of North America alone have yielded at least 1200 different kinds. Whether metazoan animals had been gradually developing over a long span of Precambrian time or whether they originated in an explosive burst of evolution late in the Pre-

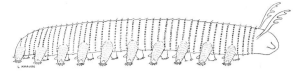

Figure 8-15 Restoration of the Middle Cambrian onychophoran *Aysheaia*. Slightly enlarged. (After G. E. Hutchinson.)

cambrian, they were present as soft-bodied forms before shells of chitin (a nitrogenous substance similar to our fingernails) or of calcium carbonate were evolved. These early animals were probably like the soft-tissued creatures of the Burgess shale, scarcely capable of preservation except under extraordinary conditions. The great variety of such forms in the Middle Cambrian, most of which are recorded nowhere else, shows clearly how abundant and how varied soft-bodied animals may have been at least in Late Precambrian time.

The cause of the relatively abrupt development of shells in many stocks of animal life at the beginning of the Cambrian has been a subject of much speculation. Brooks has suggested that before this time marine animals were chiefly small free-floating or swimming creatures like the larvae of existing types, and that near the close of the Precambrian they began to inhabit suitable parts of the shallow sea floor, then rapidly increased in size, experienced for the first time the effects of crowding and keen competition, and hence required protective armor. It has also been postulated [2] that the Precambrian oceans may have been so deficient in dissolved carbonates that limy shells could not be formed, that the lime brought to the sea by rivers may have been chemically precipitated as fast as it was supplied because of the decaying organic matter on the sea floor before scavenging types of life had developed. This is a most unlikely hypothesis since algae seem to have had no difficulty finding calcium carbonate to make stromatolites. Moreover, it does not really solve our problem because even in the absence of lime, animals could still make their shells of chitin as, in fact, most of the Lower

Cambrian types did. Protection against ultraviolet radiation, mentioned in Chapter 7, is difficult to accept as a possible stimulus to the production of hard parts in light of the existence of the soft-bodied Ediacara fauna, which occurs in beds of apparent shallow-water origin.

The development of actively predacious habits may have been the first great stimulus to the development of protective armor. It is not improbable that all Precambrian animals were herbivorous or scavenging, and that with the development of the active carnivorous habit acquisition of hard protective armor was an adaptation highly favored by natural selection.

REFERENCES

1. Brooks, W. K., 1894, *The origin of the oldest fossils and the discovery of the bottom of the ocean.* Jour. Geology, v. 2, pp. 455–479.
2. Daly, R. A., 1907, *The limeless ocean of Pre-Cambrian time.* Am. Jour. Sci., 4th ser., v. 23, pp. 93–115.
3. Hutchinson, G. E., 1930, *Restudy of some Burgess shale fossils.* U. S. Nat. Museum, Proc., v. 78, Art. 11.
4. Lochman-Balk, Christina, and Wilson, J. L., 1958, *Cambrian biostratigraphy in North America,* Jour. Paleontology, v. 32, pp. 312–350.
5. Ross, C. P., 1962, *Paleozoic seas of central Idaho.* Geol. Soc. America Bull., v. 73, pp. 769–794.

Figure 9-1 Trenton Gorge of West Canada Creek north of Utica, New York, showing about 300 feet of thin-bedded and very fossiliferous Mid-Ordovician limestones. This view shows only the middle part of the section. This is the type section of the Trenton Limestone. Within 40 miles to the east it grades laterally into black shale. (Carl O. Dunbar.)

CHAPTER **9** *The Ordovician Period*

Yes, small in size were most created things
And shells and corallines the chief of these.
T. A. CONRAD

PHYSICAL HISTORY OF NORTH AMERICA

The Great Submergence. The faunal break at the end of the Cambrian Period indicates that the continent was completely emergent for a time, but the late Cambrian lands were low and in North America the period closed quietly. The post-Cambrian hiatus apparently reflects a gentle drop in sealevel.

But with the beginning of Ordovician time another great cycle of submergence began as shallow epicontinental seas crept in from east, west, north, and south. Early in the period the Appalachian trough and much of the eastern half of the United States were flooded and an extensive sea spread across northern Canada while another covered parts of California, Nevada, Utah, Wyoming, and Colorado. Later in the period fully half of the continent was

Figure 9-2 Correlation of important Ordovician sections.

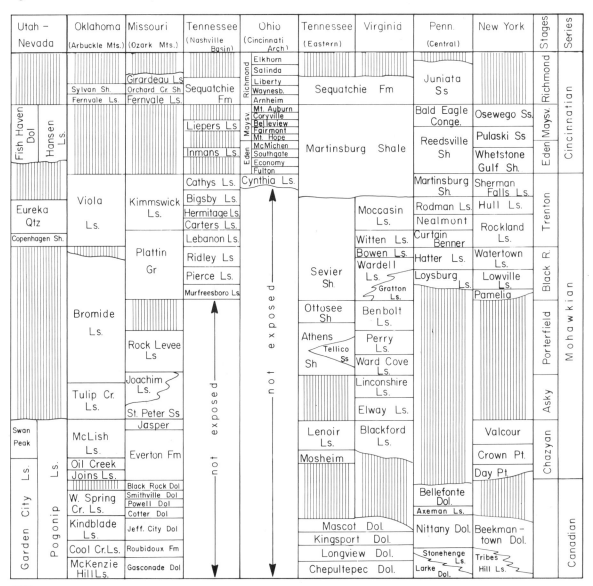

Figure 9-3 Upper Ordovician limestone resting on peneplaned surface of Precambrian granite about 100 miles west of Cape Churchill, which is on the shore of Hudson Bay. (Courtesy of L. M. Cumming.)

submerged and the land was reduced to a series of great low islands (Map III). No later submergence was quite so extensive.

The Lower Ordovician formations (Fig. 9-2) are nearly all calcareous—either limestones or dolostones—indicating that all the land surfaces were so low they suffered little erosion. The faunas also were strikingly similar over great areas and indicate that the seas were shallow and generally clear.

Later in the period similar seas spread widely across the Canadian Shield and covered much of Canada and the western United States and here, also, the seas were shallow and generally clear so that over vast areas limestone rests directly on Precambrian granite or ancient metamorphics (Fig. 9-3).

The Taconic Orogeny. In contrast to what was happening over the craton, orogeny affected the eastern side of the northern Appalachian geosyncline. Here land began to rise early in the period and before its close it was

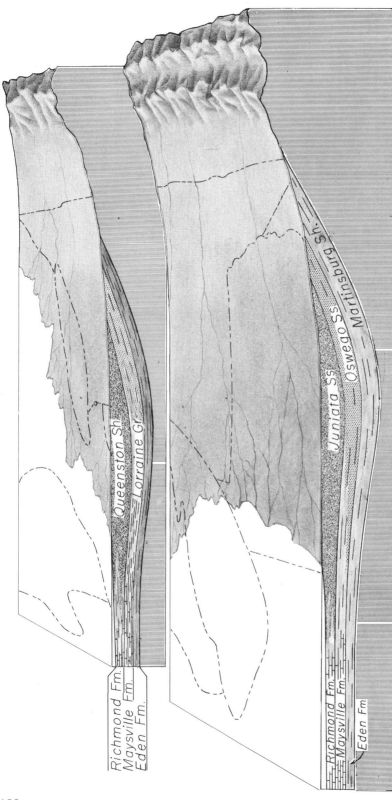

Figure 9-4 Block diagram to show relations of the Queenston Delta during its formation in Late Ordovician time. The view is north. The edge of highlands produced by the Taconic orogeny are at the extreme right. Length of front face about 600 miles, vertical scale greatly exaggerated and not uniform.

Queenston Sh.
Lorraine Gr.

Juniata Ss.
Oswego Ss.
Martinsburg Sh.

Richmond Fm.
Maysville Fm.
Eden Fm.

Richmond Fm.
Maysville Fm.
Eden Fm.

Figure 9-5 Cowhead breccia near Cow Head, western Newfoundland. It is made of blocks and slabs of limestone ranging in age from Upper Cambrian to Mid-Ordovician. They show almost no rounding and are interpreted as the accumulation of a landslide derived from submarine faulting. The hammer near the center at the top gives the scale. Some of the blocks are much larger. (C. O. Dunbar.)

the scene of an extensive mountain range from which an immense volume of detrital sediments was carried westward into the geosyncline. It also was the scene of much volcanic activity and of westward thrusting against the continent. This was named the **Taconic Orogeny** for the Taconic Mountains of eastern New York State, which are but a small remnant of this mountainous Ordovician landmass.

Evidence for the Taconic Orogeny is displayed in three ways. One is the vast **Queenston Delta** made of detrital sediments eroded from the rising mountains and carried westward into the geosyncline (Fig. 9-4). As it grew, the shoreline was pushed far to the west, in western New York and Pennsylvania, and the landward part of the delta was a broad coastal plain sloping gently westward. In this region mud began to pour into the geosyncline in Mid-Ordovician time forming the black Martinsburg shale that can be traced from

northern New York into southern Virginia. As the uplift in the marginal land grew, the black shale was followed by sandstone and conglomerate, in part marine, and this in turn was followed by nonmarine sandy redbeds. Each of these units become finer of grain toward the west and grade laterally into limestone in Ohio.

An earlier pulse of uplift farther south gave rise to another great wedge of detritals in Virginia known as the **Blount Delta** (Fig. 9-16) of Mid-Ordovician age, and from New England northward into Newfoundland black shales began to form early in the period and in places include remarkable limestone conglomerates believed to represent submarine landslides resulting from faulting. In Newfoundland the Cow Head Breccia includes blocks of limestone over 100 feet across (Fig. 9-5).

A second line of evidence of the Taconic disturbance is seen in the volcanic activity. Late in the period much of Newfoundland was covered by great lava flows (Fig. 9-6). And in Quebec the Mictaw group is made of shales and volcanic tuff of great but undetermined thickness. Small lava flows are found in the shale belt at Starks Knob in eastern New York and near Jonestown, Pennsylvania, and the Ammonoosic volcanics of New Hampshire are probably the same age. A chain of volcanoes extended as far south as Cape Hatteras on the eastern side of the geosyncline, probably in the region of the present Piedmont belt. Evidence of these volcanoes is to be seen in widespread layers of volcanic ash blown westward over the geosyncline. One such bed reached into Wisconsin, Minnesota, and Iowa, and in central Pennsylvania 15 such layers of volcanic ash have been recognized and two beds have been found in Alabama. In eastern Kentucky one bed of volcanic ash is as much as 7 feet thick.

A third line of evidence of the Taconic orogeny is seen in the structure. A great zone of thrust faulting follows the western coast of Newfoundland, is continued along the axis of the St. Lawrence River to near Quebec City where it bends south along the axis of Lake Champlain, then follows the Hudson Valley to near Kingston, New York, and then curves southwestward across Pennsylvania (Fig. 9-7). This has been called **Logan's Line** after the great pioneer geologist of Canada, Sir William Logan. To the west of this line there has been little disturbance to this day but to the east of it the Ordovician shale belt is badly crushed, and in three areas has been overthrust to the west in great masses called klippen. One of these forms the low coastal plain along much of the west coast of Newfoundland, another forms the Taconic Range in eastern New York, and the third is the Hamburg klippe of central eastern Pennsylvania. Most of the region east of Logan's Line was strongly deformed by

Figure 9-6 Flows of pillow lava exposed in the sea cliff near the mouth of Fox River, Port-au-Port Bay, western Newfoundland. (C. O. Dunbar.)

later orogenies, but in places along the Hudson Valley their effect did not reach as far west as the Taconic orogeny and here, as at Becraft Mountain near Hudson, New York, and in the Alsen quarries near Kingston, the strongly deformed Ordovician rocks are unconformably overlain by the Early Devonian, Manlius limestone (Figs. 9-8 and 9-9). Since the Ordovician folds had been truncated and the region was very low when the Manlius limestone was spread across the region of the Hudson Valley, we may infer that the Taconic orogeny reached its climax about the end of the Ordovician Period. The highlands it formed may be considered the first generation of the Appalachian Mountain system.

THE STRATIGRAPHIC RECORD

Series of Rocks and Epochs of Time

American geologists have traditionally recognized three series of the Ordovician formations and three corresponding epochs of time. These are commonly referred to as Lower, Middle, and Upper Ordovician rocks and the corresponding time units are Early, Middle, and Late Ordovician. In more technical usage the Lower Ordovician is the **Canadian Series** based on exposure in southeastern Canada; the Middle Ordovician is the **Mohawkian Series** named for exposures in the Mohawk Valley in eastern New York and western Vermont; and the Upper Ordovician is the **Cincinnatian Series** named for the richly fossiliferous sections so well exposed in the region about Cincinnati, Ohio.

The Canadian Series. The choice of a type region for this series in the early days of the New York Geological Survey has proved to be rather unfortunate because the section there represents only the lower part of the series and the outcrops are low and somewhat discontinuous. A much better section was later described on the east side of Lake Champlain opposite Crown Point. One of the finest sections is exposed in the Nittany Arch in central Pennsylvania (Fig. 9-10). This is the westernmost great open fold in the Appalachian region in front of the eastern face of the Allegheny

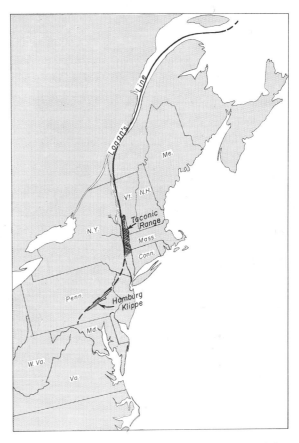

Figure 9-7 Logan's Line, a major belt of Late Ordovician thrust faulting.

Plateau. Here, about the town of Bellefonte, may be seen a simple, continuous and gently dipping succession of about 4000 feet of Lower Ordovician formations [3], every bed of which is exposed. The succession is represented in Fig. 9-2. Other thick sections are exposed in Virginia, eastern Tennessee, and northern Georgia, but there the relations are complicated by great thrust faults. One of the finest and most complete displays of the Lower Ordovician rocks occurs in the flanks of the Ozark Dome in southeastern Missouri and northern Alabama [2] (see column 2 of Fig. 9-2). Other fine sections may be seen in the flanks of the Arbuckle Mountains of Oklahoma [8] and in the Central Mineral belt of Texas [4].

A remarkable feature of the Canadian Series

Figure 9-8 Unconformable contact of Lower Devonian (Manlius) limestone on strongly folded Middle Ordovician sandstone in the Alsen quarries south of Catskill, New York. The Devonian beds strike N35°E and dip 20°NW, whereas the Ordovician beds strike N5°E and dip 55°E. (C. R. Longwell.)

Figure 9-9 Early Devonian, Manlius Limestone resting with an erosional unconformity on badly mashed Hudson River Shale (Middle Ordovician) at Becraft Mountain near Kingston, New York. (C. O. Dunbar.)

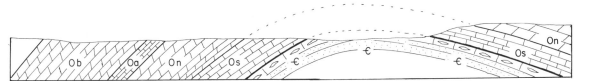

Figure 9-10 Above, a cross section of the Nittany Arch near Bellefonte, Pennsylvania, showing 4000 feet of Lower Ordovician strata resting on Upper Cambrian. Os, Stonehenge Limestone; On, Nittany Dolostone; Oa, Axeman Limestone; Ob, Bellefonte Dolostone. Length of section 2.7 miles. Vertical scale exaggerated. (After Charles Butts.)

Figure 9-11 Right, map showing the boundary between the calcareous and the black shale facies of the Lower Ordovician in the Far West. Open circles indicate sections in the calcareous facies; black dots indicate sections in the shale facies. (Adapted from Ross [10].)

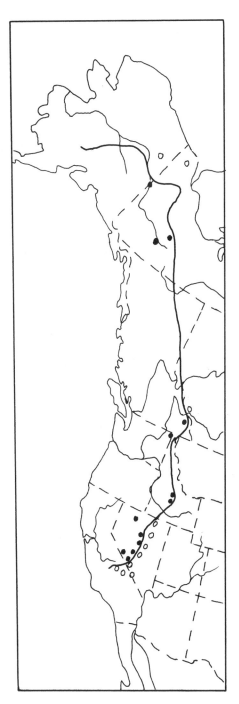

is that all across the craton and far into the Appalachian geosyncline, the formations are almost completely calcareous, indicating that the lands were then very low. It is also notable for the enormous amount of dolostone. This creates a problem. In the process of dolomitization the fossils were generally destroyed and in the central and southern part of the Appalachian geosyncline, where the sections are badly faulted, it is difficult to make correlations across the faults. In some regions the change from dolostone to limestone follows the bedding planes and formational units, but in others the change of facies crosses stratification boundaries and a limestone formation changes into a dolomite within a short distance.

From New York to Newfoundland the Lower Ordovician east of Logan's Line is in a black shale facies and bears a rich fauna of graptolites. In the Far West there is another belt of black shale bearing similar graptolites. Reuben J. Ross [10] has traced a line between the calcareous facies and the black shale facies which runs across Nevada and central Idaho and then passes northward into Canada near Banff and then roughly parallels the coast as far as Alaska (Fig. 9-11). Since the mud could not have come from the east, where calcareous deposits were forming, it must have come from marginal land on the west. But later deformation along the west coast, and the spread of younger formations, has so completely masked it that we have no further knowledge of the nature of this land. It may have been an island arc. In Nevada, where the structure is complicated by many faults, the calcareous facies and the black shale facies are distinctly separated, but in the southern Rockies of Canada the two facies intertongue, indicating the lack of a barrier between the limy and the muddy sea floors.

The Mohawkian Series. The Middle Ordovician record differs strikingly from the Canadian in both lithology and faunas. Over the craton the formations are predominantly limestones and fossils are extremely abundant. In the Appalachian trough, however, they grade eastward into black shale and siltstone, with which they commonly are deeply intertongued (Fig. 9-12). In the trough the formations thicken greatly. This reflects the steady rise of the Taconic lands and the concomitant

sinking of the geosyncline; the uplift increased to a climax late in the period. In New York and Pennsylvania the Adirondack arch formed a barrier which was not crossed by the detritals until late in the Mohawkian Epoch when the black Utica Shale spread across it and reached as far west as Lake Erie. The Trenton Limestone which is typically developed in the Black River Valley west of the Adirondacks (Figs. 9-1 and 9-13) grades eastward within a distance of 40 miles into the black Canajoharie Shale (Fig. 9-14). Exposures along the New York Throughway clearly reveal the deep intertonguing of the two facies. And in the region of Schenectady the shale intertongues with and grades into the flaggy Schenectady Sandstone (Fig. 9-15).

In Pennsylvania the Trenton Limestone equivalents likewise grade eastward into the great Martinsburg Shale belt. In the southern part of the Appalachian Valley the changes of facies reach extreme complexity, as illustrated in Fig. 9-16. Here the great wedges of sandstone (Tellico, Sevier, and Bays) indicate a sharp local rise in Appalachia, which has

Figure 9-12 Idealized west-east section of the Middle and Upper Ordovician formations along the Mohawk Valley from Oswego to Schenectady, New York. (After Marshall Kay [9].)

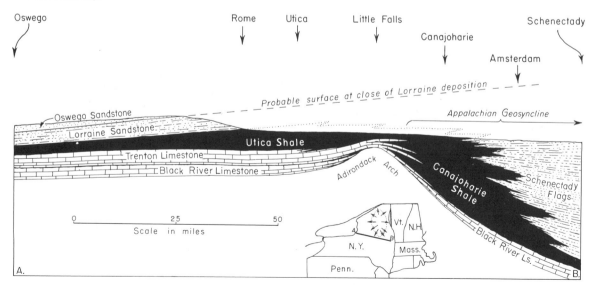

Figure 9-13 Typical outcrop of the Trenton Limestone in Trenton Gorge. (C. O. Dunbar.)

Figure 9-14 Type section of the black Canajoharie Shale above the dam on Canajoharie Creek at Canajoharie, New York. Here the shale rests on about 10 feet of the basal Trenton limestone. (C. O. Dunbar.)

Figure 9-15 Thin-bedded gray sandstone and siltstone (Schenectady Flags) one mile south of Sloansville, New York. This is the sandy equivalent of the Trenton Limestone. (C. O. Dunbar.)

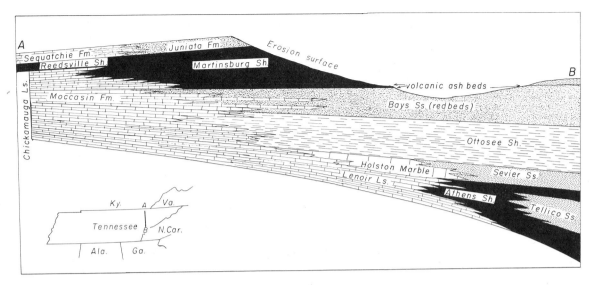

Figure 9-16 Stratigraphic diagram to show changes of facies in the southern part of the Appalachian trough resulting from the Blountian uplift in Appalachia in Mid-Ordovician time. The line of the section, shown on the inset map, runs slightly east of south across eastern Tennessee, crossing from the edge of the plateau into the axis of the geosyncline. Black shales are shown in black, redbeds in fine stippling, and sandstones in coarser stippling. The beds at the right end of the section total more than 3000 feet thick and are all detrital but their equivalents at the left end are less than half as thick and are all limestone. (After John Rodgers.)

been named the **Blount Orogeny.** The Eureka Quartzite of Utah and Nevada indicates a local area of uplift in the western geosyncline.

The Cincinnatian Series. The Late Ordovician is especially well developed in the region about Cincinnati, Ohio, where it is divisible into many limestone units and is extremely fossiliferous. From there to the south and west as far as Tennessee and Oklahoma the sections are very incomplete, indicating only temporary incursions of the sea. In the Appalachian region, on the contrary, the section is thick and is made up of sandstones and sandy shale and conglomerates derived from Appalachia, which was rising to a crescendo of the Taconic Orogeny (Fig. 9-4).

The enormous sheet of limestone spread across the Canadian Shield deserves special comment. It covered a stable region far from highlands and is nearly all limestone. It is rarely more than 300 feet thick but stretches from Baffinland over a large area west of Hudson Bay and across the Lake Winnipeg region and extends into Wyoming where it is named

the Bighorn Formation (Fig. 9-17). It carries a rich fauna that is certainly of Late Ordovician age and is generally correlated with the latest, or Richmondian, stage of the Cincinnati region, but precise correlation is still uncertain because the Bighorn formed in a northern seaway with little or no contact with the eastern sea and its fauna shows only general resemblance to those of the type Cincinnatina.

Economic Resources

The Ordovician rocks are noteworthy for four kinds of economic resources. A very large share of the slate once quarried in North America came from the "slate belt" in New Jersey and Pennsylvania where later orogeny metamorphosed thick beds of Ordovician black shale into slate. Only thick and homogenous layers yielded usable slate and about 80 percent of the rock quarried was rejected. The mountains of such waste material in the slate belt form an impressive symbol of the one-time booming industry.

Figure 9-17 Bighorn Limestone in the canyon of Bighorn Creek, Bighorn County, Wyoming. The Bighorn Limestone, about 145 feet thick, forms the lower White Cliffs. The upper limestone is the Madison Limestone of Mississippian age. (U. S. Geol. Survey.)

The white "American Carrara" marble comes from quarries in Middle Ordovician limestone in the vicinity of Rutland, Vermont, where later orogeny changed the limestone to marble (Fig. 9-18). Arlington Memorial Amphitheatre and many of the beautiful white monuments in our National Capitol came from here (Fig. 9-19). Black marble is quarried from near the same horizon on Isle Lamotte in Lake Champlain and pink or red marble comes from eastern Tennessee. These colored marbles are used only in interior decoration because they fade when exposed to the weather.

A large body of sedimentary iron ore was worked for many years on Belle Isle in Conception Bay, southeastern Newfoundland. During peak production the annual output of this ore exceeded 1,000,000 tons.

In the early days of the petroleum industry important oil and gas fields produced from the Trenton Limestone about Lima, Ohio, but these fields were exhausted early in this century. In the meantime, however, during the

Figure 9-18 Entrance to a marble quarry at Proctor, Vermont. Similar quarries in west-central Vermont follow the steeply dipping beds of pure white marble to depths as great as 300 feet. (C. R. Longwell.)

Figure 9-19 Arlington Memorial Amphitheatre in the National Cemetery near Washington, D. C. (Courtesy of the Vermont Marble Company.)

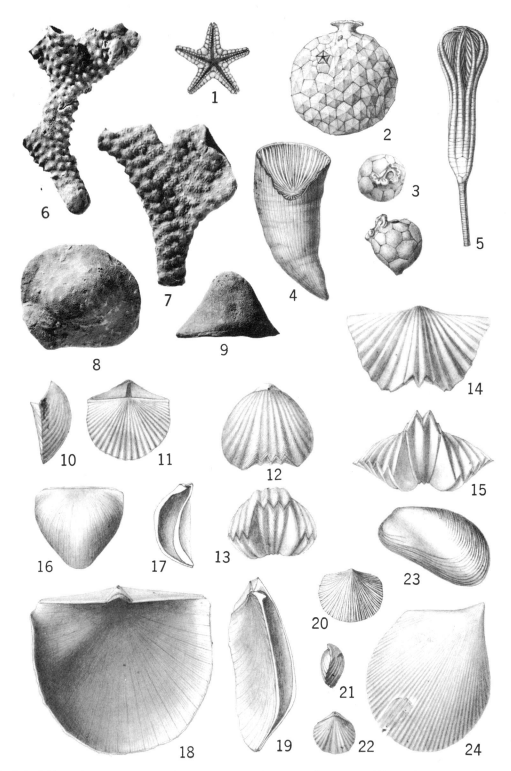

Plate 4 Ordovician Echinoderms (1–3, 5), Coral (4), Bryozoa (6–9), Brachiopods (10–22), and Pelecypods (23–24).
Figure 1, *Protopaleaster narrawayi,* one of the oldest known starfish; 2, 3, the cystoids *Echinosphaerites aurantium* and *Malocystites emmonsi* (upper and side views); 4, *Streptelasma rusticum;* 5, the crinoid *Ectenocrinus grandis;* 6, *Hallopora ramosa* (fragment of a stemlike colony); 7, *Constellaria florida;* 8, 9, *Prasopora simulatrix* (summit and lateral views); 10, 11, *Hesperothis tricenaria;* 12, 13, *Rhynchotrema capax;* 14, 15, *Platystrophia laticosta;* 16, 17, *Strophomena nutans;* 18, 19, *Rafinesquina alternata* (19, section to show flat living chamber); 20, *Resserella meeki;* 21, 22, *Zygospira modesta;* 23, *Modiolopsis concentrica;* 24, *Byssonychia radiata.* All natural size. (Drawings by L. S. Douglass.)

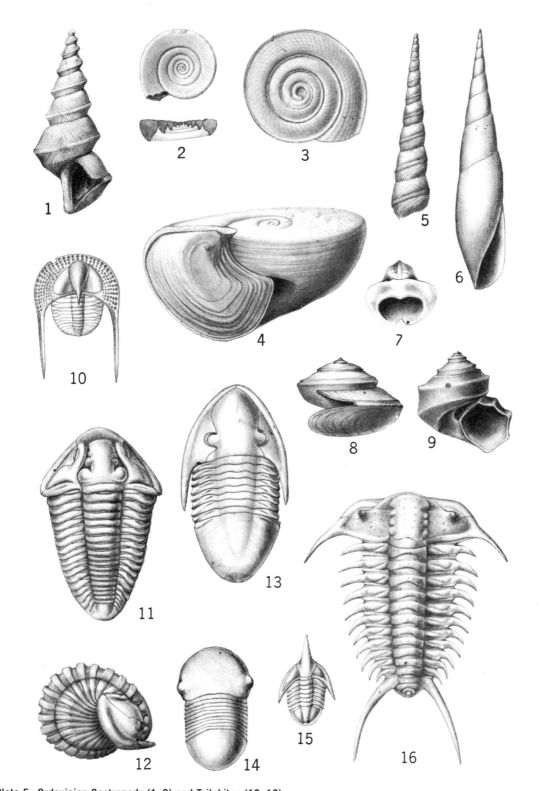

Plate 5 Ordovician Gastropods (1–9) and Trilobites (10–16).

Figure 1, *Lophospira bowdeni;* 2, *Lecanospira compacta;* 3, *Ophileta grandis;* 4, *Maclurites logani* (with operculum); 5, *Hormotoma artemesia;* 6, *Subulites canadensis;* 7, *Bellerophon troosti;* 8, *Eotomaria supracingulata;* 9, *Trochonema umbilicatum;* 10, *Cryptolithus tessellatus;* 11, 12, *Calymene meeki* (dorsal view in crawling position, and side view enrolled); 13, *Isotelus gigas;* 14, *Bumastus trentonensis;* 15, *Ampyx nasutus;* 16, *Ceraurus pleurexanthemus.* All natural size except 2 and 13 which grew to 3 times this size. (Drawings by L. S. Douglass.)

1920s, deep wells in Oklahoma and northern Texas reached the Middle Ordovician formations, which became heavy producers of petroleum.

The St. Peter Sandstone of the Mississippi Valley has been an important source of pure quartz sand used in the manufacture of glass. It is a widespread sheet of very pure, well rounded sand that spread southward from the Canadian Shield early in Middle Ordovician time and extends from Minnesota as far south as the Arbuckle Mountains in Oklahoma.

THE LIFE OF ORDOVICIAN TIMES

The Ordovician Period is truly remarkable for the enormous progress and expansion of the marine invertebrates (Plates 4 and 5).

Trilobites (Plate 5, Figs. 11-16) probably reached their peak in variety and numbers during Early Ordovician time, but nearly all of them belong to new stocks quite different from those of the Cambrian. Later in the period they began a great decline, which continued until their extinction shortly before the end of the Paleozoic Era.

Brachiopods (Plate 4, Figs. 10-22), which were small and nearly all bore thin chitinous shells in Cambrian time, now expanded into a plethora of genera and species and many families, nearly all of which bore limy shells. They were the dominant "shellfish" of the time.

A related but very distinct phylum, the **Bryozoa** (Plate 4, Figs. 6-9), now made their first appearance and are represented by four distinct orders and a host of genera and species. In many of the limestones and limy shales they and the brachiopods are far the most abundant Ordovician fossils.

Of the great phylum **Mollusca** only the **gastropods** (snails) were represented in the Cambrian and these were quite small and of few kinds; but early in Ordovician time they appeared in great numbers and bewildering variety, some with low spired shells and others with slender graceful spires of many whorls (Plate 5, Figs. 1-9). By Middle Ordovician time one genus (*Maclurites*) was making shells as much as 8 inches across.

Cephalopods were exceedingly rare and quite small in the latest Cambrian but, near the base of the Ordovician, nautiloid cephalopods appeared in abundance (Fig. 9-20) and many of these had rather specialized shells. One tribe in the Early Ordovician developed shells larger than that of the modern *Nautilus* and by the middle of the period one of the straight-shelled tribes, the endoceroids, produced shells as much as 15 feet long and over 10 inches in diameter at the living chamber. This was the largest animal on earth during the Ordovician.

Pelecypods (clams) are not represented in the Cambrian rocks. In North America they make their appearance near the base of the Middle Ordovician, and this first fauna had thin and small shells little more than an inch long, but later in the period larger shells in considerable variety were developed (Plate 4, Figs. 23 and 24).

Corals, totally unknown in the Cambrian, are represented by two distinct orders near the base of the Middle Ordovician. One of these belongs to the **Rugosa** or "horn corlas" and the other to the **Tabulata.** The earliest of the latter (*Tetradium*) were colonial forms with minute, slender, four-sided corallites but soon they were joined by the first of the honeycomb corals (*Favosites*), and the chain corals (*Halysites*). The oldest known coral reefs were made by *Lamottia,* a primitive relative of the "honeycombs" in the Middle Ordovician sea in what is now Lake Champlain.

The great Phylum **Echinodermata** was represented only by a few primitive kinds in Cambrian time. But near the base of the Lower Ordovician primitive starfish and brittle stars appear (Plate 4, Fig. 1) and before the end of the period some of the starfish closely resembled the small modern species that feeds on the oyster beds along the U. S. Atlantic coast. In the Mid-Ordovician in Scotland the oldest true sea urchin (echinoid) was found. Small crinoids of many kinds are known also in Ordovician rocks.

The **Crustaceans** are represented by several groups, notably the **Ostracoda** (Fig. 9-21), whose bean-shaped bivalved shells in places

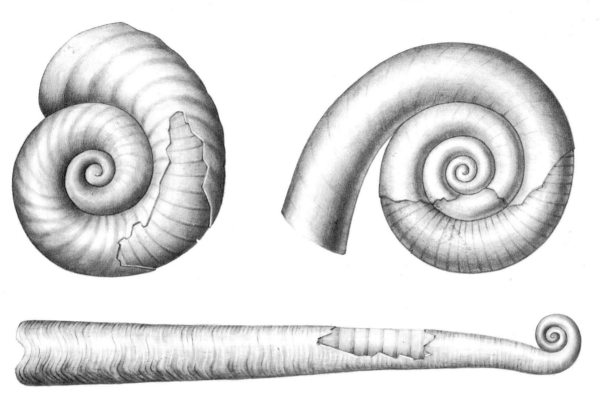

Figure 9-20 Ordovician cephalopods. Upper left, *Plectoceras occidentale;* upper right, *Schroederoceras eatoni;* below, *Lituites lituus.* In each specimen a part of the shell has been removed to show the sutures where the simple septa between chambers joined the shell. About natural size.

Figure 9-21 Large ostracods on the surface of a layer of Mid-Ordovician limestone. (Yale Peabody Museum.)

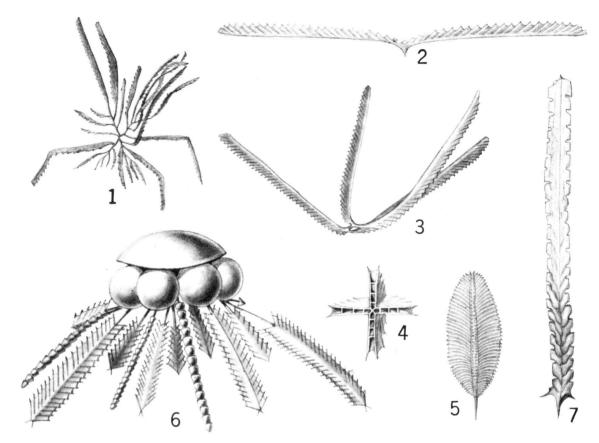

Figure 9-22 Ordovician graptolites. 1, *Clonograptus flexilis;* 2, *Didymograptus nitidus;* 3, *Tetragraptus serra;* 4 and 5, *Phyllograptus typus;* 6, *Diplograptus pristis;* 7, *Climacograptus modestus.* All are colonies of individuals. Slightly enlarged.

almost cover the bedding planes. Ostracods are still present in the modern seas where they are an important source of food for the larger animals.

Finally, the **graptolites** (Fig. 9-22), which are so abundant in the black shales, deserve note because they were mostly floaters, drifting like the sargassum does in the modern oceans, and thus spread rapidly to far places. They are thus one of the most important fossils for intercontinental correlation. They evolved rapidly during Ordovician time and mark a number of faunal zones that can be recognized in both the eastern and western shale zones in North America and in Europe, and even in far-off Australia. The biologic

relationships of the graptolites is still problematical.

But of all advances, the appearance of the oldest known **vertebrates** near the base of the Middle Ordovician is most noteworthy. At three places along the front of the Rockies the bony armor plates of an extinct tribe of primitive fishes occur in some abundance (Fig. 9-23).

Of **plant life**, blue-green algae are represented by stromatolite reefs (Fig. 9-24), which are widespread and commonly extensive. But no convincing evidence of land plants is yet known, and, of course, without land plants no animal life could have existed on the lands.

Figure 9-23 Right, fragment of the bony armor of the basal Mid-Ordovician fish *Astrapsis desiderata.* From the Harding Sandstone at Canon City, Colorado. (Courtesy of W. L. Bryant.)

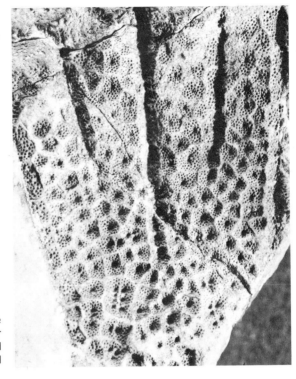

Figure 9-24 Below, surface of an extensive stromatolite reef in the St. George Dolostone (Lower Ordovician) near Port au Port, western Newfoundland. The bed was tilted to the right during post-Ordovician deformation. (Carl O. Dunbar.)

REFERENCES

1. Barnes, T. R., 1953, *Williston Basin,* Am. Assoc. Petroleum Geologists Bull., v. 37, pp. 340–355.

2. Bridge, Josiah, 1931, *Geology of the Eminence and Cardareva Quadrangles.* Missouri Bur. Geol. and Mines Bull. 24.

3. Butts, Charles, and Moore, Elwood, 1936, *Geology and mineral resources of the Belle- fonte Quadrangle, Pennsylvania.* U. S. Geol. Survey Bull. 855.

4. Cloud, P. E., Jr., and Barnes, Virgil, 1948, *The Ellenburger Group of Central Texas.* Univ. of Texas, Publ. 4621.

5. Cooper, B. N., and Cooper, G. A., 1946, *Lower Middle Ordovician Stratigraphy of the Shen- andoah Valley, Va.* Geol. Soc. America Bull., v. 57, pp. 35–114.

6. Cooper, G. A., 1956, Chazyan and Related Brachiopods. Smithsonian Misc. Coll., V. 127, pt. 1.

7. Cullison, James, 1944, *The stratigraphy of some Lower Ordovician formations of the Ozark Uplift.* Univ. Missouri School of Mines and Metallurgy Bull. v. 15, no. 2, pp. 1–112.

8. Ham, William E., 1955, Field Conference on the Arbuckle Mountain Region. Okla. Geol. Surv. Guidebook 3.

9. Kay, Marshall, 1937, *Stratigraphy of the Tren- ton Group.* Geol. Soc. America Bull., v. 48, pp. 233–302.

10. Ross, Reuben J., Jr., 1961, *Distribution of Ordovician Graptolites in Eugeosynclinal Facies in Western North America and its pale- ogeographic implications.* Am. Assoc. Petro- leum Geologists Bull., v. 45, pp. 330–341.

Figure 10-1 Niagara Falls. The gorge below the falls exposes the classical section of the American Silurian.

CHAPTER 10 *The Silurian Period*

Coral reefs were the cities of those days.
A. O. THOMAS

The early Paleozoic systems were defined and named in England and Wales, and this is an appropriate place for a brief historical review. The first general account of the geology of a large region was published by Conybeare and Phillips in 1820 and covered most of England where the rocks dip gently into the London Basin (Fig. 10-4). Beneath the Coal Measures and the "Old Red" sandstone, which they embraced in the Carboniferous Order, lay the *terra incognita* of Wales, then mistakenly thought to be a rather hopeless chaos of deformed and nearly unfossiliferous rocks.

This presented a challenge to a young Scottish gentleman, Roderick Impey Murchison, who after 6 years of public schooling had joined the army at the age of 15 and served through the Napoleonic War. With the return of peace he retired to his estate in the Highlands of northwest Scotland as a gentleman of leisure. Fortunately he soon fell under the influence of Sir Humphrey Davy, who persuaded him to come to London and take courses in chemistry and allied subjects. There the lectures in geology aroused in him an interest that was fanned into enthusiasm as he tramped the hills with two of the foremost geologists of the time, William Buckland of Oxford and Adam Sedgwick of Cambridge. Thus at the age of 32 he set himself the task of reading and gaining an education in geology. His spectacular rise to become one of the most distinguished scientists of his time, eventually to become the first Director of the Geological Survey of Great Britain, is one of the inspiring chapters in the history of geology (Fig. 10-2).

By 1831 he and his friend Professor Adam Sedgwick (Fig. 10-3) had decided to tackle the virtually unknown geology of Wales (Fig. 10-4), and here fate took a hand. To insure no conflict of interest, Murchison decided to begin at the base of the "Old Red" and work downward along the River Wye while Sedgwick plunged into the rugged region of Snowdonia and the Harlech Dome in northwest Wales.

Working westward along the Valley of the Wye, Murchison had easy going with his carriage and camping equipment and, to his delight, he found that, although somewhat deformed, the rocks below the Old Red were sedimentary beds of shale and limestone in a gently dipping sequence and that they were richly fossiliferous.

Meanwhile Sedgwick found the rocks in northwest Wales to be detrital, to be strongly deformed, and to include great thicknesses of barren sandstone and poorly fossiliferous black slate along with extensive volcanics. By 1835 Murchison and Sedgwick decided that they

Figure 10-2 Roderick Impey Murchison.

Figure 10-3 Adam Sedgwick.

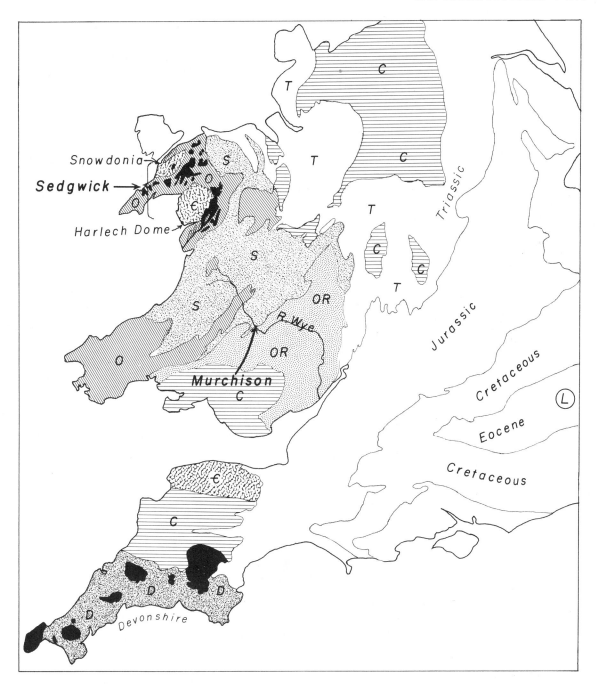

Figure 10-4 Map showing distribution of the Early Paleozoic rocks of Wales and western England.

had worked out two distinct and mutually independent new geologic systems, and choosing to give them classical names, Murchison called his the **Silurian System** for the ancient tribe of the Silures who had inhabited his region in Roman days, while Sedgwick called his the **Cambrian System** for Cambria, the Roman name for Wales.

In 1835 the two friends appeared jointly before the British Association, meeting in Dublin, to name and define the new systems. At the time, they were convinced that the Cambrian System lay entirely below the Silurian. In ensuing summers Sedgwick continued to work eastward while Murchison worked westward until they were on common ground. But by 1839, when Murchison published his great monograph on the Silurian System, it became evident that much of his Lower Silurian (the **Llandeilo** and **Caradoc** groups) was also claimed by Sedgwick as Upper Cambrian (Fig. 10-5). Thus a controversy began that estranged the two old friends and split British geologists into two camps for more than half a century. In defense of his position, Sedgwick emphasized that the Caradoc of Murchison consisted of two parts that in many places were separated by a marked unconformity. In 1854 Murchison recognized this important break but claimed it to be the boundary between his Lower and Upper Silurian. Sedgwick and his followers bitterly opposed this interpretation and insisted that the Lower Silurian of Murchison was part of the Cambrian, but after Murchison became Director of the Geologic Survey in 1855 Sedgwick's claims got little sympathy from that powerful organization.

Meanwhile Charles Lapworth (Fig. 10-6) was engaged in his epoch-making study of the graptolite faunas of southern Scotland, and this study, extended into Wales, convinced him that the Lower Silurian of Murchison was really a distinct system lying between the original Cambrian and the Silurian, and so in 1879 he "cut the gordian knot" by defining the **Ordovician System,** named after another ancient Celtic tribe, the Ordovici. The controversy had become too highly charged to be readily settled and, in deference to Murchison, the Geological Survey of Great Britain did not recognize the Ordovician System until a change of personnel occurred in 1900. American geologists, however, had a more detached viewpoint, and soon accepted the Ordovician System. When in 1894 Dr. Charles D. Walcott became Director of the U. S. Geological Survey it was officially adopted by that great organization [6] and has since been widely accepted.

PHYSICAL HISTORY OF NORTH AMERICA

The close of the Ordovician Period left rugged highlands in Appalachia and the whole of North America is believed to have stood for a short time above sealevel. Early in Silurian time three large embayments began to form as indicated in Map IV. One of these entered the Appalachian trough from the north and soon spread westward across the Great Lakes region and in Canada covered the southwestern part of Hudson Bay and the region of Lake Winnipeg [5] and the recently discovered Williston Basin. A second embayment spread up the Appalachian trough from the south, as far as New York, and spread westward over the craton as far west as Oklahoma and central Texas. Distinct differences in the faunas of these seaways suggests that they were separated by a low barrier whose limits are not clearly known. A third embayment entered far northwestern Canada across the region of the Mackenzie Valley.

By Middle Silurian time the eastern seaways had united and spread over much of the east-

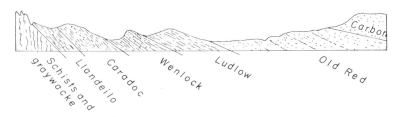

Figure 10-5 Section of Murchison's original Silurian System as published in 1839.

ern United States. In the meantime the Mackenzie Valley embayment spread southward to beyond Great Slave Lake and may have joined the southern sea in the Williston Basin. It also spread widely to the east across the Arctic Archipelago of Canada and northern Greenland. Along the west coast a small embayment covered the panhandle of Alaska, and from southern California a sea followed the axis of the Cordilleran geosyncline as far north as central Idaho, while an embayment spread eastward across southern Arizona and New Mexico and into Trans Pecos Texas (map IV).

Coral Seas of Mid-Silurian Time. Except for the Appalachian geosyncline, all these seas must have been warm, clear, and shallow since the Silurian deposits are almost exclusively limestone and abound in corals that built reefs in many places, from Tennessee across Kentucky and Indiana and along the west side of Hudson Bay and in the Lake Winnipeg region. Some of the reefs in northern Indiana were as much as a mile across and 75 feet thick (Fig. 10-7).

Figure 10-6 Charles Lapworth.

Figure 10-7 Below, restoration of a Silurian reef. A diorama in the U. S. National Museum. (Courtesy of the Smithsonian Institution.) In the foreground are three large coral "heads," the middle one a chain coral and the others honeycomb corals. Brachiopods and seaweeds cling to the reef and on the adjacent sea floor may be seen nautiloid cephalopods, trilobites, and brachiopods. The slender stalked creatures are cystoids.

In the western seaways where only the Middle Silurian is represented, the deposits are also calcareous and coraliferous (the Fusselman Dolostone in New Mexico and Texas, Lake Town Dolostone in Utah, and the Lone Mountain Limestone in Nevada). Here we have no evidence of a marginal land. In the panhandle of Alaska, where the Silurian is thick but still calcareous, it includes volcanics.

In contrast, the rapidly subsiding Appalachian trough was filled with an enormous mass of detrital sediment derived from the borderland, Appalachia.

During Late Silurian time North America seems to have been above sealevel except for a restricted basin in the region of the present Great Lakes and the middle part of the Appalachian trough (Fig. 10-16). For a time this was a low desert basin and late in the period it was reduced to two dead seas, one in central

Pennsylvania and central New York and the other in the Michigan Basin, and in these places a phenomenal amount of salt was deposited. These saline deposits involved the evaporation of an enormous volume of sea water, which must have streamed in to balance the evaporation; the position of the connection to the ocean is unknown. At the very close of the period, however, the water was freshened so that thin limestones with somewhat restricted faunas were laid down in western New York and Ontario and along the eastern margin of the Appalachian trough (Fig. 10-8).

THE STRATIGRAPHIC RECORD

Influence of the Highlands in Appalachia. A section across the Appalachian trough (Fig. 10-9) demonstrates the influence of highlands farther east. The geosyncline sank much more

Figure 10-8 Correlations of important Silurian sections.

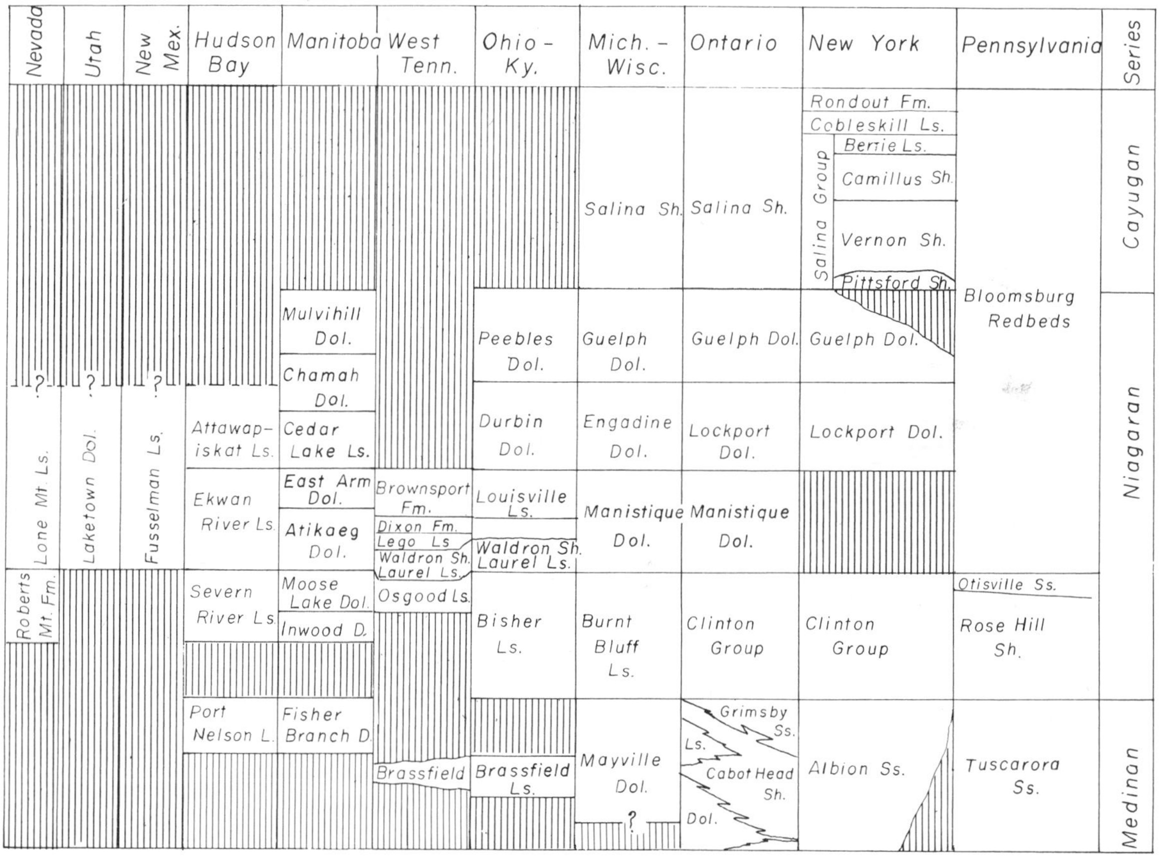

Nevada	Utah	New Mex.	Hudson Bay	Manitoba	West Tenn.	Ohio-Ky.	Mich.-Wisc.	Ontario	New York	Pennsylvania	Series
									Rondout Fm.		Cayugan
									Cobleskill Ls.		
							Salina Sh.	Salina Sh.	Berrie Ls. (Salina Group)		
									Camillus Sh. (Salina Group)		
									Vernon Sh. (Salina Group)	Bloomsburg Redbeds	
									Pittsford Sh. (Salina Group)		
?	?	?		Mulvihill Dol.		Peebles Dol.	Guelph Dol.	Guelph Dol.	Guelph Dol.		Niagaran
				Chamah Dol.							
			Attawap-iskat Ls.	Cedar Lake Ls.		Durbin Dol.	Engadine Dol.	Lockport Dol.	Lockport Dol.		
Roberts Mt. Fm. (Lone Mt. Ls.)	Laketown Dol.	Fusselman Ls.	Ekwan River Ls.	East Arm Dol.	Brownsport Fm.	Louisville Ls.	Manistique Dol.	Manistique Dol.			
				Atikaeg Dol.	Dixon Fm. / Lego Ls. / Waldron Sh. / Laurel Ls.	Waldron Sh. / Laurel Ls.					
			Severn River Ls	Moose Lake Dol.	Osgood Ls.	Bisher Ls.	Burnt Bluff Ls.	Clinton Group	Clinton Group	Otisville Ss.	
				Inwood D.						Rose Hill Sh.	
			Port Nelson L.	Fisher Branch D.			Mayville Dol.	Grimsby Ss. / Ls. / Cabot Head Sh. / Dol.	Albion Ss.	Tuscarora Ss.	Medinan
					Brassfield	Brassfield Ls.	?				

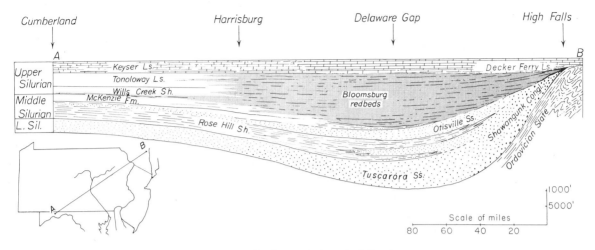

Figure 10-9 Section of the Silurian formations in the central part of the Appalachian geosyncline.

rapidly than the craton; it was filled about as fast as it sank by sand and mud which reached a maximum thickness of more than a mile. A low submarine rise, the Cincinnati Arch, running slightly east of north through the Nashville Dome of Tennessee, and the vicinity of Niagara Falls, formed a threshold separating the detrital formations from the widespread coral sea to the west.

East of this arch the Lower Silurian is represented by a vast sheet of pure quartz sand. Before its relations were well known local names were applied to it in different regions—the Clinch Sandstone of the southern Appalachians, the Tuscarora Sandstone of the central Appalachians, the Albion Sandstone of western New York. In southeastern New York where it thins to a feathered edge at the old shore line it is the Shawangunk Conglomerate. But all these are parts of one great body of sandstone that stretches fully 700 miles along the trough and extends westward for some 200 miles. It is one of the greatest bodies of pure quartz sand known anywhere. And because of its resistance to erosion it is the great ridge-maker throughout the Appalachian folded mountains.

The weathering of igneous rocks produces far more mud than sand, and in light of this the great excess of sandstone over shale in the Lower Silurian seems puzzling. But in the Shawangunk facies, where quartz pebbles as large as marbles are extremely abundant, Krynine (personal communication) found that the pebbles and the sand are made of older quartzite. This suggests that Appalachia included large areas of ancient quartzite and probably had developed a deep sandy regolith before Silurian time. By Middle Silurian time Appalachia was worn lower and the geosynclinal deposits are mostly red mud and silt, which thin and wedge out far to the west.

The Upper Silurian is represented largely by the red, sandy Bloomsburg Shale, which reaches a maximum thickness of over 3000 feet in central Pennsylvania, is nonmarine, and accumulated over the floor of a desert basin. Mud-cracked layers are common here. Figure 10-10 shows such a bedding plane near Roundtop, Maryland.

In contrast to the detrital formations of the Appalachian region, the Lower and Middle Silurian formations on the craton are relatively thin and are predominantly limestone. Figure 10-11 shows a classical section near Cincinnati. This was beyond the limits of the Late Silurian basin and here a thin Middle Devonian limestone rests directly on Middle Silurian limestone. Figure 10-12 shows about half of the section in eastern Missouri. In the Lake Winnipeg region of Canada the Silurian section is about 300 feet thick, is all limestone,

Figure 10-10 Mud cracks on a bed of impure limestone in the Bloomsburg Formation near Roundtop, Maryland. (Charles Schuchert.)

and includes coral reefs; to the west it thickens to about 1100 feet in the Williston Basin. The section west of Hudson Bay is closely similar.

The classic section exposed along the walls of the gorge below Niagara Falls (Fig. 10-13) is especially interesting because it lies upon the low barrier at the edge of the craton mentioned earlier. Here the basal tongue of the Albion Sandstone (known locally as the Whirlpool Sandstone) is an old beach deposit, as shown by its distinctive cross-bedding and swash marks about pebbles and fossils on the surfaces of the cross-bedded layers (Fig. 10-14). This is one of the few places in the early Paleozoic rocks where a beach is clearly recorded. The Middle Silurian beds hold up the Niagara Cuesta, which runs across the Great Lakes region, completely exposing along its face the transition from the detrital

Figure 10-11 Exposure of Middle Silurian limestone in the Bear Grass Quarries, south of Cincinnati where the Upper Silurian is absent and Middle Devonian limestone rests paraconformably on the Middle Silurian as it does over a large part of the craton. (Charles Schuchert.)

Figure 10-12 Section along the Mississippi River in Ste. Genevieve County, Missouri, showing about half of the Middle Silurian beds. (C. O. Dunbar.)

Figure 10-13 Silurian section exposed in the gorge below Niagara Falls. The Albion Sandstone is Lower Silurian and the rest is Middle Silurian. (Carl O. Dunbar.)

Figure 10-14 Whirlpool Sandstone at the Whirlpool in Niagara gorge showing the typical crossbedding of a beach deposit. (C. O. Dunbar.)

Figure 10-15 Stratigraphic section of the Lower and Middle Silurian formations exposed in the Niagara Cuesta showing changes of facies across the area of Niagara Falls.

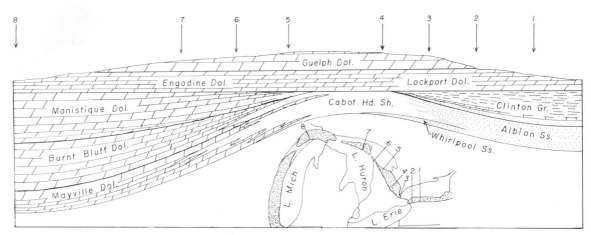

facies of the east to the limestone facies of the craton (Fig. 10-15).

The Late Silurian Desert. All the Late Silurian deposits known on our continent are confined to a limited basin indicated in Fig. 10-16. It straddled the low arch that runs through Niagara Falls dividing it into two deeper parts, the Salina Basin on the east and the Michigan Basin on the west [1]. In the Salina Basin the lower half of the Salina shale group (the Vernon Shale) is red, sandy shale like the Bloomsburg, of which it is a northern tongue. The upper half is gray and includes enormous quantities of salt. This is the basis of the large salt industry of New York State. Salt is mined about Salina (Fig. 10-17); and to the south of Ithaca, where these beds lie between 2000 and 3000 feet underground, seven beds of salt have a combined thickness of 250 feet.

In the Michigan Basin the salt deposits are immense [4]. Here deep wells reveal a series of interbedded layers of rock salt having an aggregate thickness of 1600 feet. The precipitation of such quantities of salt must have

CAYUGAN

SYRACUSE FM. THICKNESS

Contour Interval 250 feet

Figure 10-16 Isopach map showing by contours the distribution and thickness (in feet) of the Late Silurian formations. (After Alling and Briggs.)

Figure 10-17 Underground salt mine in the Salina Shale (Late Silurian) at Retsof, New York. The tunnel is cut in solid rock salt. The cars are loaded with salt on their way to the shaft. (Retsof Mining Company.)

involved the evaporation of enormous quantities of sea water, and this leads to two deductions. First, this region must have been intensely arid during Late Silurian time and these were truly dead seas. Second, the basin must have been connected to the ocean in order to permit the inflow of so much seawater. Recently discovered Silurian beds in the Lake Temisconta region in the Notre Dame Mountains south of the Gulf of St. Lawrence may indicate that the seaway was to the northeast (F. F. Osborne, personal communication).

As the inflow of sea water exceeded precipitation at the very close of Silurian time the Bertie Limestone was deposited over western New York and adjacent parts of western On-

tario (Fig. 10-18). It is only a few feet thick and in west-central New York it intertongues with the upper part of the Salina Shale. Its fauna, limited and specialized, is the chief source of the Silurian eurypterids (Figs. 10-21 and 10-22).

In eastern New York and along the eastern margin of the Appalachian trough two thin limestones were formed, first the Cobleskill Limestone and then the Rondout Limestone. The Cobleskill Limestone includes a variety of small corals but the Rondout is silty and almost unfossiliferous and its age is controversial. It may be, at least in part, of basal Devonian age. The thickest section of these Late Silurian limestones is in New Jersey (the

Figure 10-18 Exposure in the Bennett Quarries at Buffalo, New York, showing the Late Silurian Bertie and Cobleskill limestones overlain by Middle Devonian (Onondagan) limestone. (Charles Schuchert.)

Decker Ferry Limestone).

Volcanoes in Northern Appalachia. The volcanic activity in Appalachia, which began during the Ordovician, continued through Middle Silurian time from New England to Newfoundland. In a local basin about Westport, Maine, layered volcanics have the impressive thickness of some 10,000 feet. In part these were submarine, for interbedded lenses of sediments include marine fossils. These date the volcanics as Silurian. Farther to the northeast, about Black Cape on the New Brunswick coast of Chaleur Bay, the Silurian sedimentary formations have a thickness of about 8000 feet, all exposed in the sea cliffs, and in the midst of the Middle Silurian beds is a thickness of about 4000 feet of black lava flows (whence the name Black Cape) (Fig. 10-19). The basal flow poured out over the sea floor engulfing corals and brachiopods which are still well preserved. Here a geologist may have the unusual experience of collecting marine fossils in igneous rock. In north central Newfoundland, where the Silurian section is very thick and is all detrital, it includes some 1600 feet of rhyolitic and andesitic lava flows.

Climate. The arid climate of Late Silurian time was discussed previously; the mild climate of Middle Silurian time also deserves comment. The wide range of reef-building corals far to the north is truly remarkable. Similar or related species are found in Tennessee, Kentucky, Indiana, western New York, Ontario, the Lake Winnipeg area, along the west side of Hudson Bay, and within the Arctic Circle, as at Polaris Bay in northern Greenland. One highly specialized genus, *Goniophyllum* (Fig. 10-20), found in Iowa, must have come by way of the Arctic seaway from Europe.

A recent study of liquid inclusions in salt crystals indicates that in central New York the temperature in Late Silurian time ranged between 32° and 48°C. This may be compared with the surface temperature in modern oceans which at Gibraltar is about 27°C and in the Mediterranean is only 21°C. The evidence we have, therefore, indicates that during Silurian time the temperature was mild,

Figure 10-19 Silurian lava interbedded in the Mid-Silurian limestone at Black Cape, New Brunswick. (C. O. Dunbar.)

Figure 10-20 *Goniophyllum,* a peculiar coral that was an immigrant from Europe by way of the Arctic seaway.

even north of the Arctic Circle, and that climatic zones were not nearly so distinct as they are now.

ECONOMIC RESOURCES

Before the Civil War the chief U. S. source of iron was in beds of oolitic hematite in the Clinton Shale and its equivalents (basal Mid-

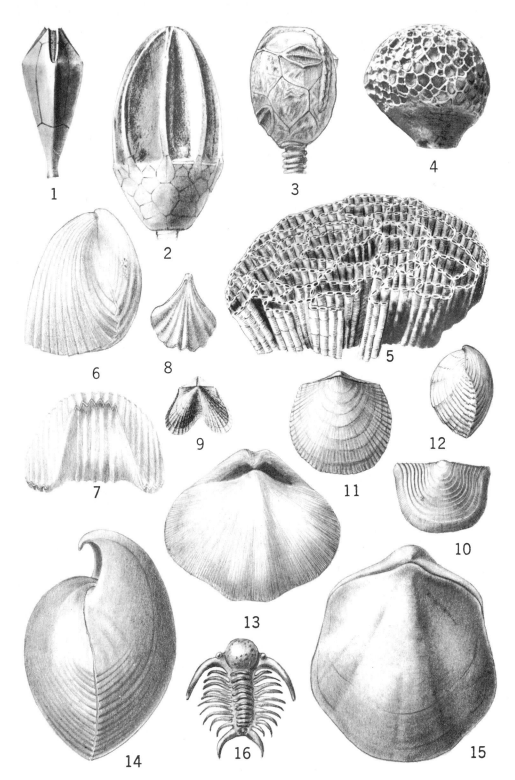

Plate 6 Silurian Blastoid (1), Crinoid (2), Cystoid (3), Corals (4, 5), Brachiopods (6–15), and Trilobite (16).
Figure 1, *Troostocrinus reinwardti;* 2, *Eucalyptocrinites crassus;* 3, *Lepocrinus manlius;* 4, *Favosites forbesi,* a honeycomb coral; 5, *Halysites catenularia,* a chain coral; 6, 7, *Uncinulus stricklandi;* 8, *Rhyncotreta? americana;* 9, *Dicoelosia biloba;* 10, *Leptaena rhomboidalis;* 11, 12, *Atrypa reticularis;* 13, *Eospirifer radiatus;* 14, *Conchidium laqueatum;* 15, *Pentamerus oblongus;* 16, *Deiphon forbesi.* All natural size. (Drawn by R. G. Creadick.)

dle Silurian). It was extensively mined about Clinton, New York, and is still the basis of an important industry about Birmingham, Alabama, where beds of ore are as much as 16 feet thick. Here the ore is close to important coal fields. It is estimated that more than 600,000,000 tons of ore are still available underground.

The salt of central New York is the basis of another big industry. During the years 1943–1945 the average annual production from this area was in excess of 2,900,000 tons, which was about 20 percent of U. S. national output and had a value of nearly $10,000,000. In 1965 the output was 5,002,000 tons, worth over $34,000,000.

The Solway process extensively used here involves pumping water down into deep wells where it dissolves the salt and is then pumped to the surface and into large evaporation pans where the salt is reprecipitated and refined.

LIFE OF SILURIAN TIME

Marine organisms developed out of the diverse Ordovician faunas with steady specialization but without the sensational innovations that marked the Ordovician. One of the major advances was in the corals, which were in their heyday and for the first time made extensive reefs. The **Rugosa** (horn corals) were legion as to both genera and species. Among the **Tabulata** the honeycomb corals (*Favosites*) were abundant and varied and were important contributors to the reefs. The chain corals (*Halysites*) were also abundant and since they died out about the end of the period are one of the most distinctive Silurian guide fossils (Plate 6, Fig. 5).

Of the great phylum **Echinodermata** the crinoids increased greatly in numbers and some were highly specialized. Cystoids (Plate 6, Fig. 3) were locally abundant, and the first typical blastoids (Plate 6, Fig. 1) made their appearance. Starfish were probably common but are seldom seen because their dermal plates are small and irregular in shape and are only held in place by flesh so they readily separate after death and are not easily recognized. Echinoids also were

present but small and are extremely rare fossils.

Brachiopods (Plate 6, Figs. 6 to 15) were still the most common "shellfish" and for the first time the spire-bearers appeared. They would become increasingly important in the next period. Bryozoans were still abundant and in places made small limy reefs.

Of the phylum **Mollusca** clams are locally common in the shales, and thick-shelled gastropods are abundant in some of the limestones. Cephalopods are less conspicuous than they were in the Ordovician rocks.

Of **Arthropods** the trilobites (Plate 6, Fig. 16) were on the decline but are locally common fossils. More noteworthy were the eurypterids or "sea scorpions" (Figs. 10-21 and 10-22). They somewhat resemble scorpions and are probably ancestors of the latter, but in the Silurian seas they were aquatic and relatively large. *Pterygotus buffaloensis* reached a length of more than 7 feet and was probably the largest animal in the world in its day. Ostracods are extremely abundant and serve as important guide fossils but most of them are small.

Graptolites, like the trilobites, were on a great decline but are still important guide fossils in the black shales.

Fishes. In the highest Silurian beds on the Isle of Oesel east of Sweden, and in equivalent beds in Norway, remarkably preserved but very primitive fish have been found (Fig. 10-23). These have no well defined jaws and appear to be ancestral to the modern hagfish or cyclostomes. Elsewhere small bony dermal plates of primitive ostracoderms are found locally, but the record of Silurian fishes is still very sparse.

Land Plants. In 1935 a single type of plant fossil from the Upper Silurian of Australia was described as a land plant. If so, it had drifted into the sea since it lies on a bedding plane amongst abundant graptolites. With this exception there is still no evidence of Silurian land plants.

Very rare specimens of scorpions and millipeds found in the Upper Silurian of Wales have been claimed to be the first air-breathing animals, but both may still have been aquatic

Figure 10-21 A large eurypterid (*Pterygotus buffaloensis*) and above, on the right, *Eusarchus scorpionis,* both from the Bertie Limestone. A diorama in the U. S. National Museum. (Courtesy of the Smithsonian Institution.)

Figure 10-22 Three genera of eurypterids from the Bertie Limestone. Left, *Eusarchus scorpionis* ($\times\frac{1}{2}$); center, *Pterygotus buffaloensis* ($\times\frac{1}{25}$); right, *Eurypterus remipes* ($\times\frac{1}{8}$). (After Clarke and Ruedemann.)

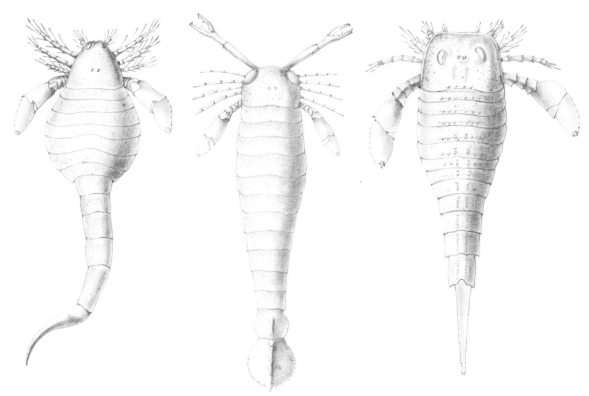

Figure 10-23 A primitive jawless fish, *Pharyngolepis oblongus,* from the uppermost Silurian, Downtonian, beds of Norway. Below, dorsal view of a specimen in the rock; above, a reconstruction from the left side view. About ⅔ natural size. (After Kiaer.)

ancestors of later terrestrial forms. The unique milliped is associated with eurypterids, which were surely aquatic.

REFERENCES

1. Alling, H. L., and Briggs, L. I., Jr., 1961, *Stratigraphy of Upper Silurian Cayugan evaporites.* Am. Assoc. Petroleum Geologists Bull., v. 45, pp. 515–547.

2. Barnes, Virgil, et al., 1966, *Silurian of central Texas.* Science, v. 154, pp. 1007–1008.

3. Fisher, D. W., 1960, New York State Museum Geol. Survey Map and Chart Series No. I.

4. Landes, K. K., Ehlers, G. M., and Stanley, G. M., 1945, *Geology of the Mackinac Straight region.* Mich. Geol. Survey Publ. 44, pp. 1–204.

5. Stearn, Colin, 1956, *Stratigraphy and paleontology of the Interlake Group and Stonewall Formation of southern Manitoba.* Canada Geol. Survey Mem. 281, pp. 1–162.

6. Walcott, C. D., 1903, 24th Ann. Report of the Director of the U. S. Geol. Survey, pp. 21–27.

Figure 11-1 "Old Red" sandstone resting on strongly deformed Silurian beds at Sicar Point, Berwickshire, Scotland. It was here that the concept of *unconformity* was first perceived in 1788 by James Hutton. (British Crown copyright.)

CHAPTER **11** *The Devonian Period*

Discovery of the Devonian System. At the close of the Silurian Period the Caledonian Disturbance had produced a bold mountain system along the western margin of Europe all the way from southern Ireland and Wales to northern Norway. In Norway it resulted in a great overthrust of the older rocks toward the east, and in the British Isles it left a series of ranges and intermont basins aligned east of north, with the thrust apparently having come from a marginal land to the northwest (Fig. 11-2). In the intermont basins red sandstones and gray silty shales accumulated throughout Devonian time, reaching exceptional thicknesses. The pioneer British geologists called these deposits "The Old Red." These beds generally lie conformably below the Carbonif-

erous System and until Murchison had defined the Cambrian and Silurian systems they formed the base of the known geologic column.

In 1836 Murchison and Sedgwick began to work in Devonshire and Cornwall, the southwestern provinces of England, which had long been known to be underlain by gray rocks considered to be Carboniferous because of the presence of fossil plants. They found that only the upper part of these rocks is plant-bearing, and the lower part they referred to the Cambrian solely because it was badly deformed and in that respect resembled the rocks of northwest Wales. However, when fossil corals found by local collectors were submitted to the paleontologist Lonsdale, he found them intermediate between corals of the Silurian and those of the Carboniferous, and suggested that these beds might be contemporaneous with the Old Red. Murchison and Sedgwick found this hard to believe, but after two years they accepted Lonsdale's view and proposed the name Devonian for a new system between the Silurian and the Carboniferous. In it they embraced these marine deposits of Devonshire, the Old Red Sandstone, and correlative formations elsewhere.

It was eventually found that in Devonshire the system is 10,000 to 12,000 feet thick and consists of graywacke, slates, and limestone, associated with lavas and tuff. The region was an unfortunate one on which to base a system, for the beds are so disturbed by folding, faulting, and intrusions that the detailed succession is still not wholly known. Equivalent but less disturbed beds had already been described in the Rhine Valley in Germany and Belgium and these became the actual standard section of the system in Europe. A still finer section in New York State is the standard of reference for America.

Figure 11-2 Basins of the Old Red Sandstone. (After Joseph Barrell.)

PHYSICAL HISTORY OF NORTH AMERICA

During Early Devonian time nearly all of North America remained low and the seas were quite limited. The axis of the Appalachian trough was flooded from Gaspé to northern Mississippi (see Map V). A thin formation

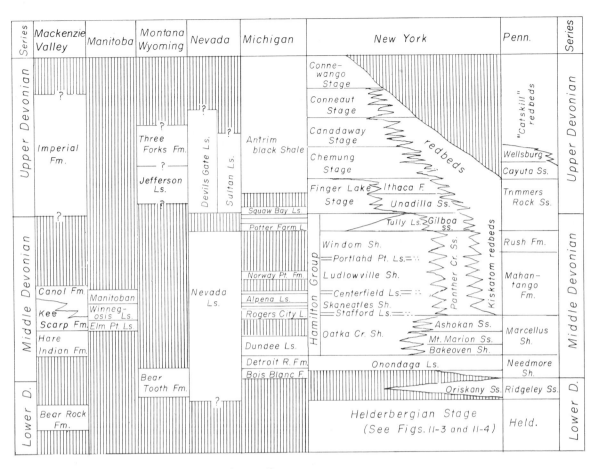

Figure 11-3 Correlation of important Devonian sections.

in Arkansas and eastern Oklahoma and tiny outliers near the center of Texas indicate a temporary incursion in this direction, and fossiliferous beds near Sault Ste. Marie indicate that a slender embayment briefly crossed the Great Lakes region. Otherwise the Lower Devonian deposits occupy a relatively narrow belt along the axis of the Appalachian trough and it is here that the full development is represented. Even here the nature of the deposits indicate that the adjacent land to the east was low. Recently, Lower Devonian deposits were identified in the northern part of the Mackenzie River Valley [1]. A few very local areas of marine Lower Devonian rocks are known in the far western United States, but they are now so isolated that their original connections are unknown.

Early in Middle Devonian time, however, submergence allowed the seas to spread as indicated in Map V,B. The broad western seaway covered the Cordilleran trough all the way from the Mackenzie delta to southern California and in the more darkly shaded areas the Middle Devonian Series attains an impressive thickness. The wide embayment over eastern Arizona, New Mexico, and Colorado and that which crossed the northern Great Plains lay upon the craton and in these areas only part of the series is represented and the deposits are relatively thin.

In the eastern United States a wide, shallow sea spread over the craton from the geosyncline to the position of modern Lake Michigan and for a time reached north to James Bay. This pattern of lands and seas was maintained until late in Devonian time over much of the area and constitutes one of the great submergences experienced by North America. Conditions were different in the northern part of the Appalachian region. Here crustal movements again brought about uplift in the northern lands of Appalachia from the New England states to the Maritime Provinces of Canada, and from this growing highland an immense volume of detrital sediment was transported westward into the geosyncline to form the great Catskill delta and related detrital deposits to the north. This was the **Acadian orogeny,** named in the Maritime Provinces of Canada where it was first recognized. Its movements increased in intensity to the close of the period and ultimately affected most of Appalachia north of the latitude of Cape Hatteras. From New England north it deformed Devonian and older formations and was accompanied by much igneous intrusion. Its effect was to destroy the northern end of the geosyncline so that the seas never again traversed it.

THE STRATIGRAPHIC RECORD

Lower Devonian. From New York southward the geosyncline subsided slowly and received but little sediment from the bordering lowlands. As a result the Lower Devonian formations are generally not over 300 feet thick, and they display with exceptional clarity the influence of local environments in the sea and in the bordering land. Thus they will be discussed in detail, for they illustrate an important stratigraphical principle [11].

To an exceptional degree these Lower Devonian formations can be studied in three dimensions. They are exposed in the Appalachian folds which parallel the axis of the geosyncline (Fig. 11-4) and along the south side of the Mohawk Valley which cuts directly across the geosyncline (Fig. 11-5) where they hold up the Helderberg Escarpment (Fig. 11-6). This is the type region of the Helderbergian Stage, the lower of two major divisions of the Lower Devonian. The upper unit—the Deer Park Stage—is represented in Fig. 11-5 only by the thin Oriskany Sandstone, which is not shown; it is much thicker in Fig. 11-4 where it is represented by the Ridgeley Sandstone and its equivalents.

In a shallow sea such as this, local bottom environments and the influence of varied conditions in the surrounding lands produce a bewildering complex of facies changes, and formational boundaries generally cross time

Figure 11-4 Section of the Lower Devonian formations along the axis of the Appalachian trough from Catskill, New York, to Clifton Forge, Virginia. Bedding planes mark time lines. Vertical scale greatly exaggerated. (Adapted from L. V. Rickard [11].)

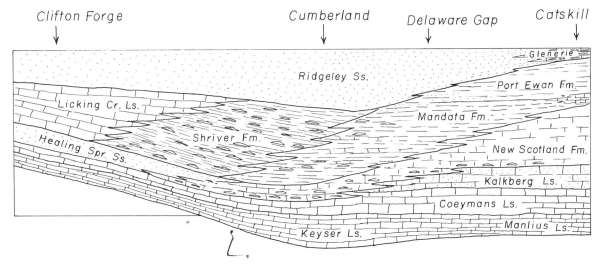

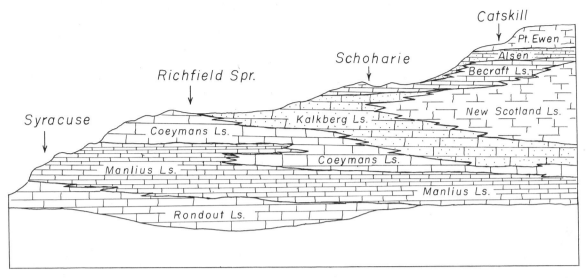

Figure 11-5 Section of the Lower Devonian formation along the Mohawk Valley from Catskill to Syracuse, New York. Time lines run horizontally. Vertical scale greatly exaggerated. (Adapted from L. V. Rickard [11].)

planes. The many local formations in our sections are named only to make this point clear.

A few of them deserve further comment. The Manlius Limestone is rather thin-bedded and shows almost no contribution of detrital sediment from Appalachia; but in New York State it includes two units called "water-limes," which tell an interesting story. Each is an impure limestone carrying an appreciable percentage of very fine siliceous silt, which can only be windblown dust carried eastward from the old Silurian desert floor in central New York. In early colonial days it was discovered that if this rock was burned in a lime kiln and then ground to fine powder, the resulting lime would set to form cement when water was added (whence the name water-lime). Until after 1824 this natural cement rock was our only source of cement and a great industry grew up about Rosendale on the Hudson River south of Catskill, New York. Here the Devonian beds are folded and the water-lime beds were quarried to depths up to 300 feet and some 300 miles of "drifts" were quarried out under the region.

Then in 1824 the Portland cement process was invented. It involves the mixing of the proper ratio of siliceous shale and pure lime-

Figure 11-6 Face of the Helderbergian Escarpment at Indian Ladder State Park overlooking the Hudson Valley. At the base is the Ordovician Shale. It is overlain by the thin bedded Manlius Limestone and that in turn by the massive Coeymans Limestone. The New Scotland Mudstone forms a gentle slope above this outcrop. (Carl O. Dunbar.)

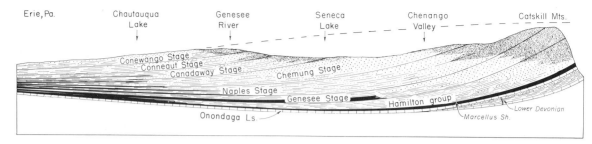

Figure 11-7 Cross section of the Catskill delta from the Hudson Valley to Erie, Pennsylvania. The broken line at the top indicates the probable position of the surface of the delta at the close of Devonian time. The missing part has been removed chiefly during the modern erosion cycle by the Susquehanna River System. Section about 250 miles long. Vertical scale greatly exaggerated. (Data from Chadwick and Cooper.)

Figure 11-8 Onondaga Limestone in a quarry south of Stafford Springs, New York. This pure limestone is extensively quarried for the manufacture of cement and for crushed stone for construction and road building. (Carl O. Dunbar.)

stone and could be used wherever they are available. With this development the industry at Rosendale collapsed and many of the shallow drifts are now used as mushroom cellars.

The New Scotland Limestone, which appears in both our sections, is really a dark, calcareous mudstone and reflects the contribution of mud from Appalachia, but it grades westward into purer limestones. The Healing Springs Sandstone clearly shows the contribution of a stream system debouching in Virginia while limestone was accumulating farther north. And the Ridgeley Sandstone must represent the contribution of a much larger stream late in Early Devonian time. Study of heavy mineral suites has shown that

the Ridgeley Sandstone came from Appalachia, whereas the Oriskany Sandstone along the Mohawk Valley came from the Adirondack region far to the north. The Ridgeley Sandstone is extensively quarried as glass sand.

Middle and Upper Devonian. Because the Middle and Upper Devonian series are so much a part of one continuing evolution it will be convenient to treat them together on a regional basis.

Figure 11-7 is a restored section of the enormous Catskill delta in which the entire Devonian System is represented. The Onondaga Limestone (Fig. 11-8) forms a continuous sheet under the geosyncline and far out onto the craton, remaining about 100 feet thick from the Hudson Valley to Erie, Penn-

Figure 11-9 Channel sand in the midst of the redbeds near Parksville, New York, showing cut and fill at the base. (C. O. Dunbar.)

sylvania, and then thinning gradually to the Cincinnati arch. It forms a clearly defined base to the Middle Devonian Series and proves that at this time Appalachia was still very low.

Then Appalachia began to rise, responding to early movements of the Acadian orogeny, and a sheet of black mud spread far across the geosyncline. This is the Marcellus Shale at the base of the Hamilton Stage in which four distinct facies were forming at the same time: (1) over the landward surface of the young delta, redbeds were deposited; (2) west of the shoreline, well sorted gray sands formed extensive sand flats on which brachiopods and clams lived in great abundance; (3) farther west and in somewhat deeper water the silt and mud winnowed out of the sand flats accumulated to form gray flaggy siltstones and shale; (4) still farther west only the finest mud

and organic debris accumulated to form black shale. These distinct facies are deeply intertongued and thus their equivalence can be proved. The silty bottoms must have been inhospitable to most benthonic organisms and in large part they are sparsely fossiliferous. The black mud bottoms were unfit for bottom-dwelling animals, but swimming creatures such as fish and pteropods lived above the bottom and, with these, tiny clams that lived attached to seaweeds make up a distinctive fauna.

As the delta grew, each of these four facies shifted progressively farther west rising in the section while individual time-stratigraphic units dip down to the west. The pioneer geologists of New York State were confused by this relation and mistook each facies zone as a time-stratigraphic unit. They believed, for

Figure 11-10 Cross-bedding in channel sandstone at the Neversink Dam south of the Catskill Mountains. (C. O. Dunbar.)

Figure 11-11 Fossiliferous Chemung Sandstone. (Yale Peabody Museum.)

example, that the redbeds exposed in the face of the Catskill Mountains were younger than the Chemung Sandstone of the center of the state. It was not until the field work of Chadwick [3] and of Cooper [4] in the 1930s that the true relations were understood.

In the redbed facies many lenses of gray sandstone represent the channel sands of old streams. The bottom contact of such lenses is commonly irregular as a result of cut and fill by the stream (Fig. 11-9) and, in some of the sands, striking cross-bedding records shifting sand bars (Fig. 11-10).

The gray, well-bedded sandstones represent the deposits of wide sand flats from which the mud was winnowed out and shifted farther west. Fossils are extremely abundant in these beds and in places were washed together in windrows before burial (Fig. 11-11). In scattered places a single bed of the sandstone has been folded and faulted for a distance of several yards, although the beds above and below are completely undisturbed (Fig. 11-12). The pioneer geologists were puzzled by such disturbed beds and called them "storm rollers," imagining that heavy storms had churned up the bottom at time of deposition. In view of the fact that the sand was then loose, such an

Figure 11-12 Flow rolls in the Chemung Sandstone at Chemung Narrows, New York. (C. O. Dunbar.)

idea now seems ridiculous. These are simply miniature landslides where the sand had been built up to an unstable position and are now termed "flow rolls."

The silty shales and flagstones (Fig. 11-13) were deposited in somewhat deeper water and must have formed an inhospitable surface for benthonic animals, since in general they are strangely barren of fossils. But in the Hamilton Stage a wide area of calcareous shales was forming in central New York and these are richly fossiliferous and include abundant corals.

The black shales (Fig. 11-14) carry a specialized fauna of tiny shells of animals that swam or were attached to seaweeds. This fauna changed little throughout Devonian time. The most abundant of these fossils is *Styliolina*, a delicate slender cone, apparently a pteropod similar to the modern *Clio*, which in the Genesee Shale accumulated by the millions to cover large areas on the sea floor.

The Devonian beds are deeply buried under the Allegheny Plateau and when they come to the surface along the Cincinnati arch the Onondaga Limestone is much thinner and passes under local names. It bears numerous coral reefs in the outcrops across New York State but the biggest one known is exposed on the Cincinnati arch near Louisville where for about three-quarters of a mile the Ohio River flows over rapids formed by a reef. This locality is commonly known as the "Falls of the Ohio." Here the corals weather free as siliceous pseudomorphs and immense collections gathered since colonial days have provided the types of almost 200 coral species [14]. In this region the rest of the Devonian is black shale.

The Michigan Basin underwent a rather independent and interesting development, being separated from the Appalachian region by a low barrier that checked the westward movement of detrital sediment from Appalachia.

Figure 11-13 Flaggy siltstones of the Naples Stage in the gorge of Genesee River near Mount Morris, New York. (Courtesy of New York State Museum.)

Figure 11-14 Genesee Shale in Taughanock State Park, about 9 miles north of Ithaca, New York. When exposed it weathers into paper-thin sheets and bleaches to a dark gray color. (Carl O. Dunbar.)

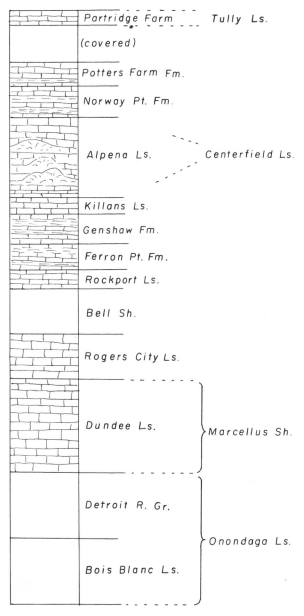

Here only the Middle Devonian is present. A vertical section is represented in Fig. 11-15 and a cross section of the Onondagan equivalents is shown in Fig. 11-16. The latter is based upon deep-well logs and shows a quite remarkable development that was not understood until 1930 [4]. These beds crop out in places around the margins of the basin but are deeply buried by younger formations across its center. In outcrops near Detroit the beds below the Dundee Limestone are only a few feet thick and consist of cross-bedded sandstone at the base overlain by dolostones of the Detroit River Group that in part carry an abundant fauna of poorly preserved corals; however, many deep wells have shown that the Sylvania Sandstone thickens basinward and intertongues with the Bois Blanc Limestone, which is as much as 400 feet thick in the center of the basin and carries a distinctive Onondaga fauna. The Detroit River Group also thickens and intertongues with an enormous body of salt and anhydrite. The overlying Dundee Limestone is about 140 feet thick and lithologically resembles the Onondaga Limestone of New York. Until the faunas were critically studied, it was supposed to be Onondagan and the underlying formations were referred to the Silurian. Only the fossils could prove that the Dundee correlates with the Marcellus Shale of New York and that the two thick formations below correlate with the

Figure 11-15 Vertical section of the Middle Devonian formations of the Michigan Basin. (After G. Arthur Cooper and Scott Warthin.)

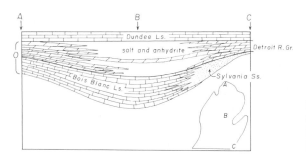

Figure 11-16 Cross section of the Onondagan equivalents and the Dundee limestone in the Michigan Basin. The inset map shows the positions of the section. (Data from Landes et al. [10].)

Onondaga Limestone. This illustrates another important principle; namely, the thickness of a formation is determined chiefly by the rate of subsidence and the supply of sediment and has little relation to the time span it represents. The Onondaga Limestone is only about 100 feet thick all across New York State but in the rapidly subsiding Michigan Basin, equivalent deposits are more than 10 times as thick. The saline deposits of the Detroit River Group are now believed not to indicate severe aridity during Onondagan time but to have been leached out of the Silurian beds uplifted around the rim of the basin and drained into the Devonian basin during Onondagan time, to be reprecipitated.

The formations above the Dundee Limestone, well exposed about Alpena on the east side of Michigan, are normal limestones and calcareous shales, and are extremely fossiliferous; but only two horizons carry faunas that can be directly correlated with the New York section. The Partridge Point Limestone is equivalent to the Tully Limestone, which is at the very top of the Hamilton Stage in New York, and the Alpena Limestone carries distinctive elements of the fauna of the Centerfield Limestone which is near the center of the Hamilton Stage. The Alpena Limestone is extensively quarried and shipped down to the steel mills in Ohio and Pennsylvania to use as flux in the iron smelters, and in these quarries numerous coral reefs are exposed (Fig. 11-17).

In the great Cordilleran seaway Middle and Upper Devonian formations were thick, but over large areas they are now covered by younger formations and over much of the region post-Paleozoic structural deformation has left them exposed in widely separated

Figure 11-17 Small coral reef in the Alpena limestone in quarries near Alpena, Michigan. (Preston E. Cloud.)

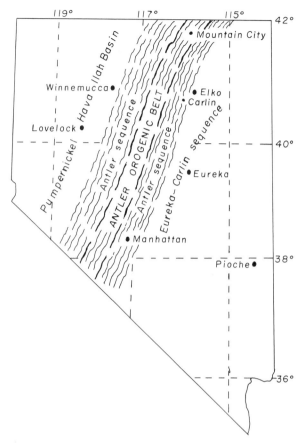

Figure 11-18 The Antlers Orogenic Belt (after Ralph Roberts). Until after Devonian time this was the western margin of the stable craton. To the east of it the Devonian sea was shallow and clear and the deposits are essentially calcareous. To the west of it lay a broad basin of much deeper water in which sediments derived from the west accumulated to great thickness. These deposits are generally detrital, are sparingly fossiliferous, and include much detrital volcanic material.

areas. Much information is available, but the formations are still under study and in Canada particularly drastic revisions in the classification and naming of stratigraphic units are still being made. It is therefore impossible to make a brief synthesis of the stratigraphy as was done for the east.

In eastern Nevada 4000 feet of thick bedded limestone is about equally divided between Middle and Upper Devonian, and in the Gold Hill district of Utah the section is also carbonate and is about 3000 feet thick, but here it consists of dolostones. In Nevada the Ant-

lers orogenic belt (Fig. 11-18) separates two totally different facies [12]. To the west of this zone, extremely thick but generally unfossiliferous detrital deposits had formed, probably ranging in age from Cambrian to Late Devonian. In Pennsylvanian time this material was greatly deformed and was thrust up onto the carbonate rocks of the craton. In the eastern arm of this sea in eastern Arizona and western New Mexico the Devonian section is relatively thin and is largely shaly.

In Middle Devonian time a large basin occupied much of Alberta, southern Saskatchewan, and southwestern Manitoba [13], but the Devonian rocks are covered by younger formations, except along the eastern margin of the basin where they were thinning toward the Canadian Shield and in the front range of the Canadian Rockies where they form towering mountain blocks (Fig. 11-19). Deep wells in the oil fields of the plains have penetrated the Devonian and showed that it reaches a maximum thickness of at least 1000 feet and that complex facies changes are involved with local coral reefs and extensive saline deposits.

In the Mackenzie Valley the Middle Devonian includes numerous limestones that crop out along the river and in isolated uplifts, but the Upper Devonian is mostly weak detrials that form low country.

Throughout the Cordilleran region there are many faunal zones of local importance but two key faunal zones range from the Mackenzie Valley to Utah and Nevada. The lower of these, in the Middle Devonian is characterized by the large "owl-head" brachiopod, *Stringocephalus* (Fig. 11-20). The second zone high in the Upper Devonian is characterized by the fat little spirifer, *Theodossia* (Fig. 11-21). Both of these genera are common in northern Europe and they almost certainly reached North America by way of the Arctic. Twice during the Devonian a slender arm of the western sea reached across the northern part of the Mississippi Valley to make temporary contact with the eastern seaway. Part of the *Stringocephalus* fauna appeared in the Rogers City Limestone in the Michigan Basin and recently *Stringocephalus* was discovered

Figure 11-19 Mount Devon in the Canadian Rockies about 20 miles north of Lake Louise, exposing about 2000 feet of Devonian limestone. (Charles D. Walcott.)

Figure 11-20 *Stringocephalus,* the "owl-head" brachiopod, distinctive guide fossil to a zone in the Middle Devonian of the Cordilleran sea.

Figure 11-21 *Theodossia,* a fat spirifer, guide fossil in a faunal zone in the Upper Devonian of the Cordilleran sea.

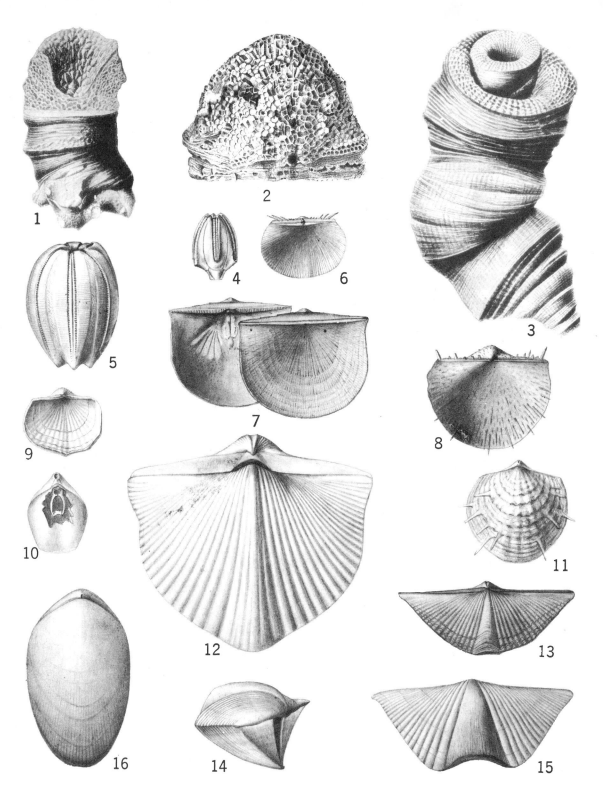

Plate 7   Devonian Corals (1–3), Blastoids (4, 5), and Brachiopods (6–16).

Figure 1, *Cystiphyllum vesiculosum;* 2, *Favosites conicus;* 3, *Heliophyllum halli;* 4, *Pentremitidea filosa;* 5, *Nucleocrinus verneuili;* 6, *?Chonetes coronatus;* 7, *Stropheodonta demissa* (dorsal view of shell and interior of pedicle valve); 8, *Productella callawayensis;* 9, *Tropidoleptus carinatus;* 10, *Cranaena sullivanti;* 11, *Atrypa rockfordensis;* 12, *Costispirifer arenosus;* 13, *Mucrospirifer mucronatus;* 14, 15, *Spinocyrtia? mesistrialis* (oblique view with ventral beak down, and dorsal view); 16, *Rensselaeria elongata.* All natural size. (Drawn by L. S. Douglass.)

244

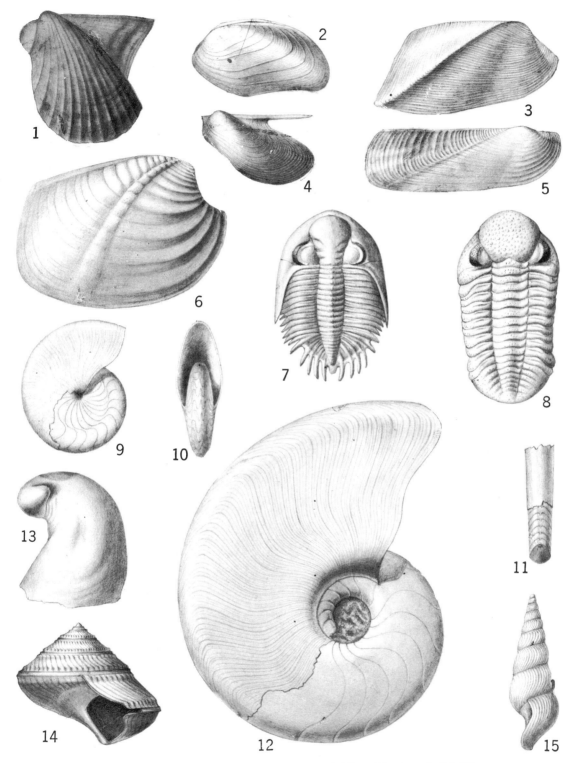

Plate 8 Devonian Pelecypods (1–6), Trilobites (7, 8), Cephalopods (9–12), and Gastropods (13–15).

Figure 1, *Cornellites flabellus;* 2, *Nyassa arguta;* 3, *Goniophora hamiltonensis;* 4, *Leptodesma longispinum;* 5, *Orthonota undulata;* 6, *Grammysia bisulcata;* 7, *Greenops (Greenops) boothi;* 8, *Phacops rana;* 9, 10, *Tornoceras uniangulare;* 11. *Bactrites arkonensis* (fragment); 12, *Agoniatites vanuxemi;* 13, *Platyceras reflexum;* 14, *Bembexia sulcomarginata;* 15, *Loxonema hamiltoniae.* All natural size. (Drawn by L. S. Douglass.)

locally in Indiana [5]. The *Theodossia* fauna is abundant in northern Iowa and appeared in central New York in the High Point Sandstone above Naples. But in neither case did the western faunas survive more than a very brief time in the eastern sea.

Special interest attaches to the Devonian section of Central East Greenland, which probably exceeds 2000 feet in thickness, consists almost entirely of detrital formations, and carries a rich fauna of fresh-water fishes closely allied to those in the Old Red of Scotland, and along with these the remains of the oldest land vertebrates (see page 252).

Igneous Activity in New England and Maritime Canada

One of the dramatic features of the Paleozoic history of North America was the widespread and oft-recurring igneous activity in New England and the Maritime Province of Canada. It began at least as early as Mid-Ordovician time, was impressive during Silurian time, and continued intermittently to the close of the Paleozoic Era. It may have reached a crescendo in the Devonian Period during the Acadian orogeny.

Great thicknesses of bedded lavas and tuffs in southern Quebec, Gaspé, New Brunswick, and Maine bear witness to active volcanoes. In much of New England and parts of New Brunswick such extrusives were later eroded away, exposing the related deep-seated plutonic rocks. The granite core of the White Mountains is one example. Other Devonian batholiths are represented by the granites at St. George and in the Little Megantic Mountains of New Brunswick, the granites that make up most of Nova Scotia, and those that form the core of Mt. Katahdin in Maine. Mount Royal, from which the city of Montreal takes its name,

Figure 11-22 Reconstruction of a bit of a Devonian coral reef showing a crinoid (a), assorted seaweeds (b), assorted corals (c), sponges (d), and a snail (e) about $\frac{1}{10}$ actual size. (A diorama in the Buffalo Museum of Natural History.)

Figure 11-23 Life of a normal Devonian sea floor. A crinoid at the left, seaweeds at the right, two trilobites in the left foreground, one of which is being attacked by a frilled cephalopod. (A diorama in the New York State Museum in Albany.)

is made of basic igneous rocks, as are the Monteregian Hills of southern Quebec, which represent old volcanic necks believed to be Devonian in age. Special interest attaches to volcanics in Lower Devonian beds (that is, New Scotland Formation of the Helderberg Escarpment) near Presque Isle, Maine, in which the rubidium/strontium ratio gives a radiometric date of 413 ± 5 million years. This places the Silurian-Devonian boundary near 400,000,000 years B.P. [2].

A widespread layer of volcanic ash near the top of the Onondaga Limestone in Pennsylvania, Ohio, and West Virginia may have been windblown from a volcano far to the east in Appalachia [7].

THE LIFE OF DEVONIAN TIME

Progress among the Marine Invertebrates. The marine invertebrates were now at the flood tide of their Paleozoic evolution and the shallow seas swarmed with animals of many kinds (Plates 7 and 8, Figs. 11-22 and 11-23). In the warm, shallow seas that spread so widely over the craton three great tribes of corals conspired to build reefs. These were the **Rugosa** (also known as horn corals or tetracorals), the **Tabulata** (especially the favosites or honeycomb corals) and the **Stromatoporoids** (of the class *Hydroidea*). Neither tribe would ever be so prolific again. The largest of all the horn corals, *Siphonophrentis gigantea*, made solitary coralla as much as 2 feet high and 3 inches across, while compound colonies formed "coral heads" as much as 8 feet across, and stromatoporoids formed even greater colonies.

Brachiopods also reached a peak of diversity at this time, no fewer than 700 species being known in North America alone. Of these the great tribe of the spirifers was especially prolific and was represented by many genera (Plate 7, Figs. 12 to 15). The first appearance of another great tribe, the spiny brachiopods

247

of the Superfamily Productacea (commonly known as "productids") should also be noted. They were still relatively few in Devonian time but would dominate the brachiopod faunas throughout the rest of the Paleozoic Era.

Of the mollusks, **pelecypods** were now more abundant than ever before, and since they prefer muddy or sandy bottoms they are chiefly found in the shales and sandstones of the Catskill delta. **Gastropods** are locally common but hardly noteworthy. **Cephalopods,** on the contrary, made a very significant advance in the evolution of the first of the ammonites, which in the Mesozoic Era would become the dominant "shellfish" and would supply the chief guide fossils for correlation. They were still relatively few in Devonian time but serve to mark several distinctive faunal zones in the Upper Devonian.

Trilobites and **eurypterids** were both on a decline that would lead to their extinction near the end of the Paleozoic Era, yet each produced a relative giant species in the Devonian. *Dalmanites,* one of the trilobite genera, was about 29 inches long and was probably the record for all time.

Among the **echinoderms,** starfishes were locally abundant in the sandy beds of the Catskill Delta and one locality near Mount Marion, New York, is of special interest. Here John M. Clarke discovered a bedding plane covered with some 400 individual starfish, some of which were buried while in the act of devour-ing clams, just as they now do on modern oyster beds. Figure 11-22 shows a reconstruction of the life of a Devonian coral reef and Fig. 11-23 shows part of a sea floor.

Lungfishes Learn to Breathe Air. Fishes underwent a great evolution during Devonian time. Several major tribes were abundant, some in the seas and others in the rivers. Sharks were common in the seas but since their skeletons are cartilaginous they are represented chiefly by spines and teeth. A distantly related group, the **Arthrodires,** had developed a cuirass of bony plates over the front part of the body and these are well preserved. One species of this tribe reached a length of 20 feet and was the largest animal in the world at this time (Fig. 11-24).

Far greater interest, however, centers in a new order of fishes, the **Choanichthys** or lungfishes. The name refers to a feature not found in any other group of fishes—a pair of openings in the roof of the mouth that permitted connection of external nostrils with the throat to permit breathing air [Gr. *Choana,* internal nostril + *ichthyos,* fish]. Fortunately, a few of this tribe have survived to the present (Fig. 11-25), and through these we know that a pair of pouches had developed, one on each side of the esophagus into which air could be drawn through the nostrils, and these pouches were covered by a plexus of blood vessels by means of which oxygen could be absorbed and carbon dioxide expelled from the blood. This, indeed, was the beginning

Figure 11-24 Model of the giant arthrodire *Dinichthys.* Actual length about 20 feet. This was the biggest animal in the world in Devonian time. (Courtesy of the American Museum of Natural History.)

Figure 11-25 The living African lungfish *Protopterus.* Left, the fish in its mud "cocoon" as it was shipped from Africa to Chicago in an open tin can. Right, the same after it had been placed in an aquarium where the fish returned to its normal aquatic life. The fish was in transit for more than 6 months, during which time it lived in its cocoon breathing air exclusively. (From Turtox News, courtesy of The General Biological Supply House, Inc.)

Figure 11-26 *Eusthenopteron,* a crossopterygian lungfish from Late Devonian shales near Escuminac, Quebec. (A model by George G. Simpson.)

of the evolution of lungs.

In Devonian time the Choanichthys had diverged into two major tribes, the **Dipnoans** and the **Crossopterygians.** The former, which includes the living lungfish, had weak fins, whereas the latter, which probably were essentially bottom dwellers, had developed stout, fleshy basal lobes in the fins which included a jointed cartilaginous skeleton that foreshadowed the skeleton of the five-fingered limbs of the tetrapods (four-footed land animals). One of the most significant of this group was *Eusthenopteron,* remains of which are abundant and exceptionally well preserved in Late Devonian shale beds near Escuminac, Quebec (Fig. 11-26). W. K. Gregory has compared one of these fins with the fore limb of a primitive amphibian (Fig. 11-27) to show how the latter may have evolved. His reconstruction of the transition from crossopterygians to amphibians is shown in Fig. 11-28.

Vertebrates Discover the Land. Gregory's insight was brilliantly confirmed in 1948 when a specimen was found in the Late Devonian of central East Greenland that had the contour of a crossopterygian fish with the long tail covered with scales, but had one tetrapod limb attached to the skeleton. The animal was described as *Ichthyostega*. Since then many specimens have been found, one of which is virtually complete (Fig. 11-29) and shows an animal, half fish and half amphibian [9]. It is the oldest land vertebrate, and marks the beginning of a primitive stock of amphibians known as **labyrinthodonts**, so-called because the enamel of the teeth was infolded so that in cross section it forms a labyrinthine pattern (Fig. 11-30). This peculiarity is shared by both the crossopterygian

fishes and all the primitive amphibians. They also share another common structure, the "lateral line system," a slender groove running along each side of the body which housed some sort of sensory organ. The pattern of dermal bones that covered the skull in both groups of animals was also strikingly similar.

It is now quite clear how the tetrapods evolved from the air-breathing fish and why the first ones were Amphibia. Members of that great class, which includes salamanders, toads, and frogs, are still incompletely adapted for life on land. They return to the water to spawn and lay small eggs like those of fishes, which hatch into tadpoles. The tadpoles breathe by means of gills and are essentially fishlike until partly grown, when legs bud out from their sides, lungs develop from their throat, and the gills are resorbed. Then they leave the water and breathe air. But they still are not quite fully adapted to land life; they must return to the water to spawn, and most of them remain in moist places and spend part

Figure 11-27 Comparison of the skeleton of the fin of *Eusthenopteron* with that of a very primitive land animal. Left, unretouched photograph of the left front fin of *Eusthenopteron;* center, diagram of the skeletal elements of the same; right, skeleton of the corresponding limb of a late Paleozoic amphibian *Eryops.* (Photo by W. L. Bryant; diagrams after W. K. Gregory.)

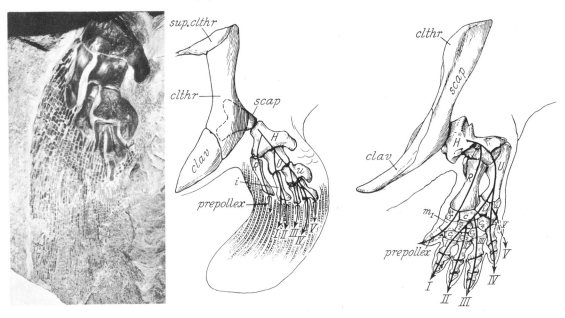

Figure 11-28 W. K. Gregory's reconstruction of the emergence of *Eusthenopteron* leaving the water to flounder about on its strong muscular fins (left), compared with the early labyrinthodont amphibian *Diplovertebron* from the lower Carboniferous of Bohemia. (Courtesy of the American Museum of Natural History.)

of the time in the water. Comparative anatomy and ontogeny both indicate that they evolved from fishes, and the geologic record indicates rather clearly when and under what circumstances that change took place.

The "Old Red" type of Devonian formations accumulated over basin floors where the rainfall was seasonal. Such conditions persisted throughout much of Devonian time in eastern North America and western Europe, and here for millions of years the fish living in the streams and evanescent lakes had to endure annual seasons of drought. Again and again the shrinking waterholes brought death and de-

struction—but always there were some that did not go dry, and there the survivors were crowded in stagnant water, starving for oxygen. A great premium was thus placed on ability to gulp air and to use the oxygen straight. Fishes with swim-bladders that could function as lungs were thus at a great advantage and were stimulated to ever greater activity as the oxygen in the water was depleted. Those with stout fins like *Eusthenopteron* could forsake their pools in the cool of the night and flounder about the banks in short forays for food or could migrate overland to other pools. Once the lungs had reached a certain efficiency and

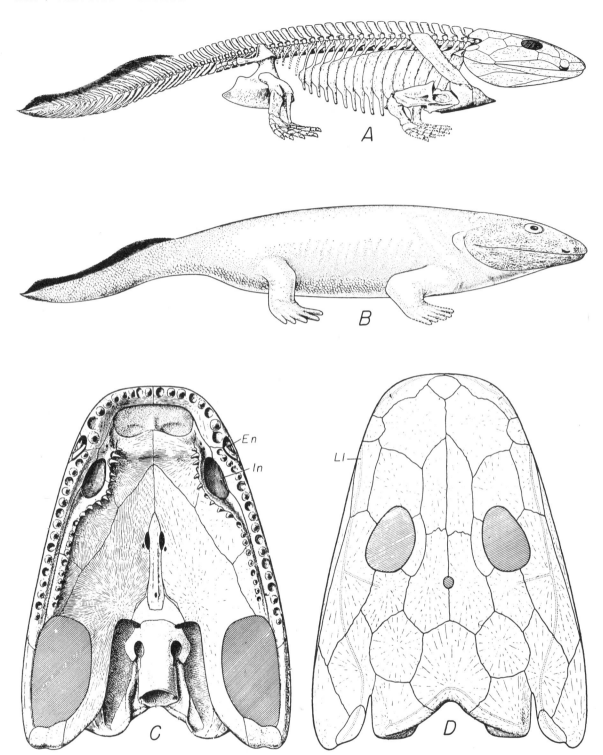

Figure 11-29 *Ichthyostega,* the oldest known tetrapod. Late Devonian beds of Central East Greenland. A and B, restoration of skeleton and of complete animal; C and D, ventral and dorsal views of the skull showing external and internal nares and lateral line. (After Jarvik, courtesy of Erik Jarvik and Lauge Koch.)

the fins had been modified into stubby limbs, land vertebrates had arrived!

The environment, of course, did not produce these modifications; it had simply selected ruthlessly those random variations that appeared from time to time and were better equipped to survive under such exacting conditions. It is almost certain, as Romer has stated, that these Late Devonian animals, as well as their Mississippian descendants, lived normally in the water and were not, strictly speaking, land animals, but they had the ability to breathe air and to leave the water when necessary.

Land Plants and the Spread of Forests. Before Devonian time the evidence of land plants is very meager and concerns only small, herbaceous types; but from this time on the record is clear and impressive. Probably the most primitive type of land plants are those well known from the Lower Devonian beds at Rhynie, Scotland (Fig. 11-31). Here the preservation is excellent and several genera have been described, all belonging to an extinct phylum, the **Psilophytales,** so called because most still had not developed leaves [Gr. *psilo*, bare + *phyton*, plant]. A typical genus *Horneophyton* is illustrated in Fig. 11-31. It has a creeping stolon from which little branching stems rose to a height of a foot or so and bore spore cases at their tips. This is about the simplest possible expression of a primitive land plant. A somewhat taller plant, *Asteroxylon*, shows the beginning of leaves as small overlapping bracts. Such plants are found also in other parts of Europe and in the Lower Devonian of North America.

By Middle Devonian time a considerable

Figure 11-30 Side view and cross section of a labyrinthodont tooth (×6). Creases along the side of the tooth are caused by the infolding of the enamel layer. Position of the cross section indicated by the black line across the tooth. (Unretouched photographs. Yale Peabody Museum.)

Figure 11-31 Very primitive land plants from the Lower Devonian at Rhynie, Scotland. A. *Asteroxylon;* B, *Horneophyton.*

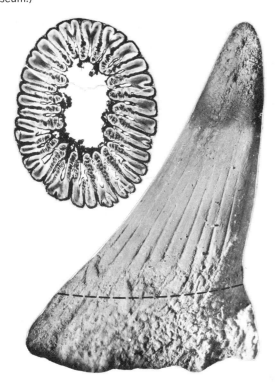

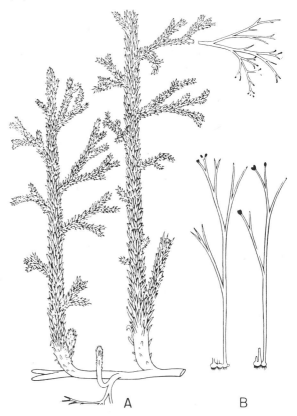

Figure 11-32 Stump of a giant tree-fern, *Eospermatopteris,* being hoisted out of its place in a quarry in a Middle Devonian sandstone bed at the Gilboa reservoir south of Schoharie, New York. A two-foot ruler lies in the hollow from which the stump is being lifted.

diversity of primitive trees had spread over the lowland of the Catskill delta. These were the forerunners of the flora so strikingly preserved in the Coal Measures of the Carboniferous (they will be discussed in more detail in that section). In quarrying sandstone with which to build the great dam for the Schoharie Reservoir on the north slope of the Cat- skills, three layers were found to include abundant stumps of trees still standing where they grew. Some of these trees had trunks more than 30 feet high and were more than 2 feet in diameter at the base (Fig. 11-32). This quarry supplied the material and the data used in the restoration shown in Fig. 11-33, which stands in the New York State Museum in Albany. This is the oldest known forest!

Western Europe and the "Old Red." While in North America the marginal land, Appalachia, had been rising periodically since Early Cambrian time and shedding vast amounts of detrital sediments westward into the geosyncline with westward thrusting during both the Taconian and the Acadian disturbances, western Europe, from southern Wales to Norway, was occupied by another geosynclinal belt in which thick sandstone formations grade from coarse in the west to fine and largely calcareous formations farther east. This is shown particularly well in Wales where Sedgwick was working in thick detrital and igneous formations in the west while Murchison was studying largely equivalent deposits farther east. Clearly the bulk of the detritals was coming from a western borderland that has now vanished. About the close of Silurian time this was the scene of the Caledonian orogeny that folded the rocks in

western Wales toward the east and caused an immense overthrust toward the east in Norway. This left chains of bold mountains running east of north across England and Scotland and as the mountains continued to rise in Devonian time five distinct intermont basins were filled with nonmarine deposits that the pioneer British geologists called "The Old Red." A large part of the Old Red is lithologically like the Catskill redbeds of the Catskill delta, but parts laid down over swampy basins are gray and have yielded large numbers of fresh-water fishes.

Here then is a striking situation—in eastern North America the margin of the continent was being repeatedly uplifted and thrust westward while the western margin of Europe was being uplifted and thrust eastward. Patrick Hurley [8] has indulged the intriguing speculation that, until the beginning of the Paleozoic Era, Europe and America had been close together, if not united, and now they

Figure 11-33 Reconstruction of the Middle Devonian forest at Gilboa, New York. e, a tree fern, *Eospermatopteris;* p, a primitive scale tree, *Protolepidendron.* The tree fern was originally interpreted as a seed fern, whence the name [Gr. *eo,* dawn + *sperma,* seed + *pteris,* fern] but later study showed that the supposed seeds were only spore capsules. [6] (A great diorama formerly in the New York State Museum at Albany.)

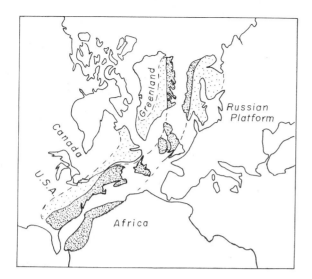

Figure 11-34 Map showing a tentative reconstruction of the relations of the continents bordering the North Atlantic Basin which was beginning to form as the continents drifted apart in Late Devonian time. Heavy shading marks the orogenically disturbed belts. (Adapted from Hurley.)

were being pushed apart by convection in the mantle, like that under the modern mid-ocean ridges, and that the Atlantic Ocean basin was coming into being or was being widened (Fig. 11-34). It is a stirring thought that can only be verified or disproved as we gather more information, especially from the distribution of the fossils and from paleomagnetism.

REFERENCES

1. Bassett, H. G., 1960, *Devonian stratigraphy, central Mackenzie River region, Northwest Territories, Canada.* In Raasch, G. O., Ed., *Geology of the Arctic,* Univ. Toronto Press, V. 1, pp. 481–498.
2. Bottino, M. L., and Fullagar, P. D., 1966, *Whole-rock rubidium-strontium age of Silurian-Devonian boundary in northeastern North America.* Geol. Soc. America Bull., v. 77, pp. 1167–1176.
3. Chadwick, G. H., 1935, *Summary of Upper Devonian stratigraphy.* Am. Midland Naturalist, v. 16, pp. 857–862.
4. Cooper, G. A., 1930, *Stratigraphy of the Hamilton Group of New York.* Am. Jour. Sci., v. 19, pp. 116–134; 214–236.
5. Cooper, G. A., and Phelan, Thomas, 1966, *Stringocephalus in the Devonian of Indiana.* Smithsonian Misc. Coll., v. 151, no. 1, pp. 1–20.
    ———, et al., 1942, *Correlation of the Devonian sedimentary formations of North America.* Geol. Soc. America Bull., v. 53, pp. 1774–1793.
6. Daugherty, L. S., personal communication.
7. Flowers, R. R., 1952, *Lower Middle Devonian metabentonite in West Virginia.* Am. Assoc. Petroleum Geologists Bull., v. 36, pp. 2036–2038.
8. Hurley, P. M., 1968, *The confirmation of continental drift.* Scientific American, v. 218, no. 4, pp. 52–64.
9. Jarvik, Erik, 1959, *De Tigida Fossilia Ryggradsdjuren* [The early fossil vertebrates]. Svensk Naturvelenskap., pp. 5–80.
10. Landes, K. K., Ehlers, G. M., and Stanley, G. M., 1945, *Geology of the Mackinac Straits region.* Mich. Geol. Survey Publ. 44, pp. 1–204.
11. Rickard, L. V., 1964, *Correlation of the Devonian Rocks in New York State.* New York State Museum and Science Service Geological Survey, Map and Chart Series: No. 4.
12. Roberts, Ralph, et al., 1958, *Paleozoic rocks of north central Nevada.* Am. Assoc. Petroleum Geologists Bull., v. 42, pp. 2813–2857.
13. Sandberg, C. A., 1961, *Distribution and thickness of Devonian rocks in Williston Basin and central Montana and north-central Wyoming.* U. S. Geol. Survey Bull. 1112-D, pp. 105–127.
14. Stumm, Erwin, 1964, *Silurian and Devonian corals of the Falls of the Ohio.* Geol. Soc. America Mem. 93, pp. 1–184.

Figure 12-1 Pennsylvanian landscape in the swampy lowlands of the eastern United States. Part of a great mural by Rudolph F. Zallinger in Yale Peabody Museum. Plants: 1, a tree fern; 2, *Lepidodendron,* a scale tree; 3, *Sigillaria,* a scale tree; 4, *Cordaites,* a precursor of the conifers; 5, *Calamites,* a scouring rush. Animals: 6, *Diplovertebron,* a primitive amphibian; 7, *Eryops,* a large amphibian; 8, *Eogyrinus,* an aquatic amphibian; 9, *Seymouria,* a primitive reptile; 10, *Limnoscelis,* a reptile; 11, *Varanosaurus,* a reptile; 12, *Meganeuron,* a giant dragonfly.

CHAPTER **12** *The Carboniferous Systems*

The coal-bearing rocks of Europe and eastern North America were among the first to attract the attention of geologists, and the fuels they contain have played no small part in the industrial development of the western nations.

As early as 1808 Belgian geologists referred to them as "the bituminous terraine," and in 1822 when Conybeare and Phillips published the first regional survey of the geology of the British Isles they coined the term "Carboniferous Order." The term **system** had not yet come into use, but after it was introduced by Murchison and Sedgwick the name "Carboniferous System" was adopted. At that time, however, it was made to include three major groups of rocks below the Coal Measures, as follows:

	Coal Measures
Carboniferous System	Millstone Grit
	Mountain Limestone
	Old Red Sandstone

After the "Old Red" was recognized as Devonian, the Coal Measures plus the Millstone Grit became the Upper Carboniferous Series and the Mountain Limestone the Lower Carboniferous Series.

In America the Coal Measures were readily recognized as Upper Carboniferous and the underlying formations, having virtually no coal, were appropriately called Subcarboniferous. In 1891 the U. S. Geological Survey recognized these as formal units of series rank and gave each a geographic name. Thus the Upper Carboniferous became the Pennsylvanian Series and the Subcarboniferous the Mississippian Series. In 1906 Chamberlin and Salisbury proposed to raise these to systematic rank, pointing out that in this country, as in Europe, they differ generally in lithology and over great areas are separated by a major hiatus. This lead is now universally followed by North American geologists, but it has not been followed in Europe and Asia where a threefold division is now preferred. Under the circumstances, although we recognize two systems, we have decided to treat both in a single chapter.

CRUSTAL UNREST

Throughout the Carboniferous periods the Appalachian geosyncline continued to subside as it was filled by detrital sediments carried west from the rising highlands in the marginal land of Appalachia; and its southern end may have swung west to join the old Ouachita geosyncline, which, with a great sigmoidal curve, extended across the southern tier of states into central Texas, then farther southwest into Trans Pecos Texas and eastern Mexico. Meanwhile the Ouachita trough was bordered on the south by Llanoria, a land of unknown extent that was rising into bold mountains and being pressed northward against the continent. About the beginning of Pennsylvanian time mountain ranges began to buckle up on the craton, which had been essentially stable since Precambrian time (Map VII, p. 126). Of these the Oklahoma Mountains ran north of west from western Arkansas across southern Oklahoma into the Panhandle of Texas. Their eroded stumps now form the Arbuckle Mountains, the Criner Hills, the Wichita Mountains, and the buried Amarillos of the Panhandle of Texas. Almost simultaneously the Colorado Mountains (sometimes called the Ancestral Rockies) formed a chain of ranges running north across central Colorado with one branch extending into south central Wyoming and another curving northwest into east central Utah while a separate range (the Zuni-Fort Defiance uplift) arose across the boundary between New Mexico and Arizona. The effect of this widespread orogeny on the Pennsylvanian stratigraphy was profound.

TWO CYCLES OF SUBMERGENCE

Out of an ever-changing panorama of shifting lands and seas we have selected five maps to illustrate the two major cycles of submergence in North America (Maps VI and VII). The earliest Mississippian seas were quite small, but by the middle of the period the Cordilleran Trough was widely flooded and most of the eastern half of the United States was covered by a vast shallow sea. This was the flood tide when vast sheets of limestone were forming all the way from the Cordilleran trough, across the region of the present Rocky Mountains, and as far east as Virginia. But during the latter half of the period the seas were much less extensive.

The earliest Pennsylvanian seas (Lower Morrowan time) were even more restricted (Map VII), but by the middle of the period they had spread over most of the United States, while they had disappeared from Canada. By this time the Oklahoma and the Colorado mountains formed extensive high islands, and

later in the period the seas were receding again.

THE STRATIGRAPHIC RECORD

Mississippian System. The type section of the Mississippian System is along the Mississippi River between southeastern Iowa, western Illinois, and western Kentucky and is shown in Fig. 12-3. At the base is a thin black shale and a little above it a local sandstone; but with these exceptions the lower half consists almost exclusively of limestones, all of which are richly fossiliferous. Three of these formations deserve special comment. The

Figure 12-2 Correlation of important Mississippian sections.

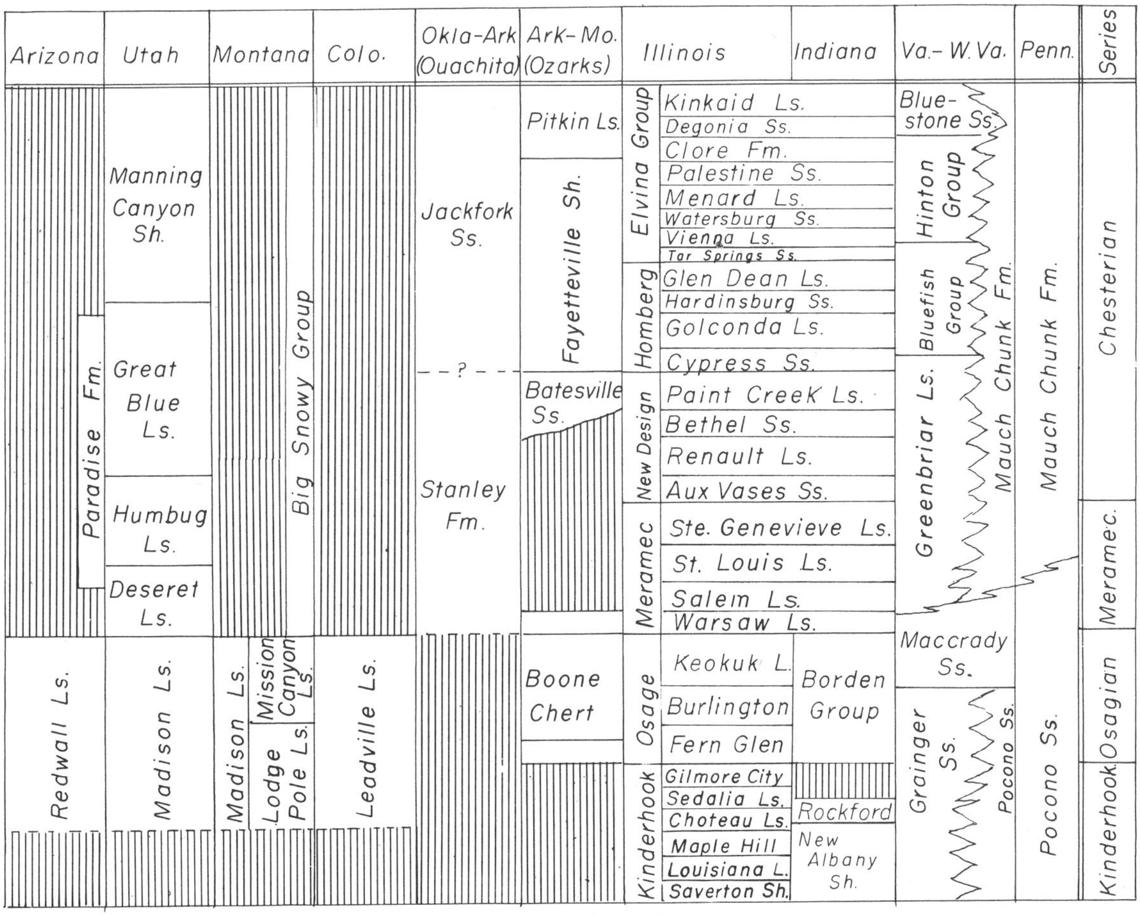

Arizona	Utah	Montana	Colo.	Okla-Ark (Ouachita)	Ark-Mo. (Ozarks)	Illinois	Indiana	Va.-W. Va.	Penn.	Series
					Pitkin Ls.	Kinkaid Ls. / Degonia Ss. / Clore Fm. / Palestine Ss. / Menard Ls. / Watersburg Ss. / Vienna Ls. / Tar Springs Ss. (Elvina Group)	Blue-stone Ss. (Hinton Group)		Mauch Chunk Fm.	Chesterian
	Manning Canyon Sh.			Jackfork Ss.	Fayetteville Sh.	Glen Dean Ls. / Hardinsburg Ss. / Golconda Ls. / Cypress Ss. (Homberg)	Bluefish Group	Mauch Chunk Fm.		
Paradise Fm.	Great Blue Ls.	Big Snowy Group		— ? —	Batesville Ss.	Paint Creek Ls. / Bethel Ss. / Renault Ls. / Aux Vases Ss. (New Design)	Greenbriar Ls.			
	Humbug Ls.			Stanley Fm.		Ste. Genevieve Ls. / St. Louis Ls. / Salem Ls. / Warsaw Ls. (Meramec)				Meramec.
	Deseret Ls.						Maccrady Ss.			
Redwall Ls.	Madison Ls.	Madison Ls. / Lodge Pole Ls. / Mission Canyon Ls.	Leadville Ls.	Boone Chert		Keokuk L. / Burlington / Fern Glen (Osage)	Borden Group	Grainger Ss.	Pocono Ss.	Osagian
						Gilmore City / Sedalia Ls. / Choteau Ls. / Maple Hill / Louisiana L. / Saverton Sh. (Kinderhook)	Rockford / New Albany Sh.			Kinderhook

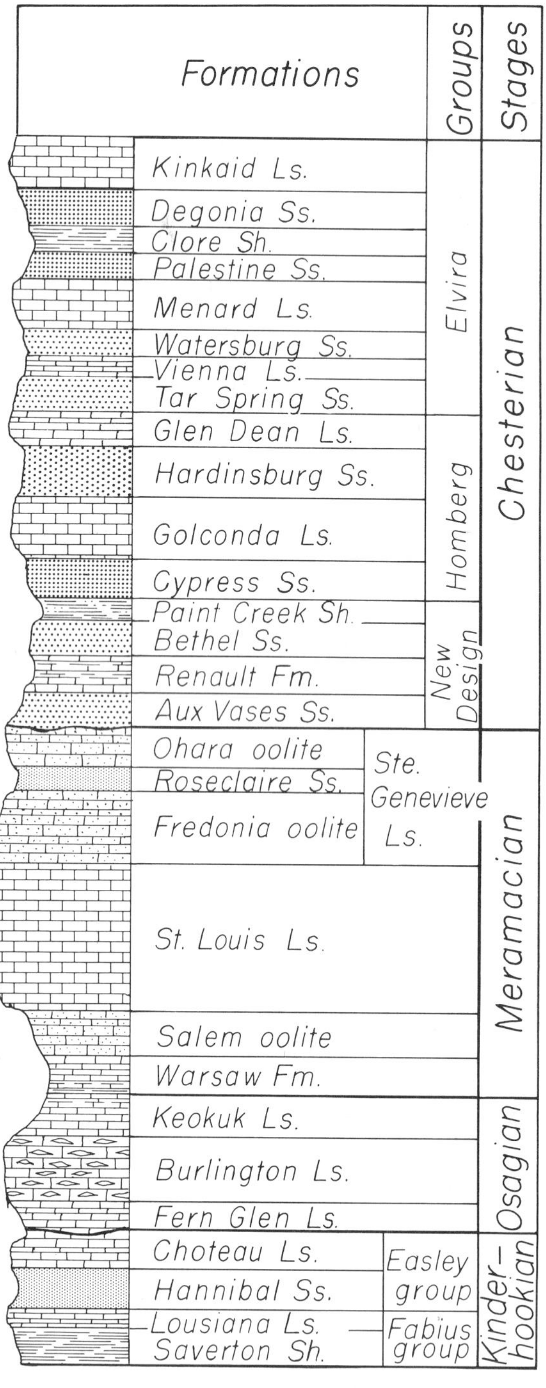

Formations		Groups	Stages
Kinkaid Ls.			
Degonia Ss.			
Clore Sh.			
Palestine Ss.		Elvira	
Menard Ls.			
Watersburg Ss.			
Vienna Ls.			Chesterian
Tar Spring Ss.			
Glen Dean Ls.			
Hardinsburg Ss.		Homberg	
Golconda Ls.			
Cypress Ss.			
Paint Creek Sh.		New Design	
Bethel Ss.			
Renault Fm.			
Aux Vases Ss.			
Ohara oolite	Ste. Genevieve Ls.		
Roseclaire Ss.			
Fredonia oolite			Meramacian
St. Louis Ls.			
Salem oolite			
Warsaw Fm.			
Keokuk Ls.			
Burlington Ls.			Osagian
Fern Glen Ls.			
Choteau Ls.	Easley group		
Hannibal Ss.			Kinder- hookian
Lousiana Ls.	Fabius group		
Saverton Sh.			

Figure 12-3 Type section of the Mississippian System in western Kentucky and southern Illinois.

Burlington Limestone is highly crinoidal and includes much chert. It passes southward into the Boone Chert of the Ozark region

where it is one of the largest deposits of siliceous rock in the continent. The Salem Oolite is extensively quarried in southern Indiana for use as trimmed stone in many of the public buildings in the eastern United States (Fig. 12-4). The St. Louis Limestone, about 400 feet thick in the river bluffs near St. Louis, is part of one of the most extensive sheets of limestone in North America.

The upper half of the section (Chesterian Series) is quite different, embracing alternate formations of limestone and of quartz sandstone, the latter representing tongues of detrital material now spread westward from the Appalachian region.

The Mississippian formations are deeply buried under the Allegheny Plateau but where they emerge in the folded belt they present a totally different picture, as represented in Fig. 12-5. In the anthracite coal fields of eastern Pennsylvania the entire system is detrital and nonmarine and is divided into two great units. The lower of these is the ridge-forming gray Pocono Sandstone, which in its lower part includes units of conglomerate made of well-rounded pebbles of quartz and old crystalline rocks, some the size of door knobs. The overlying Mauch Chunk Formation (pronounced maw chunk) is a complex of red sandstones, siltstones, and shales many layers of which display mudcracks. Land plants, mostly fragmentary, occur in the Pocono Formation, and one local coal bed has historical interest as the source of the fuel used by the **Merrimac** in her battle with the **Monitor** during the Civil War. This region was near the highlands of Appalachia and here the Pocono Sandstone is as much as 2000 feet thick and the Mauch Chunk about 3000 feet thick.

Farther down the strike of the Appalachians where subsidence was more rapid the section is nearly all marine and consists of alternating shales and sandstones. Toward the west these intertongue with more calcareous beds such as the Greenbriar Limestone of Fig. 12-5.

The Ouachita Mountains of central Arkansas and southeastern Oklahoma display a remarkable thickness of detrital deposits poured into the Ouachita trough from Llanoria (Fig. 12-6). This thickness cannot be accurately deter-

Figure 12-4 Quarry in the Salem Limestone near Bedford, Indiana. (Courtesy of Indiana Limestone Company.)

Figure 12-5 Stratigraphic diagram to illustrate complex facies relations in the Appalachian trough. (After Byron Cooper.)

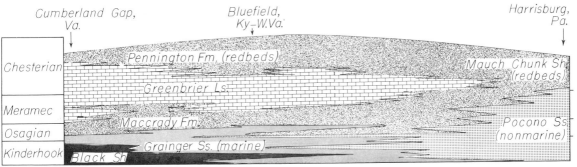

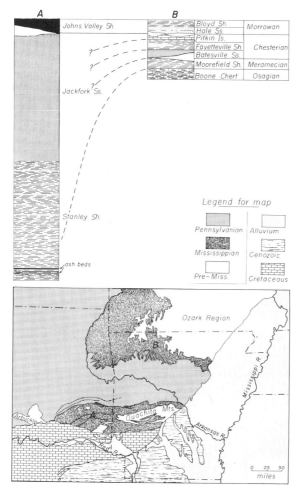

Figure 12-6 Map and columnar sections to illustrate Mississippian deposits of the Ouachita trough.

mined but has been estimated to be as much as 15,000 feet. Here the Stanley Shale consists of fissle black shale interbedded with bluish siltstone and in its basal part includes several beds of volcanic ash that range in thickness from 6 to 85 feet. The ash was certainly wind-blown from volcanoes in Llanoria. The overlying Jackfork Sandstone varies from coarse to fine and in color varies from light to dark gray. Both formations are intensely deformed for they were shoved northward in a series of large thrust plates in Permian time. They were clearly deposited much farther to the south and were overthrust upon younger rocks that had formed along the northern side of the

Ouachita trough. Fossils are sparse in both formations. The Stanley Shale has yielded radiolarians and conodonts and the Jackfork Sandstone includes macerated and poorly preserved fragments of land plants. For many years the age of these formations was the subject of controversy but they are known to be Mississippian. North of the thrust sheets where the Mississippian rocks lap up onto the flanks of the Ozark Dome they are much thinner and are richly fossiliferous.

Over the craton north of the Ouachita trough and extending across the central United States and the western states and western Canada the lower half of the Mississippian System forms one vast sheet of limestone ranging in thickness from a few hundred to more than 1000 feet. Where it crops out in isolated regions it bears local names. It is the Leadville Limestone of central Colorado, the Madison Limestone of Wyoming, and the Redwall Limestone of the Grand Canyon (Fig. 12-7). It also extends far to the north in the Canadian Rockies of Alberta.

On the contrary, the Chesterian Series is missing over most of the states west of the Mississippi River. Exceptions are northern Utah and Idaho, where it is thick. The Oquirrh Range west of Great Salt Lake shows the thickest known section of the Upper Mississippian Series; it is about 5000 feet thick and consists of limestone and shale.

Near the Pacific coast there are five isolated areas of Mississippian outcrop that are largely detrital and may represent a shelf sea or a separate trough isolated from the Cordilleran sea. In the Chichagof Island of Panhandle Alaska a thick limestone bears an Asiatic fauna including the enormous brachiopod *Gigantella*. The same fauna is known in the Yukon Valley of Alaska, in the Coffee Creek Formation of central Oregon, and in the Bragdon Formation of northern California. This fauna is not known elsewhere in America.

The Pennsylvania System. The Pennsylvanian System marks a profound change in the geologic history of much of the United States. At the beginning of the period the continent was rimmed on the east and south, all the way from Maritime Canada to west

Figure 12-7 South wall of the Grand Canyon below Grand Canyon village showing the 500 foot cliff (lower) formed by the Redwall Limestone. The other formations are identified in Fig. 1-6B. (C. O. Dunbar.)

Figure 12-8 Correlations of important Pennsylvanian sections.

Series	Colo.	Ariz. Utah Nev.	Texas west	Texas central	Okla.	Kansas–Nebr.	Illinois	Ohio– Penn.	Va. – W. Va.	Series
Virgilian		Callville Ls. / Oquirrh Fm. (lower part) / Bird Spring Ls. (lower part)	Cisco Series		Vanoss Fm.	Wabaunsee Gr.		Monongahela Series Pittsburgh coal	Monongahela Series	Virgilian
Virgilian					Ada Fm.	Shawnee Gr. Topeka L.				Virgilian
Virgilian					Vamoosa	Douglas Gr. Oread L.				Virgilian
Missourian	Hermosa Ls.		Canyon Series		Francis Fm.	Lansing Gr. Iola Ls.	McLeansboro Fm.	Conemaugh Ser. Ames Ls. Cambridge Ls.	Conemaugh Ser. =Ames Ls.=	Missourian
Missourian						Kansas City Gr. Hertha Ls.				Missourian
Missourian						Bourbon Gr. Lenapah L.				Missourian
Desmoinesian			Gaptank Fm.		Wewoka	Marmaton Gr.	Coal 8 Herrin L. Coal 6.	Freeport Coal	U. Freeport Coal	Desmoinesian
Desmoinesian			Strawn Series		Wetumka	Ft. Scott Ls.	Carbondale Fm.	AlleghenySer.	Allegheny Series	Desmoinesian
Desmoinesian	Fountain Fm.				Boggy	Cherokee Ss.	Curew L.	Brookville Coal U. Mercer Ls=	Roaring Cr. Ss	Desmoinesian
Desmoinesian					Savanna					Desmoinesian
Atokan			Haymond Fm.	Smith-wick Gr. / Big Spring Gr.	Atoka Fm.		Tradewater Fm.	Pottsville Ser.	Kanawha Group	Atokan
Atokan				Lampasan / Bendian						Atokan
Morrowan			Dimple Fm.	Marble Falls Ls.	Wapanu-cka Ls.		Caseyville Fm.	Sharon Congl.	New River Group	Morrowan
Morrowan			Tesnus Fm.		Springer Fm.				Pocahontas G.	Morrowan

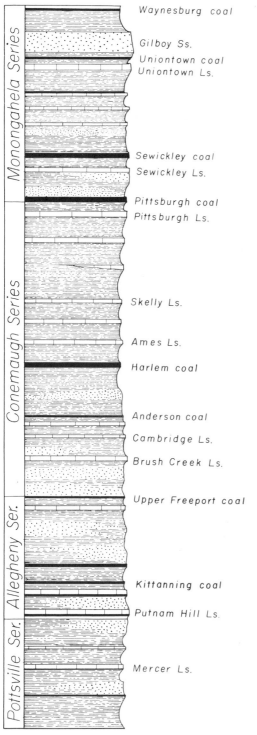

Figure 12-9 Columnar section of the Pennsylvanian System in eastern Ohio, showing repeated alternation of sandstone, shale, coal, and thin marine limestones. Vertical scale, 150 feet to the inch. (Ohio Geological Survey.)

Texas, by borderlands bearing actively rising mountains (Map VII). So far as known nearly all of Canada was emergent and between it and the mountainous border described above, a broad basin was forming that soon involved much of the United States. A vast shallow sea spread from the west across most of this basin but from Kansas and Oklahoma eastward a contest was being waged between the sea and the rivers, which were transporting vast amounts of sand and mud from the east and south, filling the basin about as fast as it sank. Where the rivers were dominant a vast swampy lowland was maintained, but when subsidence exceeded filling the sea crept eastward, drowning the lowland. Thus throughout the period the shoreline oscillated back and forth, at times advancing into Pennsylvania and again retreating westward to Kansas. This resulted in a complex intertonguing of coal-bearing sandstone and shales with thin marine limestones. A typical section in the coal fields of Ohio is shown in Fig. 12-9. In Illinois, where cyclic deposition prevailed, about 100 cyclothems are recognized, each including several members. In the Appalachian coal fields the sandstones are commonly irregularly stratified and crossbedded; they vary in thickness and in places show evidence of cut and fill. Generally the only fossils found in the sandstones and shales are land plants and these are locally abundant and well preserved. In the clay immediately under the coal beds, root stocks of plants, called **stigmaria,** mark the old land surface on which the plants were growing. In short, these formations were laid down by streams that meandered across a low aggraded plain formed of sediments carried far west from Appalachia.

Some of the coal beds are thin and lenticular, recording disconnected and rather temporary swamps, but a few such as the Pittsburgh coal (Figs. 12-10 and 12-11) cover hundreds of square miles and may be 5 feet or more thick.

The limestones are generally less than 10 feet thick but commonly extend over large areas with little change in thickness and most of them bear marine fossils. Each of them thus

Figure 12-10 Natural outcrop of the Pittsburgh coal near Riverside, Pennsylvania. (C. O. Dunbar.)

records a brief incursion of the sea, which for a relatively short time drowned the whole region and then retreated to the west.

Such thin and widely persistent units as the limestones and coal beds have made it possible to subdivide the section into many named units that are grouped into four series as indicated in Fig. 12-9. But in the geosynclines where the earliest deposits were formed five series are recognized (Fig. 12-8).

Most of West Virginia and Alabama was never reached by the sea and there the section is very thick and wholly nonmarine. Far to the north in Nova Scotia and New Brunswick where the Pennsylvanian System is as much as 13,000 feet thick, deposition was in intermont basins and is entirely nonmarine. The section exposed in the sea cliffs at Joggins near the head of the Bay of Fundy is especially interesting because the stumps and trunks of trees are preserved in the position of growth (Fig. 12-12). Here such buried forests are recorded at 20 horizons distributed through about 2500 feet of beds. Many of the standing trunks are several feet tall, some exceeding 20 feet. They indicate very rapid accumulation since in each horizon the trunks were buried before they had time to decay. It is evident, however, that deposition at any locality was intermittent since some of the

buried trees, as much as 4 feet in diameter, must have grown unhindered before their eventual burial. Deposition must have been by sluggish, meandering streams, which from time to time deserted their sediment-choked channels to burst out into new courses over timbered lowlands (compare Fig. 3-23).

In the Midcontinent region of Kansas, Missouri, Nebraska, and Iowa the marine invasions were more persistent and limestones form about 25 percent of the section. Here, also, being far out on the craton, the entire system is less than 3000 feet thick. Traced southward into Oklahoma and Arkansas, the system changes, at first gradually and then profoundly, as we enter the Ouachita trough, which was subsiding rapidly as Llanoria on the south was rising into mountains. Here the section thickens to 12,000 feet in Oklahoma and probably 23,000 feet in Arkansas; it consists largely of sandstone and sandy shale (Fig. 12-13). In basins adjacent to the Oklahoma Mountains, locally derived detritals also accumulated to great thickness and include coarse conglomerates that thin and become fine grained away from the ranges. The lower conglomerates are made of limestone fragments which can be identified with the Ordovician formations that covered the Arbuckles, but some of the higher conglomerates include granite pebbles, proving that the core of the range had then been exposed.

In central Texas the Pennsylvanian System is thick and almost entirely marine. The lower part shows the influence of the marginal land, Llanoria. In the Marathon uplift of Trans Pecos Texas the rising marginal land was very close on the south and the trough subsided rapidly and here the system is very thick and includes (in the Haymond Formation) uncommonly coarse conglomerates. About the close of the period the Marathon orogeny strongly deformed the rocks in the geosyncline and thrust them northward (Fig. 12-14).

Throughout the Cordilleran region there is so much local variation that brief synthesis is impossible. Along the flanks of the Colorado Mountains and in the basins between ranges great debris aprons of nonmarine conglomerates, sands, and red sandy shales accumulated.

Figure 12-11 Open mine in the Pittsburgh coal near Pittsburgh, Pennsylvania. (U. S. Bureau of Mines.)

Figure 12-12 Fossil tree trunks preserved in the position of growth. Left, sea cliffs at "The Joggins," Nova Scotia; right, an 8-foot trunk in exposure at Table Head, Great Bras d'Or, Cape Breton. (W. A. Bell, Geological Survey of Canada.)

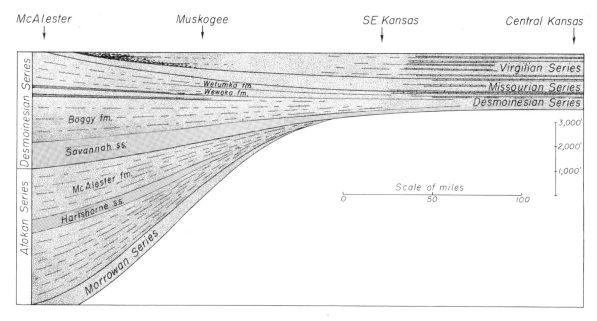

McAlester   Muskogee   SE Kansas   Central Kansas

Figure 12-13 Stratigraphic section of the Pennsylvanian System from central eastern Kansas to McAlester, Oklahoma.

Figure 12-14 Three stages in the history of the Marathon orogeny in the Marathon region of Trans Pecos, Texas. The sections run from northwest to southeast across the Marathon fold belt and includes the edge of Llanoria at the extreme right. (After P. B. King.)

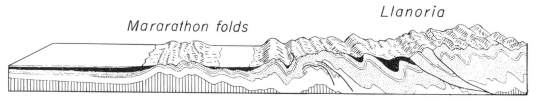

Marathon folds begin to rise out of the sea

Overthrusting at end of Pennsylvanian time

Permian overlap after peneplanation of Marathon area

Figure 12-15 The "Garden of the Gods" at Colorado Springs formed of upturned beds along the front of the Rockies. (Frank Bauer, Rapho Guillumette Pictures.)

The "Garden of the Gods" at Colorado Springs (Fig. 12-15) affords a spectacular display of this facies, but farther east, under the plains, these formations grade laterally into white limestone. In southeastern Arizona the section is thick, all marine, largely calcareous, and richly fossiliferous. In the eastern part of the Grand Canyon the Pennsylvanian System is represented by the thick, nonmarine, red Supai Formation. In the western part of the Canyon the Supai intertongues with the Callville Limestone, which thickens westward to form the lower half of the widespread Bird Spring Limestone of southern Nevada and southeastern California. In Idaho, on the contrary, the rocks are largely detrital and include thick sandstone and coarse conglomerate.

ECONOMIC RESOURCES

Coal. It was no accident that in both Europe and North America the Pennsylvanian rocks are known as the Coal Measures. No other geologic system contains so much high rank coal. In these formations lie the great coal fields of the British Isles, of the Saar Basin in France, of the Ruhr Basin of Germany and Belgium, and of the Donetz Basin in Russia. In North America it includes the vast coal

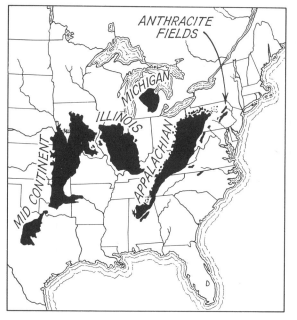

Figure 12-16 Map showing the Pennsylvanian coal fields of eastern United States.

Figure 12-17 Map showing the structural relations of the Anthracite and Bituminous coal fields of Pennsylvania.

fields that stretch from Oklahoma and Kansas to the Appalachian Mountains. These fields in Europe and America produce more than 80 percent of the world's coal. The American coal fields (Fig. 12-16) occupy an area of more than 250,000 square miles, which considerably exceeds the combined area of all the European coal fields.

Of the five well-defined fields shown in Fig. 12-16 the Anthracite Field of eastern Pennsylvania is in some respects the most interesting. Although less than 500 square miles in area, it has produced to date almost one-fourth of the total output of coal in North America. Its production reached a peak of 93,000,000 tons in 1923 and has averaged over 85,000,000 yearly for the last quarter-century. Unfortunately most of the coal has already been mined, and the supply of anthracite may well be exhausted during the life of the present generation.

The anthracite field is in reality but an eastern part of the vast Appalachian basin that was caught in the Permian folding and isolated from the rest by later erosion, which

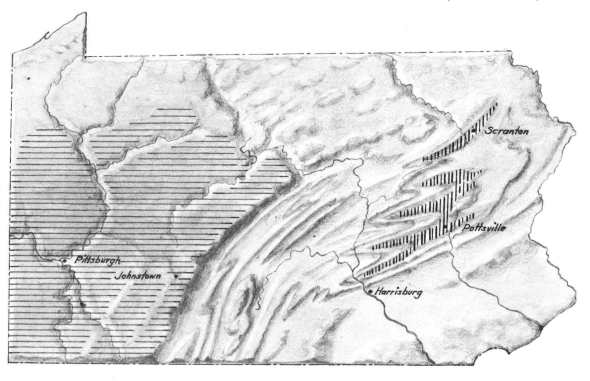

destroyed all the Coal Measures in eastern Pennsylvania except those parts preserved in deep synclinal folds (Fig. 12-17). This folding and the resulting pressure converted the bituminous coal into anthracite by eliminating much of the volatile matter. The thickness of some of the anthracite beds is noteworthy. The most remarkable bed is the Mammoth, which extends throughout the field, with an average thickness of 35 to 40 feet and in one place has a thickness of 114 feet due to overfolding.

The Appalachian Field is the largest, and much the greatest producer of the several coal regions. It underlies the Allegheny Plateau and extends from northern Pennsylvania to Alabama. Pennsylvania and West Virginia are the heaviest producers, but Ohio, Kentucky, and Alabama also yield much coal. This field alone furnishes almost one-fourth of the world's coal supply.

Throughout most of this area the strata lie quite flat, and the coal is bituminous, occurring at many levels. About 60 beds are recognized in Pennsylvania, but of these only 10 are widely mined, the rest being too thin to work profitably. The most remarkable of these beds is the Pittsburgh coal (Fig. 12-11), which is more than 13 feet thick about Pittsburgh and is known to be workable over an area of 6000 square miles in western Pennsylvania, eastern Ohio, and northwestern West Virginia, where it is estimated to have contained more than 22,000,000,000 tons of coal. Up to 1926 it had yielded approximately 3,500,000,000 tons of coal, with a value at the mines more than 20 times that of the gold produced by the greatest gold mine in the United States.

The Illinois Field is a shallow structural basin extending into southwestern Indiana and western Kentucky. It includes ten or more important producing horizons and ranks second to the Appalachian field, far outstripping the much larger Midcontinent region in production. The coal beds here do not attain as great a thickness as they do farther east, but some of them, notably the Herrin Coal, persist with remarkable uniformity over much of the state of Illinois.

The Midcontinent field embraces the coal fields of Missouri, Iowa, Kansas, Oklahoma, and northern Texas. Although the area exceeds that of the Appalachian field, the output of coal has been less than one-tenth as great. Here the beds are commonly less than 4 feet thick, and the best producing horizons are in the lower part of the system. Over extensive areas, the topography is flat and the coal so near the surface that it is mined by stripping.

The remaining fields are relatively small. In the tiny Rhode Island Basin extreme metamorphism has reduced the coal to graphite, or so nearly so that it has little fuel value.

Most of the Pennsylvanian coals of the United States include sulfur as an impurity. This is believed to indicate that the coal swamps bordered the inland seas and were brackish, since sulfur-depositing bacteria live in the sea but do not thrive in fresh-water lakes and swamps. Conversely, the slight amount of sulfur in the anthracite of eastern Pennsylvania and in the coals of the Acadian Basin indicates that in these regions swamps were entirely of fresh water.

Petroleum and Natural Gas. Pennsylvanian rocks have been an important source of petroleum and natural gas in the Midcontinent oil fields and for a number of years, from the discovery of these fields in the 1890s until about 1925, constituted their only important producing horizon. During that time Kansas and Oklahoma produced from these beds over 2,000,000,000 barrels of oil. Subsequently, however, production has been found at greater depths in the Ordovician "sands," which have given rise to the spectacular developments of more recent years in those two states and in north-central Texas.

CLIMATE

The terrestrial sediments with their plant remains are persuasive evidence of warm, moist climate during the chief coal-producing stages of the Pennsylvanian in many parts of the world. The vegetation of the coal beds clearly grew in swamps, where it accumulated under standing water, as evidenced by the spreading root systems still preserved in the fire clays that underlie the coals in many places. Moreover, the structural types of the foliage so well preserved in the roofing shales at many places indicate marked humidity. Swamp waters are required to protect the fallen vegetation from the air and thus save it from decay. The wide distribution and the repeated occurrence of coals therefore assure us that there was a persistently moist climate over vast regions of the Pennsylvanian landscape. Paleomagnetic data indicate that during Carboniferous time the equator crossed North America in the latitude of the southern tip of Hudson Bay and the Strait of Belle Isle.

This does not prove, of course, that no parts of the earth were arid. It is well to remember that at present the dripping jungles of the Amazon Valley are separated only by the narrow Andean chain from the desert coast of western Peru. Similar extremes probably existed on the continents during Pennsylvanian time, for lofty mountains must have been accompanied, then as now, by a leeward rain shadow of deficient rainfall. The salt deposits of the Paradox Formation in eastern Utah and western Colorado, like the red arkose of the Fountain Formation of central Colorado, may indicate such local aridity about the Colorado Mountains. On the contrary, the coal swamps of the eastern interior were formed where the warm, moist winds were rising up the long western slope of Appalachia. In general, it appears that humid climatic conditions were exceptionally widespread during Pennsylvanian time.

There is also much evidence that the climate was warm, even in high latitudes, during much of the period. The mere presence of abundant vegetation is no evidence, for it is well known that the most extensive modern accumulation of peat is in subarctic regions where slow growth is more than counterbalanced by slow decay; but the character of the Carboniferous vegetation indicates a lack of freezing winters, at least in the lowlands where the plants are preserved. The trees, whether tree ferns, seed ferns, cordaites, or the great scale trees, bore succulent foliage of almost unprecedented luxuriance. Not merely were the leaves large, but their texture indicates rapid growth under warm, humid

conditions. For example, the very large size of the individual cells, the arrangement of the stomata (breathing pores), the smoothness and thickness of the bark, the presence of aerial roots, and the absence of growth rings in the woody trunks are all features of significance. One of the foremost paleobotanists of our times concluded that "the climate of the principal coal-forming intervals of the Pennsylvanian was mild, probably nearly tropical or subtropical, generally humid, and equable."

The animal life of the time also seems to support this view. Insects, for example, attained an extraordinary size and, as far as known, averaged larger than in any other period of earth history. Since it is well known that the modern orders of insects have their large representatives in the tropics, with smaller and smaller species in regions of more rigorous climate, the significance of the Pennsylvanian insects is obvious. To this may be added the fact that at certain times during the period corals were able to thrive in great abundance and to form reefs as far north as arctic islands of Spitsbergen (lat. 78°N). The presence of these ancient reefs in the sea cliffs of a land now treeless and ice-covered speaks eloquently of the climatic contrast between the Present and the Pennsylvanian ages in this region. The exceptional abundance of the large fusulines (p. 274) in the limestones of the northern hemisphere, and even as far north as Spitsbergen, seems to have a significance like that of the insects. It is noteworthy that in extreme northeast Greenland where Pennsylvanian limestones peek out from under the ice cap fusulines are abundant.

The fossils thus confirm the paleomagnetic evidence that, during Upper Carboniferous time, North America and Europe were far to the south of their present position.

The record in South America is quite different. We still lack paleomagnetic evidence of its position during Carboniferous time, but early in Permian time it was part of the great Pangean landmass that was centered about the South Pole and was widely glaciated. Glacial deposits are widespread in eastern Argentina and Bolivia and across the Parana Basin of

Paraguay, Uruguay, and Brazil, and a comprehensive review by L. A. Frakes and John Crowell (*Geol. Soc. Amer. Bull.*, **80**, 1969, 1007–1042) confirms that glaciation occurred in Lower Carboniferous, Upper Carboniferous, and Early Permian time. This suggests that throughout Carboniferous time South America was part of Pangea and was far to the south of its present position.

LIFE OF THE CARBONIFEROUS SYSTEMS

Progress of Marine Invertebrates. With a few exceptions the marine life of Mississippian time is transitional between that of the Devonian and Pennsylvanian periods. The exceptions deserve comment.

Foraminifera had been present since the Ordovician but now, for the first time, one species was so abundant as to be an important rock maker. This is a small, globular form about the size of a pinhead, the genus *Endothyra* (Fig. 12-18), which locally makes up a large percentage of the Salem oolite.

Lacy **bryozoans** of the tribe of *Fenestella* now occurred in great abundance so that in places crushed specimens cover the bedding planes (Fig. 12-19). Of these the genus *Archimedes* (named for the ancient Greek in-

Figure 12-18 Shells of the small foraminiferan *Endothyra* from the Salem Oolite, near Salem, Illinois. These shells are about the size of a period on this page.

ventor of the screw) developed a twisted colony with a spiral axis. This unmistakable genus is one of the best guide fossils to the Mississippian rocks. As a student once exclaimed, "it is the screwiest fossil there is" (Fig. 12-20).

The **crinoids** had been fairly common since Early Ordovician time but in the Mississippian seas they were extraordinarily abundant and varied, and with their brilliant colors, they must have formed the most colorful sea bottoms on the earth. Their broken stems and isolated plates form a large percentage of some of the widespread crinoidal limestones of the time (Plate 9, Figs. 6 to 10).

A related group of echinoderms, the **blastoids**, are almost confined to Mississippian rocks (Plate 9, Figs. 4 and 5). They are most abundant in the Chesterian Series where they form excellent guide fossils.

Pennsylvanian Invertebrates. At this time the invertebrate animals were abundant and highly varied in the seas. One of the most notable developments was the appearance and rapid evolution of the **fusulines** (Superfamily Fusulinacea). These are foraminifera of relatively giant size (commonly about the size and shape of grains of wheat or oats) and they

Figure 12-19 Surface of a bed covered by fragments of lacy bryozoans. (Natural size.) A small spiriferoid brachiopod appears near the center. (Yale Peabody Museum.)

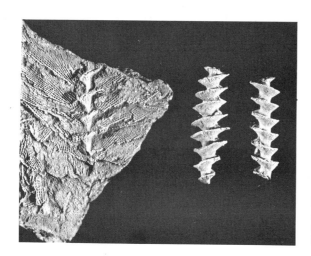

Figure 12-20 Left, fragment of a colony of the bryozoan *Archimedes;* right, the spiral axes of two colonies, one coiled to the right and the other to the left. From the Warsaw Formation in Illinois. (Yale Peabody Museum.)

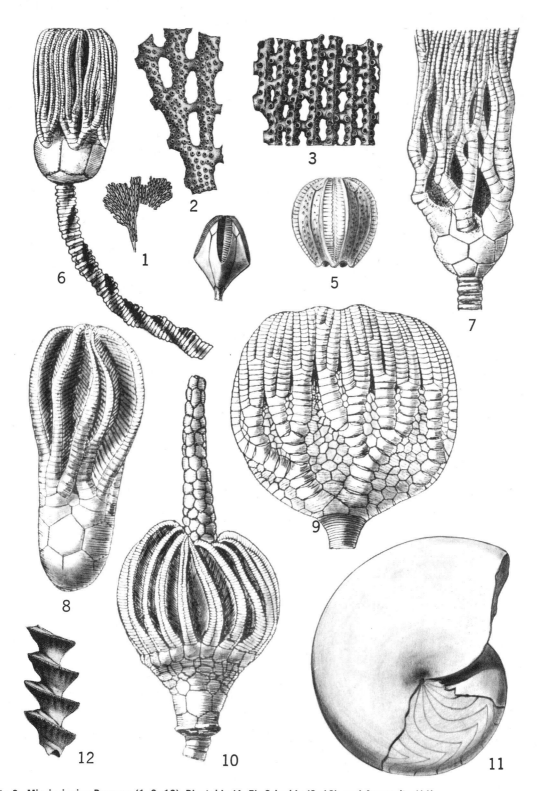

Plate 9   Mississippian Bryozoa (1–3, 12), Blastoids (4, 5), Crinoids (6–10), and Ammonite (11).
Figures 1, 2, *Polypora cestriensis* (×⅔ and ×6); 3, *Fenestella cingulata* (×6); 4, *Pentremites pyriformis;* 5, *Cryptoblastus pisum;* 6, *Platycrinites hemisphericus;* 7, *Cyathocrinus multibrachiatus;* 8, *Agassizocrinus dactyliformis;* 9, *Forbesiocrinus wortheni;* 10, *Batocrinus pyriformis;* 11, *Imitoceras rotatorium;* 12, *Archimedes wortheni,* the screw-like axis of a colony. All natural size except Figures 1–3. (Drawn by R. G. Creadick.)

occur in great abundance in many of the limestones (Fig. 12-21). Their shells were spirally coiled about an elongate axis and were divided into chambers by longitudinally arranged and commonly folded or ruffled septa. The earliest ones were scarcely bigger than a pinhead, but they increased in size and varied in shape with the progress of time and thus are most useful zone fossils (Fig. 12-22).

Brachiopods were still the commonest of the "shellfish" except on the muddy bottoms where pelecypods and gastropods were dominant. Among the brachiopods the spirifers were on the decline but the spiny productids had expanded into several families and were extremely abundant. Nautiloid cephalopods were rather on a decline but the primitive ammonoids were gradually increasing in variety, foreshadowing their expansion in the Permian Period.

Blastoids are abundant in some of the basal members of the system but are very rare thereafter. In the East Indies, however, they found an asylum where on the Island of Timos they were still common in the Permian.

In Pennsylvanian rocks two groups of microfossils—small foraminifera and ostracods—are important aids to correlation (Fig. 12-23).

Figure 12-21 Fusulines of the genus *Triticites* naturally weathered free from shaly limestone. (Yale Peabody Museum.)

Forests Cover the Swamp Lands

Forests of fast-growing, soft-tissued trees spread over the moist lowlands of the Pennsylvanian landscape. Among these were none of the deciduous forms like those of our modern forests, for they had not yet evolved. The giants of the time were strange, spore-bearing trees which today are represented only by insignificant herbaceous descendants like the ground pines and scouring rushes (Figs. 12-1 and 12-24).

Under the moist and perpetual summer of the coal swamps, shades of green must have been dominant, since flowers had not yet evolved. There was no honey to lure the insects and no sweet perfumes to scent the air—only fresh, resinous odors such as pervade the living conifer forests.

Although the seed-bearing plants were common, spore-bearing trees were even more abundant and at certain seasons must have covered the forests with a greenish-yellow or brownish dust of spores, since some of the coals (canned coals) are composed almost entirely of spore cases.

Ferns of many kinds were common, and they alone gave a modern aspect to the dells of these ancient forests. The leaves of some species attained huge proportions, single fronds reaching a length of 5 or 6 feet; and the slender, unbranched trunks grew to be as high as 50 feet.

Seed ferns resembled the true ferns in every respect save one: they bore small, nut-like seeds instead of spores appended to their fronds. They may have descended from ferns and in turn may have given rise to all the higher, seed-bearing plants. They were more common than the true ferns in Pennsylvanian time and have often been confused with them, since the two groups can be distinguished only when fruiting fronds are found.

Scouring rushes of giant size grew in solid stands like "cane brakes" in portions of the swamps. Like their humbler modern descendant, *Equisetum*, they are easily recognized by their vertically ribbed and regularly jointed stems (Fig. 12-24D). The Paleozoic forms bore at each joint a whorl of slender leaves

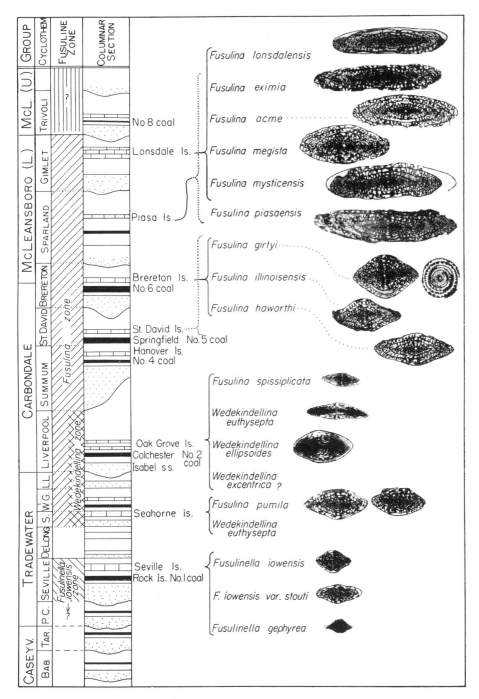

Figure 12-22 Fusulines as zone fossils. The columnar section at the left shows the succession of formations in the lower half of the Pennsylvanian beds in Illinois, and at the right are shown axial thin sections of the fusulines found in each of the marine limestones (all ×4). Obviously the faunas of the Braerton and St. David limestones are identical but are quite different from those of the higher and lower beds. Each of these faunal zones can readily be identified in a well core as well as in an outcrop. (From Dunbar and Henbest, Illinois Geological Survey.)

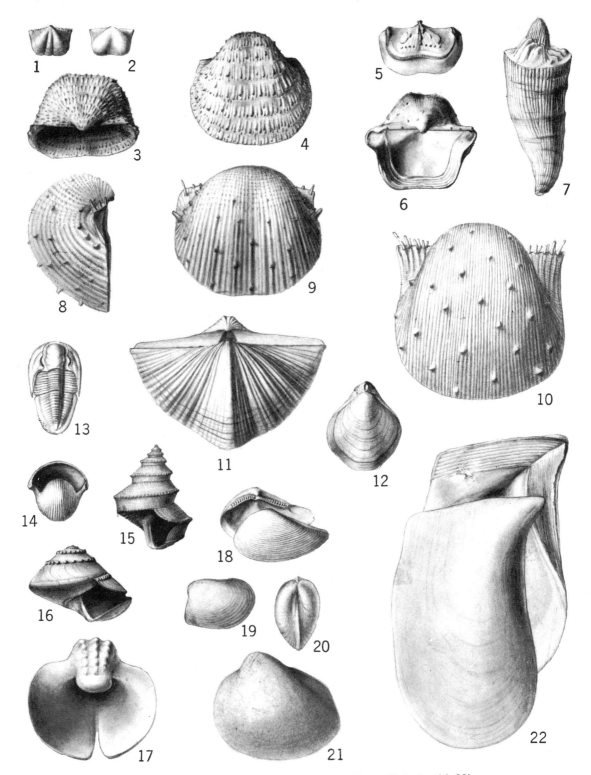

Plate 10 Pennsylvanian Brachiopods (1–6, 8–12), Coral (7), Trilobite (13), and Mollusks (14–22).

Figure 1, *Mesolobus mesolobus;* 2, *Lissochonetes geinitzianus;* 3, 4, *Juresania nebrascensis;* 5, 6, *Marginifera splendens* (5, interior view of dorsal valve); 7, *Lophophyllum profundum;* 8, 9, *Dictyoclostus portlockianus;* 10, *Linoproductus prattenianus;* 11, *Neospirifer dunbari;* 12, *Composita subtilita;* 13, *Ameura major,* one of the very last of the trilobites; 14, *Euphemites carbonarius;* 15, *Worthenia tabulata;* 16, *Trepospira sphaerulata;* 17, *Bellerophon tricarinatus;* 18, *Nuculana arata;* 19, 20, *Nuculopsis ventricosa;* 21, *Schizodus wheeleri;* 22, *Myalina subquadrata.* All natural size. (Drawn by L. S. Douglass.)

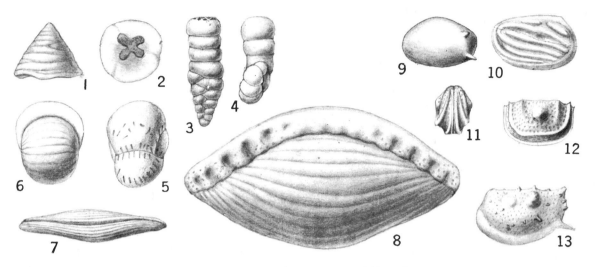

Figure 12-23 Pennsylvanian microfossils, greatly enlarged. 1–8, foraminifera; 9–13, ostracods.

Figure 12-24 Four of the dominant types of forest trees of Pennsylvanian time. A, *Lepidodendron;* B, *Sigillaria;* C, *Cordaites;* D, *Calamites.*

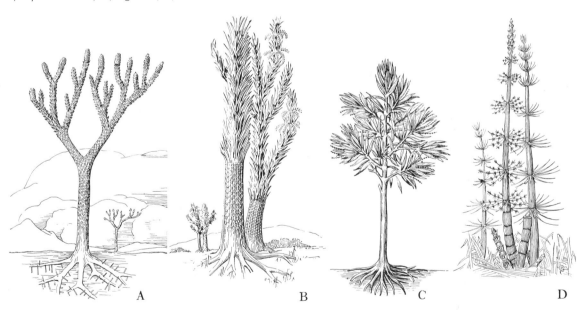

which in modern rushes are represented only by bractlike vestiges. The leaf whorls, known as *Annularia,* commonly present a false resemblance to flowers. The largest of the Pennsylvanian rushes belonged to the genus *Calamites.* Some of these exceeded 12 inches in diameter and had a height of 30 or more feet. Their trunks were not solid woody stems, but rather thin woody cylinders filled with a core of pith and surrounded by thick bark, the woody layer rarely having a thickness of 2 inches.

The **scale trees** were the most imposing plants of the forests and in many places the most common. Their name is derived from the fact that their close-set leaves left permanent leaf scars over the trunk and limbs that make them appear scaled (Fig. 12-25). So striking is this deception that twigs have been mistaken by amateur fossil hunters for petrified snake skins. The scale trees grew to a large size, their stumps reaching a diameter of 4 to 6 feet and their slowly tapering trunks an extreme height of more than 100 feet. Most of them belonged to one of two well-defined forms, **Lepidodendron** or **Sigillaria.**

Lepidodendron grew a tall slender trunk, branching repeatedly near the top to present a spreading crown of stubby twigs covered with slender, straplike leaves. These leaves, like immensely overgrown pine needles, in some species were 6 to 8 inches long and $\frac{1}{2}$ inch wide. The older leaves were shed as new ones formed at the tips of the branches, leaving sharply defined diamond-shaped leaf scars which were normally arranged in spiral rows about the limbs and the trunk (Fig. 12-25). The branching was normally dichotomous (with equal forks). Spore cases were borne as cones at the tips of the limbs.

Sigillaria (Figs. 12-24 and 12-25) possessed a thicker trunk which rarely branched and was clothed for several feet from the top with large, bladelike leaves, resembling those of *Lepidodendron* but larger. In these trees the bark was vertically ribbed, and the leaf scars were normally in vertical rows. Trunks have been found with a diameter, just above the roots, of 6 feet, and one specimen is known to be 100 feet long without a branch.

About 100 species each of *Sigillaria* and *Lepidodendron* have been described. Although many of them were large trees, some were relatively small. In all of them the structure of the trunk and limbs was peculiar in that they had a relatively large center of pith surrounded by a woody cylinder, this in turn surrounded by two very thick layers of corklike bark. The leaf scars are impressed only on the bark. The root system likewise was peculiar, the main trunk roots spreading almost horizontally without a tap root; moreover, they branched but a few times and were stubby and thick. The real rootlets sprang

Figure 12-25 Pieces of the bark of "scale trees" showing the characteristic arrangement of the leaf scars in *Lepidodendron* (right) and *Sigillaria* (left). (Yale Peabody Museum.)

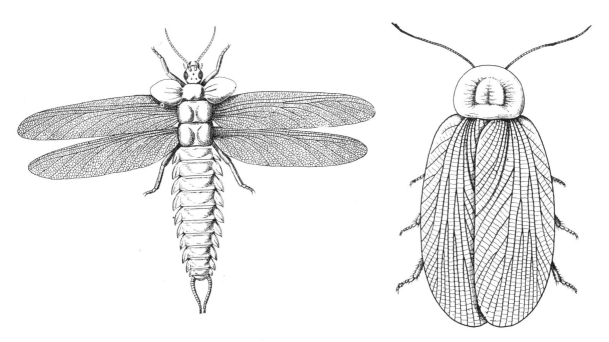

Figure 12-26 Primitive insects of the Pennsylvanian Period. Left, a primitive and generalized type; right, a primitive cockroach.

directly from the sides of these trunk roots, radiating thickly away to a distance of several inches. Such root stocks, known as **stigmaria,** are common in the fire clays under coal beds and not infrequently appear in the coal.

The **cordaites** (named after the Bohemian botanist Corda) were the forerunners of the modern conifers, which they resembled in their sturdy, softwood trunks and their parallel-veined leaves (Fig. 12-25). They differed from true conifers chiefly in two ways: (1) their leaves were not needlelike but bladelike, attaining a length of several inches to 5 or 6 feet, and (2) their seeds were borne in racemes instead of being crowded into cones. Many of them were tall, graceful trees, some attaining a height of 120 feet and a diameter as great as 3 feet. In such trees fully two-thirds of the trunk was without branches, though the top was a dense crown of branches and large simple leaves. The wood of the cordaites was much like that of modern pines, but the pith at the center was larger. They appear to have been one of the chief contributors to the vegetation that made the Pennsylvanian coal.

True conifers appeared during the period but have been found in only a few localities, probably because they lived on the uplands where they were not commonly preserved. It is obvious that our knowledge of the Pennsylvanian land plants relates almost wholly to the swampy lowlands. Possibly the most rapid advances were being made in the uplands, where the climate was more rigorous but where the chances for preservation were slight.

One of the striking features of the Pennsylvanian floras is the marked similarity of the species in different parts of the world. They were as nearly cosmopolitan as any in the earth's history. This may be attributed to their habit of reproduction by spores that are easily carried by the wind, but it also suggests the absence of well-marked climatic zones.

Animal Conquest of the Lands

Through the dank forest of Pennsylvanian time droned clumsy insects, while centipedes, spiders, and scorpions scurried about over fallen logs in search of food. This spectacular advance was a prelude to the expansion of vertebrate land animals that fed upon such small invertebrates. The insects were many and varied but nearly all belonged to primitive, extinct orders. Among the most common and the most familiar kinds were primitive

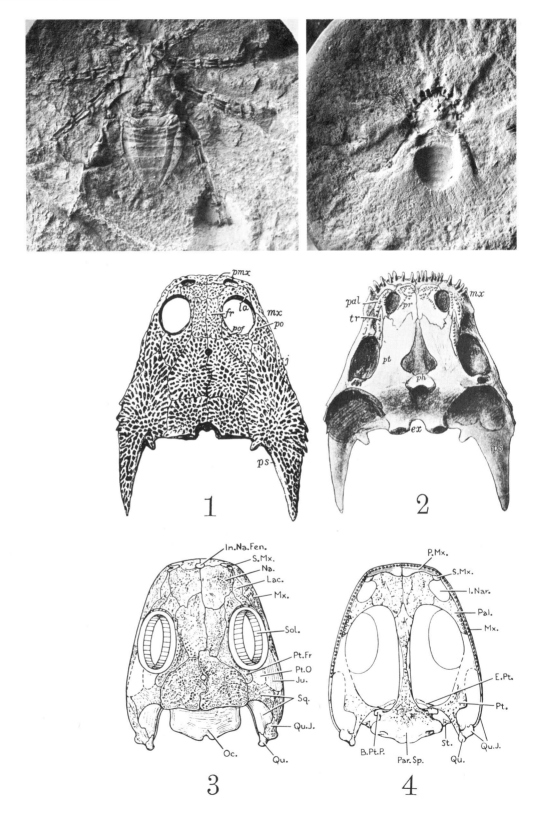

1

2

3

4

cockroaches (Fig. 12-26), most of which were larger than their living descendants. One species attained a length of about 4 inches. Close relatives of the modern dragonflies were also common and exceptionally large. One species found in the Coal Measures of Belgium had a wing spread of about 29 inches (foreground of Fig. 12-1). The insects generally were large. Out of some 400 known species more than a score exceeded 6 inches and three exceeded a foot in length. No other period has produced insects so large.

The presence of so many species and such a great variety of insects at this time is remarkable and clearly indicates a long antecedent evolution. The swamps of the Coal Measures afforded the first widespread environment favorable to the preservation of such delicate creatures.

Scorpions, remarkably like modern ones, occur with the insects, on which they undoubtedly fed. **Spiders** (Fig. 12-27) likewise occur abundantly though none of the fossils show clear evidence of spinnerets and it is not known whether they had yet learned to spin webs. One species of **Centipede** found at Mazon Creek, Illinois, was 12 inches long.

The oldest known **land snails** were found associated with the skeletons of small amphibians at "The Joggins" in New Brunswick where both had taken refuge in standing hollow stumps.

Small **fresh-water clams** of several kinds occur abundantly in the Coal Measures of Europe and in Nova Scotia, but are not found elsewhere in North America.

Vertebrate animals are represented by abundant skeletal remains of small amphibians resembling modern salamanders (Fig. 12-28). From the Coal Measures of North America alone no fewer than 7 orders, 19 families, 46 genera, and 88 species are known. But unlike modern salamanders, which are without armor, the Pennsylvanian forms bore a cuirass of bony plates over the skull (as did their ancestors, the crossopterygian lungfish). They also had labyrinthodontine teeth and represent the extinct order called **labyrinthodonts,** among which the higher verte-

Figure 12-27 Spiders preserved in concretions in Mid-Pennsylvanian shale at Mazon Creek, Illinois. Both are seen in ventral view. (After Petrunkevitch.)

Figure 12-28 Pennsylvanian Labyrinthodonts. Above, dorsal and ventral views of the bony skull of *Diceratosaurus* from Linton, Ohio (slightly enlarged); below, comparable views of the skull of *Miobatrachus,* from Mazon Creek, Illinois (×3). It is thought that *Miobatrachus* belongs to the stock from which the frogs developed at a later date.

Figure 12-29 One of the oldest known reptiles, *Eosauravus copei,* from Mid-Pennsylvanian beds at Linton, Ohio. (Slightly enlarged.) The head is missing. (Roy L. Moodie, U. S. National Museum.)

brates have their ancestors. The Pennsylvanian amphibians were nearly all small, most of them only a few inches long, and the largest known skeletons indicate animals no larger than a Florida alligator, but deep tracks in a sandstone bed near Lawrence, Kansas, indicate an animal with stubby feet over 5 inches long and a stride of about 30 inches. The right and left treads are wide apart, indicating a heavy-bodied animal weighing not less than 500 or 600 pounds.

The very first **reptiles** are known from the Pennsylvanian rocks where two genera, both small, have been found. The oldest, *Romeriscus,* was recently described from near the base of the system in Nova Scotia and the next, *Eosauravus* (Fig. 12-29), was found in Mid-Pennsylvanian beds at Linton, Ohio.

The sudden appearance of such a varied array of land animals in the Pennsylvanian rocks implies that their ancestors must have existed in Mississippian time where their apparent absence can only be due to the lack of favorable conditions for preservation. A few amphibians have, indeed, been found in Mississippian rocks but they are exceedingly rare.

Figure 13-1 The Guadalupe Mountains of West Texas and New Mexico, held up by the great reef complex. Aerial view from the southwest, looking across Salt Flat at the west face of the Guadalupes on the left and along the reef front on the right. Delaware Basin is at the right and the dissected plateau in the distance is formed of the weak lagoonal deposits. A small white cloud caps the west crest of the reef. BC, Brushy Canyon Formation; BS, Bone Spring Limestone; CC, Cherry Canyon Sandstone; C, Capitan Limestone. (Muldrow Aerial Surveys.)

CHAPTER **13** *The Permian Period*

A CRISIS IN EARTH HISTORY

Momentous changes ushered the Paleozoic Era to its close. As the mobile borderlands continued to rise, several of the great Paleozoic geosynclines were uplifted and transformed into ranges of folded mountains. While the Appalachians were forming in eastern America, the Urals were rising out of a great geosyncline in eastern Europe and other ranges were growing across southern Europe and southern Asia. By the close of Permian time all the continents were completely emergent, deserts were widespread, and the world had experienced the most severe and widespread glaciation of Phanerozoic time. For Paleozoic life, both animal and plant, it was a time of reckoning and many of the dominant groups failed to survive. Judged by the changes that occurred, the end of the Paleozoic Era was one of the great crises in the history of life on the earth.

Founding of the Permian System. The rocks on which this system was founded occupy a vast structural basin west of the Ural Mountains, measuring about 700 miles from east to west and 1000 miles from north to south. The region is divisible into two major structural elements, the Russian platform and the Uralian geosyncline. Over the platform the Carboniferous and Permian strata are nearly horizontal and have only moderate thickness, but as they pass eastward into the geosyncline they thicken greatly and undergo rapid facies changes.

Murchison visited this region briefly in 1841 and more extensively in 1842, primarily to see whether the Cambrian, Silurian, and Devonian systems recently erected in England and Wales would be recognizable and useful so far afield. He went under the patronage of Czar Nicholas II, and under the guidance of Count de Cancrine, the Imperial Minister of Finance, who provided every convenience for the field work, and he was accompanied by Lieutenant Koksharof, a trained geologist of the Imperial School of Mines.

At the end of the field season of 1842 he addressed to the President of the National Academy of Sciences in Moscow a letter proposing the Permian System, and upon returning to England he published it in the Philosophical Magazine. During the next three years he prepared his truly great monograph entitled "The Geology of Russia in Europe and the Urals." A reader of this classic will be amazed that he and his colleagues could have gained such a clear insight into the major features of so vast a region in two brief field seasons. The fact is that much of the local geology had already been worked out by Russian geologists and, by the principle of using fossils for correlation, which he had applied so successfully in England and Wales, he was able to synthesize all previous work and to interpret it in the light of his personal reconnaissance. It is also evident that he went to Russia hoping to find evidence for a new system overlying the Coal Measures. He went with the knowledge of certain formations in western Europe that contained fossil plants and reptiles that did not properly belong to the Coal Measures.

However, because the faunas in Russia were still imperfectly known, Murchison mistakenly assigned two major divisions of the Permian

Figure 13-2 Section across the Appalachian System from the Piedmont belt to the eastern edge of the Allegheny Plateau in Central Virginia. Length of section about 100 miles; vertical scale exaggerated. All this deformation is the result of the Alleghanian orogeny.

rocks to the Carboniferous System. The Artinskian detritals of the Urals (now recognized as Middle Permian) he correlated with the Millstone grit (Lower Carboniferous) of England, solely on the basis of lithologic similarity; and the limestones of the Ufa Plateau he assigned to the Upper Carboniferous. These mistakes caused confusion until as late as the 1930s but have since been resolved so that the geologists of Russia and of North America now agree on the base of the Permian System. The extensive development of the Permian System in North America was not appreciated until after 1920 when the geology of the Permian Basin of West Texas and southern New Mexico was studied. This region now affords probably the finest known section of the Permian System.

PHYSICAL HISTORY OF A CHANGING WORLD

Final Destruction of the Appalachian Geosyncline

Throughout the long Paleozoic Era the Appalachian geosyncline had subsided intermittently until the sedimentary deposits within it had attained a total thickness of some 50,000 feet. During this time Appalachia, a borderland of unknown eastward extent, had been rising intermittently and supplying most of the sediment that filled the trough. Its uplift had been accompanied by extensive orogeny and igneous activity, especially in the northern part, in every period since early in Ordovician time. Twice it had been thrust westward enough to deform the strata along the eastern margin of the trough, once during the Taconian orogeny of Late Ordovician time and again during the Acadian orogeny late in the Devonian, but neither of these seriously interrupted the continued evolution of the geosyncline. Sometime during the Permian Period, however, a much more pervasive westward compression folded all the deposits of the geosyncline. These are the great folds we now see in the Valley and Ridge Province. At the same time enormous thrust faults formed along the east side of this prov-

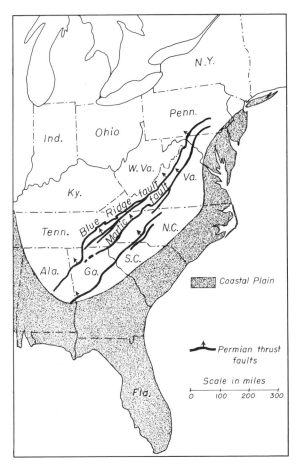

Figure 13-3 Map showing traces of the major thrust faults in the central and southern Appalachians. Overlapping Mesozoic and Cenozoic deposits of the Coastal Plain are shaded.

ince and in the region now involved in the Blue Ridge and the Piedmont belt (Figs. 13-2 and 13-3). The precise date of the folding is unknown because with this uplift the Appalachian geosyncline ceased to sink and the sediments derived from the highlands were carried far to the west. Basal Permian formations of the Dunkard Group (Fig. 13-5) remain on the plateau in western Pennsylvania, West Virginia, and eastern Ohio, and these beds were gently folded along with all the older formations. Obviously, therefore, the folding took place after Early Permian time. The mountains had been largely peneplaned before Late Triassic deposits were laid down, hence it is clear that the deformation occurred during or at the close of Permian time. This final deformation, so much greater and more widespread than the earlier ones, has commonly been called the **Appalachian orogeny** since it produced the visible Appalachian Mountain System. But since this was only the culminating phase of a long series of orogenies in Appalachia, it deserves a more distinctive name and we adopt the term **Alleghanian orogeny** proposed by Woodward [13]. With this deformation the Appalachian geosyncline ceased to exist and the eastern part of North America was never again invaded by the sea. This was the most profound change ever experienced by the eastern part of the continent, at least since Precambrian time. Can this mean that the Atlantic Ocean basin was now fully formed and that with this last push enough of the ocean floor had pushed under the geosynclinal area to thicken the continental plate and make it stable?

The closing stages of the Paleozoic Era witnessed similar great changes in many other parts of the world. The Ural Mountains were produced at this time, and western Europe was involved in Armorican orogeny. At the same time a great mountain system was arising in eastern Australia. This is the time also when Gondwanaland was breaking up and the southern continents were drifting apart. The epicontinental seas vanished from practically all the continents by the end of Permian time and they then stood higher than at any other time in the Phanerozoic except the present. With these changes deserts spread more widely than ever before, and the southern continents experienced their most extensive glaciation.

HISTORY OF NORTH AMERICA

Vanishing of the Midcontinent Sea. Early in the period shallow sea covered a large part of the western United States, as indicated in Map VIII and in the Midcontinent area the change from Pennsylvanian geography was gradual but ultimately profound. In Kansas and Nebraska, as in west-central Texas, the early Permian formations consist of alternating gray shales and thin limestones, but in successive horizons the marine faunas change progressively. At first the corals and crinoids and fusulines drop out, then the bryozoans and brachiopods, and finally the clams. The dwindling faunas were the response to increasing salinity that culminated during the deposition of the Wellington Shale and the precipitation of immense layers of salt. Then the sea disappeared completely and a thick sequence of redbeds accumulated over a desert floor. For a time near the close of deposition extensive sand dunes spread across northern Oklahoma (the Whitehorse Sandstone). The Kansas dead sea marks the last stand of the Paleozoic seas east of the Cordilleran region.

In the meantime the Oklahoma Mountains were being eroded to spread great debris aprons both north and south, largely separating the Kansas sea from that which still covered much of Texas.

The Permian Basin of the Southwest. West Texas and eastern New Mexico were occupied by a rapidly subsiding basin in which Permian deposits accumulated to a thickness of about 14,000 feet, of which all but the uppermost division is marine. This is the finest Permian section in America and probably the most important reference section for Permian rocks in the world; it involves complex facies changes of exceptional interest, discussed on page 292.

Physical Changes in the Cordilleran Region. Between the west Texas basin and the eastern Cordilleran trough that extended from northwestern Mexico to Idaho (Map VIII, left) the Permian beds are largely nonmarine, but from southeastern Arizona into southern California and northward across eastern Nevada and western Utah the Permian beds are thick and largely calcareous, indicating that no highlands were near. In central Idaho, however, the Seven Devils Formation, probably exceeding 10,000 feet in thickness, consists of clastic rocks and volcanics including andesitic flows and tuffs, rhyolitic flows, water-laid volcanic grits, and conglomerates with lenses of impure limestone, some of which bear fossils proving Permian age. In northern California the Lower Permian rocks

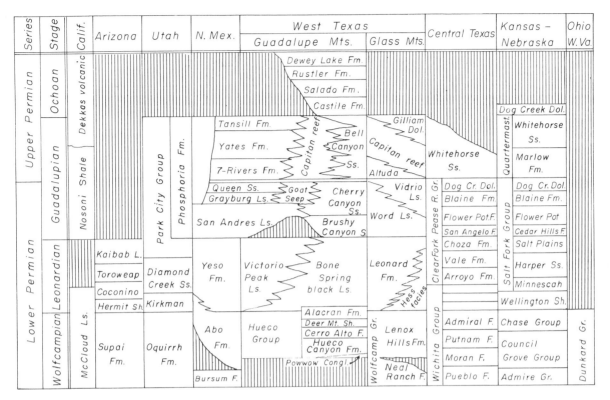

Figure 13-4 Correlation of important Permian sections.

are calcareous (McCloud Limestone) but are succeeded by thick dark shales of Late Middle Permian age (Nosoni Formation) which is overlain by thick volcanics. Here the interbedding of sediments and lava at the top of the Nosoni Formation dates rather precisely the beginning of volcanic activity. Other important areas of volcanism were the southern half of Alaska and central Mexico. This extensive volcanism marks the beginning of a profound change in the geologic history of the West Coast, which from time to time hereafter would be the scene of massive volcanism all the way from southern California to Alaska. Permian formations are locally thick and are widely distributed in British Columbia and northward into Alaska but have not yet been sufficiently studied to permit a clear understanding of the Permian history of this region.

The Ouachita Orogeny. At some time during the Permian a final northward compression shoved the thick Pennsylvanian deposits of the Ouachita trough into a series of great arcuate thrust sheets that form the Ouachita

Mountains of Arkansas and southeastern Oklahoma. The time of this thrusting cannot be closely dated, but since the Ouachita Mountains were shoved over the Arbuckle Mountain structures, which were formed during Pennsylvanian time, it seems highly probable that the movement occurred at some time during the Permian. In any event, this was the final orogeny to affect the region.

Correlation of the most important Permian formations is shown in Figure 13-4.

THE STRATIGRAPHIC RECORD

The Dunkard Group. The only Permian rocks east of the Mississippi River occupy an oval area of about 30,000 square miles in the western part of the Alleghany Plateau (Fig. 13-5). The character of these beds in the southeastern part of the area contrasts strikingly with that on the northwest side, though the two facies deeply intertongue in the middle

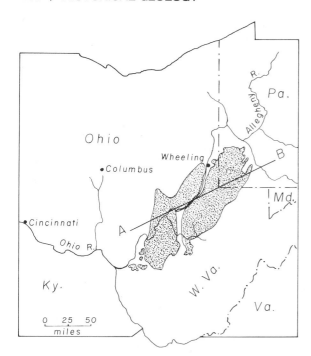

Figure 13-5 Left, the Dunkard Basin.

Figure 13-6 Below, lenticular channel sands in red shale and siltstone of the Dunkard Group near Marietta, Ohio. (C. O. Dunbar.)

Figure 13-7 Fresh-water limestone (Fishpot Limestone) with interbedded gray shale of the Dunkard Group west of Bridgeport, West Virginia. (C. O. Dunbar.)

of the basin and are certainly contemporaneous. To the southeast of line A-B of Fig. 13-5 they consist of red shales and siltstones with numerous interbedded sandstones. The sandstones are lenticular, vary rapidly in thickness, and are generally cross-stratified (Fig. 13-6). In many places they fill shallow valleys that had been cut in the underlying shales. Clearly the sandstones were deposited in and along the channels of aggrading streams and the redbeds are the deposits spread over the higher ground at times of flood. Contemporaneous deposits in the northwestern side of the basin are dark gray shales and siltstones with many local and lenticular coal beds and interbedded fresh-water limestones (Fig. 13-7). The coal beds record swamps choked with vegetation, the limestones mark shallow open lakes, and the gray shales imply poorly drained lowland in which much organic matter

was buried with the sediment.

The transition zone, where these strikingly different facies of the Dunkard Group intertongue, marks the boundary where the low, well-graded piedmont west of the Appalachian highlands met the poorly drained, humid lowlands. As the streams, heavily laden with red mud and sand from the warm, moist slope of Appalachia, meandered over the region, the sands were deposited chiefly in and near the channels, whereas the finer muds settled over higher ground during floods. Wherever the slopes were well drained during the dry season the muds remained red, but in the swampy lowland decaying organic matter destroyed the red pigment and turned the muds gray. Most of the sand was dropped on the lower slope and failed to reach the center of the basin.

By the end of Early Permian time deposition ceased here as the region began to rise, and the streams from Appalachia flowed through to spread their red muds farther west, in Kansas, Nebraska, Oklahoma, and Texas.

The Permian Basin. This area includes the standard section of the American Permian which, as in the U.S.S.R., is divided into Lower Permian and Upper Permian series. Here each series is further subdivided into two provincial stages as follows:

Upper Permian Series { Ochoan Stage
Guadalupian Stage
Lower Permian Series { Leonardian Stage
Wolfcampian Stage

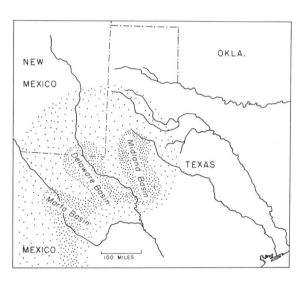

Figure 13-8 Sketch map of the deep Permian basins of west Texas in which the standard section of the Permian System in North America is preserved. (After P. B. King.)

Figure 13-9 North-south section across the Glass Mountains showing the Permian formations at the south end of the Delaware Basin.

Within a rather shallow, bowl-shaped seaway three distinct but interconnected basins sank deeply during Permian time (Fig. 13-8). Of these the stratigraphy of the Delaware Basin is best known and will afford the basis for the following discussion. Uplift long after Permian time raised the north end of the basin to an elevation of about 9000 feet and here the Guadalupe Mountains form an imposing sight, their western face (Fig. 13-1) being fully exposed in a great fault scarp facing the graben of Salt Flat. The basal stage is not exposed here but the Glass Mountains along the south side of the basin afford fine exposures of the lower stages (Fig. 13-9).

The Wolfcampian Stage was named for the fine exposures in the Wolfcamp Hills along the front of the Glass Mountains. Here the early Permian sea was lapping up against the foothills of Llanoria and present exposures, extending for some 40 miles along the face of the mountains, roughly parallel the old shoreline. As a result the Wolfcampian beds in their type region thicken and thin irregularly and show complex facies changes with thick masses of coarse limestone conglomerate grading laterally into shales. Deep wells in the basin northeast of the Glass Mountains encountered a much greater thickness of beds carrying the Wolfcampian faunas but there the structure is complicated and the section may be duplicated by faulting. The Wolfcampian sequence is particularly important because it contains a prolific fauna of fusulines that permit correlation with other regions, not only in North America but in South America, Japan, and the type region of the Permian in the U.S.S.R. This is the zone of *Pseudoschwagerina* (Fig. 13-10).

In the Hueco Mountains, which are near the western side of the Guadalupe Basin, the

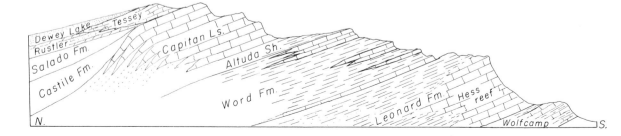

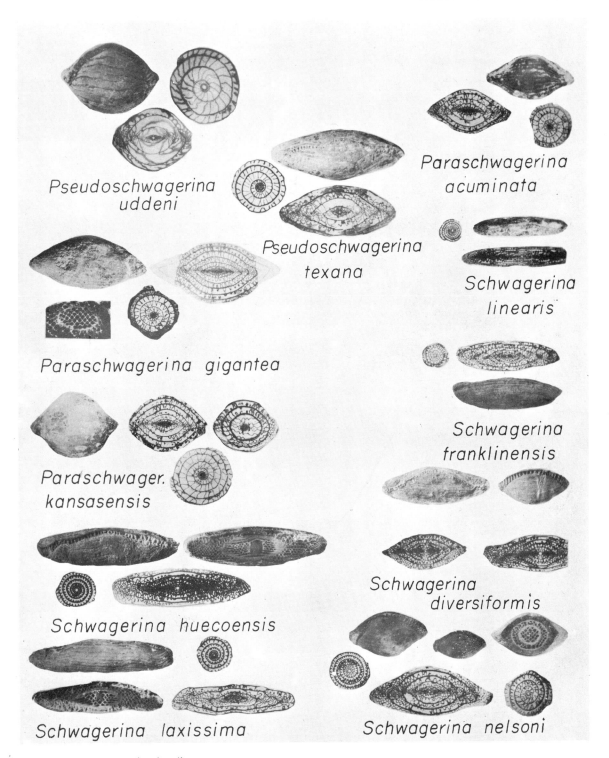

Figure 13-10 Wolfcampian fusulines.

Figure 13-11 Typical outcrop of the Leonardian strata in the Lennox Hills in the western part of the Glass Mountains. (C. O. Dunbar.)

Figure 13-12 North-south section across the Permian Reef Complex at the north end of the Delaware Basin. (Adopted from P. B. King [6].)

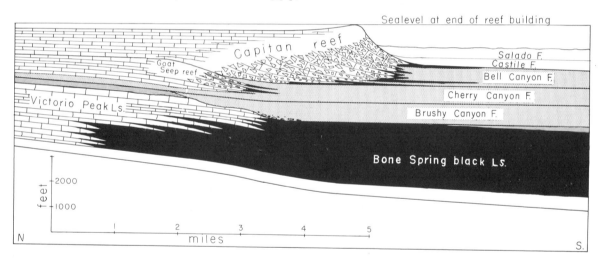

Wolfcampian strata consist of more evenly bedded limestones with interbedded shales, and reach a total thickness of about 1600 feet [12]. Here they are known as the Hueco formation. Deep wells reveal that in the center of the Delaware Basin the Wolfcampian equivalents are dark shales.

The Leonardian Stage also is based upon exposures in the Glass Mountains where it exceeds 2000 feet in thickness and consists of gray silty shales with interbedded light gray limestones (Fig. 13-11). Dipping eastward under cover in the Staked Plains, the rocks of this stage reappear in central Texas as a complex of redbeds with channel sands and local units of gray shales and several thin but widespread limestones. Here the beds constitute the upper part of the Wichita

Group and the Clear Fork Group. The channel sands and gray shales of this region have yielded a large fauna of primitive reptiles and labyrinthodont amphibians (Figs. 13-29 and 13-30).

Along the northwest rim of the Delaware Basin rocks of the Leonardian Stage are extensively exposed in the face of the Sierra Diablo and in the west face of the Guadalupe Mountains (Fig. 13-12) where they are wholly marine but lithologically unlike the type section in the Glass Mountains, consisting of two strikingly different facies, the white Victorio Peak Limestone and the thin-bedded black Bone Spring Limestone (Fig. 13-13). During Leonardian time the Marfa, Delaware, and Midland basins had subsided faster than they were filled so that the water here was several hundred feet deep while the surrounding area to the northwest, north, and east remained as a broad platform only partly awash with the sea. On the submerged area of the platform were broad, shallow lagoons and landward beyond these stretched a broad lowland over which streams were spreading red muds and sandstones. The white Victorio Peak Limestone was deposited on the platform but in the basins where the water was several hundred feet deep the bottom became stagnant and foul so that bottom-dwelling organisms died out and the organic matter settling from above produced a dark pigment of carbon and hydrocarbons. Under these conditions the black Bone Spring Limestone

(Fig. 13-13) accumulated. Around the margins of the basins these contrasting deposits of limestone intertongue.

The Guadalupian Stage has its type section in the great Permian reef complex of the Guadalupe Mountains (Figs. 13-14 and 13-

Figure 13-13 Bone Spring black limestone in road cut below Guadalupe Peak, on Highway 62. (C. O. Dunbar.)

Figure 13-14 Block diagram to show facies relations in the Permian reef complex at the end of the Capitan reef growth.

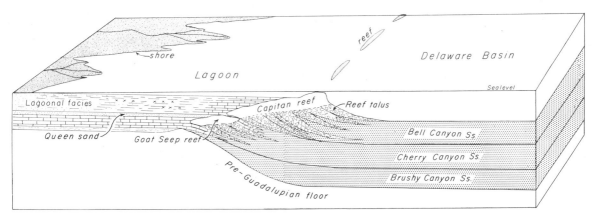

Figure 13-15 Front of the Capitan reef resting on steeply dipping reef talus at the mouth of McKittrick Canyon about 8 miles east of Guadalupe Peak. (C. O. Dunbar.)

Figure 13-16 Guadalupe Peak from the south, resting on the Brushy Canyon Sandstone. (C. O. Dunbar.)

15) where it has been comprehensively studied by P. B. King [6] and by N. D. Newell [8]. By Guadalupian time the Delaware Basin had become many hundreds of feet deep but the trough in Mexico was then deep enough to allow fresh seawater to flow in fast enough to keep the bottom suitable for a rich benthonic fauna. Meanwhile excessive evaporation in the shallow lagoons caused a sheet of water to flow in over the margin of the shelf. Here warming and evaporation caused a concentration of calcium salts that formed a perfect environment for reef-forming organisms— chiefly algae, bryozoans, and sponges. Thus a calcareous reef began to form all around the margins of the basin. Since it could not rise above sealevel, growth was chiefly along the seaward face while within a few miles back of the reef the lagoonal waters were so saline that no organisms could survive. Meanwhile, in some manner not yet well understood, a vast amount of quartz sand was transported into the basins, probably through channels crossing the reef. Thus throughout Guadalupian time three very different facies were maintained—lagoonal deposits on the shelf, reef deposits along its margin, and normal marine but deeper water sandstones in the basin (Fig. 13-14).

The reef front stood hundreds of feet above the basin floor and as storm waves continually hammered at it, debris in the form of blocks and chunks of reef rock rolled down to form a great talus slope (Fig. 13-15). In the upper part of the talus some of the blocks of reef rock are as much as 10 feet in diameter but downslope the talus is progressively finer as it tongues out into thin wedges of calcarenite, some extending many miles out into the basin. Enormous fusulines of the genus *Polydiexodina* are locally very abundant in the reef rock. In the lagoons back of the reef extensive layers of gypsiferous shale and beds of gypsum were formed.

Due to post-Permian deformation the reef dips eastward from its high point at Guadalupe Peak (Fig. 13-1) and passes below the surface near Carlsbad, New Mexico. Carlsbad Cavern lies within the reef rock. The eroded face of the reef at Guadalupe Peak (Fig. 13-16) is an imposing sight. Here the white reef rock towers above the Delaware Mountain Sandstone in a sheer cliff some 1400 feet high.

The Ochoan Age witnessed a profound change in the regimen of this region. As a result of regional warping of a threshold in Mexico, the sealevel fell below the rim of the deep basins leaving the reef exposed and the platform emergent. Under the intense aridity little detrital sediment now reached the Delaware and Midland basins and, as evaporation continued, each became a dead sea in which colossal amounts of anhydrite and salt were precipitated (Fig. 13-17). The Castile Formation, as much as 2000 feet thick in the east-central part of the Delaware Basin, consists largely of laminated anhydrite in which a number of thick salt beds are interstratified. The Salado Formation, as much as 2400 feet thick in the eastern part of the basin, is largely made of salt (halite) with interbeds of anhy-

Figure 13-17 West-east section of the Ochoan Stage in the northern part of the Delaware Basin. (Adapted from Kroenlein.)

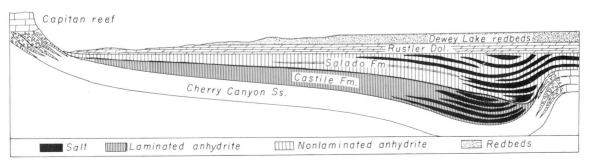

Figure 13-18 Section of a deep well core (natural size) showing the finely laminated anhydrite of the Castile Formation. From Gresham Well No. 1 in Culbertson County, Texas. This is part of the core upon which Udden's study was made. (Yale Peabody Museum.)

drite and, in its middle portion, it includes lenses of potash salts, the chief of which are

Sylvite   (KCl)
Carnallite (KCl—MgCl$_2$—6H$_2$O)
Polyhalite (CaSO$_4$—MgSO$_4$—K$_2$SO$_4$·2H$_2$O)

This is one of the world's great deposits of potassium salts.

The lamination so common in the Castile gypsum is well shown in Fig. 13-18. The light layers are nearly pure anhydrite, and the thin, dark layers have considerable organic matter and microscopic crystals of calcite. The well from which this core was taken passed through more than 1200 feet of such laminated material. Udden [10] gave reasons to suspect that these are seasonal precipitates, the purer layers representing the drier season and the darker layers a more humid season. On this assumption, and from a count of the laminae in this well core, Udden estimated that it required 306,000 years to deposit the Castile and Salado formations.

The total evaporation implied by so much salt is colossal. King has recently estimated that if the salt were precipitated from normal seawater which constantly flowed into the basin to replenish the loss, an average evaporation of about $9\frac{1}{2}$ feet per year over the 10,000 square miles of the basin would be required for a period of 300,000 years. Since the average evaporation in Death Valley is only about $11\frac{1}{2}$ feet, this is a striking commentary on the Late Permian climate of this region.

After final deposition of the salts a brief influx of normal seawater produced the Rustler Dolostone and introduced a limited fauna of marine organisms. With this exception the Ochoan Stage is entirely unfossiliferous. It is therefore a very unsatisfactory standard section for the youngest part of the Permian System. It now appears probable that marine deposits equivalent in age to the Ochoan occupy much of the western Cordilleran trough that extended from northern California along the western margin of Canada. These deposits contain an Asiatic fauna that did not get into the seaway which led to the Permian Basin in west Texas.

Powell Plateau

Figure 13-19 General view of the north wall of the Grand Canyon. At the top are four Permian formations having a combined thickness of about half a mile. These are, in ascending order, the red sandy Supai Shale (Ss), the Coconino Sandstone (C), the Toroweap Formation (T), and the Kaibab Limestone (K). These are also shown in Fig. 13-20. Below the Permian lies the Redwall Limestone (R) of Mississippian age, the Muav Limestone (M), the Bright Angel Shale, and the Tapeats Sandstone (T), all of Cambrian age. The dashed line Pu marks the base of the Paleozoic erathem, which lies with profound unconformity on the Precambrian Unkar Group (U). The lower dashed line marks the boundary between the Unkar beds and the much older Vishnu Schist (V), which is exposed in the inner gorge above the Colorado River. (U. S. Geological Survey.)

Cordilleran Region. As indicated in Map VIII, the southern part of the eastern Cordilleran trough was occupied by a shallow sea during much of Permian time. This left an imposing marine record in southern and eastern Nevada, most of Utah, southeastern Idaho, and extreme western Wyoming. Farther east, redbeds accumulated over a great area surrounding the Colorado Mountains.

Grand exposures of these Permian formations of the Far West are to be seen in the walls of the Grand Canyon (Figs. 13-19, 13-20), where they may be studied. The general relations are suggested by Fig. 13-20. The Kaibab Limestone is a key horizon rimming the inner gorge of the canyon in unscalable cliffs from 500 to 600 feet in height. It persists as a marine horizon from southern Nevada to northeast Utah, but in southeast and east-central Utah it grades laterally into sandstone like the Coconino. The Toroweap Formation, originally included in the Kaibab, consists of a marine limestone with redbed members both

above and below it. The limestone thickens and largely replaces the redbeds in southern Nevada, but thins toward the east and is replaced first by redbeds and then by the massive, cliff-forming, light gray Coconino Sandstone (Fig. 13-21). Farther east, massive windblown sands (DeChelly Sandstone) accumulated to a thickness of hundreds of feet (Fig. 13-22).

A great area of Mesozoic rocks separates the exposures of the Grand Canyon region from those of northern Utah and southeastern Idaho, where the Phosphoria Formation is the most widely distributed and best known Permian deposit. Where typically developed, it includes a basal member of black phosphatic shale and a thicker, upper member of cherty limestone, but toward the northeast it intertongues with, and is finally replaced by, red shales (lower part of the Chugwater Formation) that extend across Wyoming and into the rim of the Black Hills of South Dakota. Its marine fossils indicate that the Phosphoria

Figure 13-20 Idealized diagram showing facies changes in the Permian formations in the Grand Canyon region. (Data from McKee.)

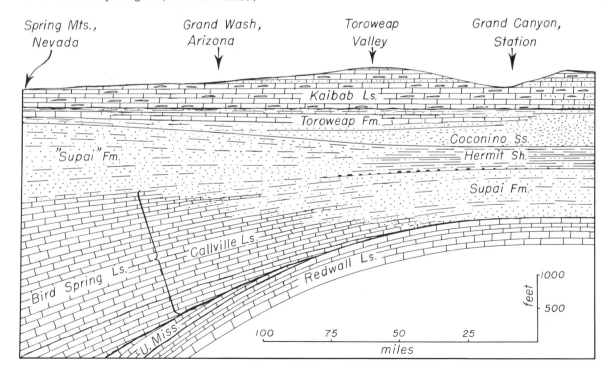

Figure 13-21 Coconino dune sandstone in the walls of Walnut Canyon near Flagstaff, Arizona. (C. O. Dunbar.)

Formation is younger than the Kaibab, the Kaibab being of Leonardian Age and the Phosphoria, Guadalupian. In the Confusion Range of western Utah and in the Provo region southeast of Salt Lake City, Kaibab equivalents are now known to underlie the Phosphoria complex.

CLIMATE

In North America the evidence for aridity is impressive. The enormous deposits of salt in Kansas (of Leonardian Age) and of the Permian basins in Texas and New Mexico have been estimated to amount to 30,000,000,000,-000 tons and would have required the evapora-

tion of 22,000 cubic miles of seawater of normal salinity. The great volume of widespread dune sand adds to the picture. Both these features suggest warmth and this may be confirmed by the presence of large fusulines, big corals of several kinds, and abundant brachiopods in Grinnell Island in Arctic Canada [4]; also by a prolific brachiopod fauna in central East Greenland, Spitsbergen, Bear Island, and Novaya Zemlya, and by abundant fusulines in extreme northeastern Greenland where the early Permian beds peek out from under the Greenland ice cap.

Europe was also the scene of intense aridity. The salt deposits of the Kungurian Formation that underlies the Permian basin of the

Figure 13-22 Massive DeChelly Sandstone and Canyon DeChelly about 130 miles east of the Grand Canyon. This is in a great area of desert sandstone generally showing large-scale cross-bedding. Here the sand is about 800 feet thick. (Spence Air Photos.)

U.S.S.R. about Solikamsk is certainly one of the three greatest known, and the Stassfurt salt deposits in the Permian of Germany is the third, exceeded only by that in the Permian basin of Texas and New Mexico.

Glaciation in the Southern Continents. The evidence of aridity and warmth in the northern continents stands in starkest contrast to the evidence of very widespread glaciation in the southern landmasses (Fig. 13-23). Early in the period a large ice sheet covered South Africa where the ice moved southwestward into the sea (Fig. 13-23). The southern end of Madagascar was also ice capped and in the Durban area of South Africa the ice moved inland, probably from Madagascar, which then was part of the mainland. Extensive glaciation occurred simultaneously in western Australia and in southeastern Australia and Tasmania where the ice was moving northward [3]. In South America a vast ice cap covered parts of Paraguay, Uruguay, and Brazil where it reached as far north as the present equator. India presents a special problem, for it, too, was heavily glaciated in Early Permian time when the ice radiated from the region of the modern Arivalle highlands. Glacial deposits of Permian age have recently been discovered also in Antarctica.

It was this remarkable spread of glacial de-

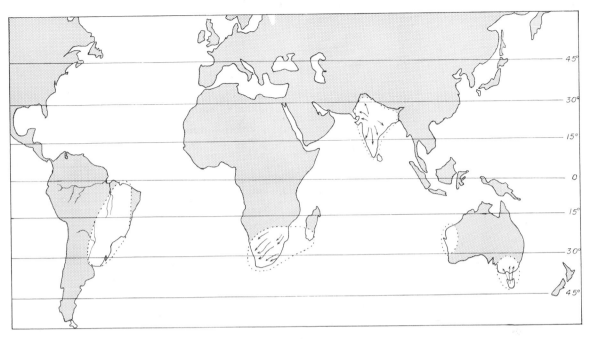

Figure 13-23 Map showing the regions of Permian ice sheets. Arrows show the direction of ice movement.

Figure 13-24 Glaciated floor overlain by the Dwyka tillite near Kimberley, South Africa. (A. P. Coleman.)

posits in the southern lands that led Wegener to postulate continental drift. He believed that at the time of glaciation these lands were all clustered near the south pole and that they later drifted apart. As supplemental evidence he cited the fact that the distinctive *Glossopteris* flora is associated with the glacial deposits in each of these regions, and is found nowhere else. According to this theory India

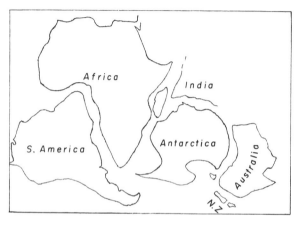

Figure 13-25 DuToit's arrangement of the southern continents during Early Permian time.

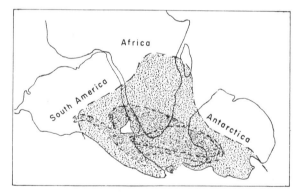

Figure 13-26 Reconstruction of a portion of Gondwanaland during the late Paleozoic glaciation, by Frakes and Crowell [2]. It is based on facies changes in the glacial and associated deposits. The lightest shading indicates the distribution of terrestrial tillites and associated sediments; intermediate shading represents shallow subaqueous tillites and interbedded terrestrial sediments; the darkest shading represents deep glacio-marine deposits and is postulated to indicate the beginning of the Atlantic Basin.

was still a part of old Gondwanaland and has since drifted far to the north to join the Asiatic mainland.

Although Wegener's theory seemed to many geologists at the time to be preposterous, it has recently gained wide acceptance and the paleomagnetic evidence appears now to offer a basis for reconciling the contrasts between the Permian climate of North America and Europe and that of the glaciation in the southern continents. As summarized by Van Hilten, North America was much farther south than at present, the equator crossing the southern tip of Hudson Bay and the north tip of Newfoundland. This would place the Permian Basin of West Texas in the Horse Latitudes (which is now the belt of the most extensive deserts). It would also bring Europe much farther south with the equator crossing just south of Germany and the Solikamsk salt basin of the U.S.S.R. in or near the northern Horse Latitudes. At the same time the southern tip of Africa was below 60°S latitude and latitude 80°S crossed the center of Australia. Many attempts have been made to arrange the southern continents as they were during the time of glaciation.

The well known reconstruction by DuToit [1] (Fig. 13-25) was recently confirmed, in part, on the basis of comparison of the bedrock masses, the structural trends, and the glacial deposits (Fig. 13-26). Admittedly more evidence from many sources is needed to establish the ideas suggested above or to disprove them, but we can feel confident that it will be available in the next few decades.

The time of the glaciation is best dated in western Australia where the ice pushed out to sea leaving thick glaciomarine deposits with interbedded fossils of Early Permian (Wolfcampian) age. Here also the glacial deposits are overlain by a limestone with an abundant marine fauna of Leonardian age [11]. In South Africa *Glossopteris* was overridden by the ice and here the Upper Permian has a great variety of reptiles that indicate an appreciably warmer climate. In India also the glacial till includes spores of *Glossopteris* and is overlain by the Lower *Productus* Limestone with abundant fusulines of Leonardian Age.

These facts indicate that the glaciation occurred early in the period.

It must be noted that if Africa was so close to the south pole in Early Permian time, its movement to the north must have been relatively rapid since during Middle and Late Permian time the prolific fauna of reptiles in South Africa could hardly have survived freezing winters. Such rapid movement may not be unreasonable. For example, on the basis of paleomagnetism, Irving [5] finds that in Early Carboniferous time New South Wales in southeastern Australia was in latitude 30°S but by the end of the Carboniferous and the beginning of glaciation it was in latitude 75°S, close to the present margin of Antarctica. This would involve a migration of about 2° latitude per million years.

THE LIFE OF PERMIAN TIME

It was once believed that the life of Paleozoic time was being greatly reduced during the Permian Period. On the contrary we now know that this was a time of extreme specialization and the marine invertebrates were legion, land plants were abundant where the environment permitted, and the amphibians and reptiles were abundant and were undergoing rapid specialization. The great decline of many Paleozoic groups of organisms came during the closing stages of the Permian.

Decline of the Carboniferous Floras. In the northern hemisphere the dominant types of Pennsylvanian plants lived on into Permian time. Lepidodendrons, sigillarias, calamites, cordaites, and seed ferns were the common forest types during the early part of the period. These swamp-dwelling plants were ill adapted to the oncoming aridity and to winter cold. With the passing of the period, therefore, hardier stocks with reduced foliage evolved, or became predominant, as the Pennsylvanian types declined. By the close of the period the great scale trees were almost extinct. The cordaites were likewise nearly gone, having first given rise to the conifers. Seed ferns were rare after the close of the period, and the race died out in the Jurassic.

True conifers rapidly sprang into the lead

Figure 13-27 Glossopteris leaves with attached reproductive organ. (After Plumstead.)

as the dominant type of woody trees, while primitive cycadeoids (allies of the sago palm) foreshadowed the expansion of higher plants in the Mesozoic.

The Glossopteris Flora of the Glaciated Regions. Throughout the glaciated regions of the southern hemisphere the most distinctive plants were the **tongue ferns** of the genera *Glossopteris* and *Gangamopteris*. These bore clusters of simple, spatulate leaves that apparently arose from creeping stolons. They produced winged spores, and some have been found with the reproductive organs attached to individual leaves by a slender petiole (Fig. 13-27), but uncertainty still exists as to whether these were spore cases or seed vesicles. Accordingly the position of these strange plants in the plant kingdom is still a subject of controversy.

Gangamopteris has been reported by Du-Toit "jammed in between the boulder beds [of the Dwyka Tillite] and the glaciated floor beneath it" near Strydenburg and at Vereening in South Africa. Elsewhere the characteristic winged spores have been found at several places in the tillite. The common association of the *Glossopteris* flora with glacial deposits may indicate that these strange plants were

Figure 13-28 A primitive relative of the dragonflies (*Dunbaria*) with color marking of the wings well preserved. Found in the Early Permian beds near Elmo, Kansas. The wing spread of this species is only about 1.5 inches but an associated form, *Megatypus,* had a wing spread of about 15 inches. (Yale Peabody Museum.)

especially adapted to a cold environment.

By Middle Permian time some members of the flora had migrated as far as northern Russia and the Altai Mountains of Siberia, but none ever reached western Europe or North America. Along with the *Glossopteris* flora there lived in the southern hemisphere many ferns, conifers, and calamites.

Animals of the Lands

Insects. Insects underwent rapid evolution during Permian time. No less than 20 orders have been described from Insect Hill near Elmo, Kansas, and most of these are also abundant in the U.S.S.R. and in Australia. Among these are forerunners of the dragonflies (Fig. 13-28) and of the beetles. One of the dragonflies had a wing spread of about 13 inches, but on the whole, Permian insects were generally smaller than those of Carboniferous time.

Sprawling Labyrinthodonts and Reptiles. Even in the semiarid regions the old labyrinthodont amphibians clung to the stream courses with surprising success. Their heavily armored skulls are locally abundant in the old channel sands in the redbeds of Texas and Oklahoma and in the fluvial deposits of Germany, South Africa, and the U.S.S.R. A

characteristic labyrinthodont is illustrated in Fig. 13-29. It was about the size of a half-grown Florida alligator. Nearly all of the amphibians had broad heads, thick bodies, and short stubby legs. As Huxley once said, "They pottered with much belly and little leg, like Falstaff in his old age."

Reptiles increased greatly in variety during Permian time. Early Permian forms are known chiefly from the redbeds of Texas and Oklahoma and from contemporaneous beds in the U.S.S.R. But Middle and Late Permian reptiles are amazingly abundant in South Africa and in the northern part of the U.S.S.R. and are found also in India and in Brazil. Before the close of Permian time reptiles had undoubtedly mastered the lands and a few had reverted to aquatic life, as have the modern crocodiles and alligators.

Most of the Permian reptiles had long bodies, long tails, and short legs. They were sprawlers. A few had short bodies, stubby tails, and stout legs. Some had blunt teeth adapted to crushing the shells of mollusks or crustaceans; others were toothless (as are modern turtles) and may have been herbivorous; but many had grasping and cutting teeth and were undoubtedly carnivores. The most bizarre were the pelycosaurs or "finbacks" (Fig. 13-30). In these the neural spines were greatly

Figure 13-29 *Eryops,* a characteristic Permian labyrinthodont amphibian from the Lower Permian of Texas. Length about 5 feet. (After a model by Dwight Franklin.)

Figure 13-30 A Permian landscape in Texas showing characteristic "finback" reptiles and associated vegetation. 1, *Edaphosaurus;* 2, *Dimetrodon;* 3, *Sphenacodon;* 4, a pelycosaur, *Ophiacodon;* 5, a primitive reptile, *Araeocelis;* 6, ferns; 7, *Lepidodendron;* 8, one of the last of the sigillarias; 9, *Walchia,* a conifer. (Part of a mural in Yale Peabody Museum, painted by Rudolph F. Zallinger.)

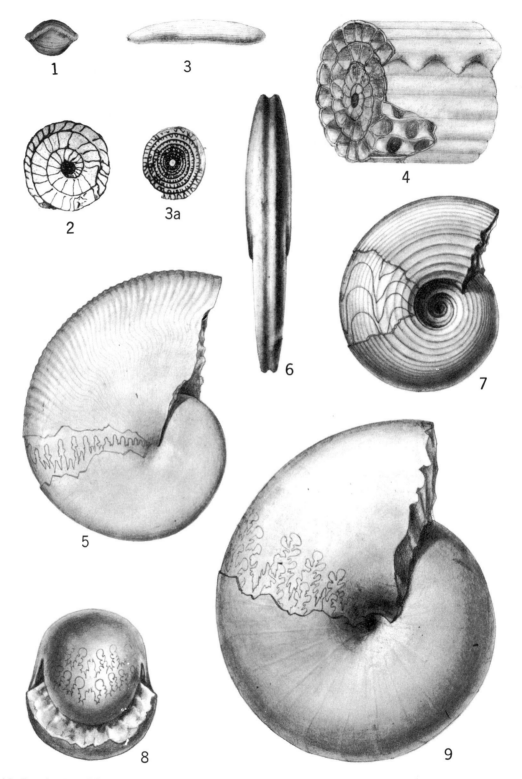

Plate 11 Permian Fusulinids (1–4) and Ammonites (5–9).
 Figure 1, *Pseudoschwagerina uddeni;* 2, enlarged section of same; 3, *Parafusulina wordensis;* 3a, enlarged section of same; 4, model of a portion of a shell showing septa; 5, 6, *Medlicottia whitneyi* (lateral and edge views); 7, *Gastrioceras roadense;* 8, *Waagenoceras dieneri;* 9, *Perrinites vidriensis.* All natural size except 2 and 4. (Drawn by L. S. Douglass.)

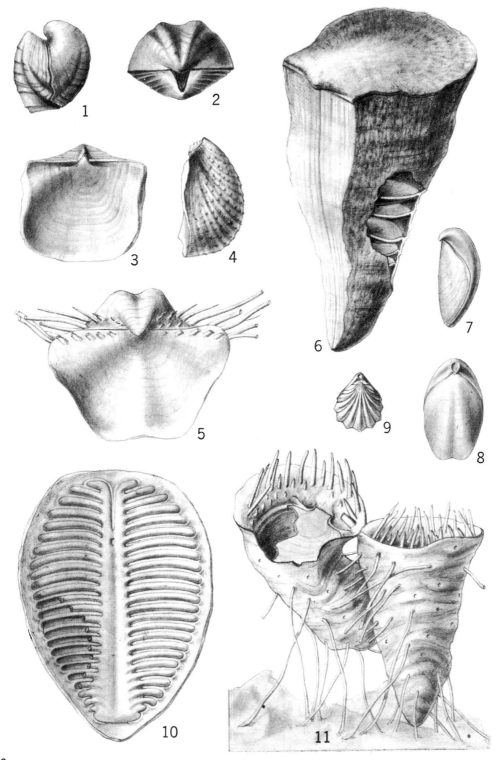

Plate 12

Figures 1, 2, *Parenteletes latesinuatus;* 3, 4, *Aulosteges medlicottianus;* 5, *Horridonia horrida;* 6, *Scacchinella gigantea* (with break in conical ventral valve showing internal septa); 7, 8, *Dielasma angulatum;* 9, *Hustedia meekana;* 10, *Leptodus americanus* (showing skeletonized dorsal valve on its seat below the overarching spines; the dorsal valve also broken); 11, *Prorhichtofenia permiana* (two specimens in position of growth). All natural size. (Drawn by L. S. Douglass.)

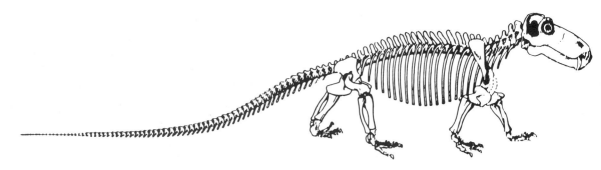

Figure 13-31 Above, skeleton of a theriodont reptile, *Titanophoneus,* from the Upper Permian (Kazanian) near Ichaevo on the Volga River south of Kazan, U.S.S.R. The animal was about 11 feet long. (Courtesy of U. A. Orlov.)

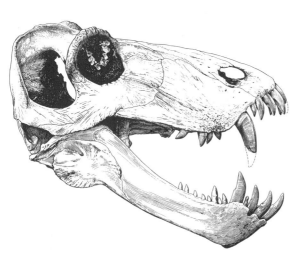

Figure 13-32 Left, skull of *Titanophoneus* showing the mammal like differentiation of the teeth. Same as Fig. 13-31. (Courtesy of U. A. Orlov.)

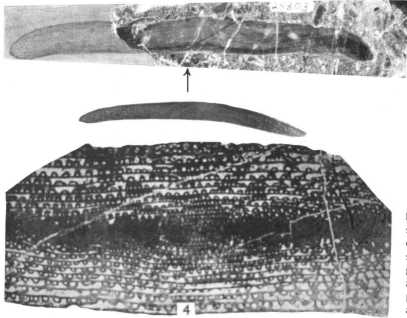

Figure 13-33 The largest known specimen of a fusulinid foraminifera. Above, a full grown microspheric individual; middle, a half grown microspheric shell; below, axial thin section of the middle part of the young individual shown above it. (Yale Peabody Museum.)

Figure 13-34 Restoration of part of a shallow Permian sea floor showing "horn" corals, many brachiopods, several nautiloid cephalopods, and several kinds of sponges. Siliceous sponges, upper right, grew to a height of 2 feet or more. A diorama in the U. S. National Museum. (Courtesy of the Smithsonian Institution.)

extended and supported a fleshy fin. Grooves on the neural spines indicate abundant blood vessels. It has been suggested that the fins served as radiators to cool the blood of these desert inhabitants!

Far greater interest centers in a group of stout-bodied flesh-eaters belonging to the reptilian order Therapsida, among which arose the ancestors of the mammals. A typical example, *Titanophoneus*, is shown in Figs. 13-31 and 13-32. The therapsids were not sprawlers but carried the body off the ground like a modern carnivore and the differentiation of the teeth into incisors, canines, and jaw teeth reveal their relations to the mammals which were to appear in the next era. The therapsids are known chiefly from the Upper Permian of South Africa and from the Permian Basin of the U.S.S.R.

Specializations among the Marine Invertebrates. The marine invertebrates evolved gradually out of Pennsylvanian ancestors. Some groups advanced steadily into progressive types while others assumed extravagant specializations before becoming extinct, and others, already on the decline, gradually died out.

The **fusulines** were amazingly abundant until near the end of the period and many of them were larger than any in the Pennsylvanian rocks. The largest individual known was approximately 4 inches long and 1 cm in diameter (Fig. 13-33) but most were less than an inch long. Previously attention was called to the inflated genus *Pseudoschwagerina*, which is an important guide to the Lower Permian formations. In spite of their extravagant evolution during Permian time not a single fusuline survived the close of the period.

Brachiopods (Plate 12) also underwent great expansion and many showed extreme specialization and made small "patch" reefs (Fig. 13-34). Among these the productids were most prolific and showed the most extreme specialization. In spite of being the most abundant "shellfish" in the Permian seas, very few brachiopods survived the end of the period and all the highly specialized types became extinct.

Bryozoans were also extremely abundant, especially the lacy types (fenestellids), but three of the great Paleozoic orders died out before the close of the period.

Crinoids too were prolific and varied during most of Permian time but all three Paleozoic orders died out by the end of the period except for a few genera that survived to start the existing order in the Triassic Period.

Blastoids had found an asylum in the region of Timor in the East Indies and here they underwent great specialization during Permian time. A few also migrated into Australia, but none of these survived the end of the Permian.

Echinoids were locally abundant but of all the Paleozoic types only a few survived to give rise to the existing stocks in Triassic time.

The history of the **mollusks** was different. Pelecypods and gastropods diversified steadily but conservatively, whereas the ammonoids diversified greatly into a number of stocks, some including fairly large species (Plate 11). Even after this rapid evolution, however, all but a few genera died out at the close of the period, but these few gave rise to another great burst of evolution during Triassic time.

Mass Extinctions. The mass extinctions of major groups of animals, both on the land and in the sea, marks the close of the Paleozoic Era as one of the great crises in the history of life, and the reason is still an enigma. Drastic changes in the environment, both physical and climatic, might seem to be a probable reason were it not for the fact that the glaciation was over near the beginning of Middle Permian time and the great expansion of the reptiles occurred later. Moreover, the marine invertebrates had made remarkable specializations adapting to the changing environment all through the period. As Newell has written:

One of the great groups of animals that disappeared at this time was the fusulinids, complex protozoans that ranged from microscopic sizes to two or three inches in length. They had populated the shallow seas of the world for 80 million years; their shells, piling up on the ocean floor, had formed vast deposits of limestone. The spiny produced brachiopods, likewise plentiful in the Late Paleozoic seas, also vanished without descendants. These and many other groups dropped suddenly from a state of dominance to one of oblivion.

By the close of the Permian, 75 percent of the amphibian families and more than 80 percent of the reptile families had also disappeared [9].

REFERENCES

1. DuToit, A. L., 1954, *The geology of South Africa.* 3rd ed., New York, Hafner Publishing Co., p. 13.
2. Frakes, L. A. and Crowell, J. C., 1968, *Late Paleozoic glacial facies and the origin of the South Atlantic Basin.* Nature, v. 217, pp. 837–838.
3. Hamilton, Warren, and Krinsby, David, 1967, *Upper Paleozoic Glacial Deposits of South Africa and Southern Australia.* Geol. Soc. America Bull., v. 78, pp. 783–800.
4. Harker, P. and Thorsteinsson, R., 1960, *Permian rocks and faunas of Grinnell Peninsula.* Geol. Surv. Canada Mem., 309, p. 189, 25 plates.
5. Irving, E., 1966, *Paleomagnetism of some carboniferous rocks from New South Wales and its relation to geological events.* Jour. Geophys. Research, v. 71, No. 24, pp. 6025–6051.
6. King, Philip B., 1948, *Geology of the southern Guadalupe Mountains, Texas.* U. S. Geol. Surv. Prof. Paper 215, pp. 1–183.
7. King, Ralph, 1947, *Sedimentation in Permian Castile Sea.* Amer. Assoc. Petrol. Geol. Bull., v. 31, pp. 470–477.
8. Newell, Norman D., et al., 1953, *The Permian reef complex of the Guadalupe Mountain Region.* W. H. Freeman and Co., 236 pages.
9. Newell, N. D., 1963, *Crises in the history of life.* Scientific American, v. 208, no. 2, p. 79.
10. Udden, J. A., 1924, *Laminated anhydrite in Texas.* Geol. Soc. America Bull., v. 35, pp. 347–354.
11. Veevers, J. J., and Wells, A. T., 1961, *The geology of the Canning Basins, western Australia.* Australian Bur. of Min. Resources, Geol. and Geophysics Bull. No. 60.
12. Williams, Thomas E., 1963, *Fusulinidae of the Hueco Group (Lower Permian, Hueco Mountains. Texas.)* Yale Peabody Museum Bull. 18, pp. 1–122.
13. Woodward, H. P., 1957, *Chronology of Appalachian folding.* Amer. Assoc. Petroleum Geologists, Bull., v. 47, pp. 2313–2327.

PART III

PART III

THE MESOZOIC WORLD

Figure 14-1 Glimpse into a Triassic landscape early in the Mesozoic Era. Plants: 1, a cycad *Macrotaeniopteris;* 2, a primitive cycadeoid, *Wielandiella;* 3, a conifer, *Voltzia.* Animals: 4, a primitive reptile ancestral to the dinosaurs, *Saltopsuchus;* 5, one of the smallest dinosaurs, *Podokesaurus* (about 3 feet long); 6, one of the largest Triassic dinosaurs, *Plateosaurus,* a probable ancestor of the great sauropods of the next period (length about 20 feet); 7, a mammal-like reptile, *Cynognathus.* (Part of a mural by Rudolph Zallinger in Yale Peabody Museum.)

CHAPTER 14 *The Triassic Period*

The Mesozoic Era. The decimation and extinction of Paleozoic plants and animals culminating at the close of the Permian period is only one aspect of the great change in the fossil record that prompted the founders of geology to distinguish a middle chapter, or **Mesozoic Era,** in the history of life. Newly evolving, generally more efficient organisms took over the dominant roles in all major Mesozoic environments, carrying evolution of life to new frontiers of specialization in the sea, on the land, and in the air. Mollusks dominated the invertebrate life in Mesozoic seas and among them the ammonoids were particularly conspicuous. Both on the land and in the sea reptiles became the ruling vertebrates and the record of their spectacular radiation, which includes the genesis of birds and mammals, is one of the best known and most fascinating family histories in the evolution of life. Land plants underwent an equally great change and undoubtedly had strong influence on the evolution of reptiles and other vertebrates.

Extinction and decline of dominant organisms marks the end of the Mesozoic as it does the beginning but the changes are perhaps not quite as drastic. Many groups of animals and plants achieved essentially modern stages of development during the Mesozoic and continued into the Cenozoic without noteworthy interruption in the evolutionary trend. But many others that had become conspicuous members of the Mesozoic scene disappeared with dramatic abruptness. The Mesozoic is thus sharply defined by the two most cataclysmic changes in the record of Phanerozoic life. These must mark times of great environmental stress, but just what the causal factors were remains unresolved in spite of a voluminous literature on the subject.

According to recent interpretations based on the isotopic age determinations of many of its rocks [8], the duration of the Mesozoic Era was about 160 million years, ending approximately 65 million years ago. The periods of the Mesozoic are unequal in length, the Triassic was between 30 and 35, the Jurassic between 55 and 60, and the Cretaceous some 70 million years in duration.

The Triassic. Colorful red sandstones and red and green marls are widespread in central Germany where they contain salt beds and, in their middle part, a conspicuous unit of shelly limestones. As early as 1800 the three distinctive parts of this sequence had been singled out and named, in ascending order, the Bunter (red sandstones), Muschelkalk (shelly limestones) and Keuper (variegated marls). Later, in 1834, Alberti grouped these three units as the Triassic System, recognizing their essential unity and the contrast of their fossils with those of the Zechstein (Permian) beds below and the Lias (Jurassic) beds above.

The threefold pattern of this type Triassic sequence, or Germanic facies, as it is more commonly called, is largely restricted to Europe north of the Alpine region and does not resemble the Triassic in most other parts of the world. Modern usage would call for a geographic rather than a descriptive name for the period, but "Triassic" became deeply intrenched in the literature during the formative years of the geologic time scale and the time to replace it has long since passed.

Redbeds of the Germanic facies formed mostly in terrestrial environments under at least seasonally arid conditions. Much of the broad basin where they accumulated was inundated by shallow seas when the Muschelkalk limes and associated evaporite beds were deposited. Locally, as much as 5000 feet of Triassic beds accumulated. Throughout its existence this shallow basin on the European continental platform was separated from the great Tethyan geosyncline flanking it to the south by a narrow barrier (Fig. 14-2). During the Muschelkalk inundation the barrier was breached in eastern Europe but there was little interchange between the distinctive faunas of the two regions with the result that the Germanic facies has proved difficult to correlate with marine Triassic deposits. Thousands of feet of marine sediments, notably limestone and dolostone, formed in the geosyncline and it is in this, the Alpine facies of the Triassic, that a succession of Triassic marine faunas was first worked out and a sequence of Triassic zones and stages established. In practice, it was the Alpine facies of the Triassic that became the standard of reference for the Triassic System, not the Germanic facies for which the system was named.

PHYSICAL HISTORY OF NORTH AMERICA

A Major Change in Continental Architecture. After the Alleghanian orogeny the eastern part of North America was emergent for two long geologic periods. Here erosion prevailed throughout Triassic time and only late in the period was any of the debris trapped under circumstances that preserved it to the present. At this time block faulting produced a chain of narrow, fault-bordered basins extending from Nova Scotia to Florida roughly parallel to the present continental margin. Predominantly red, terrestrial sandstones, silts, and shales filled the troughs and were locally interbedded with lava flows and intruded by diabase. No trace of marine Triassic rocks is known anywhere in the eastern part of the continent nor have undoubted Triassic sediments other than those in the fault troughs been encountered by deep drilling through younger formations into the coastal plain.

The Alleghanian orogeny elimated the great geosyncline that flanked the east side of the stable continental platform during the Paleozoic. In striking contrast, no orogeny disrupted the west side of the continent and the Paleozoic pattern of deposition continued into the Triassic without appreciable change. Seas occupied parts of the Cordilleran and Arctic geosynclines of North America throughout most of the Triassic (Map IX), leaving deposits rich in ammonoid faunas. The most complete succession of fossil zones on any continent, representing the entire Triassic Period, has been worked out by piecing together a number of these North American sections of fossiliferous Triassic rocks [15, 13]. Early Triassic seas advanced from the Pacific region into the eastern trough of the Cordilleran geosyncline, locally overspreading adjacent parts of the stable continental platform. During Middle Triassic time, however, a great elongate welt began to rise irregularly within

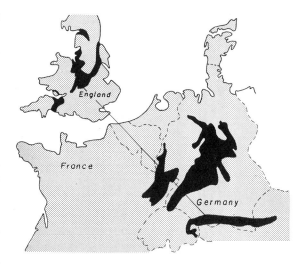

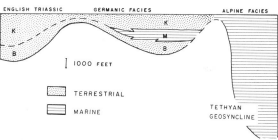

Figure 14-2 Generalized outcrop map of the Triassic and an idealized section to show the relation of the type, or Germanic facies, of the Triassic System to other Triassic deposits in Western Europe. The three original divisions of the type Triassic—Buntsandstein (B), Muschelkalk (M), and Keuper (K)—are now groups comprising a number of formations.

this eastern part of the geosyncline, paralleling the edge of the platform and extending from what is now Nevada northward into British Columbia. This major structural element, the **Cordilleran geanticline,** brought an end to the impressively long history of the eastern Cordilleran trough, which had begun accumulating its sedimentary record with the deposition of the Precambrian Belt Series (p. 150). The Cordilleran geanticline transformed the western continental architecture, effectively cutting off the western interior region over a large part of the continent from any further marine incursion from the Pacific (Map IX). West of the geanticline the unstable area of great volcanic activity, which characterized the western part of the Cordilleran geosyncline during the Permian and Early Triassic, persisted as a broad belt of dominantly volcanic islands and local sedimentary basins in which thick sequences of sedimentary and volcanic rocks were deposited. East of the geanticline lay a great interior basin, overlapping a broad area of the stable continental platform to the east and deepening abruptly along its relatively narrow western edge adjacent to the geanticline. During the latter part of Triassic time continental deposits formed in parts of this interior basin and marine waters from the Boreal region spread into it as far south as Alberta. Subsequently this interior basin played a far more prominent role in Mesozoic sedimentation.

The major Mesozoic features of the North American continent, first shaped in its eastern half by the Alleghanian orogeny, were essentially completed with the rise of the Cordilleran geanticline during Triassic time. Subsequent Mesozoic sedimentation was restricted to coastal plain and shelf deposits in the eastern half of the continent while the western half featured a great interior basin of deposition separated from a western Cordilleran volcanic trough by the tectonically active Cordilleran geanticline.

Triassic of the West and North

The North American Standard. The great orogenies of later Mesozoic and Cenozoic time in western North America have left only remnants of the Triassic marine record and much of this is metamorphosed or isolated in fault blocks. Early attempts at piecing together the scattered record using the prolific ammonoid fossils permitted a general correlation with the Alpine facies of the European Triassic so that it was possible to distinguish the major parts of the Triassic sequence and even some of the stages. However, the Alpine facies lacks most of the Lower Triassic and the sequence of some of the ammonoid faunas in the upper and middle parts is not firmly established. Other better understood Triassic marine sequences elsewhere in the world have been used to supplement the Alpine

Triassic; for example, the Lower Triassic in the Himalayas and Salt Range of India has long been used as a standard of reference in the absence of fossiliferous beds of equivalent age in the Alps. But a composite sequence of Triassic ammonoid zones based on fragments of the record from Europe and Asia has limitations as a useful reference in North America for the reason that many of the ammonoid faunas are not the same as those in Eurasia. Although ammonoid species tended to be more widespread than those of most other invertebrates preserved as fossils, they were not generally worldwide and many were narrowly provincial in distribution.

Recently biostratigraphers of the Canadian Geological Survey have assembled a standard for the Triassic of North America based on great Triassic exposures in two areas, the foothills of northeastern British Columbia and the Arctic Islands [15]. Neither locality has a complete marine section, but the sequences of their fossils can be easily related by a number of ammonoid zones common to both, and together they present the most complete Triassic faunal succession known from any one part of the world. Supplemented by additional data from Triassic formations in the western United States, this standard reference sequence for North America comprises 35 faunal zones, based chiefly on species of ammonoids, and makes possible precise correlation of the many scattered parts of the western marine Triassic.

Cordilleran Early Triassic. During Early Triassic time seas covered much of southern California, spreading northeastward across Nevada and Utah, and then northward along the edge of the stable continental platform into Montana. A comparable spread of the Boreal Sea southward down the eastern Cordilleran region through northeastern British Columbia and western Alberta apparently came very close to joining with the southern incursion from the Pacific. Today only 300 miles separate the terminal outcrops of the two terranes (Map IX). It is possible that they were indeed contiguous and that the record was destroyed by subsequent geologic events, for this particular region in western

Montana and adjacent Idaho was the site of almost continual crustal unrest throughout Mesozoic time.

In problems of this kind it is often instructive to look at the faunas of the two terranes for evidence of intermixing or the lack of it. The northern Lower Triassic ammonoids are quite clearly a boreal fauna with species identical to the Arctic Island faunas, but they have no species in common with the nearby faunas of the southern Lower Triassic in southwestern Montana and eastern Idaho. The ammonoid fauna of this latter terrane relates to those in Nevada and California and is in general much more closely allied to the ammonoids of the Alpine and Himalayan Triassic than to those in the Triassic of the Boreal region. A species of shallow-water clam is common to the two faunas, but this is a very widespread form that occurs in the Lower Triassic of the Arctic Islands and in the Alps as well. The ammonoid faunas of the two seaways in question are from distinctly different Early Triassic marine provinces and their apparent lack of faunal mixture suggests, but does not prove, that these seaways did not connect.

The eastern Cordilleran Early Triassic seaways of Canada and the United States were the site of deposition of thousands of feet of dominantly shallow-water sandstones, shales, and limestones. In both, the principal source of sediment seems to have been from the continental interior which lay to the east, but local sources to the west may also have contributed. The marine deposits in the Canadian seaway terminate abruptly in shoreline deposits at the edge of the continental platform, but in the United States the marine beds thin and intertongue eastward with a facies of dominantly terrestrial redbeds widespread on the platform. The nature of this transition is shown in the cross section, Fig. 14-3, which is based on a series of exposures of Triassic rocks beginning on the west with geosynclinal deposits in southeastern Idaho and extending eastward into contiguous platform deposits in western Wyoming. Both the Dinwoody and Thaynes formations, which comprise the marine Lower Triassic in this area, pass eastward into redbeds. A thin portion of the

fossiliferous silty limestone in the Dinwoody Formation extends for another hundred miles or so east beyond the area of the cross section, but eventually it too passes into redbeds. Beneath the Triassic lie Permian rocks deposited under similar conditions and the Park City Formation, like the Dinwoody, grades into redbeds in central and eastern Wyoming. Where this takes place it is commonly not possible to distinguish a boundary between Permian and Triassic rocks in the continuous sequence of red-colored strata.

Outside of the thick marine sequence of the Dinwoody and Thaynes formations in the Idaho-Wyoming-Utah area little is known of the Early Triassic seaway. A few exposures of similar rocks are present in northeastern Nevada and up to 3000 feet of sparingly fossiliferous Lower Triassic is contained in the Candelaria Formation and other faulted bodies of Triassic rock in western Nevada. In southern California as much as 8000 feet of Lower Triassic, much of it volcanic, has been reported but not adequately studied. Just enough is known to indicate that the southern Cordilleran Early Triassic sea connected to the Pacific region across what is now southern California. As in Idaho and Wyoming, the marine deposits of the geosyncline grade into a redbed facies southeastward onto the stable platform in eastern Utah and Arizona (p. 322).

Middle Triassic Change. In the Arctic Islands and Cordilleran trough of Canada marine Middle Triassic formations follow those of the Lower Triassic and signify continued presence of the sea. Scattered outcrops in western British Columbia reveal local areas of Middle Triassic deposits of interbedded volcanics and marine shale and limestone. These volcanic areas were apparently separated by a narrow landmass from the Canadian Cordilleran trough, where no volcanics or sediments derived from volcanics are found. Thousands of feet of associated volcanics and marine sediments also accumulated in western Nevada and parts of California during Middle Triassic time. Possibly the entire region of the present west coast states and provinces, from California through southern Alaska, was a terrane of scattered tectonic basins during the

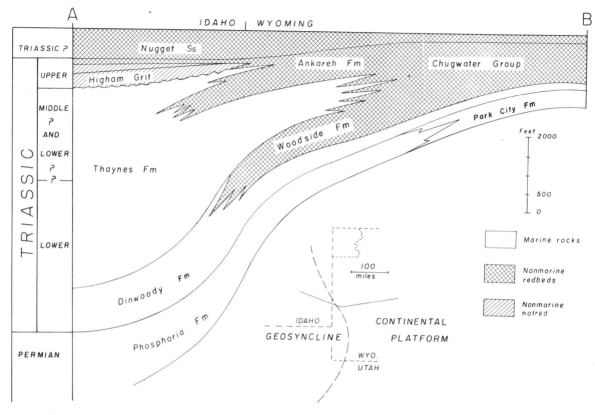

Figure 14-3 Idealized section across southeastern Idaho and western Wyoming, showing the relationship of Triassic marine deposits of the geosyncline and equivalent redbed facies of the continental platform. (Data from Kummel, U.S.G.S. Prof. Paper 254-H.)

Middle Triassic as well as the Upper Triassic. The record of the Late Triassic is more widespread in this great region, but so much of the entire Triassic record here was obliterated by later Mesozoic intrusions and enormous Cenozoic volcanic activity that we can only guess at the distribution of the western volcanic portion of the Triassic Cordilleran geosyncline.

None of the Middle Triassic features of western North America noted above represent any significant departure from the Lower Triassic pattern or from the Permian pattern of the region. The significant change in Middle Triassic time took place in the eastern Cordilleran region of the United States where the Thaynes Formation and its associated redbeds are overlain by terrestrial rocks of supposed Late Triassic age and no Middle Trias-

sic rocks whatever have been identified. This relatively sudden change has been attributed to the rise of the Cordilleran geanticline as an elongate, narrow series of uplifts trending roughly northward through eastern Nevada and western Utah, and possibly extending to join the narrow uplift which apparently separated the Canadian Cordilleran trough from the volcanic part of the geosyncline to the west. Evidence for the geanticline is largely negative in the Middle Triassic. With seas in Nevada during this time why did they not spread into the trough in Utah, Wyoming, and Idaho as they had during the Early Triassic? Presumably only a rising land barrier could block them from this interior region and it is a matter of record that no seas ever invaded this region from the Pacific after Early Triassic time.

Absence of any Middle Triassic rocks in the eastern Cordilleran region of the United States is anomalous and most likely a portion of the poorly dated nonmarine redbed facies was deposited during this time (p. 328). Similarly, hundreds of feet of beds constituting approximately the upper two-fifths of the Thaynes Formation lack ammonoid faunas or other diagnostic fossils and may be Middle Triassic. The important point, however, is that Middle Triassic time marks the end of the broad Cordilleran geosyncline open to the Pacific region. At about this time it was divided by the geanticline into a western mobile trough characterized by local deep sedimentary basins and volcanics, commonly called the **Pacific geosyncline,** and a broad interior basin, the **Rocky Mountain geosyncline.** These two great areas of deposition dominate the western North American scene throughout the remainder of the Mesozoic.

Late Triassic Inundation. Most widespread of all the marine Triassic deposits are those of the Late Triassic. Great thicknesses of sediments and volcanics were deposited throughout the Pacific geosyncline and sedimentation continued in the eastern Cordilleran trough of Canada and in the Arctic Islands and northern Alaskan regions. Similar spread of the seas in other parts of the world apparently permitted much wider distribution of ammonoid faunas, for those of the Karnian and Norian stages of the Late Triassic (see Fig. 14-4) are worldwide. During the Norian stage, however, a spectacular extermination of unknown cause swept the ammonoids leaving

Figure 14-4 Correlation of important Triassic sections.

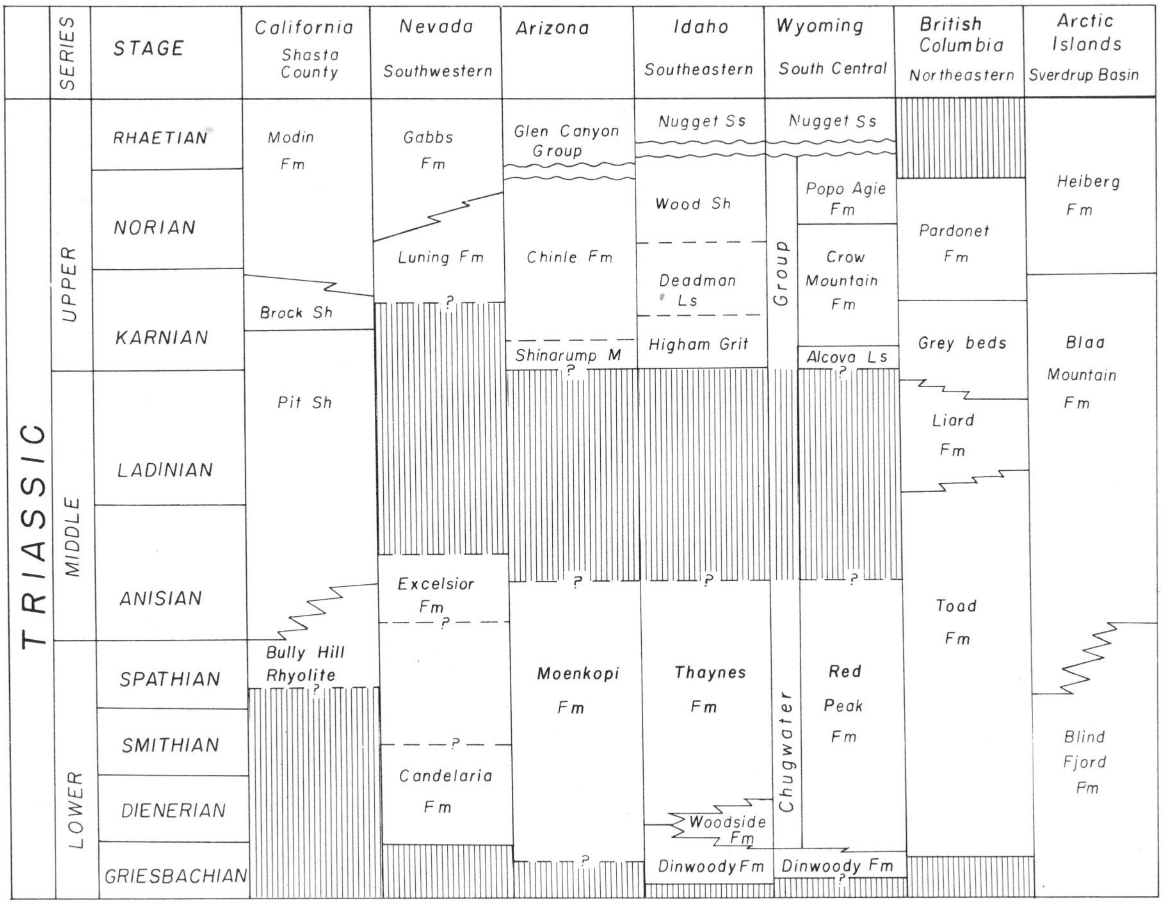

but a few forms to repopulate later Mesozoic seas. Deposits of the very latest stage of
the Triassic, the Rhaetian, contain few ammonoids.

Redbeds of the Platform. The greatest area
of Triassic terrestrial deposits in North America (Map IX) occupies much of the Rocky
Mountain region from eastern Idaho and
Wyoming southward across Utah, Colorado,
Arizona, New Mexico, and western Texas.
Here red-colored sediments predominate,
though marine members of the Lower Triassic
locally interfinger from the west. Bright red
or maroon shales and cross-bedded red sandstones make colorful landscapes such as the
Painted Desert of Arizona and the broad red
valley of the "Race Track" in the rim of the
Black Hills.

Throughout the Triassic the Colorado
Ranges, which formed during Pennsylvanian
time (p. 258), stood as highlands on the stable
continental platform and shed sediments into
shallow surrounding basins where the red-

beds slowly accumulated. Other areas on the
adjacent platform also contributed debris. In
parts of Utah and Arizona the Triassic redbeds
attain a thickness of more than 4000 feet and
are divisible into five principal units (Fig.
14-4); in ascending order, they are the Moenkopi, Chinle, Wingate, Moenavi, and Kayenta
formations.

The Moenkopi consists of reddish brown
shales, silts, and sands deposited across a
broad, low-lying coastal plain sloping gently
toward the Early Triassic Cordilleran geosyncline, which lay to the northwest. In its thicker
parts adjacent to the geosyncline the formation
is commonly about 1000 feet thick and consists predominantly of thin, even beds of red-
brown shale and siltstone (Fig. 14-5) that are
abundantly ripple-marked. These very regularly bedded fine sediments are believed to
have been deposited on broad mudflats [11].
Interbedded with the Moenkopi are one or
two thin limestone and gypsum beds locally
with marine fossils, which are thin tongues of

Figure 14-5 Moenkopi Formation at the edge of Paria Plateau near Navajo Bridge
over the Colorado River. (C. O. Dunbar.)

Figure 14-6 Chinle Formation in Petrified Forest National Monument, Arizona. Petrified logs in the foreground. (Joseph Muench.)

the Thaynes Formation and represent brief periods when the sea spread over the Moenkopi coastal flats. The lowermost tongue, called the Timpoweap Limestone in Arizona and the Sinbad in Utah, contains a fauna characterized by the ammonoid *Meekoceras*, which also occurs in the basal limestone beds of the Thaynes Formation. Somewhat higher and just below the middle of the Moenkopi a second tongue, the Virgin Limestone, contains another fauna found in the Thaynes Formation. These relationships establish the Lower Triassic age of much of the Moenkopi adjacent to the geosyncline.

To the east and southeast away from the coastal area the Moenkopi thins and grades into muddy sands and silts with less regular

bedding. Here it shows abundant evidence of deposition by aggrading streams meandering over a low alluvial plain. Sand-filled channels are common and the finer silts and clays of the flood plains are marked by dessication cracks and locally by tracks of sprawling reptiles and large amphibians that then lived along the stream courses as alligators now do in subtropical regions [12].

The **Chinle Formation** (Figs. 14-6 and 14-7) is a complex of variegated, highly colored shales and siltstones well exposed in the Painted Desert of eastern Arizona. Its colors range from red to pink and yellow, and locally to ashen gray and white. It contains much volcanic ash derived from volcanoes in California and Nevada. It reaches a maximum thickness

Figure 14-7 Chinle Formation overlain by Wingate Sandstone in the face of Paria Plateau west of Navajo Bridge, Arizona. (C. O. Dunbar.)

of 3000 feet but, like the Moenkopi, thins eastward to a feathered edge in western Colorado.

A basal member of the Chinle, known as the **Shinarump Conglomerate**, is persistent over a great area though generally less than 150 feet thick. It includes well-rounded pebbles of older rocks commonly as large as door knobs. Maximum coarseness occurs along the Mogollon Rim in central Arizona, where cobbles range up to 6 inches in diameter; but the coarseness decreases generally toward the north and east, and in Monument Valley in southeastern Utah, the conglomerate grades into sandstone. This suggests that the gravel

was coming largely from the west or southwest. The contact of the Shinarump Member with the underlying Moenkopi Formation is an erosion surface with local relief of 100 feet or more, and the conglomerate is thickest where it fills old valleys. The relief and sharp change in lithology imply a hiatus between Moenkopi and Chinle deposition, but they do not indicate how long a break it was.

Chinle sediments are more widespread than those of the Moenkopi and extend far eastward beyond the latter, across New Mexico into the western parts of Texas and Oklahoma where they are called the **Dockum Group.**

Here and there the generally barren terrestrial deposits of the Chinle contain fossil vertebrates. The fauna consists largely of river dwellers and is characterized by large, flat-headed amphibians such as *Eupelor* (Fig. 14-8) and by **phytosaurs** (Fig. 14-9), a stock of Triassic reptiles remarkably like crocodiles in structure and general appearance, and most likely their ecological parallel. In addition other kinds of reptiles have been found including a small bipedal dinosaur. Fossil wood is locally abundant in the Chinle, as in the Petrified Forest area of eastern Arizona (Fig. 14-14). On the basis of the phytosaurs, some of which are similar to those in the Keuper beds of Germany, the Chinle is considered to be of Late Triassic age.

Above the Chinle Formation lies the Glen Canyon Group, a complex of intertonguing terrestrial deposits comprising four formations, in ascending order, the Wingate Sand- stone, Moenave Formation, Kayenta Formation, and Navajo Sandstone. The Triassic-Jurassic boundary lies somewhere within this complex. Formations of the Glen Canyon Group were deposited in and around a large depression or basin of internal drainage that occupied much of northeastern Arizona and adjacent areas [9]. Gradual uplift on the east caused progressive shift of this basin westward. The Wingate, Moenave, and Kayenta formations include a variety of red-colored sedimentary facies generally totalling about 750 feet in thickness that represent different environments in and around the basin. Much of the **Wingate Sandstone** (Fig. 14-7) is an even-bedded silty sandstone laid down in a body of standing water but it grades eastward into cross-bedded, fluvial sands and local dune sands of the river systems flowing into the basin. The Moenave and Kayenta formations consist chiefly of red-colored sands, silts, and

Figure 14-8 Prepared slab of skulls, jaws, and other skeletal parts of the Triassic amphibian *Eupelor* from the Chinle Formation near Lamy, New Mexico. (United States National Museum.)

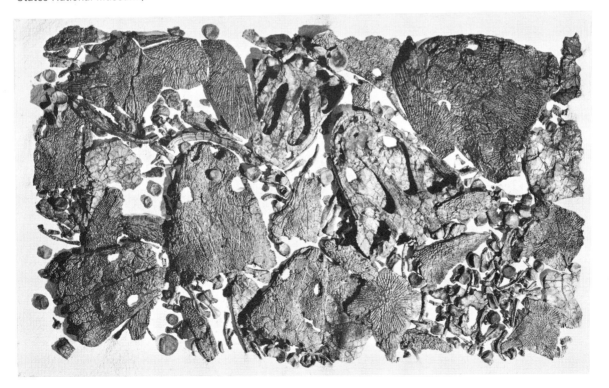

Figure 14-9 A Triassic phytosaur, *Rutiodon*. These reptiles closely mimicked the crocodiles and had similar habits. Length 10 to 12 feet or more. (After S. W. Williston.)

muds of fluvial origin and represent episodes of renewed uplift on the east and the shift westward of the basin of deposition. By this time the only standing water appears to have been in small bodies on river flood plains where lenses of freshwater limestone were deposited. These and sand dune deposits in both the Moenave and Kayenta formations suggest that the broad alluvial basin either had marked wet and dry seasons or was surrounded by generally arid conditions. The **Navajo Sandstone** is a great wedge of dominantly eolian sands that is thickest on the west, where it commonly rests directly on the Chinle Formation (Fig. 14-10). Eastward it is gradational with underlying units of the Glen Canyon Group. Apparently it began to spread across the entire basin of deposition while the Kayenta was being laid down, for its dune sands interfinger with Kayenta fluvial deposits as it overlaps them eastward. Prevailing winds from the west and northwest spread the great Navajo dune field eastward, gradually overwhelming the Kayenta drainage system. Since the Navajo Sandstone is generally considered to be Jurassic in age, we will have more to say about it in Chapter 15.

The Glen Canyon Group bridges the passage from Triassic to Jurassic time: the Chinle Formation below is Late Triassic; and above, unconformable on the Navajo Sandstone, lie marine beds of undoubted Jurassic age. The remains of the life of the rivers and plains where Glen Canyon sediments accumulated do not clearly indicate where in the sequence the change from Triassic to Jurassic takes place. In part this is because the fossils are sparse, but in addition, many that have been found are new to science and cannot be compared directly with established fossil sequences elsewhere. Vertebrate remains are the most common and most useful fossils for dating the Glen Canyon rocks. Dinosaurs, primitive crocodiles, mammal-like reptiles, phytosaurs, and fish are the principal kinds. Phytosaurs, which elsewhere in the world are known only from Triassic rocks, are represented by a few scattered teeth in the lower part of the Wingate and could have been reworked from the underlying Chinle. The very scantily fossiliferous Kayenta beds have yielded two unusual collections. One includes several skeletons of tritylodonts, mammal-like reptiles, together with remains of a primitive crocodile also found at another site in the underlying Moenave Formation. Both kinds of reptiles are new species, indeed neither tritylodonts nor primitive crocodiles had previously been known from North America. Structurally similar forms are known from South Africa and China where they occur in beds considered to be Late Triassic in age. The other Kayenta discovery is that of a primitive flesh-eating dinosaur, whose relationships and probable position in the evolutionary lineage of these beasts is still in doubt. It

Figure 14-10 West Temple of the Virgin, Zion National Park, Utah. A 4000 foot wall of Triassic and Jurassic rocks carved out of the edge of the Colorado Plateau by the Rio Virgin. The ranch houses at the bottom indicate the scale. (Union Pacific System.)

could be either Late Triassic or Jurassic. The scattered dinosaur fossils from the Navajo Sandstone are not well enough known to be useful in age determination.

Obviously there is not enough evidence to indicate with certainty where, within the Glen Canyon Group, the Triassic-Jurassic boundary falls. The occurrence of tritylodonts in Late Triassic rocks on other continents suggests

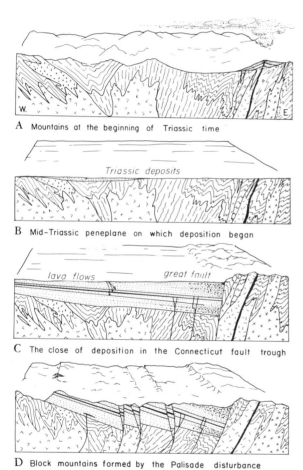

A Mountains at the beginning of Triassic time

B Mid-Triassic peneplane on which deposition began

C The close of deposition in the Connecticut fault trough

D Block mountains formed by the Palisade disturbance

Figure 14-11 Four stages in the development of the Triassic basin of central Connecticut. Length of sections about 30 miles. Block A shows the complex structure and rugged topography inherited from the Alleghanian orogeny; block B shows the beginning of Newark deposition after the region was largely peneplaned; block D shows the Newark deposits complexly faulted subsequent to deposition. The modern structure of the region dates from this time. Triassic strata stippled. (Modified from Joseph Barrell.)

that the Kayenta may be Triassic. And since the Kayenta interfingers with the lower part of the Navajo, it would follow that at least a part of it is Triassic. The problem has yet to be resolved, but it illustrates well the difficulties encountered in establishing period boundaries within terrestrial sequences. A nearby Triassic-Jurassic marine sequence offers a striking contrast. About 300 miles west of the magnificent exposure of Navajo Sandstone in Zion Park, Utah (Fig. 14-10), a continuous section of marine beds in southwestern Nevada bridges the Triassic-Jurassic boundary. Here the change from Triassic to Jurassic can be demonstrated to fall within about 35 feet of shaly beds that separate two collections of distinctive ammonoids, the lower containing the key ammonoid of the highest marine zone of the Alpine Triassic and the upper containing ammonoids of the basal zone of the standard European Jurassic sequence.

Redbeds of the stable platform north of the Colorado Ranges in Wyoming have major divisions similar to those in the southwest. Within the Chugwater Group (Fig. 14-3), the Red Peak Formation is considered equivalent to the Moenkopi. Although it lacks marine tongues, it is at least in part a lateral phase of the lower Thaynes Formation. The formations in the upper part of the Chugwater Group are generally considered to be Late Triassic in age but only the Popo Agie Formation contains undoubted Late Triassic fossils. The age of the Nugget Formation is unknown, it lacks fossils and is overlain by undoubted Jurassic rocks.

In both the Wyoming and southwestern areas of Triassic redbeds some of the lower strata are dated by association with marine Lower Triassic of the geosyncline to the west and some of the upper strata are dated Upper Triassic by vertebrate remains. Both areas contain strata between these dated intervals that could be Middle Triassic, but it is also possible that general uplift of the region may have accompanied the rise of the Cordilleran geanticline preventing preservation of any substantial Middle Triassic record.

TRIASSIC FAULT TROUGHS OF THE EAST

We do not know the origin of the tensional stress in the crust that caused great rifts to form in eastern North America during the latter part of the Triassic. Some have looked at these faults as evidence of a relaxation of the compressional stress of the Alleghanian orogeny, but tens of millions of years separate these events and there is no direct evidence of a relationship. The growing support for the idea of continental drift permits the speculation that the Triassic faulting may have been an early effect of the tensional stresses that caused the separation of the North American continent from those of Europe and Africa during the Mesozoic. There are no Triassic coastal plain deposits to indicate the presence of an Atlantic Ocean, yet the alignment of the Triassic fault basins rather closely follows the eastern continental border.

Whatever their cause, the great normal faults produced a narrow chain of block mountains bordered by downfaulted troughs (Fig. 14-11) in the metamorphic terrane of the crystalline Appalachians. The height and extent of these new block mountains are conjectural, but the preserved remnants of the structural troughs, which were filled as they sank, still retain a rich record of the time. The northermost basin lies in Nova Scotia, and others are distributed southward into North Carolina, a distance of about 1000 miles. The Triassic strata formed in these troughs have been named the Newark Group (Fig. 14-12) for the exposures near Newark, New Jersey, where they probably exceed 20,000 feet in thickness; and the structural troughs are known as the Newark basins. Deep wells in the panhandle of Florida and adjacent corners of Georgia and Alabama have penetrated redbeds with diabase flows and dikes so similar to the Newark rocks that they are believed to be contemporaneous. Here the redbeds dip westward suggesting that they also occupy a fault trough similar to the Newark basins farther north.

The Connecticut Trough. The Triassic

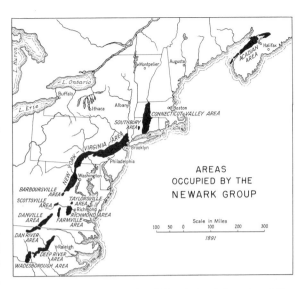

Figure 14-12 Map showing Triassic fault troughs of the Appalachian region in which the Newark Group is preserved. (After I. C. Russell, U. S. Geological Survey.)

trough of central Connecticut, standing near the middle of this chain of basins, is a good example. It extends northward across Connecticut and most of Massachusetts, its length nearly 100 miles and its greatest breadth about 25 miles.

The Triassic rocks of the basin are conglomerates, sandstones, siltstones, and shales, with interbedded flows of basic lava. Perhaps half of the sediments are gray, but the most prominently exposed beds are red, and these give the impression that the whole group consists of redbeds.

The sediments are poorly sorted and irregularly bedded, sandstones grading laterally into siltstones or conglomerate. All the coarse deposits are arkosic, and much of the feldspar is remarkably fresh. The conglomerates are thickest and coarsest along the eastern margin of the basin and clearly represent fans built where torrential streams debouched from the highlands to the east. In Connecticut the Newark Group has an estimated thickness of 10,000 to 15,000 feet, and in New Jersey it may reach more than 20,000 feet, but it diminishes to 2000 or 3000 feet in the southern basins.

Three extensive trap sheets lie near the middle of the sequence in Connecticut, each separated from the next by several hundred feet of sedimentary beds (Fig. 14-11), many of which are dominantly gray in color. The middle sheet attains a thickness of about 500 feet and now forms conspicuous ridges such as The Hanging Hills of Meriden. Each of the trap sheets has a scoriaceous and amygdaloidal upper surface; the underlying redbeds are bleached near the contact as a result of the heat from the lava; but the overlying redbeds show no similar effect and, on the contrary, include fragments of the scoriaceous lava. It is clear therefore that these are surface flows rather than sills—they represent lavas that welled out to flood the Triassic Basin. The eruptions were remarkably free of explosive violence, however, for ash and bombs are known in only one small area (near the Holyoke Range in Massachusetts). The lava must have been very fluid to spread in such flat and extensive sheets. Many dikes cut the underlying strata.

No marine fossils have been found anywhere in the Newark Group, but land plants and freshwater fishes are locally abundant in the darker gray beds, and dinosaur tracks in the redbeds are more plentiful than anywhere else in the world (Fig. 3-18). Ripple marks are common, and mudcracks cover many of the bedding planes. The imprints of Triassic raindrops are in places associated with the footprints and mudcracks (Fig. 14-13).

With these facts in mind, it is not difficult to reconstruct the events that transpired here during the latter part of Triassic time. Repeated movement along the great fault depressed the basin and elevated the adjacent highlands. Meanwhile the streams that reached the basin from the uplands dropped most of their sediment here, building fans along the eastern border and fluvial deposits over the floor. Before the igneous activity began, the basin was well drained by through-flowing streams, and much of the mud was carried away, while gravel, sand, and silt were dropped. The uplands must have had plenti-

Figure 14-13 Slab of Newark Group sandstone from Turner Falls, Massachusetts, bearing dinosaur tracks and the imprints of raindrops. These are photographed from the under side of the beds so that the tracks stand in relief; they are natural molds of the tracks and raindrop imprints which were made in the underlying mud. For other dinosaur tracks from the Newark strata see Fig. 3-18. (Yale Peabody Museum.)

ful rainfall to develop such a quantity of red mantle as is represented in these sediments, but the rains were probably seasonal. Thus the mud that spread over the floodplains during wet seasons lay exposed to the sun during the dry months. Dinosaurs crossed and recrossed, leaving their tracks in the mud. The last spatter of passing showers also left imprints of raindrops where the mud was exposed and still soft. During the dry season the mud shrank and cracked and then was sunbaked and hardened so that it could hold these surface features until they were buried by a new layer of sediment.

Where the drainage was good and the ground water not close to the surface, the iron-stained sediments remained red. After the first lava flow, however, the drainage was impounded, and for a considerable time swampy conditions obtained over parts of the lowland [10]; here vegetation accumulated with the sediments, and in its decay reduced the iron oxide, producing gray or dark colors. Such conditions held until after the last flow, and then, with better drainage, redbeds again accumulated over the basin.

Other Newark Basins. The geology of the other Triassic fault troughs is, in general, like that outlined above. In the Acadian area the sediments are predominantly red and include one large trap sheet. The dip here is northwest, and the bounding fault is on the west side. In the New York-Virginia and Danville areas also the dip is westward, whereas in the Deep River and Wadesboro areas it is to the east. As we follow the basins southward from New Jersey, more and more of the sandstones and shales are greenish-gray instead of red and, finally, in the Danville area, there are interbedded coals that locally reach a thickness of 26 feet. This may imply that the rainfall was more evenly distributed through the year in the southern basins, or that it was greater, or that the basins were not so well drained.

In the New York-New Jersey area there are three great lava flows that may be equivalent to those in the Connecticut trough. A thick sill of dolerite occurs near the base of the group and its eroded margin now forms the Palisades of the Hudson River from New York City northward for a distance of nearly 50 miles.

Age of the Newark Group. The fishes, reptiles, and plants of the Newark beds indicate that the entire group belongs to the upper half of the Triassic System. The group rests on a surface of generally low relief cut across the complex structures of older Paleozoic rocks in the crystalline Appalachians. It is difficult to tell how long a period of erosion preceded the formation of the Triassic basins, for the history of the erosion surface is complex.

The deformation of the older crystalline Appalachians was largely accomplished by the close of the Devonian, from which time they became the source of most of the sediment deposited in the Appalachian region during carboniferous and Permian time. The surface on which the Triassic basins lie may thus be the result of well over 100 million years of erosion.

Post-Triassic Faulting. Slight tilting of the Triassic deposits in the Newark basins was accomplished by continued movement along the major faults as deposition progressed. Much more pronounced tilting and more complex normal faulting of the basin deposits took place after the filling of the basins (Fig. 14-11). Subsequent erosion has beveled these tilted sediments and etched out the resistant trap sheets into the prominent ridges of today, such as the Palisades of the Hudson River, the Holyoke Range in Massachusetts, and the Watchung Mountains of New Jersey.

TRIASSIC OF OTHER COUNTRIES

In a general view of the earth, one of the most remarkable features of the Triassic is the almost universally emergent condition of the continents, and the extensive spread of non-marine deposits, largely redbeds. Great geosynclines were present on most continents but seas remained largely confined to their relatively narrow limits and the vast lowlands bordering them were chiefly the site of terrestrial deposition. An associated feature of Triassic history is the immense volcanic activity characterized by intrusion and extrusion of basaltic magmas. These aspects of the Triassic are nowhere better illustrated than in the "Gondwanaland" continents of the southern hemisphere. Some examples from these lands follow.

South Africa. Here nonmarine formations of great thickness (upper Karroo) are overlain by volcanics and shot through with basic intrusions of extraordinary magnitude (Drakensberg Volcanics). The lower part of the series includes gray sandstones, siltstones, and shales with thin beds of coal and abundant plant remains, but the middle part consists of thick redbeds with mudcracks and a very interesting reptilian fauna. Overlying the redbeds come purer, wind-blown sands varying in thickness up to 800 feet. The succession of formations is interpreted to imply a growing aridity that resulted in desert conditions over a considerable area in South Africa before the close of .the Triassic [7]. The basic igneous rocks intruded into this series have a present area of fully 220,000 square miles, and before their erosion covered at least 330,000 square miles in a great belt between latitudes 26° and 33°S, extending from the east coast probably to the Atlantic. With a volume estimated as between 50,000 and 100,000 cubic miles, this constitutes one of the greatest known masses of basic intrusives. The time of its intrusion is either Late Triassic or more probably Early Jurassic.

Brazil. In the Paraná Basin of southern Brazil, Late Triassic redbeds with reptiles similar to those of Africa are also overlain by enormous lava flows which still cover an area of some 300,000 square miles to a depth ranging from 400 to 2000 feet [1]. These lavas, like those of South Africa, may be dated as either Late Triassic or Early Jurassic.

CLIMATE

We have noted evidence of semiarid climate in North America, South America, South Africa, and western Europe. On the whole, semiarid climate seems to have been remarkably widespread during the Triassic. Perhaps this was partly due to the size of the emergent landmasses, since the interiors of the continents,

dependent for their moisture upon evaporation from the seas, include the chief deserts of the world. At the same time, parts of the Triassic lands were well watered, just as parts of the present continents are humid.

In view of the widespread Permian glaciation, it is noteworthy that no glacial deposits have been found in the Triassic rocks. The temperature had become mild long before the close of the Permian. This we may infer from the distribution of Late Permian and Triassic vertebrates, all of which were cold-blooded, that is, without a device to keep their bodily warmth above the temperature of their environment. Modern reptiles and amphibians, without exception, become torpid and helpless when the temperature drops to near freezing. Small species may take refuge in holes and hibernate, but all large species are confined to regions without frost. For example, the alligators and crocodiles, the great land tortoises, the large lizards and boas, all live in the tropics or subtropics. It is therefore highly probable that the sprawling reptiles and the large labyrinthodonts of the Triassic, as well as the primitive dinosaurs, could not endure freezing weather. By Middle Triassic time corals had reestablished themselves and were making small reefs in the seaways along the Pacific coast of America as far north as Alaska. However, since these are of few kinds and the reefs are small, it is probably not safe to infer that the water was subtropical so far north.

LIFE OF TRIASSIC TIME

Land Plants. Unfavorable conditions for preservation where redbeds were forming most likely accounts for the relatively sparse record of Triassic plant life. In America we get two glimpses of the Triassic flora, one in the foliage preserved in the dark shales of the Newark Group, particularly in the southern areas (Virginia and the Carolinas), the other in the petrified logs of the western redbeds, as at Petrified Forest, Arizona.

The first is a swamp flora of ferns and scouring rushes, to which are added, where streams entered the swamps, the transported leaves of

Figure 14-14 Agatized logs from the Chinle Formation in the Petrified Forest National Monument, Arizona. (Joseph Muench.)

conifers and cycads that formed the forests on the uplands and slopes.

The Petrified Forests of Arizona, on the contrary, has yielded chiefly petrified logs, although foliage has been found in several places [6] recording cycadeoids and ferns that grew along the stream courses. The logs are of conifers, not unlike the great pines that now stand in stately grandeur upon the rim of the Colorado Plateau. Many of the logs are of noble size, some attaining a diameter of 10 feet at the base and a length exceeding 100 feet. It has been estimated that some of these trees stood nearly 200 feet high. They now lie imbedded in the Chinle shale, petrified as agate (Fig. 14-14).

These two occurrences give a fair representation of what we know of Triassic land plants of the world as a whole. The forests (Fig. 14-1) were then predominantly of conifers much like our modern evergreens, and of cycads (14-15). The undergrowth consisted of ferns, tree ferns, and scouring rushes. The chief groups of Paleozoic plants were extinct, or nearly so. The seed ferns so characteristic of the Coal Measures had largely vanished, and the great scale trees are known only from rare specimens of *Sigillaria* in the Early Triassic and a few other doubtful representatives. *Lepidodendron* is not represented, and the cordaites were no longer conspicuous.

Leaves believed to be those of a primitive

Figure 14-15 A living cycad. The short trunks heavily armored by leaf bases and the graceful pinnate leaves are characteristic of the cycads. (Yale Peabody Museum.)

Figure 14-16 Leaves of the Late Triassic plant (*Sanmiguelia lewisi*) interpreted by R. W. Brown to be a palm and the oldest known angiosperm foliage. From the Dolores Formation, southwestern Colorado. One-fifth natural size. (After R. W. Brown, U. S. Geological Survey.)

palm (Fig. 14-16) have been found in Triassic redbeds of southwestern Colorado [3]. If true, this is the earliest known angiosperm, or flowering plant. The resemblance to palm leaves is great, but angiosperms can only be identified with certainty by their fruiting organs.

The Beginning of a Reptilian Dynasty. As might be expected, the great change in Triassic plant life over that of the Paleozoic is matched by equally significant changes in the vertebrate faunas during Triassic time. Labyrinthodonts became highly specialized to an aquatic life during the Triassic and were abundant in the stream systems of most of the world. Even Antarctica has yielded a fragment of a labyrinthodont jaw from rocks of probable Triassic age—the first fossil of a land animal found on that continent [2]. While the labyrinthodonts attained their culmination in size and variety, they were already being surpassed by the reptiles, which had adapted themselves to all conditions of life on the lands, and early in the period began to invade the seas and compete with the fishes as do the modern seals and whales. **Phytosaurs** (Fig. 14-9) were common in the streams, and several other orders of reptiles, now extinct, were adapted to various modes of life on the land. The phytosaurs resembled the modern gavials in appearance and habits but were not closely related to crocodiles. Their bones are found in association with river clams, amphibians and lungfishes. One species from western Texas had a length of 25 feet. All the phytosaurs were confined to the Triassic Period.

Conquest of the land by **reptiles,** begun in the Permian, was accomplished during the Triassic. Stemming from the sprawling pelycosaur tribe of the Permian, the mammal-like, or therapsid, reptiles became adapted to a running locomotion, carrying their bodies up off the ground as mammals do with their limbs under the body rather than projecting from its sides. This significant advance won dominance of the land for the therapsids before the close of the Permian. They remained common and widespread throughout the Triassic but had their leadership wrested from them during the period by a rising reptilian stock aptly named the **archosaurs,** or "ruling reptiles." This competitive reptile group had become differentiated late in Permian time and, in a parallel to the earlier therapsids, had also developed a running stance with the limbs under the body. During the Early Triassic certain archosaur groups took a further adaptive step, perfecting a bipedal stance in which the hind limbs were strongly developed for running while the front limbs were essentially freed from locomotive function. By Late Triassic time swift, bipedal, carnivorous archosaurs had become the lords of the land. Together with other successful archosaur tribes, both bipedal and quadrupedal, they dominated the terrestrial faunas of the earth for remainder of Mesozoic time, well over a hundred million years!

The principal groups of archosaurs are popularly known as **dinosaurs** [Gr. *deinos,* terrible + *sauros,* reptile]. The name was coined long ago when the large bones from Mesozoic rocks were first recognized as reptilian and before it was learned that they constituted not a single order of reptiles but two great orders that are only distantly related. The name is useful in that it embraces the two main groups of carnivorous and herbivorous archosaurs participating in the spectacular radial evolution of reptiles in Mesozoic time. Indeed, the dinosaur has become the trademark of the Mesozoic Era.

Compared with the giants of later Mesozoic ages, the Triassic dinosaurs were hardly "terrible reptiles," for nearly all of them were slender of build and few reached a length of more than 10 or 15 feet. Almost all the known Triassic species were bipedal (Fig. 14-1) and shaped somewhat like a kangaroo, with powerful hind legs and a thick, powerful tail which aided in balancing the body as they ran. Abundant trackways prove that they ran like ostriches instead of leaping.

The side toes on the hind feet were already vestigial in most of the Triassic species, so that they made three-toed footprints that were for a long time mistaken for bird tracks. Although the dinosaurs were very numerous in the eastern United States, skeletal remains are extremely rare, because the redbeds were

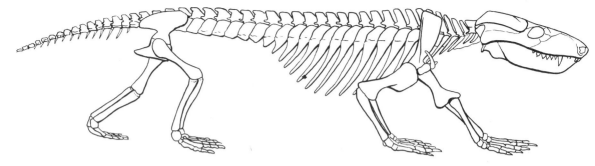

Figure 14-17 A mammal-like reptile, *Thrinacodon,* from a zone low in the Triassic part of the Karoo Series of South Africa. (After Brink.)

Figure 14-18 A mammal-like reptile, *Cynognathus,* from the Triassic beds of South Africa. (From a painting by F. L. Jaques under the direction of W. K. Gregory, American Museum of Natural History.)

a poor environment for the preservation of bones. As the dinosaurs crossed and recrossed the mud flats of the Connecticut trough, however, they left an amazing record, not of dead but of living creatures, now hurrying in search of food or water and again stopping to rest and to leave in the soft mud an impression of the body and the tiny front feet.

In 1947 a rich deposit of small Triassic dinosaur skeletons was discovered near Abiquiu, New Mexico [5]. Dinosaurs similar to those of America were also present in Europe, and nearly complete articulated skeletons of dinosaurs are known from the Upper Triassic of China.

The mammal-like reptiles (order Therapsida) continued from the Permian through the Triassic, and became extinct in Early Jurassic time, meanwhile having given rise to true mammals. The therapsids are best known from the Traissic of South Africa, China, and from European U.S.S.R. where many different kinds occur. Figure 14-17 shows the skeleton of a well-preserved form from near the base of the Triassic in South Africa and a reconstruction of a similar form from the same region is shown in Fig. 14-18. Another interpretation of what *Cynognathus* looked like appears in the foreground of Fig. 14-1.

A remarkable occurrence was discovered near the base of the Chinle formation at St. Johns, which is southeast of Petrified Forest in eastern Arizona [4]. Here, from a single layer, some 1600 bones were collected representing at least 39 individuals of a single species of therapsid, *Placerias gigas* (Fig.

14-19), along with fragments of five other kinds of reptiles. The carcasses had been dismembered by carnivorous reptiles before burial, but they had not been concentrated by currents. The reason for such mass mortality at a single spot remains an enigma.

Return of Reptiles to the Sea. Marine reptiles are known in the Lower Permian rocks of South America and South Africa, but they did not become common until late in Triassic time. Dolphinlike reptiles called **ichthyosaurs** [Gr. *ichthys*, fish] appeared in the Late Triassic seas and developed rapidly into one of the dominant groups of marine animals of the Mesozoic Era. They were already abundant in the Late Triassic of California and Oregon, where the largest species was about 60 feet long, and they were probably the largest animals in the world at that time. The ichthyosaurs had a fishlike contour with a laterally compressed tail and flipperlike limbs (Fig. 3-12). They were undoubtedly fast swimmers, able to capture fish or the ancient squids (belemnites) of their time.

The structure of their limbs indicates clearly that they had descended from terrestrial ancestors, their legs being modified into flippers similar to those of a seal.

The First Mammals. During Triassic time the evolving therapsid reptiles became more and more mammal-like in structure. One group in particular, the **ictidosaurs**, had progressed so far in this direction that separation of some of its (Late Triassic) forms from those of the earliest mammals depends on very fine structural distinctions. Unfortunately the animals in question were very small and their skeletons delicate; the teeth and jaw fragments commonly preserved are insufficient in many cases to determine conclusively whether a specimen is an ictidosaur or a mammal. In addition, precise dating of the terrestrial deposits in which these tiny remains occur is always difficult and often not possible. As the example of the Glen Canyon Group demonstrated, dating problems are the rule rather than the exception in terrestrial sequences that include the Triassic-Jurassic boundary.

Some of the principal discoveries of primi-

Figure 14-19 A mammal-like reptile, *Placerias,* from near the base of the Chinle Formation at St. Johns, eastern Arizona. (After C. L. Camp.)

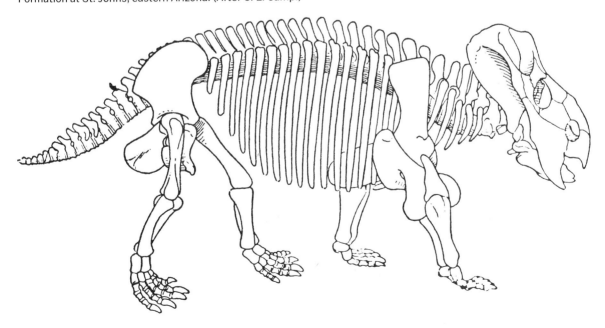

tive mammal remains have been made in ancient crevice fillings in the Lower Carboniferous limestone of South Wales and Somerset, England. The field relations suggest that these tiny creatures lived in the crevices in rocky terranes where they could evade their reptile enemies. Several orders of primitive mammals are represented in the fragmentary remains from the crevice fillings. Relationships with younger strata indicate that the deposits could either be latest Triassic (Rhaetian) or earliest Jurassic in age.

Mouse-size animals that were in general more like mammals than reptiles had arrived on the scene by the close of the Triassic and somewhat primitive mammals were certainly present by earliest Jurassic time. Our state of information about the two transitions involved here, the one from reptile to mammal, the other from Triassic time to Jurassic time, does not permit a more definitive statement.

Marine Invertebrates. The Triassic seas swarmed with **ammonoids** (Plate 13, Figs. 6 to 14) of many kinds, some far larger than any in the Permian. They were not only the most beautiful and characteristic shelled animals of the Mesozoic seas, but they have also proved the most useful marine invertebrates for defining the fossil zones used in the correlation of strata. The entire Mesozoic is zoned largely on the basis of ammonoids, and for this reason they have received a disproportionate amount of attention. In the Triassic more than in any other period they seem to dominate most marine faunas (Fig. 14-20). This is perhaps due to the apparent scarcity of hospitable shallow-water marine environments where the more abundant and diverse marine bottom faunas tend to congregate.

The rapid expansion of the ammonoids during the Triassic continued nearly to the close of the period, when a very rapid dying out almost caused their extinction. However, one genus (*Phylloceras*) with several species managed to survive into the Jurassic and to give rise to another great proliferation of forms in that period. The decline of so great and adaptive a group is difficult to explain. The ammonoids suffered great exterminations three times — at the end of the Devonian Period, at the end of the Permian, and at the end of the Triassic. Each time a few survived to start another great radiation; however, they failed to survive a fourth great decline and at the close of the Cretaceous Period became totally extinct.

Even as the ammonoids enjoyed their heyday, other cephalopod cousins that had overgrown their shells were undergoing a radiation that would lead later in the Mesozoic to essentially modern, shell-less types like the squid and octopus, which are the highest expression of invertebrate evolution in agility and strength. Of these, the **belemnites** are common in Triassic seas although more abundant and diversified later in the Mesozoic.

Less conspicuous than the ammonoid expansion but even more significant in terms of the population of the seas was the beginning of a great diversification among the **bivalved mollusks.** During the Triassic many new major groups of bivalves appeared that continued to proliferate throughout the Mesozoic and Cenozoic and today include some of our most familiar species. The stocks of the oysters and the myas, for example, both appear in the rather scanty bivalve faunas of the Triassic. Using the relationship of bivalve shell form and structure to their life habits, Stanley [14] has shown that the great Mesozoic radiation of the bivalves is probably linked to the evolution of siphons. These are the fleshy tubes which clams, like the quahogs, use to circulate water and nutrients to their bodies while they lie buried in bottom muds. The development of this habit of siphon-feeding opened up ways of life in bottom sediments that had not been possible for siphonless Paleozoic bivalves or brachiopods. In this and a number of other ways the bivalves far exceeded the limited environmental scope of the brachiopods. The latter are locally common in Triassic rocks, particularly those of the Tethyan geosyncline, but are everywhere greatly reduced in variety and numbers.

Modern kinds of **reef-building corals** appeared in Middle Triassic time and contributed to the thick limestones and dolostones of the Alps and the Himalayas. They were widely distributed through the Tethyan geo-

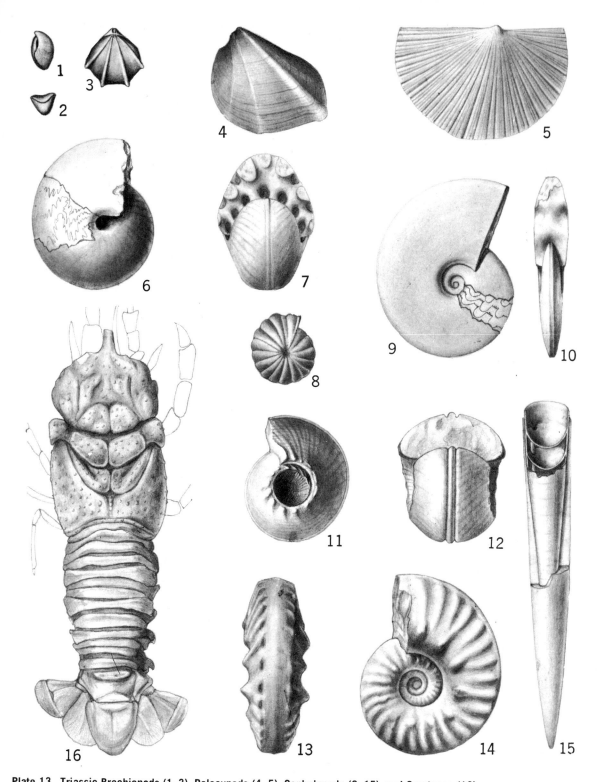

Plate 13   Triassic Brachiopods (1–3), Pelecypods (4, 5), Cephalopods (6–15), and Crustacea (16).
   Figures 1, 2, *Aulacothyris angusta* (side and front views); 3, *Tetractinella trigonella;* 4, *Myophoria kefersteini,* a forerunner of the Trigonias; 5, *Daonella americana;* 6, 7, *Paratropites arnoldi;* 8, *Leconteiceras californicum;* 9, 10, *Meekoceras gracilitatis;* 11, 12, *Tropites subbullatus;* 13, 14, *Ceratites spinifer;* 15, *Atractites macilentus,* the shell of a primitive belemnoid broken to show internal chambered portion; 16, *Pemphix sueuri,* the oldest known lobster. All natural size. (Drawn by L. S. Douglass.)

0   3 CM

Figure 14-20 A chowder of ammonoid shells from the Upper Triassic of Nevada. Fossil assemblages made up largely of ammonoids are common in Triassic marine deposits. (Yale Peabody Museum.)

syncline of Eurasia and also spread along the volcanic islands of the Pacific geosyncline of North America from California to Alaska.

The general scarcity of preserved Triassic shallow marine-shelf deposits, like that of the Muschelkalk, greatly restricts our knowledge of Triassic marine life. We know from the aspect of the succeeding Early Jurassic marine faunas that other major invertebrate groups underwent great evolutionary changes relative to their Paleozoic condition during Triassic time, but we have little direct knowledge of these innovations.

REFERENCES

1. Baker, C. L., 1923, *The lava field of the Parana Basin, South America.* Jour. Geology, v. 31, pp. 66–79.

2. Barrett, P. J., Baillie, R. J., and Colbert, E. H., 1968, *Triassic amphibian from Antarctica.* Science, v. 161, pp. 460–462.

3. Brown, R. W., 1956, *Palmlike plants from the Dolores Formation (Triassic), southwestern Colorado.* U. S. Geol. Survey, Prof. Paper 274-H, pp. 205–209.

4. Camp, C. L., and Welles, S. P., 1956, *Triassic dicynodont reptiles, Part 1: The North American genus Placerias.* Univ. Calif. Mem., v. 13, no. 4, pp. 255–348.

5. Colbert, E. H., 1947, *Little Dinosaurs of Ghost Ranch.* Natural History, v. 56, pp. 392–399; 427–428.

6. Daugherty, L. H., 1941, *Upper Triassic flora of Arizona.* Carnegie Inst., Washington, Publ. 526.

7. DuToit, A. L., 1953, *Geology of South Africa,* 3rd ed., New York, Hafner Publishing Co., 566 pp.

8. Harland, W. B., Smith, A. G., and Wilcock, B., Eds., 1964. *The Phanerozoic time-scale; a symposium.* Quart. Jour. Geol. Society of London, v. 120 s, 458 pp.

9. Harshbarger, J. W., Repenning, C. A., and Irwin, J. H., 1957, *Stratigraphy of the Uppermost Triassic and the Jurassic rocks of the Navajo country.* U. S. Geol. Survey, Prof. Paper 291, 74 pp.

10. Krynine, P. D., 1950, *Petrology, stratigraphy and origin of the Triassic sedimentary rocks of Connecticut.* Conn. Geol. and Nat. Hist. Survey Bull. 73, 247 pp.

11. McKee, E. D., 1954, *Stratigraphy and history of the Moenkopi Formation of Triassic age.* Geol. Soc. America, Mem. 61, 133 pp.

12. Peabody, F. E., 1948, *Reptile and amphibian trackways from the Lower Triassic Moenkopi Formation of Arizona and Utah.* Dept. Geol. Sci. Univ. California, Bull., v. 27, no. 8, pp. 295–467.

13. Silberling, N. J., and Tozer, E. T., 1968, *Biostratigraphic classification of the marine Triassic in North America.* Geol. Soc. America, Spec. Paper 110, 63 pp.

14. Stanley, S. M., 1968, *Post-Paleozoic adaptive radiation of infaunal bivalve molluscs—a consequence of mantle fusion and siphon formation.* Jour. Paleontology, v. 42, no. 1, pp. 214–229.

15. Tozer, E. T., 1967, *A standard for Triassic time.* Geol. Survey Canada, Bull. 156, 103 pp.

Figure 15-1 Jurassic dinosaur skeletons against a background of a mural by Rudolf Zallinger in Yale Peabody Museum. *Brontosaurus* (middle) was 67 feet long and 18 feet high.

CHAPTER **15** *The Jurassic Period*

Shallow seas teeming with life occupied western Europe during much of Jurassic time and the record they left is uncommonly clear. Fossiliferous formations of limestone and shale succeed one another over large areas of England, France, and Germany, where they now dip gently and crop out in low scarps that are easily traced. Stratigraphic geology has its roots in these Jurassic terranes, for it was here that the pioneer geologists of the early nineteenth century discovered the principle of faunal succession (p. 10) and developed the methods of correlation by fossils on which our entire chronology of Phanerozoic history is based.

The critical breakthrough was made in England by William Smith in the late 1790s. As a surveyor Smith was concerned with mapping estates, draining swamps, and laying out routes for canals across southern England where the Jurassic rocks are so well exposed.

Figure 15-2 William Smith (1769–1839), the father of English geology.

The limestones were extensively quarried for use in construction and were well known to Smith. Indeed, he gave names that are still in use to many of the formations.

As a hobby he collected fossils, not with any idea of their significance but merely because they were interesting curios. However, as his familiarity with both the strata and their fossils increased he realized that each rock formation yielded certain fossils that were distinctive of it wherever he encountered it. This being true, he reasoned, it should be possible to recognize a formation in any outcrop by its fossils. With this insight he visited a friend, Rev. Richardson, also a collector of fossils, whom he amazed by identifying the source of each lot of his fossils. Richardson, with Smith's consent, publicized the idea among prominent geologists of the day and it soon became common knowledge. Thus was born the idea of using fossils to date and correlate rock formations. It was, of course, an "open sesame," permitting synthesis of local studies into regional interpretation. It is no accident that within the next few decades all the geologic systems had been worked out and named, and a geologic time scale for the entire earth had been established. Such rapid advances would have been impossible without the principle discovered by William Smith, who has justly been called the father of English geology.

Although Smith made the first detailed study of Jurassic rocks he did not coin the name itself. On his famous map of the geology of England and Wales (1815) ten or more individual formations make up the interval that later became known as the Jurassic. By 1823 the term "Jurassique" was in use in the Jura Mountains, lying on the border of France and Switzerland and extending into southwest Germany, where several of the great continental geologists were carrying out stratigraphic studies, and by 1850 the use of Jurassic was established.

Study of the rich ammonoid faunas of the European Jurassic rocks also led to such milestones in stratigraphic geology as the concept of the **stage** by the French paleontologist Alcide D'Orbigny, and of the **zone** by the Ger-

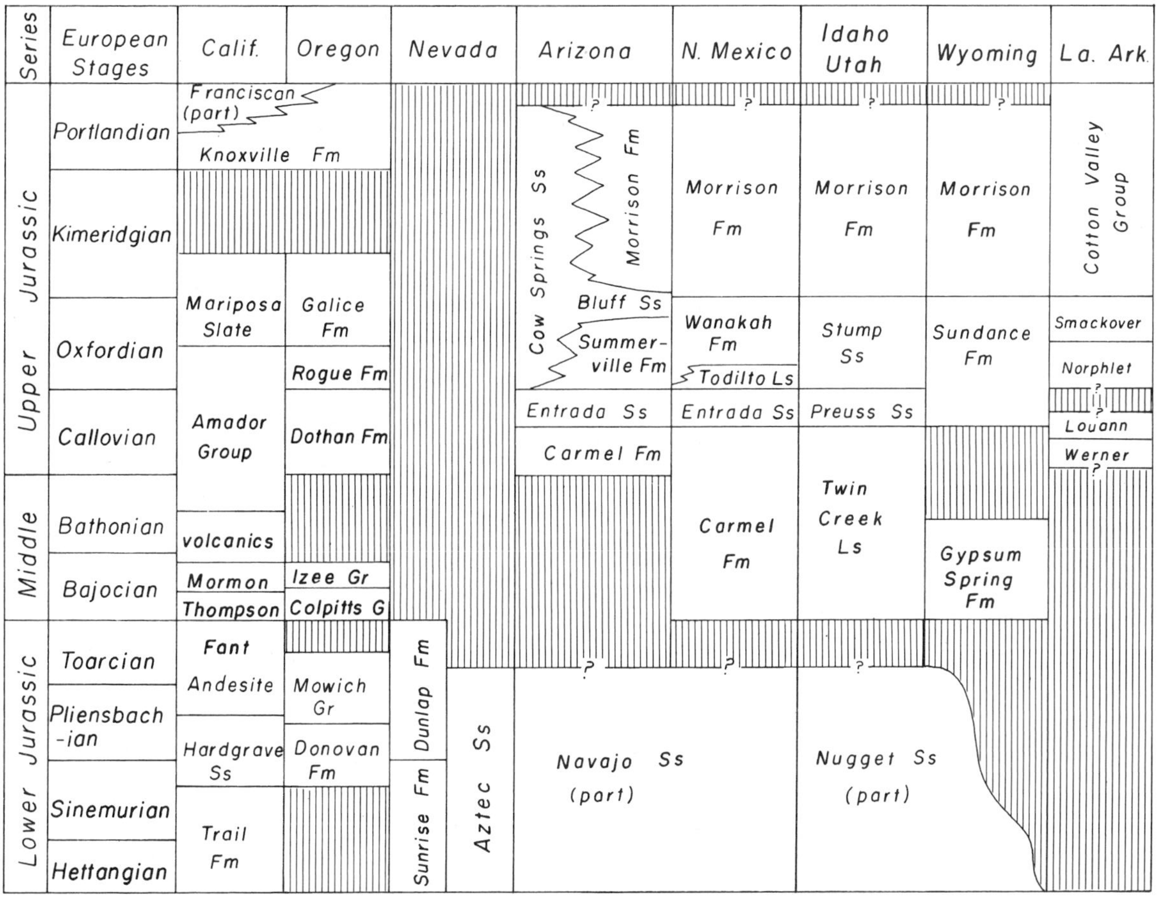

Series	European Stages	Calif.	Oregon	Nevada	Arizona	N. Mexico	Idaho Utah	Wyoming	La. Ark.
Upper Jurassic	Portlandian	Franciscan (part) / Knoxville Fm			Cow Springs Ss / Morrison Fm	Morrison Fm	Morrison Fm	Morrison Fm	Cotton Valley Group
Upper Jurassic	Kimeridgian								
Upper Jurassic	Oxfordian	Mariposa Slate / Amador Group	Galice Fm / Rogue Fm / Dothan Fm		Bluff Ss / Summer-ville Fm	Wanakah Fm / Todilto Ls	Stump Ss	Sundance Fm	Smackover / Norphlet
Upper Jurassic	Callovian				Entrada Ss / Carmel Fm	Entrada Ss	Preuss Ss		Louann / Werner
Middle	Bathonian	volcanics				Carmel Fm	Twin Creek Ls		
Middle	Bajocian	Mormon Thompson	Izee Gr / Colpitts G					Gypsum Spring Fm	
Lower Jurassic	Toarcian	Fant Andesite	Mowich Gr	Dunlap Fm					
Lower Jurassic	Pliensbach-ian	Hardgrave Ss	Donovan Fm	Sunrise Fm	Aztec Ss	Navajo Ss (part)	Nugget Ss (part)		
Lower Jurassic	Sinemurian	Trail Fm							
Lower Jurassic	Hettangian								

Figure 15-3 Correlation of important Jurassic sections.

man Albert Oppel, both during the mid-1800s. Of D'Orbigny's ten original stages of the Jurassic seven (Toarcian through Portlandian) have survived as part of the modern subdivision of Jurassic time shown in Fig. 15-3. Jurassic stages are widely useful because they are for the most part easily recognized by their characteristic ammonoid fossils. An exception to this is found in latest Jurassic time when the ammonoid faunas became markedly restricted in distribution and distinct faunal provinces were formed. This has made difficult the widespread correlation of Late Jurassic rocks. For the Jurassic as a whole, however, the relative ease of correlation on an intercontinental scale led early workers to compose the first worldwide paleogeographic maps.

JURASSIC HISTORY OF NORTH AMERICA

In marked contrast with the extensive distribution of richly fossiliferous strata in Europe, the Jurassic is the least widespread system in North America and only rarely do its outcrops abound in fossils. No outcropping Jurassic strata are known in the eastern part of the continent and presumably the Appalachian region was undergoing erosion throughout the period.

From deep wells in the Gulf Coastal Plain, Late Jurassic formations are known to underlie younger (Cretaceous) coastal plain deposits in all the Gulf border states except Florida. The Gulf embayment apparently dates from this time. No comparable deposits have been found as yet beneath the Atlantic Coastal Plain but one well at Cape Hatteras encountered strata which may be Jurassic beneath more than a mile of Cretaceous rocks.

The Rocky Mountain Geosyncline. Throughout most of the Jurassic the Rocky Mountain geosyncline was accumulating either marine or continental deposits. This great area had a rather complex geography (see Map X, p. 129). The deeper, or more rapidly subsiding part of the geosyncline consisted of two narrow troughs lying adjacent to the Cordilleran geanticline on the west, the Alberta trough in Canada, and the Twin Creek trough in southeastern Idaho, western Wyoming, and central Utah. A generally positive area, known as Belt Island, separated the two troughs in western Montana. On the stable platform east of the troughs the broad Williston Basin was the most pronounced depression and farther south the low hills of the ancient Colorado Ranges persisted from the Triassic.

The entire region appears to have been a great basin of interior drainage. Persistent dune fields occupied its southwestern part where the great Navajo dune sands, which began to form in the Triassic, continued to accumulate during Early Jurassic and were succeeded by other eolian deposits throughout the period. Marine incursions from the Boreal sea spread down the Alberta trough four times during the period. In the Early Jurassic the sea was confined to this trough, but the second incursion in Middle Jurassic time spread around Belt Island into the Twin Creek trough and also extended across the platform to flood the Williston Basin. The most widespread marine deposits of the Jurassic in the Rocky Mountain geosyncline formed during two marine invasions of Late Jurassic time which followed each other in quick succession during the Callovian and Oxfordian stages. At the maximum marine spread (Map X, p. 129) shallow seas covered much of the Rocky Mountain states, their deposits intertonguing with nonmarine mud and sand that accumulated over the bordering lowlands to the south and east and along the foot of the rising Cordilleran geanticline to the west.

As the Oxfordian sea withdrew northward, streams flowing into the basin from its margins, particularly those flowing eastward from the rising geanticline, spread a vast sheet of fluvial mud, sand, and gravel across the interior lowland to form a great fluvial plain extending from Arizona and New Mexico well into Canada. In the United States these widespread nonmarine deposits are mostly included in the Morrison Formation.

The Pacific Coast Geosyncline. The region west of the Cordilleran geanticline was even more widely submerged than in the Triassic Period and here detrital formations of great thickness accumulated throughout much of Jurassic time. Volcanics generally accompanied these formations, coming from submarine flows, volcanic islands, and in part from volcanoes in the Cordilleran geanticline. In British Columbia submarine volcanics range up to 3000 feet in thickness and in California immense flows of basic lava occur in the Mariposa Formation.

This region has been intensely deformed since Jurassic time and its early history and paleogeography are still very imperfectly known. The Jurassic rocks now crop out in relatively small and disconnected areas and are structurally involved and in large part strongly metamorphosed. In several places thick sequences of fossiliferous strata indicate submergence throughout much of Jurassic time but most sequences indicate shallowing toward the end of the period and intertonguing of progressively coarser sediments.

Nevadan Orogeny. Growing crustal unrest in the Pacific geosyncline was evident during Triassic and Jurassic time in the renewed subsidence of its depositional basins, the immense amount of volcanic activity, and the progressive rise of the Cordilleran geanticline on its eastern flank. About the close of the Jurassic Period this culminated in the first really great paroxysms of orogeny in the western part of North America since Precambrian time. As ranges of fold mountains arose, the deposits of the Pacific Coast geosyncline were crushed and in places tightly folded and complexly faulted. Shales were metamorphosed into slate and limestones into marble. Intense deformation of the older rocks at this time spread eastward across nearly all of Nevada and Idaho.

Beginning late in the Jurassic and culminating in the Early Cretaceous this episode of deformation, the Nevadan orogeny, was the first in a series of great orogenies that built the mountain ranges of the central Cordillera of North America. In addition to the deformation of large parts of the Pacific geosyncline, the Nevadan saw the emplacement, in some areas, of granitic batholiths.

JURASSIC STRATIGRAPHY

The Gulf Geosyncline Begins. By Late Jurassic time a sea occupied the present site of the Gulf of Mexico and had spread to the limits indicated in Map X. The deposits formed here are now deeply buried but they are known from many wells drilled to tap the oil which they contain. They underlie a large crescent-shaped area from Alabama to southern Texas and are in the shape of a great wedge thickening Gulfward from a featheredge updip where they are overlapped by Cretaceous deposits. Although their probable area of greatest thickness has never been penetrated by drill, more than 7000 feet of strata have been recorded from wells in Louisiana.

The Upper Jurassic rocks of the Gulf subsurface are divided into a number of units unlike usual formations in that each varies in rock type from thin, partially nonmarine redbed sequences of gravel, sandstone, and shale at their inland edge and thickens Gulfward into a succession of dark-colored, marine limestones and shales. A dominantly limy lower unit, the Smackover Formation, contains marine fossils of Oxfordian age (see Fig. 15-3) and the overlying Cotton Valley Group has yielded both Kimeridgian and Portlandian fossils. Thus a large part of Late Jurassic time is represented by this sequence, which gives evidence of open marine conditions in the Gulf region and is thus the first indication in the geologic record of a Gulf embayment or a Gulf geosyncline. The shoreward redbed facies and the shallow marine beds into which they

grade both contain evaporites locally, indicating hot and dry conditions along the coast.

Other formations whose age is not certainly known underlie the Upper Jurassic units [2]. The Smackover Formation grades downward into the gravelly redbeds of the Norphlet Formation, a thin unit that probably represents a marginal phase of the Smackover marine incursion. The Norphlet uncomformably overlies an enormous formation of salt, the Louann Salt, which ranges in thickness from 600 to 1300 feet and grades downward into a thin conglomeratic redbed unit, the Werner Formation, at its base. The closely related Louann and Werner formations contain no diagnostic fossils but similar units crop out in northeastern Mexico at approximately the same stratigraphic position and these partially overlie Lower Jurassic strata. This relationship, which has yet to be proven by continuous tracing in the subsurface, suggests a Middle Jurassic age for the Louann and Werner. Considerable interest centers around the Louann Salt inasmuch as it is believed to be the primary source bed for the famous Gulf Coast salt domes, structures that have served as petroleum traps and yielded millions of dollars' worth of oil, salt, and sulfur.

More redbeds lie beneath the Louann and Werner formations but their relationships are poorly known and there is no basis for suspecting that they are Jurassic.

Shallow Seas of the Rocky Mountain Region. The Sundance Formation of older literature was a widespread Jurassic marine unit lying between the Triassic redbeds in Wyoming and adjacent states and the overlying nonmarine Morrison Formation. In the Black Hills region where it was first named, the Sundance is about 400 feet thick and it extended without great variation in thickness over much of Wyoming, Montana, and the Dakotas. Intensive stratigraphic studies during World War II revealed that the Sundance Formation was not a single marine deposit but, together with a few feet of underlying gypsiferous shale previously considered Triassic, contained a detailed record of three separate marine invasions during Jurassic time. The old concept of a single "Sundance Sea" has

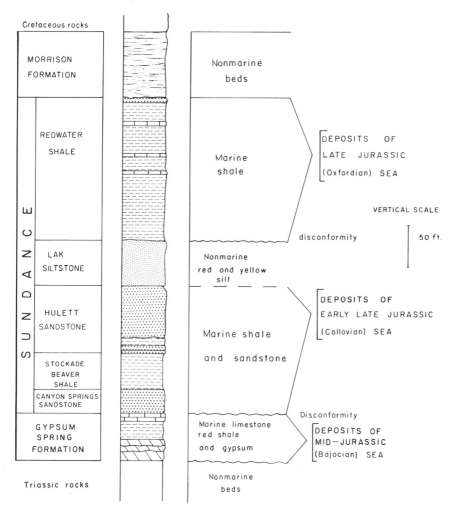

Figure 15-4 Stratigraphic interpretation of the Jurassic rocks included in the old Sundance Formation and Gypsum Spring Formation.

thus given way to a more complex history and the Sundance Formation has been subdivided into many units. Figure 15-4 shows the interpretation of the Black Hills Jurassic section.

It is remarkable that so long and complex a history has been recorded in the relatively thin veneer of Jurassic sediments preserved in Wyoming and Montana. Here formations that rarely exceed 100 feet in thickness extend over thousands of square miles, indicating great stability over this area of deposition. But the stable platform in this area was not without local activity. Formations thicken into the Williston Basin and a number are absent across the Belt Island region. In addition

there were other confined warpings of the platform which account for local absence of particular units and the presence of unconformities between many of the Jurassic units.

Normal marine conditions were rare in the Jurassic seas that spread past Belt Island, for this shallow area at the south end of the Alberta trough inhibited normal marine circulation to the south. During the Middle Jurassic, for example, the platform area in Wyoming was covered by intermittent shallow marine waters in which gypsum, reddish-brown silt and clay, and gray, sparingly fossiliferous limestone were deposited to form the Gypsum Spring Formation. The gypsum and rare fossils point to unusual conditions on the platform. Traced westward into the Twin Creek trough, the Gypsum Springs Formation grades into the Twin Creek Limestone, which locally is as much as 6000 feet thick. Relatively few fossils have been found in the Twin Creek and it is apparent that no part of the Middle Jurassic sea south of Belt Island was hospitable enough to attract the abundant shallow-water life of the Jurassic.

The one exception to this was the last and most widespread of the four Jurassic marine invasions, that during the Oxfordian stage. At this time Belt Island was apparently submerged and had little effect on circulation. The sea spread across the platform and down the Twin Creek trough, pushing southward into southern Utah and eastward into Colorado. Along these southern shores marine life was only locally abundant but in the shallow sea of the platform across Wyoming remains of a rich fauna of clams, oysters, crinoids, ammonoids, and belemnoids, along with the bones of ichthyosaurs, were preserved in the glauconitic sands and shales that now form the uppermost units of the old Sundance Formation.

Arid Borderlands of the Southern Interior. In and adjacent to the Colorado Plateau the Jurassic rocks are several thousands of feet thick and are largely nonmarine. At the base is the widespread Navajo Sandstone, in

Figure 15-5 "Frozen dunes" in the Navajo Sandstone near the entrance to Zion National Park, Utah. (C. O. Dunbar.)

Figure 15-6 Towering cliffs of Navajo Sandstone in Zion National Park, Utah. In the distance at the left, the Great White Throne, also made of Navajo Sandstone. (C. O. Dunbar.)

part Triassic, which characteristically weathers to dome-shaped masses or is dissected to form steep canyons with cliffs several hundred feet high. Formed of uniformly fine sand, the Navajo commonly weathers to natural arches and other spectacular natural monuments. Many outcrops show the steep, large-scale cross-bedding characteristic of dunes (Fig. 15-5). This is the formation impressively displayed in the walls of Zion Canyon (Fig. 15-6), in Glen Canyon of the Colorado River, in Natural Bridges National Monument, and in Monument Valley.

Overlying the Navajo Sandstone is the San Rafael Group, a complex of variable deposits totaling about 1500 feet thick, which consists of alternating or intertongued sandy marine formations, nonmarine redbed silts and sands, and eolian sands. The marine units represent tongues of the several invasions of the Jurassic sea from the north. The basal Carmel Formation contains marine fossils that show definitely that it is of Middle Jurassic age and a sandy southern correlative of much of the Twin Creek Limestone. It locally truncates the dune bedding of the Navajo Sandstone, indicating that the Middle Jurassic sea encroached on the Navajo dune fields. Partial return of the dune fields is recorded in the nonmarine Entrada Formation which overlies the Carmel and consists of fluvial sands and silts that grade into dune sands to the south and southwest. During the two invasions of the Late Jurassic sea, marked by marginal marine deposits of the Curtis and Summerville formations, the dune fields persisted south and west of the shoreline in what is now northern Arizona. Over much of this shoreline the withdrawal of the Jurassic sea was followed by a readvance northward of the dunes to form the Bluff Sandstone which appears as a tongue of the main body of Late Jurassic dune sands, the Cow Springs Formation. Apparently throughout the whole of Jurassic time and during much of the Late Triassic as well, a great area of desert sands accumulated and shifted about in Arizona and southeastern Nevada. The entire deposit constitutes perhaps the greatest single accumulation of dune sands known in the entire Phanerozoic record.

The Morrison Formation and Its Dinosaurs. All the gigantic dinosaurs of the American Jurassic have come from a single formation (Figs. 15-8 and 15-9) that was deposited as a blanket of fluvial sediments over the rest of the Jurassic formations in the Rocky Moun-

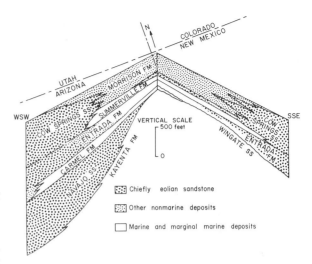

Figure 15-7 Diagrammatic cross sections illustrating the lateral interfingering of eolian sands with fluvial and marine sediments in the Jurassic deposits of the southwest. Each panel is about 125 miles long. (Data from Harshbarger, Repenning, and Irwin; U. S. Geol. Survey Prof. Paper 291.)

Figure 15-8 Upper part of Morrison Formation near Cisco, Utah, comprised of variegated clay and sandstone. (C. H. Dane. U. S. Geological Survey.)

Figure 15-9 Excavating a dinosaur skeleton in the Morrison Formation near Shell, Wyoming. (Barnum Brown. American Museum of Natural History.)

tain geosyncline following the withdrawal of the Oxfordian sea. Named the Morrison Formation for exposures at Morrison, near Denver, Colorado, the nonmarine beds extend from Arizona, where they grade in part southward into the eolian Cow Springs sands, to north-central Montana where they grade into sandy coal-bearing deposits of the Kootenay Formation, which filled the Alberta trough as the sea withdrew northward. At its maximum extent the Morrison covered almost the entire basin of Jurassic deposition in the Western Interior of the United States, an area of at least half a million square miles. In spite of its great extent it is generally less than 400 feet thick.

The Morrison Formation consists of shales, siltstones, and sandstones with local conglomerates, all of which intergrade laterally, as is the habit of fluvial deposits, so that it is impossible to follow a single bed over a considerable distance. The coarser sediments are commonly irregularly bedded and cross-laminated. The color varies locally from greenish-gray to lavender or pink or even white. No marine fossils have ever been found, but more than 150 kinds of terrestrial animals and land plants are known from these beds. These include the greatest of all dinosaurs. Of the species, 69 are dinosaurs, 25 are tiny primitive mammals, 3 are crocodiles, 24 are river clams and land snails, and 23 are plants.

From these facts we can picture the region at the time of deposition in a setting not unlike that of the present basins of the Amazon or Paraná rivers. It was a low alluvial plain crossed by sluggish streams heading in the distant highlands to the west and south, whence came heavy loads of mud and sand. Here and there swamps or small lakes interrupted the courses of the braided streams. The shifting sands and gravel bars along the channels gave rise to deposits of cross-bedded sandstone and conglomerate, while the finer mud and silt settled over the floodplains to produce the varicolored shales. Vegetation spread over much of the landscape reclaimed from the sea and animal life was abundant.

The precise geologic age of the Morrison

Formation has been difficult to prove and until the 1930s it was the basis for one of the major controversies in American stratigraphic geology. Actually two separate problems have been involved, the age of the spectacular dinosaur fauna which occurs in the lower half of the formation in most areas and the age of the Morrison Formation as a whole. Curiously, the best evidence on the age of the Morrison dinosaurs is found nearly halfway around the earth, in East Africa, where beds bearing dinosaurs much like those of the Morrison are interbedded with marine zones carrying undoubted Jurassic ammonites. Although it is possible that the Morrison and the East Africa (Tendaguru) formations are not strictly equivalent, the balance of the evidence favors the assignment of the Morrison fauna to either the Kimeridgian or Portlandian stages of the Late Jurassic. The lower part of the Kootenay Formation in Canada, which almost certainly is the lateral correlative of the dinosaur-bearing part of the Morrison, contains a few marine beds with gigantic ammonoids said to be Portlandian or possibly Kimeridgian.

It has become customary to regard the top of the Morrison Formation as marking the top of Jurassic deposits in the Rocky Mountain region, but this cannot be demonstrated. Other similar fluvial deposits overlie those of the Morrison and it is impossible to pick a consistent top for the formation throughout the Rocky Mountain region. Above the widespread Jurassic dinosaur faunas fossils are rare; without them the age of the upper part of the Morrison as well as that of overlying nonmarine beds remains in doubt. Not until late in Early Cretaceous time did a sea again invade the Rocky Mountain geosyncline and bring an end to the long interval of intermittent fluvial deposition that began with the Morrison sediments. The passage from Jurassic to Cretaceous time apparently went unrecorded in the geosyncline as continued Nevadan uplift in the Cordilleran geanticline afforded a continuing supply of sediment for streams to distribute and redistribute over the vast alluvial plain.

Rocks of the Pacific Geosyncline. Jurassic

rocks are extensively exposed in California, but only in widely separated areas where they have escaped the several later periods of uplift and deep erosion. In most places they are deformed and metamorphosed to such an extent that fossils are obscure. With these fragments of the record, it is impossible to restore a complete picture of the Jurassic history of California. Suffice it to say that the sediments were very thick and except for included lavas were almost entirely detrital.

In eastern California the Mariposa Slates and interbedded volcanics reach a thickness of possibly 10,000 feet. Mount Jura at the northern end of the Sierra Nevada presents a very complete Jurassic section, with 15 formations ranging from Lower to Upper Jurassic and containing volcanic tuffs and agglomerates of various ages throughout the period; and a still finer and more complete section is present in eastern Oregon. In Shasta County, California, the Jurassic is represented by the Knoxville Group, which may total 10,000 feet in thickness, and includes continental beds with land plants, as well as marine zones with the peculiar arctic clam *Aucella* (Plate 14, Fig. 6).

Western Canada has equally spectacular thicknesses of detrital and volcanic Jurassic beds. The Pacific geosyncline in this area has Lower, Middle, and Upper Jurassic rocks aggregating on the order of 50,000 feet of deposits. Most of these are marine in origin but toward the close of the period parts of the geosyncline were undergoing nonmarine sedimentation [3].

Southern Alaska also has a great development of Jurassic rocks, though they have not been studied in detail. Marine formations of Early, Middle, and Late Jurassic age are known and much volcanic material is associated with them.

The great thickness and detrital nature of all the Jurassic formations of the Pacific region points to highly active tectonic basins and island chains within the broad geosynclinal area. Here areas of rapid subsidence were probably linked to island arcs, many of them volcanic, which furnished the great quantities of sediment. Apparently the eastern margin lay

against a continuous, narrow Cordilleran geanticline in eastern British Columbia, western Idaho, and central Nevada. Whether this barrier was complete in Canada during all of Lower and Middle Jurassic time has been questioned, but it was in the United States and by Late Jurassic the Nevadan orogeny was obviously broadening and locally heightening this narrow land which separated the Pacific and Rocky Mountain geosynclines.

CLIMATE

The generally widespread land floras and faunas and the abundant shallow-water marine faunas have long marked the Jurassic as a period of broadly uniform, warm to temperate climate, appreciably milder and more equable than at the present. No evidence of Jurassic polar ice caps exists, no matter where one chooses to place the poles of the time, and Jurassic glacial deposits are unknown. Presence of a warm climate over much of the earth is indicated by widespread coal deposits of Jurassic age and, in Middle and Upper Jurassic time, by the existence of reefs of corals, sponges, and bryozoa 2000 miles north of the present range of similar forms.

Today the larger forms of many kinds of externally shelled animals are found in the tropics and this observation has been used to attempt to show changes of climate within the Jurassic. For example, Late Jurassic insects are larger than Early Jurassic insects and this has suggested to some that Jurassic climates increased in warmth into the latter part of the period. But until a more complete knowledge of Jurassic faunal distribution is at hand, generalization on a few observations of this kind can be very misleading. The Portlandian ammonoids illustrate this point. In the boreal region of Europe, North America, and Greenland the giant ammonoids of Portlandian rocks are widely distributed, but these great forms are unknown south of the latitude of northern France [1].

That local factors exert great influence on the climate of certain provinces is well shown by the great Jurassic desert of the southwestern

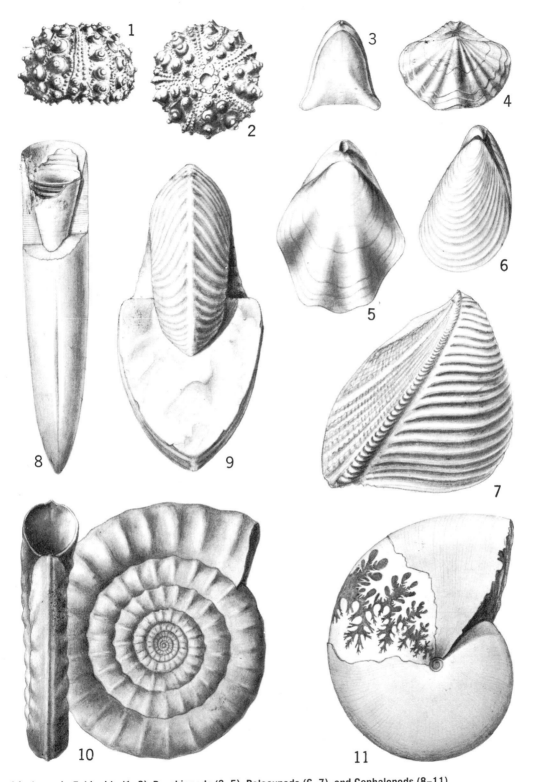

Plate 14 Jurassic Echinoids (1, 2), Brachiopods (3–5), Pelecypods (6, 7), and Cephalopods (8–11).

Figures 1, 2, *Hemicidaris intermedia* (side and upper views); 3, *Digonella digona*; 4, *Spiriferina walcotti,* last of the *spiriferoids*; 5, *Goniothyris phillipsi*; 6, *Aucella piochii;* 7, *Trigonia costata*; 8, *Pachyteuthis densus* (broken to show chambered shell); 9, *Cardioceras cordiforme*; 10, *Paltechioceras raricostatoides;* 11, *Phylloceras heterophyllum.* All natural size. (Drawn by L. S. Douglass.)

United States. As we have seen there was almost continuous accumulation of dune sands from the Late Triassic to the Late Jurassic in this area. Apparently the region lay in a local "rain shadow" of the highlands of the Cordilleran geanticline. Extensive deposits of gypsum in Wyoming and the salt deposits of the Gulf region point to general aridity for adjacent local areas. In many other parts of the world the climate was decidedly more humid than it had been in the Triassic and gray or dark gray sediments with coal beds accumulated in the lowlands. The coal-bearing Kootenay beds in British Columbia and Jurassic coal on Vancouver Island and in Mexico demonstrate pronounced differences in climate within the North American continent alone.

ECONOMIC RESOURCES

Coal. The Jurassic is an important coal-producing system, if we consider the world at large. Extensive areas in Siberia are underlain by Jurassic coal of economic importance, and in Tasmania and Australia the chief coal measures are the Jurassic rocks. There are also important coals in Spitsbergen and smaller deposits in various parts of Europe and southern Asia. In North America, coal of high quality occurs in the Kootenay Formation where individual coal seams may be as much as 20 feet thick. Billions of tons of reserves of Kootenay Coal are still available in British Columbia and Alberta. Some of it is Lower Cretaceous coal as the system boundary falls within the formation [3].

Gold. The gold that attracted the "Forty-Niners" and led to the rapid settlement of California has its source in the gold-quartz veins formed in the Jurassic slates along the western slope of the Sierra Nevada. The placer gold that the early settlers panned from the streams was concentrated in the river gravels during Cenozoic times, but it came originally from the Jurassic veins and was freed during their erosion. A long, narrow belt of Jurassic rocks containing gold veins and extending for 120 miles along the western foothills of the Sierra Nevada is the famous "Mother Lode" belt of the early gold seekers.

In the century following the Gold Rush of 1849, California produced more than one-third of the nation's gold—slightly over 103,000,000 ounces valued at almost $3,300,000,000. Production declined to an all time low of 170,000 ounces, valued at $5,950,000 in 1957, yet California still ranks fourth among the states in gold production.

Valuable metals of many kinds are associated with the great batholithic intrusions of the Nevadan orogeny in the United States and Canada. Some of these deposits are Jurassic, but most originated in late phases of intrusion and are Early Cretaceous in age.

Uranium. In recent years a large source of uranium ore in the United States has been in the Triassic and Jurassic rocks of the Colorado Plateau, but the lead-uranium ratios indicate that the ores were introduced between 60,000,000 and 75,000,000 years ago, near the end of the Mesozoic Era, and are not related to Jurassic history.

LIFE OF THE JURASSIC PERIOD

Forest of Evergreens. The Jurassic forests consisted of pines and other conifers with a mingling of gingkos and tree ferns and an undergrowth of herbaceous ferns and scouring rushes. No grasses were yet present and the drier slopes and plains must have supported an open growth of ferns and cycadeoids.

The cycadeoids were an important and characteristic group of Mesozoic plants, widely distributed over the world in both Jurassic and Cretaceous periods. Like their modern descendants, the true cycads (Fig. 14-15), they had short trunks heavily armored with leaf bases (Fig. 16-14), and their leaves were large and pinnate; but unlike the modern cycads, the cycadeoids of the Mesozoic bore large and showy flowerlike cones (Fig. 15-10).

Medieval Insects. About a thousand kinds of insects are known from the Jurassic rocks, and among these are representatives of most of the modern orders. Caddisflies, scorpionflies, dragonflies, and beetles were common;

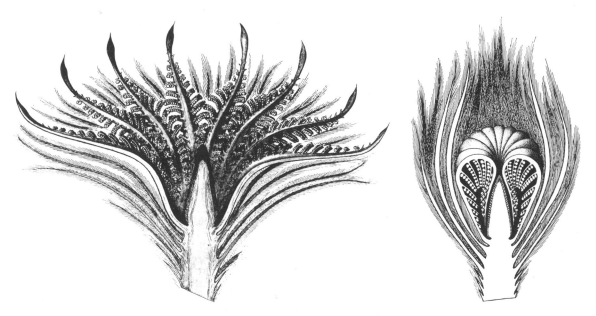

Figure 15-10 Flowerlike cone of a cycadeoid. Median section of an open and un-opened cone of *Cycadeoidea ingens,* about ½ natural size. (G. R. Wieland.)

grasshoppers, cockroaches, and termites were also present; moths and flies (Diptera) made their appearance at this time. Since the last two groups represent highly specialized stocks of the insects, it is clear that the significant features of insect evolution had already appeared by the middle of the Mesozoic Era.

Reptile Hordes. By Jurassic time the **Age of Reptiles** was in full swing. Not content with a complete domination of the lands, some of the reptiles anticipated the birds in flight, and others competed with fishes in the sea. They also attained their greatest size in the ponderous sauropod dinosaurs, so enormous that a parade of 60 to 75 individuals would have been a mile long. Never before, or since, was the earth so completely dominated by reptiles.

Dinosaurs were the conspicuous land animals and four of the five great tribes of these reptiles were represented (Fig. 15-11). **Sauropods** were the largest. *Brontosaurus* (Fig. 15-1), one of the best known American forms, reached a length of about 65 feet, but the more slender *Diplodocus* was nearly 85 feet long; the brain of each of these huge ani-

mals, however, weighed less than a pound. In spite of their great size the sauropods were probably inoffensive creatures, for they were all herbivorous.

It was once thought that these great animals were too large to support their own weight on the land and must have inhabited swamps where they were buoyed up by water. They have commonly been restored in such a setting, devouring the lush aquatic vegetation. This is almost certainly unjustified. Sauropod remains such as those in the Morrison are not found in swamp deposits but rather in flood plain and river channel deposits where they are associated with remains of smaller dinosaurs that would have drowned in water deep enough to support a sauropod. They were most likely dwellers of broad flood plains and river margins, living close enough to the water to use it as a refuge from carnivores. Anatomically, their great limbs were undoubtedly functional and there is no reason to believe they had difficulty getting about on land. They are known throughout the Jurassic and Cretaceous nonmarine rocks of the world, thus they must be counted a highly successful

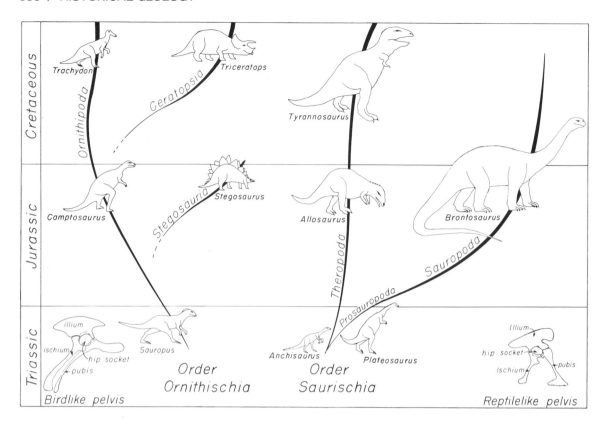

Figure 15-11 Diagram to show the relations and geologic range of the main groups of dinosaurs. The two orders stemmed from a common ancestral stock. The Saurischia [Gr. *sauros,* reptile + *ischios,* hip] had a triradiate pelvis such as is normal for reptiles. Originally they were bipedal and carnivorous, but before the close of Triassic time they were diverging into 2 suborders. The Theropoda remained bipedal and carnivorous and culminated in the great carnivore, *Tyrannosaurus,* late in the Cretaceous Period. The Sauropoda became ponderous before their front limbs had been greatly reduced and during Jurassic time they reverted to a quadrupedal habit and a herbivorous diet, culminating in the greatest of all dinosaurs such as *Brontosaurus.* The Ornithischia [Gr. *ornithes,* bird + *ischios,* hip] had a birdlike pelvis in which the pubis lay nearly parallel to the ischium. They were small, bipedal, and herbivorous in Triassic time. The main line, the Ornithopoda, increased in size but remained bipedal, culminating in the duckbill dinosaur, *Trachodon.* During Jurassic time an offshoot of this line returned to a quadrupedal habit and developed armor plates along the back, culminating in forms such as *Stegosaurus.* Early in Cretaceous, another branch reverted to a quadrupedal habit and developed bony horns. These were the Ceratopsia. All of the ornithischians were herbivorous.

Figure 15-12 *Stegosaurus,* a herbivorous dinosaur. (Courtesy of the United States National Museum.)

tribe of animals.

The plated dinosaur, *Stegosaurus* (Fig. 15-12), with a $2\frac{1}{2}$-ounce brain for 10 tons of weight, was equally distinctive of this time. In addition, there were bipedal carnivores of large and small size, one of which, *Compsognathus,* must have been as agile and slender as a small kangaroo, for it was only $2\frac{1}{2}$-feet long. Of the larger carnivores of this period, *Allosaurus* (Fig. 15-13) was perhaps the greatest, having an overall length of more than 30 feet. Still larger species lived in the Cretaceous.

The herbivorous bipedal dinosaurs (ornithopods) (Fig. 15-13) are known in several genera, but were less common here than in the Cretaceous.

Among the most bizarre animals of the Mesozoic were the **pterosaurs** (Fig. 15-14), or winged reptiles. With wings of skin they were more batlike than birdlike, though the structure of their wings shows the resemblance to be superficial. The bat has all its digits extended to bear the weblike wing membrane, whereas the pterosaur had only the fourth finger greatly extended to support the wing, leaving the other digits free to serve as claws. The Jurassic species had sharp, slender teeth, and heads that were decidedly reptilian. Some had long tails with flukes, which probably aided in keeping the balance during flight, but other forms were tailless. In the Jurassic pterosaurs the front and hind limbs were not greatly disproportionate in size, and it is clear that these winged dragons developed from quadrupedal reptiles. Upon alighting, they certainly walked on all fours (Fig. 15-15); indeed their tracks have been identified [4].

During the Jurassic period, the pterosaurs ranged in size from minute species with a wingspread equal to that of a sparrow up to others with a wingspread of 3 or 4 feet. The greater and more highly specialized forms followed in the Cretaceous Period.

In the Jurassic seas both **ichthyosaurs** and **plesiosaurs** were common. The former, with their streamlined contour and powerful, fishlike tail, must have been powerful swim-

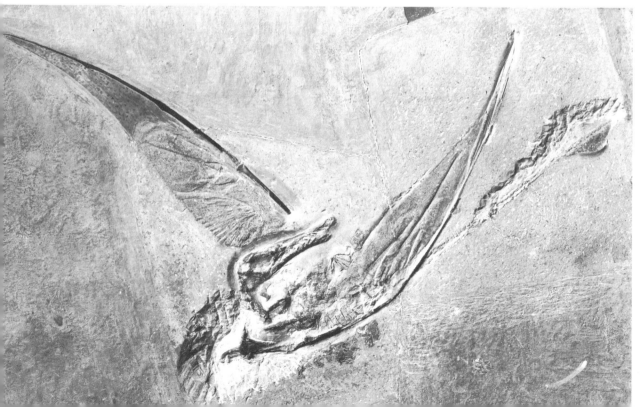

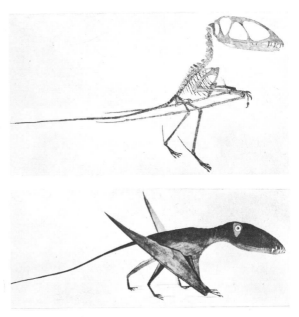

Figure 15-15 A small pterosaur, *Dimorphodon,* skeleton at the top and restoration at the bottom in a walking position. From the Lower Jurassic shales at Lyme Regis, England. This animal was slightly more than 3 feet long from nose to tip of tail. (From H. G. Seeley, *Dragons of the Air,* D. Appleton and Company, 1901.)

Figure 15-13 Jurassic dinosaurs and associated plants. At the center, *Allosaurus* feeding on the carcass of one of the sauropods. Another individual appears in the distance at the left. At lower right is a small, plant-eating dinosaur, *Camptosaurus.* At extreme lower right is a cycadeoid plant with leaves and blossoms. *Allosaurus* was about 35 feet long. (From a mural by Rudolph F. Zallinger, in Yale Peabody Museum.)

Figure 15-14 A flying reptile, the pterosaur, *Rhamphorhynchus,* from the Upper Jurassic lithographic limestone at Solenhofen, Bavaria. The delicate wing membranes and the tail fluke are preserved as an impression in the rock. (Yale Peabody Museum.)

mers (Fig. 15-16). They resemble the modern porpoise to a remarkable degree except that their tail flukes were in a vertical plane so that they swam with a fishlike motion. The ichthyosaur most likely filled much the same ecological niche as his mammalian counterpart, the porpoise.

In the Lower Jurassic black shales of Germany, remarkably preserved specimens of ichthyosaurs are found with the entire skeletons articulated and surrounded by a carbonized film that outlines the contour of the flesh (Fig. 3-12). Some of these, moreover, have been found with unborn young inside the rib case, proving that they were viviparous. In this respect the ichthyosaurs show a more perfect adaptation to aquatic life than any other known reptiles, for even the great marine turtles come ashore to lay their eggs. The Jurassic species were rather smaller than some of the Triassic forms, rarely attaining a length of 25 feet. Many were mature at a length of 5 to 10 feet. They fed on fish and cephalopods.

Figure 15-16 The marine reptile *Ichthyosaurus* with a brood of young. The adult was about 10 feet long. See also Fig. 3-12. (From a painting by C. R. Knight, American Museum of Natural History.)

A remarkable skeleton found with some 200 belemnite shells inside suggests that they were especially fond of these squidlike animals.

Plesiosaurs (Fig. 15-17) propelled themselves by means of large paddlelike flippers, as marine turtles do. In these marine animals the tail was not fluked and the body tended to be less streamlined and broader than high. Long-necked and short-necked plesiosaurs both existed and although there was some variety of form among each of these groups they all seem to have been fish-eaters equipped with long, sharp teeth for catching their prey. The largest Jurassic species attained a length of about 20 feet.

Slender-snouted **crocodiles** much like the modern gavial of India were abundant in the seas as well as in the rivers. Marine turtles also were present, though less common than the groups of reptiles already mentioned.

First Birds. Birds appear as fossils for the first time in Upper Jurassic rocks and represent one of the most remarkable advances in the life of this period. As yet only four specimens are known, and these are from the famous lithographic stone quarries about Solenhofen, Bavaria. Two of these are fine skeletons with impressions of the feathers.

To the first-discovered bird was given the appropriate name *Archaeopteryx* [Gr. *archaios*, ancient, + *pteron*, wing]. It was a strange creature, more reptile than bird, and yet because of its feathers distinctly to be classed as a bird. It would be difficult to find a more perfect "connecting link" between two great

groups of animals, or more cogent proof of the reptilian ancestry of the birds.

Archaeopteryx (Fig. 15-18) was about the size of a crow. Three remarkable features strike one at the first glance:

1. The jaws were set with a row of small teeth. These were not mere serrations on a horny beak but true conical teeth set in individual sockets like those of many reptiles.
2. In the wings the digits were not completely fused, and the first three still functioned as claws.
3. The tail was long and slender, with the feathers diverging pinnately from its axis and not fanwise as in modern birds.

The plumage was thoroughly birdlike, but the teeth, clawed wings, and long tail betray

Figure 15-17 The marine reptile *Plesiosaurus*. A pair of *plesiosaurs* in the foreground with ichthyosaurs in the right background. (After E. Frass, Stuttgart Museum.)

Figure 15-18 A. The skeleton of *Archeopteryx,* the earliest known bird, as it lies in the rock with impressions of the feathers clearly preserved. From lithographic limestone of Late Jurassic age at Solenhofen, Bavaria. This specimen is preserved in the Museum at Berlin.

B. Reconstruction of the dead bird in the position in which it was preserved. Note the toothed jaws and the claws of three digits at the front edge of the wings. (Drawing by Shirley Hartman under the direction of Joseph T. Gregory.)

reptilian affinities.

Jurassic Mammals. Although there is some doubt whether true mammals appeared in the latest Triassic or earliest Jurassic, they unquestionably became well established on the lands and underwent an initial radiation of primitive forms during the Jurassic Period. These primitive mammals are known primarily from teeth and jaws and small bits of skeleton, mostly from the Middle Jurassic Stonesfield Slate and Late Jurassic Purbeck Beds of England and the Morrison Formation of the United States.

Four principal types, or orders, of these first mammals are commonly recognized. Most were alike in being very small and possibly they were superficially similar in appearance. The chief basis for separating them into distinct orders is the different patterns of arrangement of the cusps on the jaw teeth, as illustrated in Fig. 15-19. The **triconodonts** are characterized by three simple cones in a linear series; most were the size of mice but by Late Jurassic some were as large as a small cat. In the **symmetrodonts** three cones form a triangular pattern in which the middle cone is the largest. The **multituberculates** had elongate teeth with many small cones arranged in linear rows. They were a highly specialized group that, judging from their dentition, adopted a rodentlike way of life (Fig. 15-20). The **pantotheres** had jaw teeth of roughly triangular to almost square shape with a broad and fairly complex crown. From this latter pattern all the varied tooth patterns of higher mammals were ultimately derived.

The triconodonts, symmetrodonts, and pantotheres barely survived the Jurassic, although the pantotheres gave rise to all later mammals. The multituberculates were also a domi-

Figure 15-19 Principal orders of Jurassic mammals, represented by right lower jaws and upper cheek teeth. (Adapted from George Gaylord Simpson.)

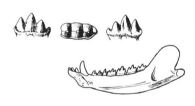

Triconodonta

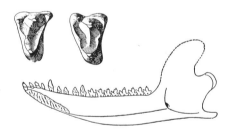

Pantotheria

Symmetridonta

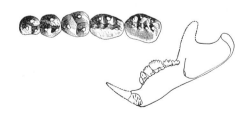

Multituberculata

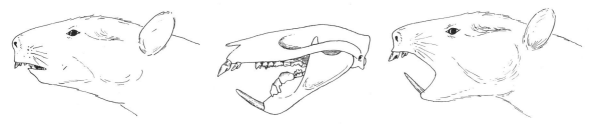

Figure 15-20 One of the common Jurassic mammals, the multituberculate *Ctenacodon,* from the dinosaur quarries in the Morrison Formation at Como Bluff, Wyoming. The skull and two views of the head. About natural size. (Restoration by F. G. Simpson.)

nant Cretaceous mammalian group and survived to the Eocene before becoming extinct.

Marine Invertebrates. The profusion of marine invertebrates and the richness of their remains in the Jurassic rocks have already claimed our attention. In many respects these faunas were essentially modern. For example, corals of the modern families were then extensive reef-makers, and abundant **pelecypods** (Plate 14, Figs. 6 and 7) and **gastropods** resembled modern forms in general features. True oysters had already become common, though strongly plicate species were more prominent than now. Lobsters and shrimplike **crustaceans** were present in numbers, and one depressed form (*Eryon*) foreshadows the evolution of the crabs (Fig. 15-21). The **crinoids,** locally abundant, resembled either of two modern types: the large-stalked forms were closely allied to the *Pentacrinus,* which still lives in deep water off the Japanese coast; small, stemless species were like *Comatula.* **Sea urchins** of modern aspect were well represented (Plate 14, Figs. 1 and 2), and among these were the first of the "heart-urchins." **Sponges** were in places important reef-makers.

More prolific and more distinctive than all other kinds of shellfish, however, were the **ammonoids** (Plate 14, Figs. 9 to 11), represented by a vast number of species, some large and some small, but all possessing delicate, pearly shells. The intricacy and the variety displayed in the fluting of the ammonoid septa during this period are remarkable, and the modification of bodily form of the living animals is an eloquent commentary on the plasticity of animal life. Some species, with slender shells coiled like a rope and with the living chamber occupying more than an entire volution, must have had bodies of eel-like

Figure 15-21 Crustacea from the Upper Jurassic limestones at Solenhofen, Bavaria. A, a shrimp; B, a lobster; C, a flattened, crablike form, *Eryon.* (After A. Oppel.)

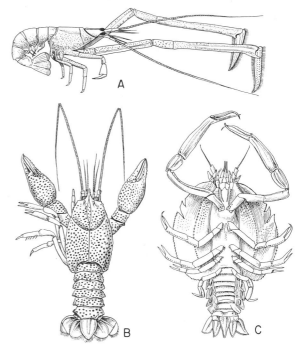

proportions, whereas others with broadly rounded, globular shells had bodies as short as that of the octopus. The most remarkable of all modifications must have existed in those species with laterally compressed and deeply involuted shells wherein the penultimate whorl of the shell was so deeply impressed

Figure 15-22 Reconstruction of a belemnite darting backward and discharging a smoke screen of ink. Such animals ranged from a few inches to 5 or 6 feet in length. The internal shell is shown in Fig. 8 of Plate 14. (Drawn by Lathrop Douglass.)

in the animal's back as to divide it almost in two.

The **belemnites,** some of them 5 or 6 feet long, were also at their climax at this time, and their cigar-shaped internal shells are extremely common fossils (Fig. 15-22, and Plate 14, Fig. 8). Rare specimens found in the black shales of the Lower Jurassic in England and Germany show the form of the body and arms, preserved as a carbonized film about the shells. From these it is certain that the belemnites were squidlike cephalopods with six instead of ten arms and with corneous hooks instead of sucking discs on the arms. True squids were also evolving from common ancestors with the belemnites, but their shells were too perishable to have left an imposing record like that of the belemnites. It is interesting to note that an ink sac exactly like that of modern squids was present in the Jurassic forms and that the pigment is sufficiently preserved in some of the Lower Jurassic speci-

mens mentioned above so that ink can still be made of it.

Solenhofen, a Remarkable Fossil Locality. In the region about Solenhofen, Bavaria, Germany, the Upper Jurassic rocks include circular reefs of sponges and corals within which there are deposits of very pure, fine-textured limestone. For generations this stone has been quarried and shipped to all parts of the world for the engraving of etchings and lithograph prints. During this time the quarries about Solenhofen and Eichstadt have yielded more remarkable fossils than any other locality in the world except possibly that of the Burgess Shale in the Cambrian (p. 184).

The flawless, fine texture of the Solenhofen stone, so essential for the reproduction of the lights and shades in lithographs, lends itself equally well to the preservation of the delicate impressions of organic tissues. It has therefore given us a knowledge of many soft-tissued Jurassic animals, such as jellyfish, and has preserved impressions of the fleshy bodies of creatures otherwise known only from their skeletons or shells. From these quarries, for example, came all the known specimens of Jurassic birds. The faithful impression of their delicate feathers is a fortunate thing, for without this evidence it would be difficult to prove that *Archaeopteryx* was not a reptile. Here also have been found specimens of pterosaurs in which the form of delicate wing and tail membrane (Fig. 15-14) are preserved with remarkable fidelity. Among other things rarely preserved elsewhere are 8 kinds of jellyfish and more than 100 species of insects, including moths and flies. Finally, very good specimens of the horseshoe crabs (*Limulus*) occur here. A total of 450 species of animals has been recovered from these quarries.

Evidently these are the deposits of lagoons within atolls that lay not far from the mainland. The fossils include a dinosaur, 29 species of pterosaurs, and 3 birds. On the other hand, there are no freshwater animals, and marine fishes and marine invertebrates (mainly crustaceans and ammonoids) comprise nearly all the fauna. One remarkable feature of the deposit is that most of the organisms were not dismembered before burial. There were certainly no scavengers living on the bottom, and the entombed creatures were either dead when they were washed over into the lagoon or died soon thereafter and were quickly buried by the fine, limy ooze that spread in from the fronts of the reefs. It is because of this quick burial that the animals are so well preserved. The floor of the lagoon may have been in part permanently submerged and in part only a mud flat covered twice daily by the tides.

REFERENCES

1. Arkell, W. J., 1956, *Jurassic geology of the world.* Edinburgh, Oliver and Boyd Ltd., 806 pp.
2. Murray, G. E., 1961, *Geology of the Atlantic and Gulf Coastal Province of North America.* New York, Harper and Brothers, 692 pp.
3. Springer, G. D., MacDonald, W. D., and Crockford, M. B. B., 1964, Chapter 10, *Jurassic,* in *Geological history of Western Canada.* Alberta Soc. Petroleum Geologists, Calgary, Alberta, pp. 137–155.
4. Stokes, W. L., 1954, *Pterodactyl tracks from the Morrison Formation near Carrizo Mountains, Arizona.* Geol. Soc. American Bull., v. 65, p. 1309.

Figure 16-1 Chalk cliffs at St. Margarets on the Strait of Dover, England. (C. O. Dunbar.)

CHAPTER **16** *Cretaceous Time and the End of an Era*

The Cretaceous Period. The name Cretaceous [L. *creta*, chalk] was first applied to the white chalk that forms impressive cliffs along the Strait of Dover in England (Fig. 16-1). It was later extended to embrace allied strata of other types, and in 1822 the Belgian geologist Omalius d'Halloy formally proposed including all the rocks lying between the Jurassic System and what we now call the Cenozoic. It thus became one of the largest of the geologic systems, widely distributed and commonly thick. For the world as a whole the Cretaceous submergence was probably the greatest the earth has known.

While remarkable for its vast deposits of chalky limestone in Europe and America, it embraces the full gamut of sedimentary rock; thus Cretaceous is really a misnomer. Furthermore, modern usage would prefer a geographic rather than a descriptive name. Nevertheless, the term Cretaceous has never had a rival and no useful purpose would now be served by a change.

An attempt early in this century to subdivide it into two distinct systems in North America was later abandoned. In most countries it is usually divided into two series, Lower and Upper Cretaceous. Twelve stages (Fig. 16-2) are widely employed in the study of Cretaceous rocks, six in each of the two series.

Figure 16-2 Correlation of important Cretaceous sections.

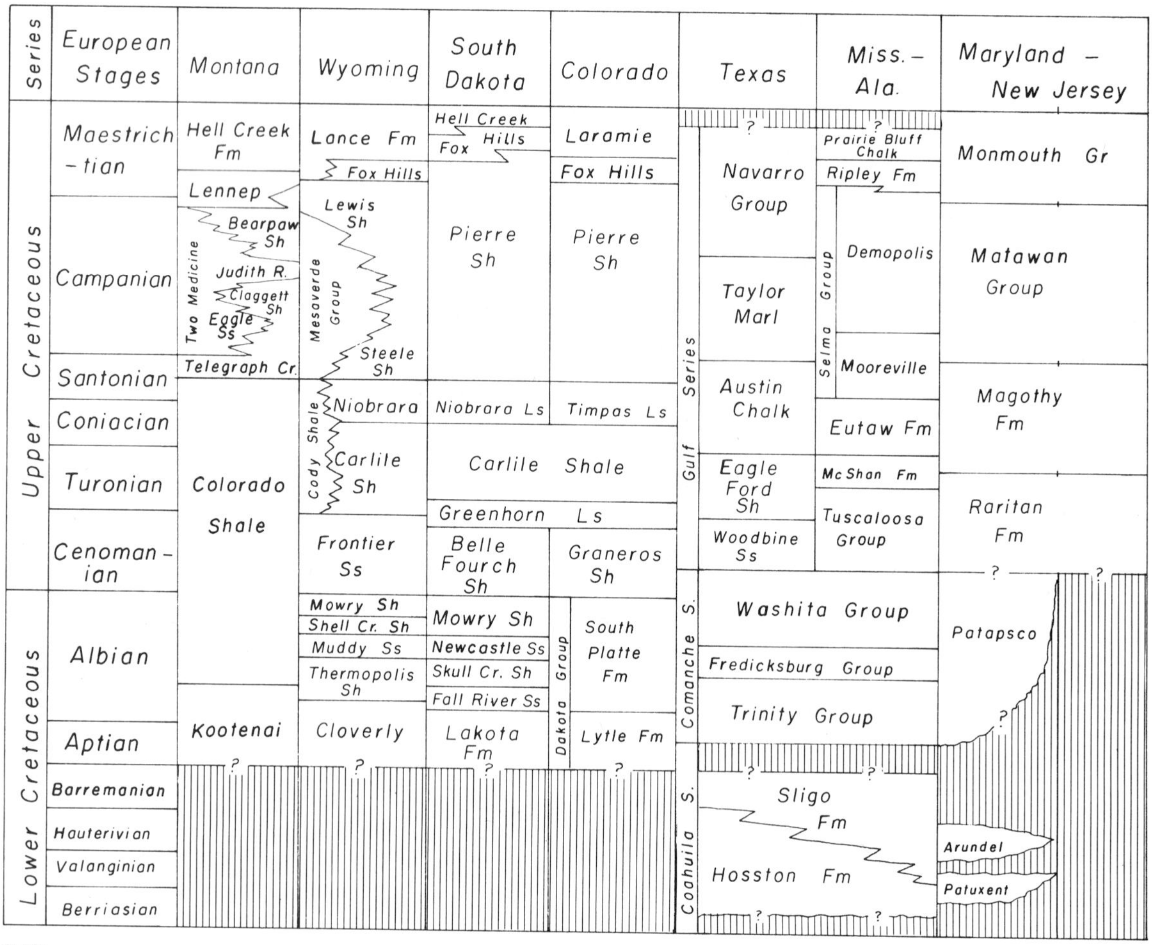

Series
European Stages
Montana
Wyoming
South Dakota
Colorado
Texas
Miss.–Ala.
Maryland – New Jersey
Upper Cretaceous
Lower Cretaceous
Maestrichtian
Campanian
Santonian
Coniacian
Turonian
Cenomanian
Albian
Aptian
Barremanian
Hauterivian
Valanginian
Berriasian
Hell Creek Fm
Lennep
Bearpaw Sh
Judith R.
Claggett Sh
Eagle Ss
Two Medicine
Telegraph Cr.
Colorado Shale
Kootenai
Lance Fm
Fox Hills
Lewis Sh
Mesaverde Group
Steele Sh
Niobrara
Carlile Sh
Cody Shale
Frontier Ss
Mowry Sh
Shell Cr. Sh
Muddy Ss
Thermopolis Sh
Cloverly
Hell Creek
Fox Hills
Pierre Sh
Niobrara Ls
Carlile Shale
Greenhorn Ls
Belle Fourch Sh
Mowry Sh
Newcastle Ss
Skull Cr. Sh
Fall River Ss
Lakota Fm
Dakota Group
Laramie
Fox Hills
Pierre Sh
Timpas Ls
Graneros Sh
South Platte Fm
Lytle Fm
Gulf Series
Comanche S.
Coahuila S.
Navarro Group
Taylor Marl
Austin Chalk
Eagle Ford Sh
Woodbine Ss
Washita Group
Fredicksburg Group
Trinity Group
Sligo Fm
Hosston Fm
Prairie Bluff Chalk
Ripley Fm
Demopolis
Mooreville
Selma Group
Eutaw Fm
McShan Fm
Tuscaloosa Group
Monmouth Gr
Matawan Group
Magothy Fm
Raritan Fm
Patapsco
Arundel
Patuxent
?

PHYSICAL HISTORY OF NORTH AMERICA

The Last Great Submergence. Late in Early Cretaceous time submergence was renewed on a grand scale as the continental margins were gradually flooded and a vast interior sea divided the continent into two landmasses (Map XI). For a time almost 50 percent of the present land surface was submerged. Then the seas began a final retreat and by the end of the period North America had assumed essentially its present size and shape.

Atlantic Coastal Submergence. From New Jersey southward the Atlantic Coastal Plain is underlain by Cretaceous formations that dip seaward with increasing thickness, but thin to feathered edges where they lap out against the Piedmont Belt. The oldest of these are nonmarine deposits of varicolored gravels, sands, and clays representing fluvial deposition on a low coastal plain during Early Cretaceous time. They are followed by Late Cretaceous marine formations that in places grade landward into brackish or freshwater facies. All this detrital material clearly came from the present land and was spread eastward onto the continental shelf. This forms a striking contrast with the Paleozoic history of the region for, so far as they are known, the older sedimentary formations of the eastern United States were transported westward into the Appalachian geosyncline. The orogeny at the close of the Paleozoic Era had thickened the granitic crust under the Appalachian chain, transforming it into a permanently positive axis. The old Appalachian geosyncline had thus ceased to exist and long erosion during the Triassic and Jurassic periods had reduced the marginal land so that with the beginning of Cretaceous time the waves of the Atlantic were for the first time breaking along the eastern seaboard essentially as they do today.

Gulf Coastal Submergence. During Cretaceous time the entire Gulf border was likewise submerged and, for a time, an embayment reached as far north as Cairo, Illinois. Here, as on the Atlantic border, the Cretaceous formations dip gently seaward with increasing thickness, and toward their landward margins grade locally into nonmarine sands and gravels. Deep wells reveal thick Cretaceous formations under much of Florida where the oldest ones are nonmarine.

The Gulf Coast Cretaceous is a more complete sequence than that of the Atlantic Coastal Plain; it includes an appreciable thickness of Lower Cretaceous deposits, both marine and nonmarine in origin. The Gulf embayment, particularly its southern part in eastern Mexico, was flooded with Atlantic waters at an earlier time than was the Atlantic Coastal region.

The Rocky Mountain Seaway. In the latter part of the Early Cretaceous, seas began to invade the Rocky Mountain geosyncline, advancing down the Alberta trough from the boreal region as they had during the Jurassic and at the same time spreading from the Gulf embayment northward through Mexico. These embayments advanced slowly over the broad fluvial plains of the western interior, meeting late in Early Cretaceous time (Late Albian) somewhere in southeast Wyoming or northeast Colorado. Thereafter, with minor fluctuations, this throughgoing seaway spread to form a vast strait nearly 1000 miles wide dividing the continent into two widely separated landmasses, the eastern land broad, low, and stable, the western narrow, mountainous, and the scene of constant orogenic and volcanic activity. During almost all of the approximately 35 million years of Late Cretaceous time this great seaway occupied the Rocky Mountain geosyncline, but gradually it became more restricted as enormous quantities of sediment poured into it from the western highlands of the Cordilleran geanticline. Gradually it withdrew from both the north and the south so that its youngest marine beds are found today in eastern Wyoming and the Dakotas. The record of its final withdrawal has been removed by erosion and we can only guess at its path. The fossils indicate that its last connections were with the boreal sea across the Canadian Shield but there is no sedimentary record to substantiate this.

The swampy coastal regions over which

dinosaurs roamed were broadened as streams spread mud and sand into the withdrawing sea to form great deltaic plains. In the swamps of this lowland, vegetation accumulated to form lignite and coal over much of the interior region.

Pacific Coastal Basins. In spite of the deformation late in Jurassic and early in Cretaceous time, large parts of the Pacific coastal region, in the western parts of California, Oregon, Washington, and British Columbia, were receiving great quantities of sediment in deep marine troughs. Volcanism locally accompanied these deposits, whose source was from highlands to the west that have since completely disappeared leaving only the sedimentary record of their presence.

Growth of the Cordilleran Geanticline. Foreshadowed in Late Triassic time, the Cordilleran axis of the continent had become a well-established feature during the Jurassic and must have been a mountainous region during the Nevadan orogeny. It continued to rise and broaden during the Cretaceous Period as the pulses of orogeny were almost continuously active along its eastern side. The volume of gravel, sand, and mud carried eastward into the Rocky Mountain trough has been estimated [5] to be more than a million cubic miles. To derive such a colossal amount of sediment from the area of the Cordilleran geanticline would require the erosion of an average thickness of 5 miles of rock from this narrow landmass. Of course this does not mean that the mountains of the geanticline ever stood 5 miles above sealevel; the distribution of gravel and sand in the Cretaceous formations preserved in the geosyncline clearly shows that the source area was repeatedly and irregularly elevated while it was being eroded throughout Late Cretaceous time.

During much of the Cretaceous active volcanoes were present on the Cordilleran geanticline. A great volcanic field persisted throughout most of the period in the region of southwestern Montana and from it clouds of volcanic ash periodically were blown eastward into the Cretaceous sea as far as Nebraska and Kansas. These ash falls, chemically altered by seawater, now occur as layers of light-colored, soapy clay, called **bentonite,** in the dark marine Cretaceous shales. There were undoubtedly volcanoes in other parts of the Cordilleran highlands but none have left a clearer or more spectacular record of their presence.

Birth of the Rocky Mountains. Orogeny along the Cordilleran geanticline and, subsequently, to the east of it in the Rocky Mountain geosyncline combined to leave a record of continuous tectonic activity in North America throughout Cretaceous time. Out of this activity rose the majestic Rocky Mountain system stretching from Alaska to Central America, with a breadth of nearly 500 miles in the Rocky Mountain states. This was the most profound episode of mountain building experienced by western North America since Precambrian time, and, with its counterparts in the Andean chain of South America and in the island arcs of the Caribbean region, it ranks as one of the great orogenic episodes of the Phanerozoic.

At one time it was thought that the Rockies were raised in one great paroxysm of mountain making at the close of the Cretaceous period. The name **Laramide orogeny** was applied to this event, after the Laramie Range which forms the Rocky Mountain front in Wyoming. But as the Cretaceous deposits and mountain structure of the region became better known, it was realized that the deformation had been continuous since the Nevadan orogeny, lasting throughout the Cretaceous and well into the Eocene. In addition, the deformation that took place along the geanticline was earlier and distinctly different from the deformation that built the ranges to the east in the geosynclinal area. Today this fact is recognized [1] by restricting the name Laramide orogeny to the building of the eastern ranges in very Late Cretaceous and Early Cenozoic time. The name **Sevier orogeny** is given to the episode of deformation along the geanticline which lasted from very Early Cretaceous time until the close of the period. The name is taken from a paleogeographic feature in southwestern Utah, the Sevier arch, a structurally active area of uplift that contributed great quantities of clastic sediment to the adjacent part of the Cretaceous geosyncline.

The Sevier Orogeny. The Sevier orogenic belt (Fig. 16-3) formed along the east flank of the Cordilleran geanticline and was characterized chiefly by great thrust sheets driven eastward as much as 25 miles, in places carrying rocks as old as Cambrian over the Cretaceous deposits. This great belt of Cretaceous low-angle thrusting extended from southeastern Nevada well into Idaho. Nearly continuous similar orogenic belts extended southeastward from it into Mexico and northward from Western Montana into Canada, but their relationships to the Sevier belt remain to be worked out and they were only in part contemporaneous with the Sevier orogeny.

Along the Idaho-Wyoming border and extending into Utah, a series of roughly parallel thrust faults have been mapped that appear to be successively younger in age eastward. Some of these extend for as much as 250 miles and have been displaced more than 20 miles. In this area it has been estimated [9] that there has been 75 miles of shortening of the outer crust by Cretaceous thrusting. Elsewhere along the Sevier orogenic belt in Utah and Nevada the shortening has been estimated at between 40 and 60 miles [1].

Detailed studies of the Cretaceous sequence in central and eastern Utah by Spieker [10] and his students have clearly revealed the stratigraphic evidence of the Sevier orogeny. Here repeated orogenic movements are recorded in thick piedmont gravels that pass seaward to the east into progressively finer clastics and eventually into marine shale (Fig. 16-4). The Indianola Group, for example, locally consists of more than 10,000 feet of nonmarine conglomerates equivalent to more than half (Cenomanian to Santonian) of the Upper Cretaceous marine sequence to the east. These conglomerates pass eastward within a short distance into marginal marine sands that in turn intertongue with the marine Mancos Shale. Clearly the gravels were coming from the west and their source was not far away since they formed a piedmont slope running to the edge of the sea. Pebbles of Lower Paleozoic rocks are abundant in the gravels and structural relation-

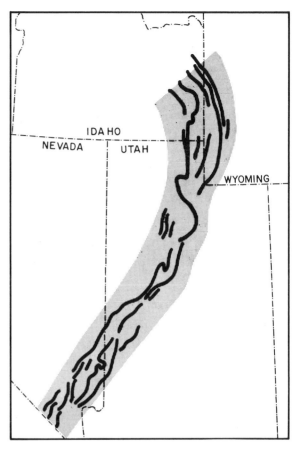

Figure 16-3 The Sevier orogenic belt. Black lines indicate the major thrust faults. (Data chiefly from Armstrong) [1].

ships to the west show that their source was the Paleozoic formations carried into the area over Triassic and Jurassic rocks on the great thrust faults of the Sevier orogenic belt. Individual tongues of Indianola gravels, each hundreds of feet thick, indicate pulses of movement on the thrusts.

Eventually uplift and continued thrusting locally involved the Indianola gravels and they were partially eroded and succeeded by younger gravels (Fig. 16-4). The great quantities of sediment pouring into the geosyncline from the Sevier belt filled in the coastal areas and the spread of coal swamp deposits eastward over marine shales in eastern Utah and western Colorado records the eastward shift of the shoreline. Uplift continued into Pale-

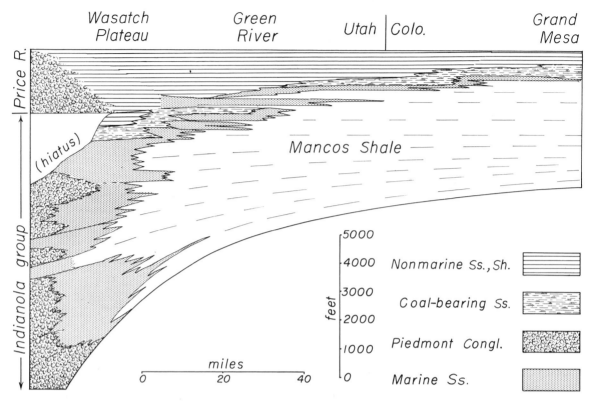

Figure 16-4 Generalized diagram showing the sedimentary relationships of Cretaceous deposits in Utah and northwestern Colorado. (After Spieker [10].)

Figure 16-5 Section across the Front Range near Denver, Colorado, showing the arched structure and suggesting the amount of erosion that has occurred since uplift began. (After W. T. Lee, U. S. Geological Survey.)

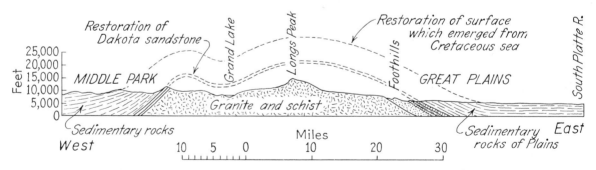

ocene time along the Sevier belt but the Late Cretaceous Price River Formation locally covers the thrust front, indicating that activity along the faults had largely ceased by the close of the period.

The Laramide Orogeny. The eastern ranges of the Rocky Mountains rose from within the geosyncline during Latest Cretaceous, Paleocene, and Eocene time, chiefly as great vertical displacements. The broad anticlinal uplifts and deep intervening synclinal basins produced by the dominantly vertical movements attained a structural relief of 30,000 feet and more at some places. The structures along the flanks of the uplifts are chiefly sharp flexures and high-angle faults.

The Front Range of Colorado (Fig. 16-5) and its northward extension, the Laramie Range in Wyoming, is a great arch from 15 to 30 miles wide and more than 200 miles long in which the Precambrian granite floor now rises to a height of 11,000 feet and the eroded Cretaceous and older formations dip off its flanks in bold hogbacks (Fig. 16-6). It is paralleled on the west by similar great arches forming the Park and the Sawatch ranges of Colorado and is separated from these by an equally large synclinal trough now occupied by North, Middle, and South Parks. The tops of these great arches were completely reduced by ero-

sion during the next era, but the Precambrian floor, still deeply buried under the Parks and under the Plains in front of the mountains, if restored as in Fig. 16-5, shows a differential uplift of the great arches of from 3 to 6 miles. In Wyoming the Bighorn Mountains rise similarly like a great island out of the plains with the Bighorn basin on the west and the Powder River basin on the east, each filled with thousands of feet of sediments. The low, gentle arch of the Black Hills is a smaller but similar structure, the easternmost of all Laramide uplifts.

Laramide uplift and warping in some areas involves strata of Paleocene and Eocene age. Its effects are spread over an appreciable interval of time, although not as long a time as the Sevier orogeny which it overlaps. Early Laramide movements in the geosyncline began during the Campanian Stage (Fig. 16-2)

Figure 16-6 East flank of Colorado Front Range west of Denver. Hogbacks of Cretaceous and older rocks arched up during Laramide orogeny. Great hogback in foreground formed by Dakota Group. Small one to right by Niobrara Limestone; those to left by Pennsylvanian Fountain Formation which here form Red Rocks Park. The mountains, left, are of Precambrian granite and metamorphics. An aerial view looking north. (T. S. Lovering and F. M. Van Tuyl, U. S. Geol. Survey.)

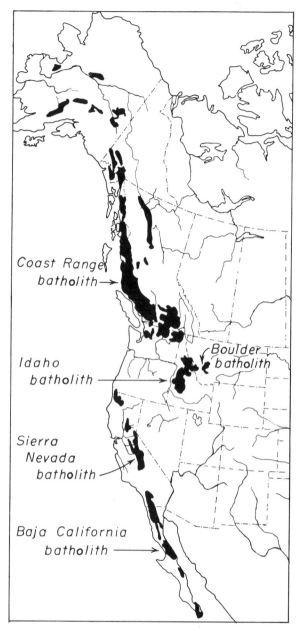

Figure 16-7 Cretaceous batholiths.

when deformation along the Sevier belt was dying out.

Granitic Intrusion. Vast areas of the surface of western North America expose granitic rocks which were intruded into the upper crust as great plutons, or batholiths, and subsequently were unroofed by erosion (Fig. 16-7). In most places the granite is intruded into Jurassic or older rocks and it was once thought that the batholiths formed during the Nevadan orogeny. In recent years isotopic dating of the batholiths has revealed that while a few have dates in keeping with a Late Jurassic-Early Cretaceous age, the great majority have dates that fall within a span of about 30 million years in the middle of the Cretaceous period [5], approximately from Early Albian to Santonian in age. Possibly the isotopic ages are somewhat misleading, for they may indicate the date of final cooling and crystallization of the great plutons rather than their date of intrusion. But even allowing for cooling, the great period of intrusion must have occurred largely in the Cretaceous. At a few places the granites cut fossiliferous rocks and this affords independent evidence on the age of intrusion. In Baja California, for example, granites cut rocks with fossils of Albian age. A few other similar occurrences tend to support the Mid-Cretaceous emplacement of many of the granites.

The great Coast Range batholith, largest of all, is exposed over an area 1100 miles long. The Sierra Nevada batholith is about 400 miles long and 80 miles wide. The eroded summit of the Idaho batholith is now exposed over an area of some 16,000 square miles. Each of these is a complex of many intrusions of somewhat varied lithology suggesting that injections of magma followed one another over a considerable span of time or that the magma differentiated as it gradually cooled.

Some batholiths in the western states formed subsequent to the principal episode of intrusion. The Boulder batholith in western Montana cuts sedimentary rocks of Late Cretaceous age and isotopic dating of minerals in the granite substantiates the stratigraphic evidence, indicating ages ranging from 87 to 62 million years, or Latest Cretaceous to Early Paleocene time. A number of Mid-Cenozoic batholiths are also known but neither these nor the ones of Latest Cretaceous age are significant in volume compared with the great Mid-Cretaceous plutons.

James Gilluly has pointed out [5] that the

emplacement of the Cretaceous batholiths was not directly associated with Cretaceous orogeny, which was essentially continuous throughout the period and mostly east of the areas of intrusion. Moreover, the granites represent a vast volume of sialic material, the building matter of continents, emplaced in the Pacific geosynclinal area, which, since Late Paleozoic, had been the site of enormous outpourings of basaltic rock more characteristic of the simatic material of the Pacific Ocean basin. Gilluly concludes that this great concentration of granitic intrusion in the middle of the Cretaceous marks a major event in continental history, such as the foundering of sialic borderlands beneath the edge of the continent where the material, remobilized, was intruded into the upper crust as great granitic plutons. This implies westward movement of the continent relative to the mantle. The evidence of compressional forces from the west afforded by the flat, eastward-moving thrusts of the Sevier belt during the Cretaceous fits such a general pattern of crustal stresses, but much still remains to be learned about the events leading to the great intrusion of Cretaceous granites.

OROGENY IN CENTRAL AND SOUTH AMERICA

During both the Paleozoic and Mesozoic there existed along the western side of South America a great Andean geosyncline, which accumulated thick deposits of sediments from mobile borderlands to the west. During the last half of Cretaceous time this geosyncline was being gradually folded and rising into a great mountain chain, largely completed by the close of the Mesozoic. Beginning east of Trinidad, off Venezuela, these mountains extended southwestward into Columbia and thence southward to beyond Cape Horn, a distance of nearly 5000 miles. They were the South American counterpart of the Rocky Mountain System.

Standing athwart the course of the Rockies and Andes is the Antillean Mountain System of Central America, also formed at this time, following the trend of the Greater Antilles.

STRATIGRAPHY OF CRETACEOUS ROCKS

Atlantic Coastal Plain. Cretaceous formations underlie the Atlantic Coastal Plain from New Jersey southward and form a wide belt of outcrop where not overlapped by Cenozoic beds. They are mostly sands and clays, disposed in nearly flat-lying beds that dip very gently seaward. For the most part, they are only slightly indurated, the clays being soft and the sands loose and friable.

The entire system is less than 1000 feet thick in the outcrop belt, where it thins out against the older rocks of the Piedmont. Down dip it thickens and probably reaches its full thickness some distance offshore, as shown by a well drilled in 1946 at Cape Hatteras, approximately 100 miles east of the Cretaceous outcrops, which penetrated some 6800 feet of Cretaceous strata. Both Lower and Upper Cretaceous formations are marine at this locality [8].

Lower Cretaceous formations are all nonmarine on the outcrop, and exposures are largely limited to the region around Chesapeake Bay. The Lower Cretaceous shoreline was clearly farther east than the present outcrop belt. The name Potomac Group is used to include the 400 to 600 feet of outcropping Lower Cretaceous strata. The beds consist chiefly of variegated clays and sands with some gravel in the lower beds. Locally a middle part of dark gray lignitic clay is present and where it is, the Potomac sequence is divided into three formations. Fossils include fragmentary crocodiles and dinosaurs and plants, which indicate the Early Cretaceous age of the beds.

Upper Cretaceous formations overlap and conceal whatever Lower Cretaceous may be present north and south of the Chesapeake Bay region. These younger formations are predominantly marine but commonly include a nonmarine formation at the base. The alternating sand and sandy clay units in the outcrop area record fluctuations in water depth and sediment supply; all appear to be shallow shelf deposits. The upper formations commonly contain grains of glauconite, a

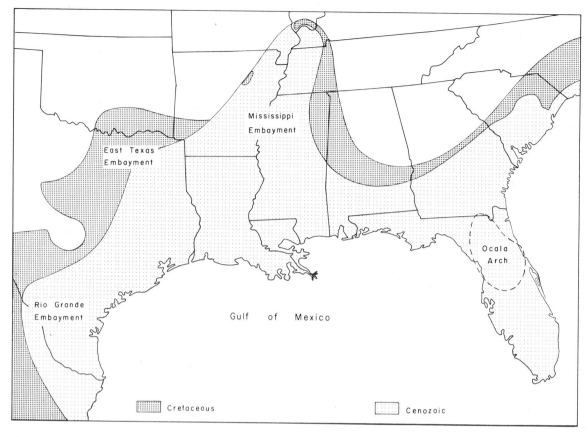

Figure 16-8 Cretaceous outcrops and major features of the Gulf Coastal Plain.

silicate mineral rich in potassium and iron. Locally it is present in quantity, forming deposits of "greensand" in both the Cretaceous and overlying Paleocene marine beds. At one time the New Jersey greensands were extensively quarried for the extraction of potassium used in the manufacture of fertilizer; in recent decades it has been utilized as a water softener. Curiously, greensand, like chalk, is more abundant in Cretaceous rocks than in any other systems, though by no means confined to them. In the Lower Cretaceous of western Europe glauconite is so common that a considerable sequence of strata is commonly called the "Greensand."

The Gulf Region. Thick deposits of both Lower and Upper Cretaceous rocks are present throughout the region of the Gulf geosyncline in the southern United States and

Mexico. The belt of Cretaceous outcrop trends southwestward across Georgia from the Atlantic Coastal Plain into the Gulf area. Here it follows the inner edge of a broad apron of coastal plain sediments (Fig. 16-8) that dip gently Gulfward under Cenozoic deposits. In three areas sediments of the Gulf Coastal Plain extend inland in broad embayments; the greatest of these is the Mississippi embayment, which reaches up the Mississippi Valley bringing the Cretaceous outcrop as far north as southern Illinois. Less extensive are the east Texas and Rio Grande embayments.

Only Upper Cretaceous strata crop out east of the Mississippi, but to the west, chiefly in Texas, a complex series of Lower Cretaceous deposits are exposed along the inner margin of the coastal plain and over much of central Texas. These rocks belong to the

Comanche Series and consist of three great divisions, or groups (Fig. 16-2), each of which represents a sedimentary cycle and grades seaward from shoreline sands through offshore limy clays into limestones containing reefs of rudistid clams (p. 396). All units thicken Gulfward in the subsurface where still older Cretaceous formations come in beneath the Comanche Series; these are included in the Coahuilan Series. Together the Comanche and Coahuilan rocks attain thicknesses of more than 7000 feet where penetrated by drill in the subsurface of the Gulf Coastal Plain of the Gulf border states, and more than 8000 feet under the peninsular tip of Florida [6].

Upper Cretaceous deposits of the Gulf Region follow a similar pattern, with the non-marine or shallow marine sands and clays grading Gulfward into dominantly carbonate rocks. This is well illustrated by the Upper Cretaceous formations of the East Gulf Coastal Plain where the marginal sands are extensively preserved. The sequence, shown diagrammatically in Fig. 16-9, is exposed in an arcuate outcrop the central part of which, along the Alabama-Mississippi state line, was deposited farthest from the shoreline and has the thicker sequence, totaling more than 2000 feet and thickening down dip. In this area the succession includes a lower sandy part (Tus-

Figure 16-9 Section of the Upper Cretaceous formations of the eastern Gulf Coastal Plain, showing facies changes along the strike. As indicated by the inset map, the section runs nearly south from western Tennessee to east-central Mississippi, and then swings eastward into Georgia. The central part was deposited farthest from the Cretaceous shoreline and deposition began there earlier than at either end. Vertical scale greatly exaggerated. (Adapted from L. W. Stephenson.)

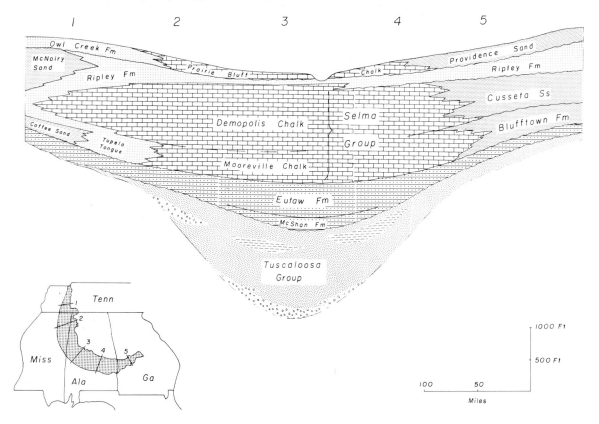

Figure 16-10 Exposures of chalk of the Selma Group, Jones Bluff on the Tombigbee River, Alabama. (L. W. Stephenson, U. S. Geol. Survey.)

caloosa Group) consisting largely of nonmarine strata with marine tongues that thicken Gulfward and thin out landward. Above the basal Tuscaloosa sands come the shallow marine glauconitic clays and sandy clays of the McShan and Eutaw formations marking the spread of the Cretaceous sea into the area. These are succeeded by a thick series of soft clayey limestone, or chalk, formations of the Selma Group (Fig. 16-10), which continue to the top of the Cretaceous sequence. Eastward across Alabama and into Georgia the chalks become sandy and grade laterally into sands (Blufftown-Cusseta-Providence) and clayey sand (Ripley Formation). A similar pattern occurs northward along the outcrop up the Mississippi embayment, the chalks grading laterally into clayey sands of the Ripley and Owl Creek formations and these in turn into the Coffee and McNairy sands.

Fossils are common throughout much of this sequence, tending to be more varied and abundant in the clayey sands and glauconitic sand bodies than in the chalk formations. Most of the sands are shallow marine deposits and the nonmarine strata into which they must all have graded shoreward are rarely preserved. Much of the record of the Cretaceous coastal plain deposits has been removed by erosion during Cenozoic time.

All of Florida now has Cenozoic strata at the surface, but deep wells indicate the presence of Upper Cretaceous formations resting on those of the Lower Cretaceous. Wells in western Florida have penetrated over 3000 feet of Upper Cretaceous strata in which are all of the principal divisions recognized elsewhere in the Gulf region. In northern Florida the Ocala uplift, a persistent structural high area of older rocks forming the backbone of the peninsula (Fig. 16-8), has only a thin veneer of Cretaceous strata across its top indicating that it was emergent during much of the period.

Outcrop of Upper Cretaceous strata in the western Gulf region is more extensive and the stratigraphy more complex than in the eastern Gulf region. Much of the complexity results from the presence of local uplifts and basins that had strong influence on sedimentation. The Gulf Series, which embraces the Upper Cretaceous strata of the Gulf region, is traditionally subdivided into five units in the West Gulf Coastal Plain (Fig. 16-2). Originally these units were employed as formations, then, as further study revealed their complexity, they became groups. The wealth of stratigraphic information from oil well drilling has made it possible to trace these units Gulfward in the subsurface; they change facies in this direction, as they also do, less abruptly, southward along the outcrop into Mexico. Experience has shown that the boundaries of the original five divisions can be projected across these changes in facies by using fossils as markers and this is how the division boundaries are commonly recognized today. In this way the original formations of early workers have come to be used as provincial stages. In general, the basal Woodbine division includes dominantly sandy formations similar to the equivalent Tuscaloosa Group of the east Gulf succession. Beyond this the stratigraphy of the west Gulf Upper Cretaceous is too complex to make useful generalizations. A complex array of fossiliferous clays, glauconitic sands, marls, and chalks interfinger and intergrade with one another, attaining thicknesses of 4000 feet in the East Texas embayment and more in the Rio Grande embayment. Local highs were common sites of reef formation and chalks and marls were widespread, particularly during Austin, Taylor, and Late Navarro deposition. Unconformities are common between different parts of the Upper Cretaceous sequence along the outcrop but traced Gulfward into the thicker deposits of the subsurface most of these, including that between Upper and Lower Cretaceous, tend to disappear, indicating essentially continuous deposition over most of the Gulf geosyncline with some fluctuation of sealevel and local crustal warping about its margins.

The Western Interior Region. Cretaceous formations underlie all of the Great Plains and much of the Rocky Mountain region, preserving a remarkably complete record of the Rocky Mountain geosyncline and the vast epicontinental sea that occupied it from late in Early Cretaceous time until just before the close of the period. In this great trough, which joined a major Gulf embayment in central Mexico with the boreal sea in the MacKenzie delta region of Canada, the bulk of sediments were supplied from the Cordilleran geanticline along its western flank. Relatively little sediment was derived from the lowlands to the east. As a result the deposits in the eastern plains commonly do not aggregate much more than 2000 feet in thickness, whereas in certain areas along the geanticline, such as western Wyoming, they reach a thickness of as much as 20,000 feet. As might be expected, the stratigraphic pattern of the deposits also reflect the dominance of a western sediment source. Along the western highlands of the geanticline a narrow marginal belt of thick and coarse piedmont and coastal plain deposits accumulated. As we have seen in the example of the Indianola Group, these deposits formed at the edge of the Sevier orogenic front. From these dominantly nonmarine marginal deposits, great clastic wedges of sediment thin eastward and intertongue with marine shale. Still farther east in the plains states the sand wedges have disappeared and the sequence is dominantly shale.

The chief exceptions to this pattern of stratigraphic relationships (illustrated by Figs. 16-11 and 16-12) are the dominantly sandy deposits at the base and top of the Cretaceous sequence. These represent, respectively, the nonmarine and near shore deposits marking the initial invasion of the Cretaceous sea and the shoreline and coastal plain deposits that signify its final withdrawal from the geosyncline.

The Gulf embayment into Mexico crossed northern Sonora and entered southeastern Arizona early in the period, and here the Lower Cretaceous is thick and largely detrital. In southeastern Arizona it constitutes the Bisbee Group and is as much as 7000 feet thick. At about this same time the boreal sea had spread

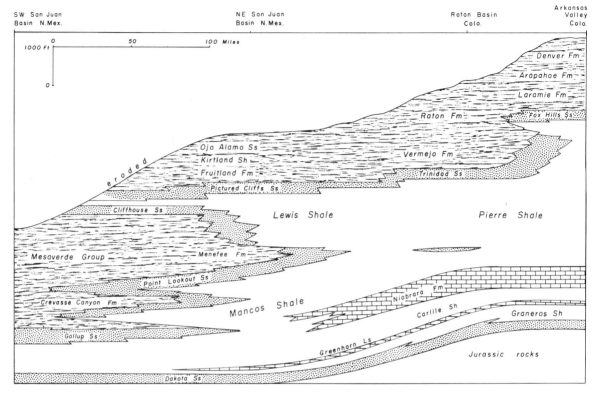

Figure 16-11 Generalized stratigraphic section of Cretaceous formations from west-central New Mexico to central Colorado.

down the geosyncline as far as central Alberta. Over the large area of the Rocky Mountain geosyncline between these two embayments continental deposits were accumulating, particularly along the front of the Cordilleran highlands.

During the latter part of Early Cretaceous time (Albian Stage) the Gulf embayment spread northward and the Boreal embayment southward, the seas reworking the surficial alluvial deposits as they advanced. The two arms of the sea met before the close of the Early Cretaceous and subsequently spread laterally to the east and west reaching a maximum spread early in the Late Cretaceous (Turonian). In the United States the sandy shoreline deposits of the spreading sea and the nonmarine coastal plain deposits laid down in advance of it are included in the Dakota Group. The Dakota generally consists of 300 feet or more of alternating sandstone and shale

formations. At many places a flat surface of disconformity within the group marks the marine transgression, separating irregularly bedded, nonmarine sands and variegated clays below from thin-bedded marine sands and black shales above. A very sparse marine fauna is present at some places in the marine shale and plant fossils are common in the sandstone.

The two idealized sections across the Cretaceous deposits of the geosyncline (Figs. 16-11 and 16-12) illustrate the general stratigraphic relations above the initial deposits of the Dakota Group. In the Great Plains belt a continuous marine succession consisting chiefly of shale with two widespread limestone units extends to the continental beds that mark the withdrawal of the Cretaceous sea from the geosyncline. The Niobrara Limestone is an important key formation in this sequence and early in the study of these deposits its upper contact with the overlying

Pierre Shale was used to divide this marine Upper Cretaceous succession into the Colorado Group below and the Montana Group above. Recognition of a marked change in the marine faunas at this horizon prompted the subdivision and the two units can be separated by their faunas throughout the Western Interior, whether or not the Niobrara Limestone is present. Consequently the Coloradoan and Montanan have come to be used as broad provincial stages rather than groups of formations, in a manner similar to the major units of the Texas Cretaceous.

Westward from the Great Plains the limestone units grade into shale and the great sand wedges intertongue with the fine-grained marine sediments. Although there were many minor fluctuations of the western shoreline there were only two or three major episodes during which great bodies of coastal plain deposits built seaward and subsequently were inundated again by readvance of the sea.

In Montana (Fig. 16-12), for example, the Eagle and Judith River formations represent times when wide coastal plains and great deltas were built seaward and they most likely signify increased tectonic activity and uplift in the adjacent highlands of the Cordilleran geanticline to the west. Various combinations of subsidence in the geosyncline, rate of sediment supply, and shifts in sealevel can be called on to account for the formation and preservation of the great sand wedges. At times great increase in the supply of sediments apparently exceeded the ability of subsidence to accommodate their deposition below sealevel and coastal deposits built seaward, essentially forcing the shoreline eastward. Along the lower or seaward-growing edge of the wedges the presence of marine sands representing barrier bars and beaches suggests that rivers were dumping great quantities of sand into the sea. The absence of a comparable thickness of coastal deposits on

Figure 16-12 Generalized stratigraphic section of Cretaceous formations from northwestern Montana to the Missouri River in South Dakota.

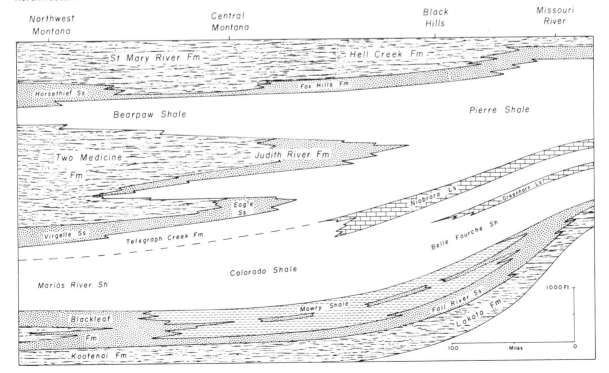

the tops of these same wedges (Fig. 16-12) indicates abrupt fall-off in sediment supply together with relatively rapid readvance of the sea.

Throughout the Upper Cretaceous the wedges of coastal plain deposits built progressively farther eastward and the readvance of marine deposits over them extend progressively less farther westward. The final spread of continental deposits eastward over the geosyncline occurred at somewhat different times in different parts of the interior region. Apparently the sea withdrew southward from the Canadian part of the trough and northward from the southwestern part of the Rocky Mountain region. In a late restricted phase it occupied a wide embayment extending east-west along the Colorado-Wyoming line and broadened eastward into Nebraska and the Dakotas, probably extending over the Canadian Shield to the Boreal sea, although there is no direct evidence remaining of such a connection. The coastal plain deposits that accompanied this withdrawal and the subsequent final retreat of the sea across the region of the present Great Plains comprise a formation known locally by a number of names but most commonly as the Lance, or Hell Creek. These beds are locally coal-bearing and include fossil plants and the last of the dinosaurs, among them ceratopsians, and the great *Tyrannosaurus* (Fig. 16-19). In parts of Wyoming the Lance Formation attains a thickness of several thousand feet but it thins eastward in the northern Great Plains to a few hundred feet.

Cretaceous of the Pacific Region. Cretaceous marine deposits in the Pacific Region of the United States are restricted to the western parts of California, Oregon, and Washington where clastic sediments accumulated, commonly with volcanics, in deep troughs. In this region orogenic pulses were scattered both geographically and in time throughout the period and locally affected the depositional troughs. However, the principal deformation of these deposits resulted from Late Cenozoic orogeny.

The aggregate thickness of sediment in some of the local basins may be as much as 50,000 feet; the Cretaceous deposits in the Great Valley of California alone have been estimated to amount to more than 13,000 cubic miles. The scattered fossil record from the different Cretaceous basins indicate that all the stages of both the Lower and Upper Cretaceous are represented. The local, discontinuous nature of lithologic units is in marked contrast to the widespread, continuous formations of the Cretaceous of the Western Interior and Gulf Coastal Plain. In spite of intensive study, no simple pattern of stratigraphic history has emerged from the west coast Cretaceous and the indications are that there was little regional uniformity of events, each basin of deposition having essentially a separate history. The great plutonic intrusions of the western part of the continent add to the complexity of its Cretaceous record.

CLIMATE

Early Cretaceous climates apparently were generally mild, although perhaps not as warm as in the preceding period, for reef corals are more restricted in the Early Cretaceous than they had been in the Jurassic. Footprints of a plant-eating iguanodont dinosaur were recently described from Lower Cretaceous rocks of Spitsbergen [4]; this implies adequate soft vegetation to support animals of this type, but such vegetation could be had in temperate as well as tropical regions.

With the great spread of the seas early in Late Cretaceous time, the climate was mild and equable over most of the land surface of the earth. Upper Cretaceous rocks preserve, even in high latitudes, abundant remains of land plants belonging to genera now restricted to warm-temperate or subtropical regions. In central-western Greenland, for example, the Cretaceous beds contain figs, breadfruits, cinnamons, laurels, and tree ferns, and in Alaska they have yielded cycads, palms, and figs. Although most of these are commonly thought of as strictly tropical trees, each of the genera has representatives in the temperate zone, and it is not necessary to conclude, as some have done, that tropical climate extended into polar latitudes. Nevertheless, there is sufficient evidence for maintaining that Greenland was without an ice cap and that the climate there was then temperate rather than frigid.

The abundance of the great dinosaurs in Alberta and Mongolia during part of the Cretaceous seems to imply a very mild temperate climate at least that far north. But there is growing evidence that dinosaurs may not have been cold-blooded like modern reptiles and their presence may not indicate as warm a climate as has usually been assumed.

LIFE OF CRETACEOUS TIME

Spread of Modern Plants

Deciduous trees suddenly became conspicuous late in the Early Cretaceous, and long before the close of the period they dominated the landscape in all the continents, just as they do today. Among the oldest of these were the magnolia, fig, sassafras, and poplar, all of which appear in the Lower Cretaceous deposits. By earliest Late Cretaceous (Cenomanian) the forests were essentially modern, including such trees as beeches, birches, maples, oaks, walnuts, planes, tulip trees, sweet gums, breadfruit, and ebony, along with shrubs like the laurel, ivy, hazelnut, and holly (Fig. 16-13).

With these, of course, there were evergreens, just as now, but they no longer dominated the landscapes as they had done during the earlier Mesozoic. Among the conifers a conspicuous type was the sequoia; although none as large as the modern giants of California have been found, it is interesting to note that the race was widely distributed over the northern hemisphere at this time.

As in the Jurassic Period the cycadeoids were the most distinctive plants of the upland slopes and their short stocky trunks are preserved in abundance and extraordinary perfection in the lower part of the Dakota Group (Lakota Formation) in the Black Hills region, and in the Upper Cretaceous (Mesaverde Group) in the San Juan Basin of Colorado.

In both regions the woody tissue has been replaced by quartz, yet all the details of woody structure are faithfully preserved (Fig. 16-14).

Most deciduous trees belong to the **angiosperms,** the highest order of the plant kingdom and the one that includes all the true flowering plants. Besides the hardwood trees mentioned above, this order includes the grasses and cereals as well as the seed- and fruit-bearing shrubs, annuals, and our common vegetables. The strong deployment of this group of plants in the Cretaceous was one of the most significant advances in the whole evolution of plant life. On their own account, this was an important milestone in the history of life, for the angiosperms are the most highly specialized of all plants and the ones destined to dominate the Earth during later geologic time. In addition, their indirect effect upon the evolution of the higher animals can hardly be exaggerated, for they supply nearly all the plantfood for the mammals that now dominate all other life upon the earth. Angiosperms provide the nuts and fruits of the forest, the grasses of the prairies, the cereals that furnish fodder and grain for man and his domestic animals, and all the vegetables and fruits that man has cultivated, to say nothing of the flowers that add so much pleasure and inspiration to human surroundings. It would hardly be too much to say that the great expansion of the mammals and the birds had to wait for the evolution of the flowering plants.

Fossil foliage and wood of angiosperms appear in some abundance in the basal sands of the Comanche Series of the Gulf region, the Potomac Group of the Atlantic Coastal Plain, and the Dakota Group and correlative Canadian strata in the Western Interior region. Interestingly, neither the lowermost beds in the Potomac Group (Patuxent Formation) nor those in the Dakota Group of the northern plains states (basal Lakota Formation) have yielded angiosperms, although both have abundant remains of ferns, cycads, and conifers. A few angiosperms appear in the middle parts of these units and in their upper

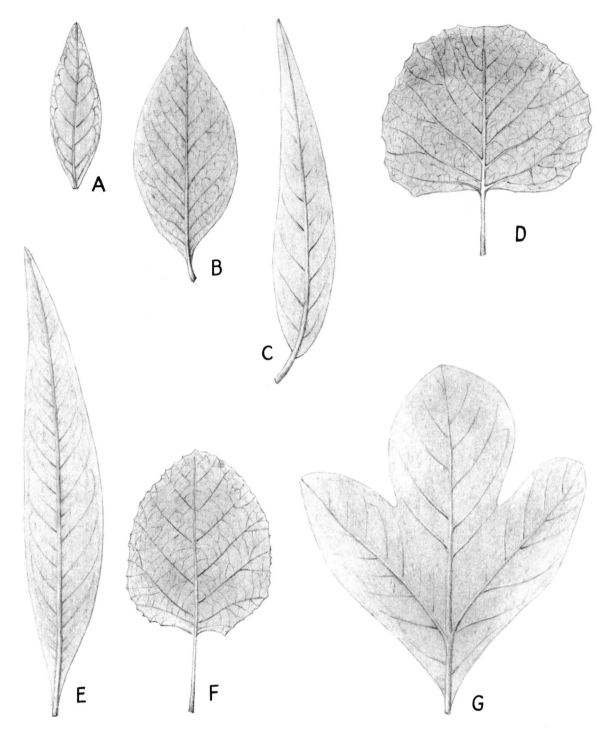

Figure 16-13 Cretaceous plants. Leaves of deciduous trees. A, *Andromeda*, a relative of the *Rhododendron;* B, *Magnolia;* C, *Salix,* a willow; D, *Populites,* a poplar; E, *Ficus,* a fig; F, *Betula,* a birch; G, *Sassafras.* (After E. W. Berry and L. F. Ward.)

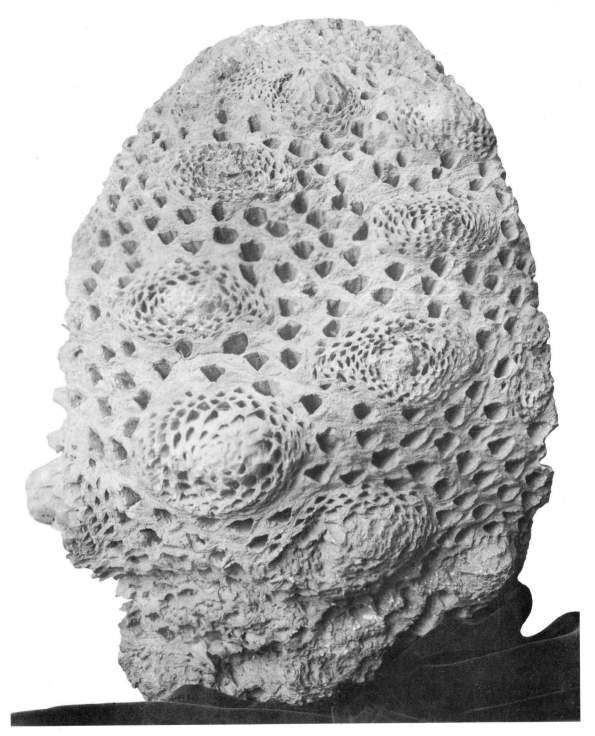

Figure 16-14 Silicified trunk of a cycadeoid, *Cycadeoidea,* from the Lakota formation in the rim of the Black Hills, South Dakota. The nodes are areas where flower buds were about to appear and the deep pits were formed by the shedding of old leaves. See Fig. 15-13 for a restoration of the living plant, and Fig. 14-15 for a modern cycad. (G. R. Wieland, Yale Peabody Museum.)

Figure 16-15 Late Cretaceous landscape in Wyoming, with dinosaurs amid flowering plants. Animals: 1, *Tyrannosaurus rex;* 2, *Trachodon;* 3, *Triceratops;* 4, *Struthiomimus;* 5, *Ankylosaurus;* 6, *Pteranodon.* Plants: 7, sedges; 8, a conifer; 9, *Ginkgo;* 10, a palm; 11, willow; 12, live oak; 13, dogwood; 14, sassafras; 15, laurel; 16, *Magnolia.* (Part of a mural by Rudolph F. Zallinger in the Yale Peabody Museum.)

parts constitute well over 50 percent of the fossil flora. In the succeeding basal Upper Cretaceous (Cenomanian) formations 90 percent of the fossil plants are angiosperms. Similar records of angiosperm appearance are found on other continents. Apparently they took over the coastal plain habitats in a remarkably short time, essentially within the Albian Stage, and remained dominant thereafter.

Neither the source of origin of the angiosperms nor the reason for their relatively abrupt rise to dominance is known. The resting stage represented by the ripening of the seeds and the shedding of the foliage of these plants clearly seems to be an adaptation to seasonal climatic change, presumably either

warm to cool or wet to dry. It has been suggested that the evolution out of older plants took place on the highlands where seasonal change was more pronounced than in the milder coastal areas and that their sudden appearance in the sedimentary record marks the migration into the lowlands of an already well-advanced angiosperm flora. In the absence of upland deposits this can be neither proven nor disproven. It is quite possible, however, that angiosperms were actually present on the uplands before Cretaceous time. The palmlike foliage from the Triassic was mentioned earlier and similar foliage of angiosperm aspect has been reported from Jurassic rocks. However, it is the fruiting organs, not the foliage, which mark the attainment of the angiosperm level of evolution, and the earliest known angiosperm fruit is from Lower Cretaceous rocks [3].

Culmination of the Reptilian Dynasty

Dinosaurs continued to hold the center of the terrestrial stage until the curtain dropped on the last scene of the Mesozoic drama (Fig. 16-15). The great **sauropods** persisted on most continents throughout the period although in North America their remains are not as common in the Cretaceous as they are in Jurassic rocks. **Stegosaurs** had a few representatives in the Early Cretaceous but did not survive it, being the first major group of dinosaurs to be-

come extinct. A new ornithischian group of short-legged, heavily armored, quadrupedal dinosaurs, the **ankylosaurs,** apparently took over the stegosaur way of life, feeding on soft plants and depending on its armor-plating and remarkable clublike tail for protection (Fig. 16-15).

The dominant herbivores of Cretaceous time were the bipedal **ornithopods** and quadrupedal **ceratopsians.** In the European Lower Cretaceous, *Iguanodon* is a common ornithopod genus and was the first dinosaur known to science. The most prolific and varied group of ornithopods, however, were the Late Cretaceous **trachodonts,** or "duck-billed" dinosaurs (Figs. 16-15 and 16-16), some of which were

Figure 16-16 Restoration of the duckbill dinosaur *Trachodon mirabilis,* from the Late Cretaceous (Lance) deposits of Niobrara County, Wyoming. (From a painting by Charles R. Knight, American Museum of Natural History.)

as much as 45 feet in length. This highly successful group developed a variety of head crests but basically were quite similar in form. Their mouths were equipped with a great battery of teeth forming a broad, rough grinding surface for crushing and shredding the vegetation on which they fed. Many had laterally compressed tails like that of the crocodile, and they may have been powerful swimmers. Mummified skeletons of trachodonts which preserved the skin (Fig. 3-15) show that some had webs between their toes. Evidently they were adapted to both terrestrial and aquatic life.

The **ceratopsians** were horned dinosaurs. These were ponderous quadrupedal herbivores of rhinoceroslike form. Their massive skulls typically bore three stout, bony horns, one on the nose and one over each eye (Fig.

16-15), but the earliest forms were hornless and in some of the later ones the nasal horn was much the largest.

The earliest known ceratopsians are from the Lower Cretaceous of Mongolia (Fig. 16-17) where many skeletons were found along with nests of unhatched eggs that had evidently been covered and preserved by drifting sand during Cretaceous time (Fig. 16-18). With the exception of this small primitive genus, all known ceratopsians are from North America. Some of these from the latest Cretaceous in Wyoming were as much as 20 feet long and more than twice as bulky as a large modern rhinoceros. Pieces of some unidentified dinosaur egg shell were found in the Lance Formation near Red Lodge in Montana.

Ceratopsians and duck-billed dinosaurs lived until the very end of the Cretaceous

Figure 16-17 The primitive ceratopsian dinosaur, *Protoceratops,* with nest of eggs. This species attained a length of only 8 to 10 feet. (Painting by Charles R. Knight, based on skeletons and nests from the Gobi Desert in Mongolia. Chicago Museum of Natural History.)

Figure 16-18 Dinosaur nest with broken eggs weathering from the rock. The eggs are about 4½ inches long, and are believed to be those of *Protoceratops.* Lower Cretaceous, Gobi Desert, Mongolia. (American Museum of Natural History.)

Period and are particularly common in the Lance and equivalent formations. Curiously, no dinosaurs were known from the Pacific Coast region until bones of a single trachodont were found near Patterson, California.

The carnivores (Theropoda) now reached their greatest size in *Tyrannosaurus rex,* the largest terrestrial carnivore of all time. This beast attained a length of some 45 feet and could carry its massive head 20 feet from the ground (Figs. 16-15 and 16-19). Many of the carnivores were much smaller, however, and a few such as *Struthiomimus* (Fig. 16-15, no. 4) had lost their teeth and had jaws rather like those of an ostrich.

Reptiles of the Sea. Ichthyosaurs passed their heyday before the close of the Jurassic and were unimportant in the Cretaceous seas.

On the other hand, the **plesiosaurs,** though less numerous than before, attained their greatest size. One species of these (*Elasmosaurus*), found well preserved in the Niobrara chalk of Kansas, reached a length of 40 to 50 feet, about half of which was a very slender, agile neck. Short-necked plesiosaurs also reached great size; in *Kronosaurus* from the Cretaceous of Australia the skull alone is 12 feet long.

The dominant group of marine reptiles was a newly evolved tribe of marine lizards, the **mosasaurs** (Fig. 16-20), which made its appearance in Early Cretaceous time. They had short, paddlelike limbs and a powerful, laterally compressed tail. The mosasaurs were obviously rapacious carnivores and were probably the most ruthless pirates of the Meso-

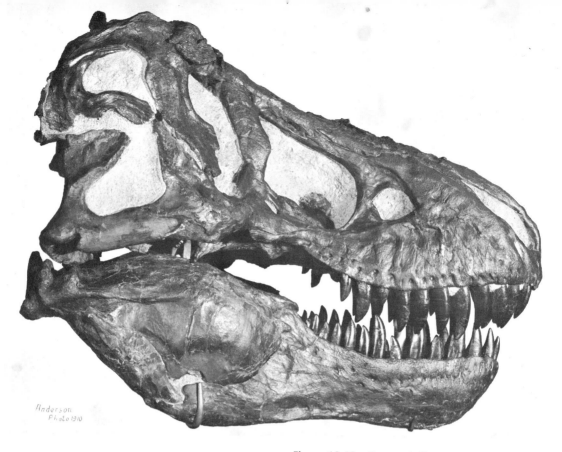

Figure 16-19 Above, skull of *Tyrannosaurus rex*, greatest of all known carnivorous dinosaurs. Length of skull, 4 feet 3 inches. (American Museum of Natural History.)

Figure 16-20 Left, a mosasaur, *Clidastes*, from the Niobrara Limestone in Kansas. The species was 12 to 15 feet long. (After S. W. Williston.)

zoic seas. The largest reached a length of about 35 feet.

Marine **turtles** were present, and one specimen of phenomenal size (*Archelon*), found in the Pierre Shale of South Dakota, measures 11 feet in length and 12 feet across the flippers (Fig. 16-21). In the rivers both broad-nosed and narrow-snouted **crocodiles** were common.

Last of the Winged Dragons. Pterosaurs, though less varied and numerous than before, were large and remarkably specialized. The largest known form, *Pteranodon*, had the amazing wingspread of 23 to 25 feet, greatly

Figure 16-21 *Archelon,* the largest turtle of all time, with its collector, the paleobotanist G. R. Wieland. From the Pierre Shale on the South Fork of Cheyenne River about 35 miles southeast of the Black Hills in South Dakota. (Yale Peabody Museum.)

Figure 16-22 *Pteranodon,* the greatest of the flying reptiles. Wingspread between 23 and 25 feet. (U. S. National Museum.)

exceeding any other winged creature of all time. Even so, the body was not larger than that of a wild goose, and with its delicate, hollow bones the creature must have been almost as light and fragile as a kite. The remains of *Pteranodon* are found in the Niobrara Limestone in Kansas (Fig. 16-22), and it was clearly adapted to soaring over the waves like the modern albatross. In fact, this great reptile would have been quite helpless on the ground, for the hind limbs were so small and degenerate that they probably could not have borne even its light weight; everything had been sacrificed to achieve sustained flight.

Unlike the Jurassic pterosaurs, the known Cretaceous species were toothless, their horny beaks displaying a remarkable parallelism with those of the post-Mesozoic birds.

Birds with Teeth

Fossil birds are still extremely rare in Cretaceous rocks. Best known is the large flightless bird *Hesperornis regalis* (Fig. 16-23), of which numerous specimens have been found in the Niobrara Limestone of Kansas. It was adapted, like the modern penguin, to life in the sea where it could swim and dive. Only vestiges of wings were present and its legs were articulated so as to move laterally as efficient swimming organs but could hardly have been used for walking. Furthermore, all known specimens have been found in marine deposits. The most remarkable feature of *Hesperornis* was its long bill armed with slender conical teeth, not unlike those of the Jurassic bird, *Archeopteryx.*

Several of the modern orders of birds are

now known to have existed in Cretaceous time—ancestral ducks, grebes, and pelicans —all of which were probably toothless, but their remains are still rare and fragmentary.

Mammals in Transition

The Cretaceous embraces the end of the first radiation of primitive mammals and the beginning of the second, in which modern orders are introduced. The four orders of mammals known from Jurassic rocks are each represented by at least one genus in Lower Cretaceous rocks, but the triconodonts, symmetrodonts, and pantotheres are extremely rare and have not been found in Upper Cretaceous rocks. Only the multituberculates carried on through Cretaceous time, diversifying somewhat in the later part of the period.

Until recent years collections of Cretaceous mammals were small and knowledge of them limited to a few prolific formations, like the Lance Formation of Wyoming. Nevertheless, teeth and jaws of two new orders of mammals had been discovered, the pouch-bearing **marsupials** and the **insectivores,** the group to which modern moles and shrews belong (Fig. 16-24). Today a mass-production technique of quarrying and washing great quantities of sediment through screens is revealing a far richer fauna of the tiny Cretaceous mammals than was suspected. Much of this new material remains to be studied; it apparently includes at least one major group of mammals not pre-

Figure 16-23 *Hesperornis,* the great diving bird of the Cretaceous seas. Overall length about 4½ feet. (Yale Peabody Museum.)

Figure 16-24 A Cretaceous insectivore, *Zalambdalestes,* from Mongolia. Flesh restoration (left) and composite reconstruction of the skull (right). Slightly enlarged. (After G. G. Simpson.)

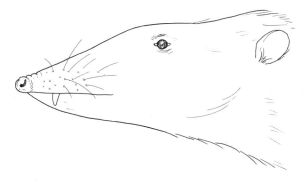

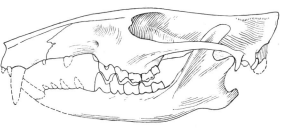

viously known from the Cretaceous; a species of early **primate** has been described from the Hell Creek Formation in Montana [12].

Modernization of the Invertebrate Fauna

By Cretaceous time the marine invertebrate faunas had a decidedly modern aspect that was, however, obscured by the dominance of a number of characteristic forms which did not survive the period. The proliferation of modern types of **corals** continued in the warm seas and by the close of the period most of the major families of modern seas had appeared. Modern forms of the small, colonial "moss animals," or **bryozoans**, appear first in Early Cretaceous time and begin a great diversification that eventually reached its peak in Eocene time.

Bivalves (Plate 15, Figs. 3 to 9; Plate 17, Figs. 4 to 8) and **gastropods** (Plate 17, Figs. 9 to 11) of many kinds and of essentially modern appearance were abundant. By Cretaceous time bivalves appear to have occupied all the principal environmental niches that they do today and, in addition, a group known as **rudistids** (Plate 15, Figs. 3, 4, and 6) became remarkably specialized for reef dwelling and their shells lend a distinctive element to Cretaceous organic reef environments. Several types of rudistids were attached by the beak of one valve, which grew to be a deep conical shell while the opposite valve served as a lid, or operculum. One form, found in the Cretaceous reefs of Jamaica, grew 5 feet tall. The **oyster** family was also very common in Cretaceous seas and more diverse than today; two stocks, the **exogyras** and **gryphaeas**, were particularly abundant and are among the most distinctive invertebrates of this time.

Among the mollusks the **cephalopods** were responsible for giving the Cretaceous faunas a medieval aspect, for both **ammonoids** and **belemnoids** were abundant throughout the period but became extinct at its close. Variation in form characterizes Cretaceous ammonoid faunas; although many kinds of normally coiled shells are found, many forsook symmetrical coiling to uncoil in adult stages. Some became spiral, like a snail's shell, a few straighten, many became loosely coiled, and a few lost all semblance of regularity or symmetry (Plate 16). This interesting phenomenon, once attributed to the decadence of "racial old age," most likely reflects attempts to adapt to new modes of existence, possibly under conditions of stress that led to extinction by the close of the period. Modern cephalopod types, the **squids** and **octopuses**, were present in Cretaceous seas and probably plentiful, but the fossil record of these largely soft-bodied creatures is sparse. The only octopus specimen in the entire fossil record is from the Late Cretaceous of Lebanon.

Sea urchins, so common in warm seas today, were in transition and several Mesozoic stocks died out in the Cretaceous; but modern types, in particular the **heart urchins**, were abundant then as now. **Brachiopods** were no more common than they are today, and are abundant only locally in Cretaceous rocks. Modern types of crustaceans were common and traces or fragments of crabs are common in strata of the shore zone of Cretaceous seas.

Insects are not common fossils simply because suitable deposits have not been found, but it is fairly certain that all the chief modern types were already represented. Probably early in the Cretaceous the insects adapted themselves to feeding upon the nectar of the newly arisen flowering plants and thus gradually assumed their important role in pollenization. Among remarkable discoveries is a fossil wasp nest from the Upper Cretaceous of Utah, and fossil ants in amber from the New Jersey Cretaceous (Fig. 3-17) [13].

WIDESPREAD EXTINCTIONS

The end of the Cretaceous, like the close of the Paleozoic, proved to be a great crisis in the history of life. Several stocks of animals declined markedly during the period; others flourished till near its end only to become extinct. For example, the dinosaurs were highly varied and apparently adaptive right up to the end of Cretaceous time, yet not one is known to have lived to see the dawn of the Cenozoic Era. The **pterosaurs** specialized

Plate 15 Lower Cretaceous Echinoids (1, 2), Clams (3–9), and Ammonite (10).
Figure 1, *Holectypus planatus;* 2, *Heteraster texanus;* 3, *Monopleura pinguiscula;* 4, *M. marcida* (a cluster of three); 5, *Exogyra arietina;* 6, *Toucasia patagiata;* 7, *Exogyra texana;* 8, *Alectryonia* cf. *A. carinata;* 9, *Gryphaea tucumcari* (inner view of larger valve); 10, *Oxytropidoceras acutocarinatum.* All natural size. (Drawn by L. S. Douglass.)

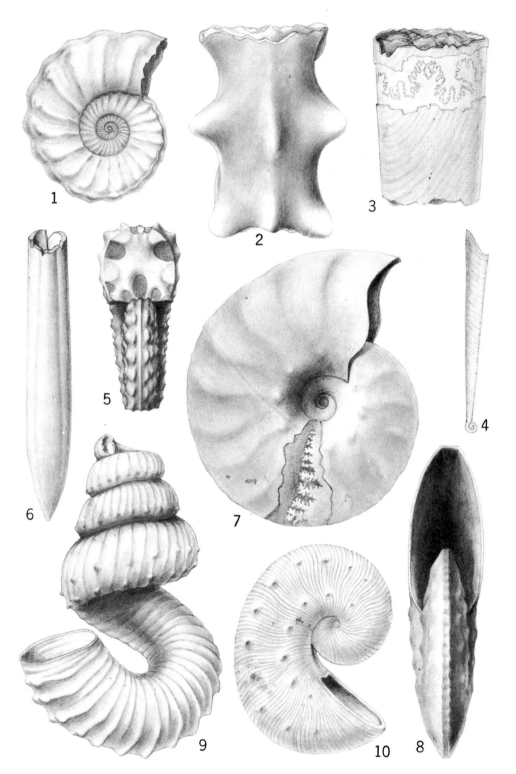

Plate 16 Upper Cretaceous Cephalopods.

Figures 1, 2, *Collignoniceras woolgari* (young shell and ventral view of fragment of large shell); 3, 4, *Baculites compressus,* a straightened ammonite (a juvenile shell, ×3, showing initial coiled stage, and a section from an adult shell showing sutures); 5, *Mortoniceras texanum* (edge view of a shell with living chamber broken away revealing last septum); 6, *Belemnitella americana,* one of the last of the belemnites; 7, 8, *Placenticeras placenta* lateral and edge views; 9, *Didymoceras* sp., an ammonite irregularly coiling at maturity; 10, *Scaphites nodosus.* All natural size, except Fig. 3. (Drawn by L. S. Douglass.)

perhaps too far, attaining their greatest size only to die out considerably before the close of the period. Among the great marine reptiles, the **ichthyosaurs** and **plesiosaurs** were already on a marked decline, while the **mosasaurs** underwent a meteoric evolution, yet all these died out and only the marine turtles survived. The decline and extinction of the **ammonoids** and **belemnites** at the very close of the period, and the passing of the reef-dwelling **rudistid** bivalves and other very common Cretaceous groups like the **exogyras, gryphaeas,** and **inoceramids** (Fig. 16-25 and Plate 17) show that the marine invertebrates did not escape the crisis. Indeed a close survey of many invertebrate groups would reveal that a number of common Mesozoic stocks became extinct.

It is difficult to account for the approximately simultaneous extinction of animals so diverse in relationships and in habits of life. This is particularly true in light of some of the more apparent exceptions to the pattern of extinction. Land plants, for example, do not appear to have been greatly affected, nor do corals, and the bony fishes began, in Cretaceous time, their great modern radiation and evidently were not measurably affected by whatever brought about the extinctions and exterminations in other groups. Such a selective pattern obviously points to environmental stress or stresses, which some organisms were adapted to meet and others were not.

As was true for the Late Paleozoic extinctions, speculations about cause are numerous. The great restriction and final disappearance of the epicontinental seas at the end of the era, the rise of highlands from Alaska to Patagonia, a sharp drop in the temperature accompanying the Laramide uplift, the vanishing of the swampy lowlands, and the vastly changed plant world have all been invoked to account for the extinctions. More recently, cosmic radiation from supernovae passing near the earth [11] and a failure in the supply of nutrients to the sea causing great extermination in the plankton—the basis of the marine food chain [2]—have been offered as possible explanations. These illustrate the range of

Figure 16-25 A block of the Jetmore Chalk Member of the Greenhorn Limestone showing a bedding surface crowded with shells of the bivalve *Inoceramus labiatus,* Russel County, Kansas. (C. O. Dunbar.)

hypotheses on the subject, but are only a small sample. The complexity of the problem is great and recent reviews of the subject [7] suggest that no single catastrophic event is likely to have been the cause. All the accumulated knowledge about the Mesozoic and its life appears to be inadequate, as yet, to provide a satisfactory answer to the great organic changes that mark the end of the era.

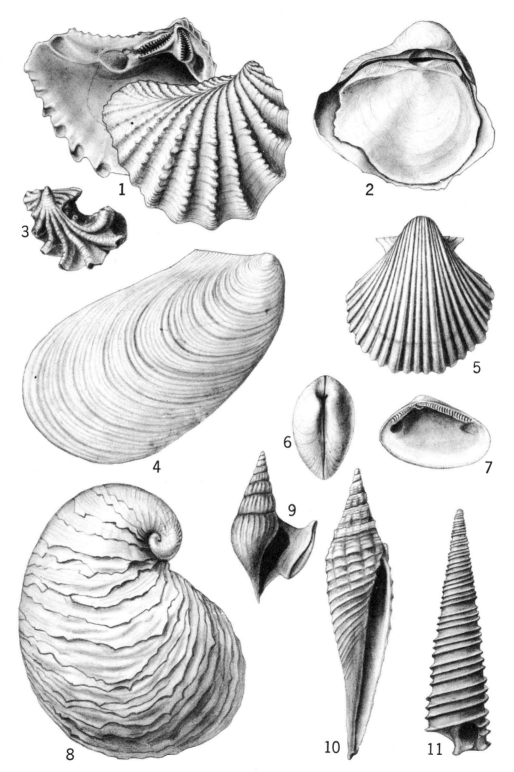

Plate 17 Upper Cretaceous Pelecypods (1–8) and Gastropods (9–11).

Figure 1, *Trigonia thoracica* (valves parted to show interior); 2, *Gryphaea convexa* (showing unequal valves); 3, *Alectryonia falcata* (a ribbed oyster); 4, *Inoceramus labiatus;* 5, *Neithea quinquecostata;* 6, 7, *Nucula percrassa* (dorsal view and interior of right valve); 8, *Exogyra ponderosa* (×⅓), an oysterlike clam with spirally twisted beak; 9, *Anchura lobata;* 10, *Volutoderma appressa;* 11, *Turritella trilira.* All natural size, except Fig. 8. (Drawn by L. S. Douglass.)

REFERENCES

1. Armstrong, R. L., 1968, *Sevier orogenic belt in Nevada and Utah.* Geol. Soc. America Bull., v. 79, pp. 429–458.
2. Bramlette, M. N., 1965, *Massive extinctions in biota at the end of Mesozoic time.* Science, v. 148, pp. 1696–1699.
3. Chandler, M. E. J., and Axelrod, D. J., 1961, *An Early Cretaceous (Hauterivian) fruit from California.* Am. Jour. Sci., v. 259, pp. 441–446.
4. Colbert, E. H., 1964, *Dinosaurs of the Arctic,* Natural History, v. 73, no. 4, pp. 20–24.
5. Gilluly, James, 1963, *The tectonic evolution of the western United States.* Quart. Jour. Geol. Soc. London, v. 119, pp. 133–174.
6. Murray, G. E., 1961, *Geology of the Atlantic and Gulf Coastal Province of North America.* New York, Harper and Brothers, 692 pp.
7. Newell, N. D., 1963, *Crises in the history of life.* Scientific American, v. 208, no. 2, p. 77.

———, 1962, *Paleontologic gaps and geochronology.* Jour. Paleontology, v. 36, pp. 592–610.
8. Richards, H. G., 1947, *Developments in Atlantic Coast states between New Jersey and North Carolina in 1946.* Am. Assoc. Petroleum Geologists Bull., v. 31, pp. 1106–1108.
9. Rubey, W. W., and Hubbert, M. K., 1959, *Overthrust belt in geosynclinal area of western Wyoming in light of fluid-pressure hypothesis.* Geol. Soc. America Bull., v. 70, pp. 167–206.
10. Spieker, E. M., 1946, *Late Mesozoic and Early Cenozoic history of central Utah.* U. S. Geol. Survey Prof. Paper 205-D, pp. 117–161.
11. Terry, K. D., and Tucker, W. H., 1968, *Biologic effects of supernovae.* Science, v. 159, pp. 421–423.
12. Van Valen, Leigh, and Sloan, R. E., 1965, *The earliest primates.* Science, v. 150, pp. 743–745.
13. Wilson, E. O., Carpenter, F. M., and Brown, W. L., Jr., 1967, *The first Mesozoic ants, with the description of a new subfamily.* Psyche, v. 74, no. 1, pp. 1–19.

PART IV

PART IV

THE MODERN WORLD UNFOLDS

Figure 17-1 Eocene (Wasatch) strata in Bryce Canyon, Bryce Canyon National Park, Utah. (C. O. Dunbar.)

CHAPTER **17** *The Modern World Unfolds*

Every feature of the modern landscape was shaped during this last short era. The Rocky Mountains were peneplaned and then uplifted to their present height. The Appalachian ridges and valleys were etched out of a rising peneplane. The Alps and the Himalayas came up out of the sea. It was time of great crustal unrest resulting in the vanishing of almost all the epicontinental seas and it left the continents as high as they have ever been and gave our world such scenic beauty as it may never have possessed before. In the meantime it witnessed the mammals' rise from small forebears with the mentality of opossum to one that can contemplate the mysteries of time and space and can journey to the moon. For these reasons the Cenozoic history will be treated in more detail than that of earlier periods.

THE CENOZOIC TIME SCALE

We live in the first period of this latest era, which is divided into seven epochs as indicated in Fig. 17-2. The name Tertiary Period is an anachronism inherited from the mistaken classification used by Arduino in 1760, but it is now so deeply intrenched in geological literature that it will probably persist.

The scheme of subdivision into epochs of time and series of rocks was introduced by Sir Charles Lyell (Fig. 17-3) in 1833 and was based on richly fossiliferous formations in the Paris and London basins. By 1833 these faunas had been largely described and had been analyzed by the great French paleontologist Deshayes, who recognized in them 3000 extinct and 5000 living species of invertebrate shells. He also observed that in the oldest formations the species were nearly all extinct and progressively younger formations included an increasing percentage of living species. On this basis Lyell proposed and named three series based on the percentage of living species, as follows:

Pliocene
 [Gr. *pleios*, more + *kainos*,
   recent]   30–50% living
Miocene
 [Gr. *meios*, less + *kainos*]   ± 16% living
Eocene
 [Gr. *eos*, dawn + *kainos*]   ± 3.5% living

The percentages would indicate that the sequence was not complete, but Lyell was

	Holocene Epoch	←11,000 B.P.
	Pleistocene Epoch	1,000,000±B.P.
	Pliocene Epoch	
		13,000,000 B.P.
	Miocene Epoch	
		25,000,000 B.P.
	Oligocene Epoch	
		36,000,000 B.P.
	Eocene Epoch	
	Paleocene Epoch	58,000,000 B.P.
		65,000,000 B.P.

(Tertiary Period)

Figure 17-2 The Cenozoic time scale.

Figure 17-3 Sir Charles Lyell.

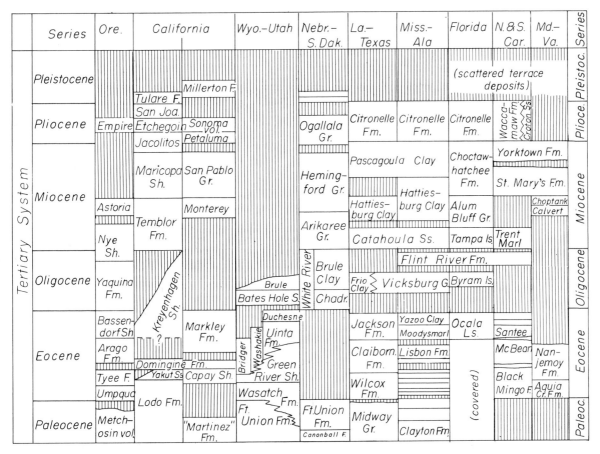

Figure 17-4 Correlation of important Cenozoic formations in North America.

no catastrophist for he wrote, "In regard to distinct zoological periods the reader will understand that we consider the wide lines of demarcation that sometimes [i.e., in places] separate different Tertiary epochs, as quite unconnected with extraordinary revolutions of the surface of the globe and arising, partly, like the chasms in the history of nations, out of the present imperfect state of our information and partly from the irregular manner in which geological memorials are preserved." He foresaw that in the future his series might need to be subdivided or other series added. And he cautioned that the proportion of living species is only one criterion for subdivision, since each series includes certain species that distinguish it from all others. His foresight was vindicated by later workers who over the years added four additional series.

The youngest of these has commonly been called Recent, but the term Holocene is now preferred. It began only about 11,000 years ago. The dates indicated for the beginning of the other series are only approximate.

CENOZOIC HISTORY OF NORTH AMERICA

At the close of Cretaceous time North America had assumed approximately its present size and shape (see Map XII). Marginal seas covered the continental shelves during Cenozoic time and at times spread over parts of the present Atlantic and Gulf Coastal plains and the area of the Coastal Ranges along the Pacific Coast, but on the average covered only about 3 percent of the present land surface. Nonmarine deposits accumulated over large areas in the western half of the continent and much of the Cenozoic history of the continent is recorded in these deposits and in the extant topographic features. Remnants of uplifted peneplanes, for example, tell of erosion cycles and give a measure of uplifts in both the Appalachian and the Rocky Mountain regions. The history can best be told by natural regions.

Eastern North America

The Atlantic Coastal Plain. Marginal seas covered the continental shelf all the way from the Grand Banks southeast of Newfoundland

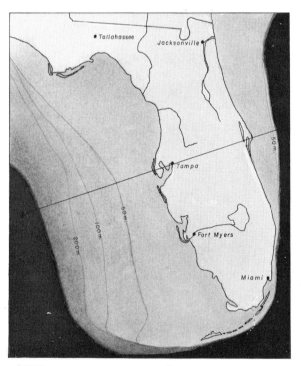

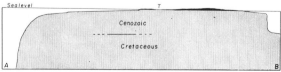

Figure 17-5 The great Florida bank. Top, map; bottom, cross section of the bank.

to the Gulf Coast of Mexico during Cenozoic time and at times spread over the present coastal plains. Along the Atlantic Coast each series is a wedge of sediment thinning to a feathered edge at its landward margin and thickening gradually to the continental margin (Fig. 4-14), and the older ones dip more steeply than the next younger, proving that the continental shelf was progressively tilted seaward. A deep well at Cape Hatteras revealed a thickness of about 3000 feet of Cenozoic formations.

Florida is the exposed summit of an enormous calcareous bank built up as the region slowly subsided throughout Cretaceous and Cenozoic time. Deep wells off the west coast penetrated 1800 feet of Cenozoic formations underlain by more than 7000 feet of Cretaceous beds, practically all of which are calcareous (Fig. 17-5). From the west coast the sea floor slopes very gently out to a depth of 200 meters, then more steeply down to 1000 meters where it drops very steeply down to the floor of the Gulf of Mexico at a depth of about 3000 meters. Until recently it was suspected that this steep face of the bank was a fault scarp along which the Gulf had subsided, but recently discovered evidence indicates that the bank is entirely constructional and that the steep western wall is due to the growth of marginal reefs of corals or algae during Cretaceous time [13].

The Gulf of Mexico. The origin and history of the Gulf of Mexico has been the subject of much speculation but extensive research during the last 10 or 15 years has for the first time laid the foundations for a convincing interpretation. This new information now has been synthesized by Uchupi and Emery [13]. Among the basic facts is the discovery by seismic research that normal oceanic crust underlies the Gulf. Another is that extensive salt deposits, of Triassic, or more likely Jurassic age, underlie much of at least the northern half of the Gulf where it has been rising in the form of salt domes into the younger sediments. Neither Jurassic formations nor older rocks crop out, but they have been penetrated by deep wells in the coastal plain north of the Gulf (p. 345). The Yucatan peninsula

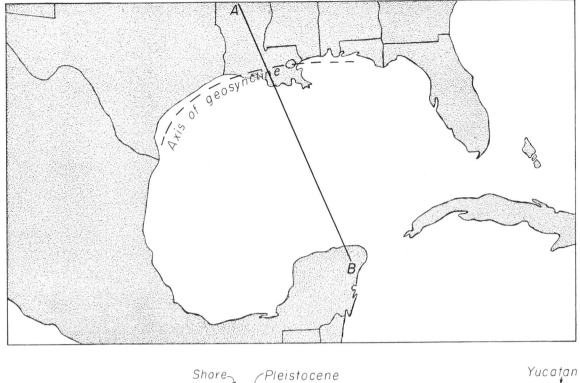

Figure 17-6 Map of the Gulf of Mexico with a section showing the great thickness of Cenozoic deposits in the Gulf Coast geosyncline. (Adapted from Fisk, 1956.)

on the south side of the Gulf is flanked on the north by a broad submarine plateau, the Campeche Bank, which is a counterpart of the Florida Bank. Between the banks a relatively narrow channel connects the Caribbean and the Gulf. The escarpment of the Campeche Bank was almost certainly formed as was that along the west side of the Florida Bank by the growth of marginal reefs during early Cretaceous time.

Three considerations indicate that the region of the Gulf was a very shallow sea at the beginning of Late Jurassic time and has been slowly and progressively sinking ever since. First, the pre-Late Jurassic salt beds must have been deposited in a shallow and

nearly landlocked sea. Second, the reefs that formed the great Cretaceous scarps could only have formed near sealevel. And third, disconformities at several levels in the Cretaceous and Cenozoic deposits on the Florida Bank indicate temporary emergences.

The wide coastal plain north of the Gulf is formed of Cenozoic sediments largely supplied by the Mississippi River system, an early stage of which was well established by Paleocene time. For a time in the Eocene Epoch the sea spread northward almost to Cairo, Illinois, but thereafter it was pushed back by the fluvial deposits. Meanwhile the Gulf Coast Geosyncline was subsiding along the north side of the Gulf (Fig. 17-6). Deep

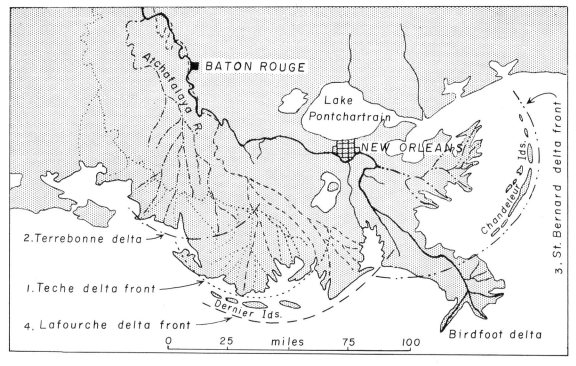

Figure 17-7 Map of the Mississippi River Delta. (Adapted from Fisk et al. [7].

wells in southern Louisiana have failed to reach the bottom of the Cenozoic beds at depths as great as 25,000 feet and it is estimated that along the axis of the trough the thickness exceeds 30,000 feet. This is about 3 times as deep as the Sigsbee Abyssal Plain in the center of the Gulf.

The present great delta of the Mississippi was formed during the Pleistocene and Holocene epochs and its history has been worked out in great detail by H. W. Fisk and his colleagues [7]. Some features of its complex history are indicated in Fig. 17-7. After the Teche subdelta had formed, the river found a shorter outlet by shifting to the west where it built the Terrebonne subdelta. It later shifted far to the east to build the St. Bernard subdelta, the front of which extended beyond the present Chandeleur Islands. It then shifted to the west to build the Lafourche subdelta. The Birdfoot subdelta is known from archeologic data and radiocarbon dating to have formed within the last 450 years. If the river had not been restrained artificially it would by now have been diverted into the Atchafalaya distributary above Baton Rouge, which would have shortened its journey to the sea

by more than 100 miles. As each subdelta in turn was abandoned its front was drowned by regional subsidence and eroded back by the sea. The Chandeleur Islands, for example, are a chain of wave-built sand barriers made of sand reworked from the front of the St. Bernard subdelta. The Holocene delta deposits are thin over the exposed area, but reach a thickness of about 850 feet at 20 miles offshore and probably exceed 2000 feet farther out, as indicated by offshore drilling.

Sculpturing of the Appalachians. At the beginning of Cenozoic time nearly all of the Appalachian region had been peneplaned. A chain of low mountains 2000 to 3000 feet high remained along the Tennessee-North Carolina border, and several monadnocks persisted in New England. These unreduced areas form the crest of the modern Great Smokies, the summit of the White Mountains, and scattered peaks such as Mount Katahdin in New Hampshire and Cadillac Mountain in Maine; they show no evidence of ever having been reduced to a level summit. Elsewhere in the Appalachian region, remnants of a widespread and remarkably flat erosion surface may be seen in the even crests of the highest ridges in the folded belt (Fig. 17-8), and in the summit of the Allegheny Plateau. This is the **Schooley Peneplane.** When it was formed the surface must have been near sealevel and the region was obviously a low plain.

The present mountains are therefore due almost entirely to Cenozoic changes. They are not, however, the result of either folding or faulting, for such structures were inherited from the Alleghanian orogeny late in the Paleozoic Era. The Cenozoic history involved the gentle regional upwarp in the form of a broad arch, and erosion which etched out the weaker rocks and created the local relief. The erosional history is summarized in Fig. 17-9. After completion of the Schooley peneplane (block 1), at a time not yet accurately dated, the region was arched up a few hundred feet (block 2). The streams then incised themselves to a new base level and opened out extensive lowlands on the weak formations in the folded belt while the resistant formations remained as ridges (block 3). Remnants of these lowlands are still conspicuous about Harrisburg, Pennsylvania, and record the **Harrisburg surface.** A second gentle upwarp caused the streams to incise their valleys into the Harrisburg lowlands and to excavate new lowlands of more local extent (block 4). Still younger and more frequent uplifts caused the major streams to carve strath terraces along their present deep valleys. Thus by a series of gentle uplifts the Schooley peneplane has been elevated to its present altitude of about 4000 feet along the crest of a broad simple arch which is near the eastern edge of the Allegheny Plateau. From this axis it slopes gently away both east and west declining but a few feet per mile. It is preserved only on the most resistant rocks in the folded belt and has been completely destroyed throughout most of the Piedmont belt. If it were preserved here it would undoubtedly pass under one of the sedimentary series in the coastal plain, for it is clear that as the Appalachian arch was elevated the continental shelf was depressed and tilted down toward the east. If stages of uplift in the mountains could be correlated with the sedimentary breaks in the coastal plain, the erosional history could be dated in detail, but unfortunately such correlation is not yet secure.

Figure 17-8 Peneplaned summit of the ridges in the Appalachian folds seen in the Susquehanna Gap near Harrisburg, Pa. k, Kittatinny Mountain; s, Second Mountain; p, Peters Mountain. (George Ashley.)

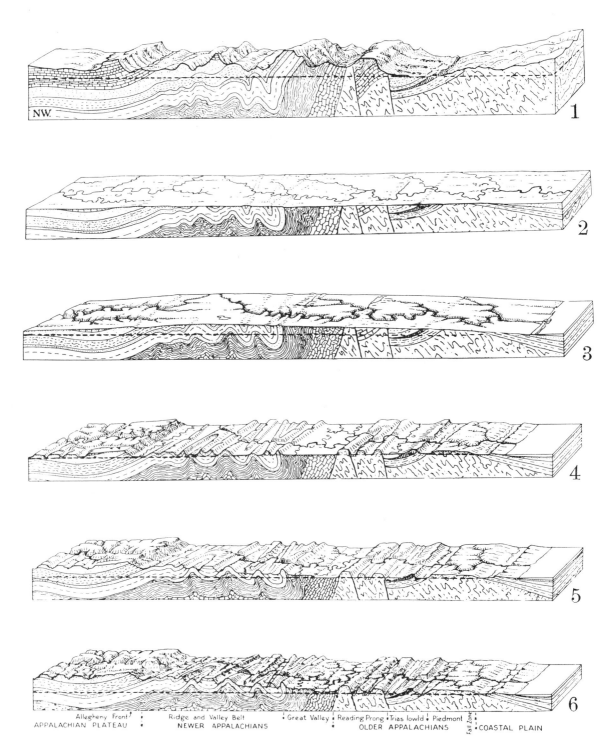

Figure 17-9 Diagrams illustrating the evolution of the Appalachian topography. 1. The rugged topography during Mesozoic time, after the Alleghanian orogeny and the Triassic faulting. 2. The Late Cretaceous peneplane. 3. Uplift as a great simple arch in early Cenozoic time, with master streams beginning to incise themselves. 4. A later stage after development of local peneplanes on the weak formations. Remnants of the Schooley peneplane remain on the ridges. 5 and 6. Later stages leading to the modern topography. (Modified slightly from Douglas Johnson.)

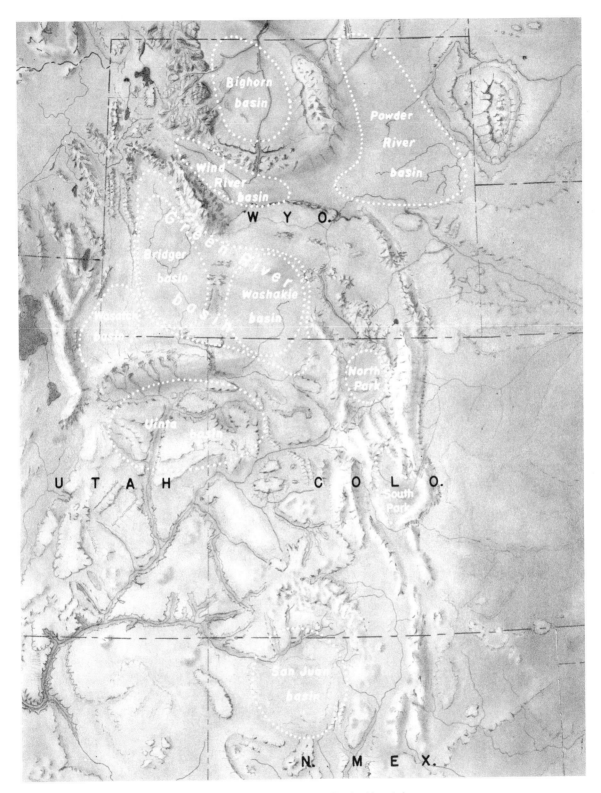

Figure 17-10 Basins and ranges of the middle and southern Rocky Mountains. The topography is modern but the structures were inherited from the Late Mesozoic orogenies. (Drawing by Milton Wallman.)

Decay and Rebirth of the Rockies

The widespread orogenies that produced the Rocky Mountain System continued locally into the Cenozoic Era, finally dying out in the Eocene Epoch. This left a vast highland of bold mountains and deep structural basins (Fig. 17-10). As erosion carved away at the mountains much of the debris was trapped in the adjacent basins where it was spread by meandering and overloaded streams in nearly horizontal layers. Thus within almost 30,000,000 years before the beginning of the Oligocene Epoch the basins had been filled and the mountains largely peneplaned leaving only scattered monadnocks to mark the highest parts of the old ranges (Fig. 17-11). Along the axis of the continental divide this nearly flat surface may have had an elevation of 2000 to 3000 feet and the streams radiating away spread over it a mantle of Oligocene silts and muds that masked the ranges except for the monadnocks. During Miocene time this region began to rise as a very broad gentle arch, similar to that of the Appalachian region.

This uplift started a new erosion cycle in which the major streams found themselves superposed upon several of the buried ranges [1], which they now cross in spectacular canyons. Meanwhile as uplift continued the weak fill of the basins was rapidly carved out to form relatively flat plains and much of the debris was carried beyond the ranges to form a broad debris apron which now forms the High Plains (Fig. 17-12). The Bighorn River crosses the Owl Creek Range in a deep canyon south of Thermopolis, then flows about 150 miles north between low banks as it crosses the Bighorn Basin, before turning northeast to cut through the bounding range in spectacular Bighorn Canyon. Green River flows south across Bridger Basin and then cuts through the eastern end of the Uintas in Flaming Gorge. Laramie River flows north across the floor of Laramie Basin before turning east to cut through the Front Range in Laramie Canyon. South Platte River heads in the basin of South Park but flows northeast to cut through the Front Range in Platte Canyon. The Arkansas River flows out of South Park to cut through the Front Range in spectacular Royal Gorge, whose sheer walls tower 1400 feet above the stream.

The evidence of peneplation followed by uplift and erosion as described above is recorded in four ways: first in remnants of the summit peneplane; second in the superposition of the major streams on the mountain ranges; third in remnants of the basin fills which still remain in horizontal beds high up on the flanks of several of the ranges; and fourth the great debris apron of the adjoining plains, which is graded to the summit of the peneplaned Front Range.

These features are illustrated in Fig. 17-12. A particularly fine peneplane remnant may be seen along the summit of the Laramie Range in southeastern Wyoming. Here, at an elevation of 8000 feet, is a nearly flat surface 10 or 12 miles wide, cutting across granite and other types of igneous rock of the mountain core (Fig. 17-13). Roads run in almost every direction; and the Union Pacific Railroad crosses the divide, not through a deep pass but across an open plateau. At the station of Sherman one may look for miles in almost any direction, and it is with difficulty that he realizes that his viewpoint is 8000 feet above sealevel, or as high as the summit of many of the rugged mountains of the Northwest [4].

East of the Laramie Range lie the High Plains, capped by Miocene (and locally Pliocene) beds. Their surface is extraordinarily flat and slopes gently eastward to an elevation of not over 2000 feet in central Nebraska and Kansas. This great flat area is the remnant of the aggraded plain that existed when the mountains were peneplaned. Generally its most elevated western margin has been dissected and eroded back several miles from the mountain front, especially where large streams cross it, but locally in southeastern Wyoming it extends up to the mountains rising to the level of the summit peneplane. Here it forms the "Gangplank" by which the Union Pacific Railroad crosses the mountains from the plains. Even where the High Plains beds have been eroded back several miles it can be seen

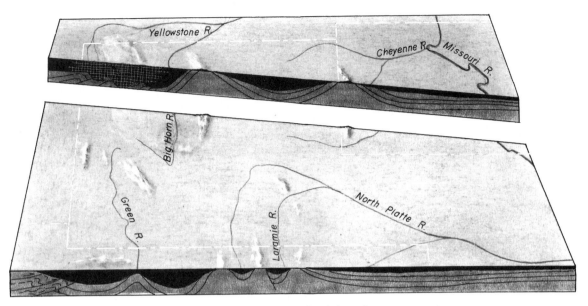

Figure 17-11 Block diagram of part of the Rocky Mountain region after peneplanation and cover by Oligocene formations. The basin fill is shown in black with an overlay of white lines to emphasize that the beds in the basins are essentially horizontal. The vertical white lines mark the great mass of volcanics in the region of Yellowstone Park. (Adapted from Atwood and Atwood [1 .)

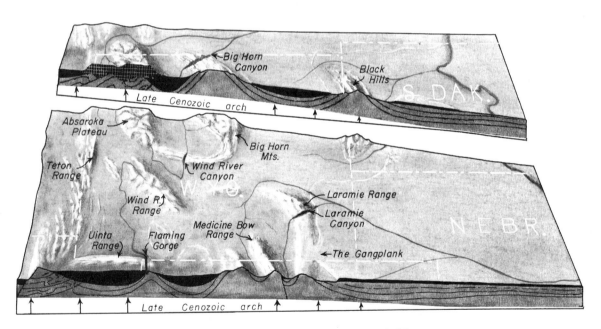

Figure 17-12 The same region shown in Fig. 17-11 as it appears at present. After it was uplifted to form a broad gentle arch, much of the basin fill was removed, leaving the mountains in bold relief. The streams superposed on the ranges cut through them in spectacular gorges. The great debris apron of the High Plains has been eroded back from the flank of the Front Range except for the "Gangplank" in southeastern Wyoming where they are still graded to the peneplaned summit of the Laramie Range. The peneplaned summit is shown in Fig. 17-13. (Adapted from Atwood and Atwood [1].)

Figure 17-13 Peneplane on the granite summit of the Laramie Range near Buford, Wyoming. The view is northward toward distant monadnocks that rise above the peneplane. (C. O. Dunbar.)

Figure 17-14 Idealized block diagram to show the relation of the High Plains deposits to erosion surfaces in the Rocky Mountains. It is drawn on the assumption that the Flattop peneplane was completed at the end of Oligocene time and the Rocky Mountain peneplane was completed by the end of Miocene time. Although these correlations appear probable, the precise dates are not certainly proved. (Adapted from an unpublished figure by R. F. Flint.)

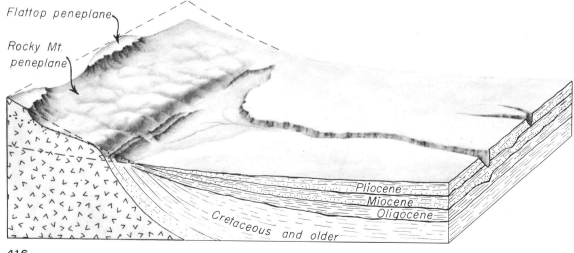

that, if they were projected toward the mountains, their surface would meet the peneplaned summit.

Stratigraphy of the Basin Fills. The fluvial deposits spread as an immense debris apron east of the Rockies have been partly destroyed by later erosion along both their eastern and western margins, but their central portion now forms the monotonously flat High Plains (Fig. 17-14). Here, as in the intermont basins, they consist of clay, silts, and sands which generally are only partly consolidated and commonly weather into badlands (Figs. 17-15 and 17-16).

The Paleocene Epoch is represented by the Fort Union Formation which occupies a large area in the northern Great Plains, chiefly in the Dakotas, Wyoming, Montana, and Alberta, and in the Laramie Basin. It includes friable yellow sandstones, somber gray shales, and enormous deposits of low-grade coal.

During Middle Eocene time the Green River Basin of Wyoming and Colorado was occupied by a vast shallow lake in which fine, evenly-laminated shale accumulated to a thickness of about 2000 feet. In these lake beds are included vast deposits of "oil shale" from which petroleum can be distilled. At present only pilot plants are in operation but as the oil fields decline these shales will become an important source of petroleum. Bradley has concluded that fine laminae in these shales, like the growth rings of trees, mark seasonal accretions, and on the basis of representative laminae he estimated that deposition of the Green River shales required about 6,500,000 years [5]. The Green River Shale has yielded an amazing number of beautifully preserved

Figure 17-15 Big Badlands east of the Black Hills in South Dakota, exposing the Oligocene, Brule Clay. These beds have yielded abundant fossil mammals, notably the extinct oreodonts and titanotheres. (N. H. Darton, U. S. Geol. Survey.)

Figure 17-16 Lower Eocene (Willwood) beds along Gray Bull River in the Bighorn Basin. (Am. Museum of Natural History.)

fish skeletons as well as leaves, microscopic plants, insects, and a bat (Fig. 3-25). Similar lakebeds (Elko Shale) in northeast Nevada are illustrated in Fig. 17-17.

A smaller lake existed near Florissant, Colorado, in Oligocene time when ash falls from nearby volcanoes buried rich deposits of insects and plants. The Oligocene basin of central Oregon was also encircled by active volcanoes and there the ash falls, reworked by streams, buried one of the most important known assemblages of fossil mammals.

One of the surprising discoveries in the northern plains was the Cannonball marine member in the Fort Union Formation in the western parts of the Dakotas (Fig. 17-18). It has yielded a fauna of about 150 species of marine and brackish water fossils, of which about 80 are mollusks and 64 are foraminifera. Of the foraminifera 38 species occur also in the Paleocene beds of the Gulf Coast [8], but the general aspect of the fauna is more like that of the Paleocene of northern Europe. Since there is no vestige of a former connection with either the Gulf or the boreal region, the geographic relationships of the Cannonball sea remain a mystery. This is the last vestige of the interior seaways in America.

Exceptional interest attaches to Yellowstone Park and the adjacent Absaroka Range that bounds it on the east. This was the scene of enormous lava flows and associated volcanic

Figure 17-17 Elko lake beds near Elko, Nevada. (C. O. Dunbar.)

Figure 17-18 Stratigraphic section showing the relations of the marine Cannonball Member of the Fort Union Formation. Length of section about 250 miles; vertical scale greatly exaggerated.

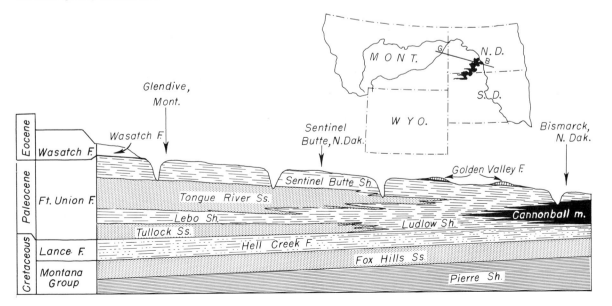

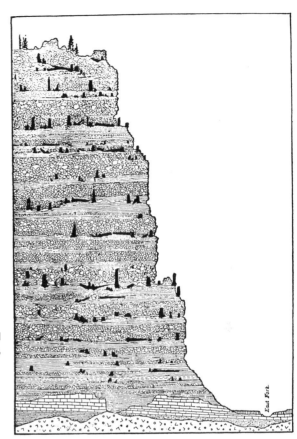

Figure 17-21 Profile section of Amethyst Cliff showing successive forests, each killed and buried by volcanic debris. (U. S. Geol. Survey.)

Figure 17-19 Lava flows of the Absaroka Range viewed across Shoshone Canyon about 25 miles west of Cody, Wyoming. (C. O. Dunbar.)

Figure 17-20 Fossil tree stump and log exposed in the face of Amethyst Cliff, Yellowstone Park. (J. P. Iddings, U. S. Geol. Survey.)

breccias. The igneous activity began in the Absaroka Range in Eocene time and continued intermittently through the Oligocene. The lavas poured out in a highly fluid condition and spread widely in thin sheets (Fig. 17-19) without explosive violence. From the west side of the range, flows spilled down into the basin that is now Yellowstone Park, where they commonly burst into volcanic breccia that killed and buried the timber. Lamar River has cut through to expose a section of some 2000 feet of these beds in which the stumps and trunks of 27 successive forests are preserved [6]. This is Amethyst Cliff (Figs. 17-20 and 17-21). Sedimentary beds interstratified with the volcanics yield fossil plants that prove the igneous activity to have ranged from Middle Eocene time into the Oligocene. In the basin

Figure 17-22 Relief model of the Cordilleran region.

of Yellowstone Park igneous activity continued into Miocene time and the geysers and hot springs of the Park show that volcanic heat still smoulders.

The Central Cordilleran Region

Figure 17-22 shows the relations of three major structural units in this region: the Basin and Range Province, the Columbia Plateau, and the Colorado Plateau.

The **Basin and Range Province** was the region of the great folds and overthrusts produced by the Late Mesozoic orogenies, and in Early Cenozoic time it was the keystone in a broad arch that extended from the Rockies on the east to the Sierra Nevada region on the west. It was then a bold, rugged highland standing higher than the rest of the Cordilleran region and had exterior drainage [12]. It therefore contains virtually no pre-Miocene sedimentary deposits.

A profound change began early in Miocene time as the arch gradually collapsed and subsided between great normal faults on both the east and west sides. Meanwhile the whole region was riven by many north-south normal faults that separated large blocks of the crust which rotated so that the higher edge of each formed a small range of mountains and the depressed side a structural basin. The subsequent history is recorded in the sedimentary deposits that accumulated in the intermont basins.

The Miocene deposits begin with coarse conglomerates that range in thickness up to 3000 feet near the adjacent range and contain angular fragments of all the older formations in the range. They are obviously the coarse debris of piedmont fans for they grade laterally into finer sediments in the center of the basins, clays, and silts that include local beds of gypsum and other salines. This indicates that the basins, lying in the rain shadow of the ranges, were deserts as they are now. The Miocene formations are overlain with angular discordance by Pliocene formations of similar character. In southern Nevada, for example, Pliocene formations as much as 1800 feet thick include local beds of gypsum and salt as much as 100 feet thick. It is evident, therefore, that faulting began early in Miocene time and continued actively throughout the Pliocene. Indeed, well-defined fault scarps and historically dated movements prove that faulting is still continuing. The net result is that the general level of the basins in this province (about 6000 feet above sealevel) is now more than a mile below the summit of the Sierra Nevada on the west and of the Colorado Plateau on the east. The structural relations and history just described extend across southern California and southwestern Arizona into northern Mexico.

The **Colorado Plateau** was scarcely disturbed by the Late Mesozoic orogenies and is still capped by nearly horizontal Paleozoic and Mesozoic formations (Figs. 17-22 and 17-23). Essentially it consists of a broad, flat-topped dome now eroded into a region of remarkable tabular plateaus that rise steplike to the north of the Grand Canyon, the upper one having an elevation of about 10,000 feet. Cliffbound mesas and deep canyons, all of the most impressive magnitude, are characteristic of the region. It is crossed by a few north-south normal faults and a few monoclinal flexures and is transected by the spectacular Grand Canyon of the Colorado River.

Mesozoic formations once covered the area but have since been stripped back to its northwestern flank. Until recently it was suspected that the stripping was largely the work of the Colorado River, but a 1967 symposium involving some 75 of the chief students of the region developed a new and more convincing interpretation, namely, that the stripping preceded the appearance of the Colorado River [12]. According to this interpretation the Basin and Range country stood higher than the Colorado Plateau until about the end of Miocene time and the drainage over the Plateau was generally to the northeast and east, probably with two or more river systems working headward into the Mesozoic formations, which were thus stripped back to near their present

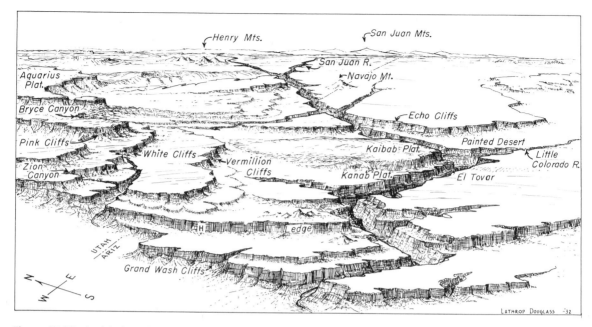

LATHROP DOUGLASS '32

Figure 17-23 Aerial view of the Grand Canyon District, looking east-northeast up the Grand Canyon. The Grand Canyon proper extends from Grand Wash Cliffs to the mouth of Little Colorado River. From there to Echo Cliffs the gorge is known as Marble Canyon and from there upstream it is Echo Canyon.

position. Then, in Pliocene time, as the Basin and Range Province subsided by some thousands of feet, a small southwest flowing stream (the present lower Colorado River) cut rapidly downward and headward until it captured the east-flowing steams, thus reversing the drainage and creating the present Colorado River System.

The evidence for this interpretation is too complex for adequate treatment here but two facts deserve explanation. The first is that where the Colorado now leaves the Plateau through Grand Wash cliffs it flows over the Muddy Creek Formation of Miocene age which is not made of material that could have been brought in by the Colorado River but includes conglomerates made of old crystalline rocks that must have come from the highlands of central southern Arizona. Its saline deposits also prove that it was a closed basin without exterior drainage. In short, the Colorado River did not then exist. Lavas interbedded in the Muddy Creek Formation have been dated at 11.8 ± 0.7 million years. This gives a maximum date for the beginning of the cutting of the Grand Canyon as near the beginning of Pliocene time.

A second line of evidence concerns Late Miocene and Pliocene deposits that still lie in places on the eastern part of the Plateau. For example, in the valley of the Little Colorado River the Bidahocki lake beds and fluvial deposits overlie a widespread erosion surface that could not have been formed under present drainage conditions; and the Chuska Sandstone along the east side of Arizona (of probable Pliocene age) was deposited by south-flowing streams.

The **Columbia Plateau** region subsided, as did the Basin and Range Province, but was progressively covered by flows of basic lava that welled up quietly through fissures and spread widely in thin layers. Such flows continued intermittently throughout Miocene and Early Pliocene time. Recently secured dates based on potassium-argon ratios indicate that the oldest layers are about 21 million years old and the youngest ones about 12 million years old [10]. Locally, as in the Valley of the Moon area in southern Idaho, the activity continued until near recent time. One very destructive flow, the Mesa Basalt, so called because it caps so many of the mesas in parts of 3 states, has recently been stated to have covered approximately 100,000 square miles, although it is generally less than 40 feet thick [14]. The mechanism by which it was able to spread so widely before congealing is unknown, but as a molten lake of this size it must have presented a lurid scene such as the earth has seldom seen.

Where the Snake and Spokane rivers have now cut through the lavas they reveal a prebasalt floor of schists, granites, and other old rocks with a relief of some 2500 feet. The early basalt flows, seeking the lowest places, gradually filled the basins and eventually buried the hills to a depth of some 2500 feet. As the flows spread over an area of some 200,000 square miles they interrupted the drainage, damming streams and giving rise to lakes and swamps around the margins of the province in which sedimentary deposits interbedded with

the lavas include abundant and well-preserved plant fossils. Examples are the Latah Formation near Spokane and the Payette Formation some 250 miles farther east.

During the time of extrusion (and possibly because of it) the region was gently warped so that its present surface ranges in elevation from 3000 feet or so to more than 8000 feet above sealevel.

The Pacific Border

As shown in Fig. 17-22, two parallel ranges of mountains separated by long structural basins parallel the western margin of the United States. These ranges differ notably both in structure and in history.

The Sierra Nevada represents part of a colossal fault block more than 100 miles wide and between 300 and 400 miles long, the eastern margin of which has been uplifted to an elevation of about 13,000 feet (Figs. 17-24 and 17-25) and its western margin has been de-

Figure 17-24 Idealized block diagram to show the relation of the Sierra Nevada to the Basin and Range Province on the right and to the California trough and the Coast Ranges on the west. *Gm,* Mesozoic granite; *J,* Jurassic; *M,* Miocene; *P,* Pliocene; *Ple,* Pleistocene; *Pl,* Lower Paleozoic; *Pu,* Upper Paleozoic.

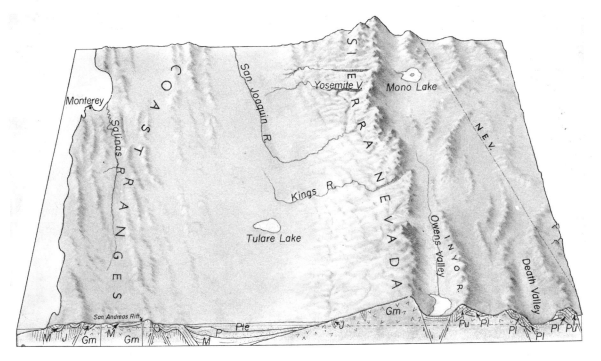

Figure 17-25 The east face of the Sierra Nevada, a fault scarp about 2 miles high. Telephoto view from Owens Valley, California. (F. E. Matthes, U. S. Geol. Survey.)

pressed to perhaps 25,000 feet below sealevel. The uplifted side forms the Sierra Nevada and the depressed half the California trough. The Sierra Nevada has been sculptured from this single fault block of Jurassic and older rocks and a complex of Cretaceous granite batholiths. Its Early Cenozoic history is obscure but by Miocene time the region had been peneplaned. This means that its surface was then almost horizontal.

Fossil plant remains preserved locally at many places on the range and in Owens Valley to the east give critical evidence of the history of uplift [2, 3]. They prove that the major uplift of the Sierra Nevada came after Pliocene time. Miocene floras from the summit region that are now at altitudes of 5000 to 9000 feet resemble modern forests in areas of low to moderate relief, chiefly less than 2000 feet. Deciduous hardwood forests of Miocene date on the summit of the range are similar to those in the

foothills area, a relation that could not have existed if the relief were then high. Late Pliocene and Early Pleistocene floras from altitudes near 9000 to 10,000 feet resemble modern forests at only moderate altitude. Both the plant fossils and the topography indicate that the major uplift of the Sierra Nevada occurred in three pulses during Pleistocene time. The first uplift of some 3000 feet came early in this epoch, when the central part of the range was about 6000 feet above sealevel and the highest peaks along its eastern margin were about 8000 feet high. A second uplift occurred in mid-Pleistocene time, and the final uplift to some 13,000 feet came late in the epoch. While the region of the Sierra Nevada was still low in Pliocene time, Owens Valley had moderate rainfall and supported forests similar to those on the Sierra Nevada block, but as the mountains rose, intercepting the westerly winds, the Basin and Range Province lying in its rain shadow took on its modern desert character.

Following the second uplift the west-flowing streams began to cut canyons and the profiles across their valleys confirms the plant evidence of three pulses of uplift as indicated in Fig. 17-26.

The **Cascade Range** which borders the Columbia Plateau is covered by lavas and is crowned by a chain of magnificent volcanoes which were still active in Pleistocene time (Fig. 17-27). The lavas hide the underlying rocks.

The **Coast Ranges** have a very complex structure involving Mesozoic and Cenozoic sedimentary formations and granitic batholiths all of which have been folded and faulted. Movements along some of the faults, such as the San Andreas Fault, still continue and are responsible for the earthquakes that occasionally disturb this region.

The **California Trough** contains a long sequence of Eocene to Pleistocene formations with a maximum thickness of some 25,000 feet. The stratigraphy is complicated by large-scale facies changes resulting from the movements in the Coast Ranges. Some of the arches in the basin are the site of the major oil fields of California.

Figure 17-27 Mount Hood, a volcano on the Cascade Range in Oregon. Mount Rainier, Washington, appears in the background. (Stephen Green-Armytage, Photo Researchers.)

Figure 17-26 Cross-profile of the Merced Valley below Yosemite Park, showing evidence of successive uplifts. *a*, an early broad valley; *b*, a deeper and narrower inner valley; *c*, the modern deep gorge. Contrary to the earlier opinion of Matthes, it is now known that all canyon cutting occurred in Pleistocene time.

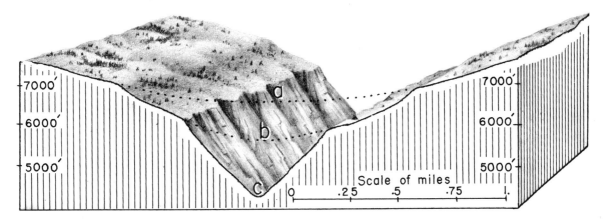

CLIMATIC HISTORY

Fossil plants throw much light on Cenozoic climate. The forest trees have come down to us with only trivial change since early Cenozoic time and probably have not changed appreciably their preferred habitat. Moreover, it is well known that most of them are restricted in their distribution by rainfall and temperature, each preferring a definite environment. Accordingly, the vegetation of a subtropical lowland like Florida has little in common with that of a desert basin, or with the subarctic barrens. Thus it is possible to infer the climatic conditions under which a Cenozoic flora lived.

A comprehensive study of such material throughout western North America has shown a striking change of climate since early Cenozoic time. During the Eocene and Oligocene epochs subtropical types of trees, now restricted to moist lowlands, ranged widely over the United States and Europe. Palms and alligators were then common as far north as the Dakotas, suggesting a climate like that of modern Florida and Louisiana. At the same time a moist temperate forest existed in high northern latitudes, notably in Alaska, Greenland, Spitsbergen, and northern Siberia. It was dominated by the giant redwood and included such deciduous trees as the basswood, beech, chestnut, and elm. Even cycads, magnolias, and figs then lived in Alaska.

The O^{18}/O^{16} ratio in fossil shells from cores in the deep ocean floor strikingly confirms the evidence from land plants. The reasoning is as follows. Water from the polar regions settles to the bottom and spreads over the deep ocean basins. As a result, under present conditions, the temperature below a depth of 15,000 feet is nearly constant and is barely above the freezing point even in low latitudes. But shells of bottom-dwelling foraminifera in cores from the Pacific Ocean floor west of South America indicate that the bottom temperature was about 51°F at a date of 32,000,000 B.P. The polar regions must then have had a warm, temperate climate.

The climate was not only milder and more humid, but also more uniform than now over the far western United States, evidently because the region was generally low and the mountains were not lofty enough to interfere seriously with the moisture-bearing westerly winds.

After Eocene time there was a slow but general southward migration of the various plant assemblages, indicating a gradual cooling of the climate that became more marked in Pliocene time and culminated in the Pleistocene glaciation. Meanwhile, in Miocene and Pliocene time, the climate became more diversified in the western United States as the rising mountains intercepted the winds, producing moist western slopes with arid regions in their lee. The diversity was not nearly so extreme or so widespread in Miocene time as it is today, though the salt and other precipitates entombed in the Miocene deposits of southern Nevada indicate rather intense local aridity in the basins then forming. Even as late as Early Pliocene time, however, a flora like that of southern California was still living in western Nevada (the Esmeralda flora), indicating a rainfall of 12 to 15 inches a year in a region where the present rainfall is only 4 inches. The final uplift of the Cordilleran ranges in Late Pliocene and Pleistocene time gave the intermont basins and the Great Plains their present degree of aridity.

For a time it was believed that the Eocene of southwestern Colorado contained local glacial deposits (the Ridgeway Formation) but recent work has shown that two quite distinct formations were confused. The Ridgeway Formation includes tillite but it is of Pleistocene age and the Eocene Gunnison "tillite" represents landslide and mudflow deposits [11].

ECONOMIC RESOURCES

Petroleum. Two of the major American oil field regions draw their production predominately from Cenozoic rocks. The Gulf Coast pools of Louisiana and southeast Texas are in small domes associated generally with plugs of rock salt that have pressed up from below into Miocene sands and clay. These fields have now been extended seaward under the northern edge of the Gulf of Mexico. The

California oil fields draw about 30 percent of their oil from Pliocene beds and 65 percent from the Miocene.

Although the greatest American oil fields are in Paleozoic or Mesozoic rocks, it is a very striking fact that nearly all the foreign fields are in Cenozoic formations. For example, the rich Baku fields of Russia produce from the Miocene, the Galician fields from the Eocene, Oligocene, and Miocene, the Rumanian fields from Oligocene to Pliocene, those of Burma, Sumatra, Java, and Japan from the Miocene, and those of the Persian Gulf chiefly from the Miocene.

Coal. Lignite occurs in the Eocene of the Gulf Coast but is not commercially exploited. In the Puget Sound region, however, the strong folding of the coal-bearing series has advanced the Eocene coals to a subbituminous rank. These coals are now being extensively used west of the Sierra Nevada and Cascade ranges. Production declined to 55,000 tons in 1965. About 2,732,000 tons of lignite was produced in North Dakota in 1965 chiefly from the Fort Union Formation. In most parts of the world, Cenozoic coals are of lignite or low-grade subbituminous rank, and therefore of little present value.

Placer Gold. The placers, which now yield about two-thirds of the annual gold of California, were formed during Cenozoic time by streams degrading the Mother Lode belt in the Sierra Nevada. During the early decades nearly all the gold was secured from placers, and they still yield about 92 percent of California's gold. In 1965 the placer gold amounted to 58,571 ounces and was valued at more than $2,000,000.

Metalliferous Veins. The fabulous wealth of gold, silver, and copper so widely distributed throughout the Rocky Mountain region is for the most part a byproduct of the intrusions of Cenozoic time. The mineral-bearing solutions of various sorts formed the vast copper deposits of Bingham, Utah, of Morenci, Arizona, and of Santa Rita, New Mexico. The silver of Park City, Utah, as well as of the great Comstock Lode, Tonopah, and other famous localities in Nevada, was similarly formed in middle and later Cenozoic time. The gold of many of the spectacular mining camps of the West, such as Goldfield, Nevada, and Cripple Creek in the Rockies, had a similar date of origin.

Mexico, Central America, Peru, and Bolivia also provide notable examples of the great mineral wealth we owe to the crustal disturbances and intrusive activity of the Cenozoic.

Diatomite. This is a white porous rock formed of the siliceous shells of microscopic pelagic plants called **diatoms.** It has extensive use in industry, for insulation, as a filter used to purify water, alcoholic beverages, antibiotics, sugar, oil, and solvents, and as a filler in paper, paints, plastics, and soap. All of it comes from Cenozoic deposits and the chief American source is in the Miocene formations near Lompoc, California. In 1967 the production in the United States amounted to 580,278 tons worth over $29,000,000.

REFERENCES

1. Atwood, W. A., and Atwood, W. A., Jr., 1938, *Working hypothesis for the physiographic history of the Rocky Mountain Region.* Geol. Soc. America Bull., v. 49, pp. 957–980.
2. Axelrod, D. I., and Ting, W. S., 1961, *Early Pleistocene floras of the Chagoopa surface, southern Sierra Nevada.* Univ. Calif. Publ. *in* Geol. Sciences, v. 39, no. 2, pp. 119–194.
3. Axelrod, D. I., 1962, *Post-Pliocene uplift of the Sierra Nevada, California.* Geol. Soc. America Bull., v. 73, pp. 183–198.
4. Blackwelder, Eliot, 1909, *Cenozoic history of the Laramie region, Wyoming.* Jour. Geology, v. 17, pp. 429–444.
5. Bradley, W. H., 1929, *The varves and climate of the Green River epoch.* U. S. Geol. Survey Prof. Paper 158, pp. 87–110.
6. Dorf, Erling, 1960, *Tertiary fossil forests of Yellowstone National Park, Wyoming.* Billings Geol. Soc. Eleventh Ann. Field Conference, pp. 253–260.
7. Fisk, H. N., McFarlan, Edward, Jr., Kolb, C. R., and Wilbert, L. J., Jr., 1954, *Sedimentary framework of the modern Mississippi Delta.* Jour. Sed. Petrology, v. 24, pp. 76–99.
8. Fox, S. K., and Ross, R. J., Jr., 1942, *Foraminiferal evidence for the Midway (Paleocene) age of the Cannonball Formation in North Dakota.* Jour. Paleontology, v. 16, pp. 660–673.

9. Gilluly, James, 1949, *Distribution of mountain building in time.* Geol. Soc. America Bull., v. 60, pp. 561–590.

10. Gray, Jane, and Kittleman, L. R., 1967, *Geochronometry of the Columbia River Basalt and associated floras of eastern Washington and western Idaho.* Am. Jour. Sci., v. 265, pp. 257–291.

11. Mather, K. F., and Wengerd, S. A., 1965, *Pleistocene age of the "Eocene" Ridgeway Till, Colorado.* Geol. Soc. America Bull., v. 76, pp. 1401–1408.

12. McKee, E. D., et al., Eds., 1967, *Evolution of the Colorado River in Arizona.* Museum of Northern Arizona Bull., v. 44, pp. i–x; 1–67.

13. Uchupi, Elazar, and Emery, K. O., 1968, *Structure of Continental margin off Gulf Coast of United States.* Am. Assoc. Petroleum Geologists Bull., v. 52, pp. 1162–1193.

14. Wheeler, E., and Cooms, H. A., 1968, *100,000 Square Miles of Burning Rock.* Saturday Review, Oct. 5, pp. 60–63.

Figure 18-1 Margin of a continent still in the Ice Age. Shore zone north of Gneiss Point, Antarctica. At the right glacier ice is flowing into the sea; in the left foreground a tabular iceberg is surrounded by floe ice. (Official U. S. Navy photograph.)

CHAPTER **18** *Ice Sculptures the Final Scene*

During the Pleistocene Epoch vast ice sheets, like that of modern Antarctica, covered nearly one-fifth of the land surface of the earth, not once but four times, and the last ice sheet vanished from North America and Europe less than 10,000 years ago. The effects of the glaciation are universal. Over large parts of the Northern Hemisphere human culture and industry are profoundly influenced by the glaciation, which stripped away all the soil from large areas, made swampland of others, deposited coarse boulder till in some, and over large areas spread the materials of an uncommonly deep, rich soil (Figs. 18-2 and 18-3). In the glaciated regions ice has given the final touches in the shaping of the modern landscape.

The major ice fields of the Northern Hemisphere are indicated in Fig. 18-4. The Scandinavian sheet spread across the North Sea to cover all but the southern margin of the British Isles, across the Baltic Sea to cover the low countries and Germany to the foot of the Alps, and extended far to the east across Siberia. A separate sheet occupied far eastern Siberia and, of course, Greenland was covered then as now. The maximum of the ice in North America is shown in more detail in Map XIII. The limits of the ice and the direction of movement have been determined by mapping the distribution of the glacial till, especially the marginal moraines that loop across the northern states in great festoons, by plotting the direction of striae on the bedrock, and by noting the distribution of glacial boulders made of distinctive types of rock that can be traced back to their source. The striking fact is that the ice did not spread southward from the polar region but radiated from centers in the latitude of Hudson Bay.

The Laurentide ice sheet, centered over the Hudson Bay lowlands, spread northeastward into the sea and at its maximum may have joined the icecap of Greenland. It spread east

Figure 18-2 Barren lands of the Canadian Shield stripped of soil by glacier ice that gouged out basins now occupied by a maze of lakes. Looking south from Tookaret Lake. (Royal Canadian Air Force.)

Figure 18-3 Fertile wheatland of the Red River Valley about 20 miles south of Fargo, North Dakota. This is a part of the floor of glacial Lake Agassiz. (North Dakota Agricultural Experiment Station.)

Figure 18-4 Map of the major ice fields of the Northern Hemisphere.

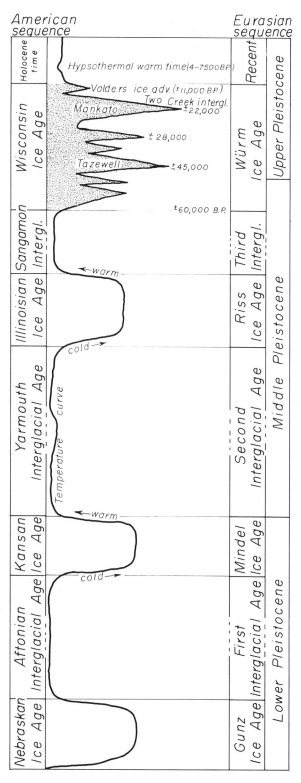

American sequence

Eurasian sequence

Holocene time — Hypsothermal warm time (4–7500 B.P.)

Valders ice adv. (±11,000 B.P.)
Two Creek intergl.
Mankato ±22,000
±28,000
Tazewell ±45,000
±60,000 B.P.

Wisconsin Ice Age — Würm Ice Age — Upper Pleistocene

Sangamon Intergl. — Third Intergl.

Illinoisian Ice Age — Riss Ice Age — Middle Pleistocene

warm ←
cold →
Temperature curve

Yarmouth Interglacial Age — Second Interglacial Age

warm ←

Kansan Ice Age — Mindel Ice Age

cold →

Aftonian Interglacial Age — First Interglacial Age — Lower Pleistocene

Nebraskan Ice Age — Gunz Ice Age

Recent

across Labrador and Newfoundland, overriding the Torngat Mountains of eastern Labrador and the Long Range of Newfoundland, and it spread southeastward over all of New England, leaving a big marginal moraine which is now Long Island. It spread southward for a distance of some 1600 miles across the Great Lakes Region and into the Ohio and the lower Missouri River valleys where summer temperatures now range occasionally up to 100°F. To the west it flowed up the long slope to the foothills of the Canadian Rockies at an elevation of about 5000 feet. Since the direction of flow is determined by the surface slope of the ice, it is evident that in the Hudson Bay region the ice was at least 10,000 feet thick (as it is now in Greenland and Antarctica).

In the Cordilleran Highlands of western Canada a separate center of accumulation covered the mountains, filled the intermont basins, and spilled westward into the ocean and eastward to meet the Laurentide ice sheet.

Smaller ice fields widely scattered over the mountains of western United States are recorded by cirques and great U-shaped valleys. Magnificent Yosemite Valley is only one, for example, on the western slope of the Sierra Nevada.

GLACIAL AND INTERGLACIAL AGES

When erratic boulders strewn over the plains of northern Europe and New England were recognized as the work of a former ice sheet, they were at first quite naturally attributed to a single glaciation. But as the study of the glacial deposits was carried westward into Illinois, Wisconsin, and Iowa, two distinct sheets of drift were found at many places to be separated by old soil, beds of peat, or layers of till that had been leached and decayed (Fig. 18-10). Here the uppermost drift, like that in New England, appeared fresh, but

Figure 18-5 Pleistocene chronology. The space allotted to the several ages is not to scale. The ice advances in the Wisconsin Age are representative but too simple. The terminology differs from state to state and the radiocarbon dates vary somewhat depending on the location. Dates for the earlier ages are still very uncertain.

the buried drift sheet showed the effect of chemical decay and was obviously much the older. Moreover, in places, the soil and peat, or gravels, between two such sheets of till included fossil wood, leaves, or bones, recording the existence of animals and plants of temperate climate. Thus it came to be realized, about 1870, that a continental ice sheet had developed more than once, and that warm interglacial ages had intervened. During these warmer periods the ice melted away, forests grew over the glaciated regions, and plants and animals from warm temperate climates migrated northward even beyond their present limits. Eventually four glacial and three interglacial ages were recognized in the upper Mississippi Valley. Each glacial age was named after a state in which its deposits are well displayed. These are, from oldest to youngest, the Nebraskan, Kansan, Illinoian, and Wisconsin Ice Ages (Fig. 18-5). The interglacial ages are named for localities where their records are well displayed.

While this chronology was being worked out in America, four glacial and three interglacial ages were recognized also in Europe and were given local European names. It is now certain, however, that the glacial ages were essentially synchronous on both sides of the Atlantic.

Records of the Wisconsin Ice Age are so fresh and so widespread that its history has been worked out in great detail, revealing that the ice sheet receded (i.e., melted back) a considerable distance to the north and then readvanced on three successive occasions. As a result, the Wisconsin glacial age is now divided into four glacial and three interglacial subages, as indicated in Fig. 18-5.

In detail the history of the ice sheets is exceedingly complex because the ice did not advance on a simple front but pushed out in great lobes that followed the lower places and in its retreat it paused repeatedly and readvanced locally, leaving an amazingly complex array of marginal moraines. Meltwater repeatedly formed shallow lakes between the ice and the last-formed moraine and local soils formed before different lobes have individual characteristics that depend on the bedrock in

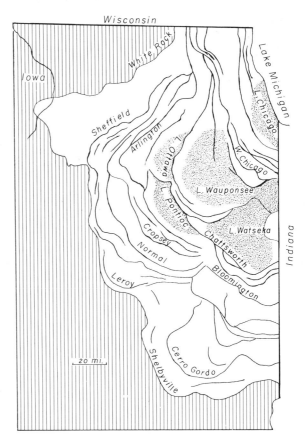

Figure 18-6 Map showing the complexity of moraines and proglacial lakes of a Late Wisconsin drift sheet in northeastern Illinois. Each line represents a marginal moraine. (After Frye et al. [6].)

the source area. Figure 18-6 indicates the complexity created by a single lobe during its final retreat [6]. Meanwhile the record left by another lobe of the ice in another state is equally complex but different in detail.

It should be noted here that since these moraines were formed in sequence as the ice moved northward no two are of precisely the same age and even if precise radiocarbon dates are secured in different places they may show a considerable spread.

METHODS OF STUDY

Because the glacial drift still lies essentially unaltered at the surface, it can be studied in several ways that are scarcely applicable to older formations and it is now known in great detail. A recent summary of Pleistocene geology [11] comprises a volume larger than this book. Our purpose is to present a simplified and more generalized account of the glacial history, but first a brief review of the chief methods of study will be useful.

Radiocarbon Dating. For the last 50,000 years or so the ratio of carbon-14 to carbon-12 will establish the date **in years** at which the organic material (fossils, peat, and muck) associated with the several sheets of till were deposited. These are fairly precise back to 30,000 years or so and less accurate for the older deposits. This gives a detailed chronology for the Wisconsin Ice Age [4]. For the earlier deposits the dates are estimates based on a variety of considerations and are still the subject of controversy.

The absolute time scale for the Wisconsin age is one of the most significant develop-

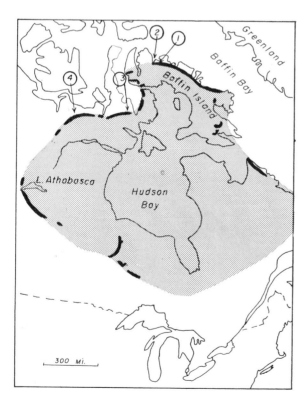

Figure 18-7 Map showing the limits of the Laurentide ice sheet about 8000 years ago. (After Falconer et al. [3]).

ments of recent decades. For example, a log in the base of the last sheet of till near Meriden, Connecticut, represents a tree that was pushed over by the ice on its last advance; radiocarbon proves that the tree was living approximately 32,000 years ago [1].

Radiocarbon dating of varved clay deposited in a lake bordering the ice near Hartford, on the contrary, indicates that the ice had receded this far by approximately 18,000 years ago [8]. And organic matter under the edge of the Mankato till at Two Creeks, Wisconsin, proves that the ice front was still farther north at 11,000 years ago. Similar evidence clearly shows that the ice disappeared very rapidly between 10,000 and 11,000 years ago. For example, a peat bog at Plissey Point in Aristook County, Maine, shows that it was free of ice about 6000 years ago; and radiocarbon dating proves that pine forests had spread over England by 9000 years ago and prehistoric people with a Mesolithic culture were then living there. Figure 18-7 shows the last stand of the Laurentide Ice Sheet. The marginal moraines, shown in black, were mapped in part on the ground and in part from aerial photographs. Fossils at localities 1 to 4 give radiocarbon dates of 7930 ± 300, 8350 ± 300, 8870 ± 140, and 8160 ± 140 years, respectively [3].

Palynology, the Study of Pollen. Leaves and fossil wood are found locally in interglacial deposits but pollen is much more abundant. In the peat and muck associated with the drift pollen can be used to identify the vegetation that existed in the area when they were deposited. This gives a clue to the local climate. For example, the pollen in a bog near Quincy, Illinois, indicates that when it was formed the region was forested with fir, tamarack, pine, and birch—an assemblage that now lives 200 to 300 miles farther north. Evidently the climate in Illinois was colder then than now. In Iowa the sedimentary deposits of the Aftonian (second) interglacial age contain three pollen zones. The lower one resting on the Nebraskan till records a pine forest, indicating climate colder than the present; the middle zone shows that Iowa was then a grassland as at present; and the upper zone shows a return of pine forest before the advancing Kansan ice

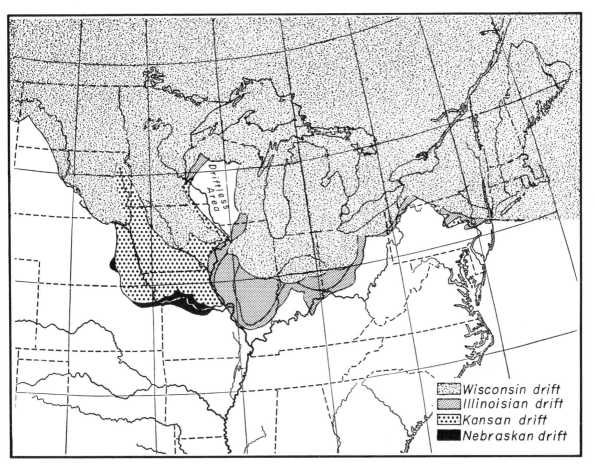

Figure 18-8 Map showing the southern limits of the four major ice advances. (After R. F. Flint.)

sheet. Comparable data are now known from many places but these examples will suffice to illustrate the principle of determining the fluctuations of climate as the ice sheets advanced and retreated.

Where larger fossils are found they supplement the evidence secured from pollen. A particularly interesting case is presented by the Don beds (second interglacial) at Toronto, Canada, which include leaves of the pawpaw and other plants indicating climate appreciably warmer than that of the present.

Proglacial Lake Shores. When the ice was retreating into Canada it still dammed the northward drainage and a series of large proglacial lakes were formed before it, standing long enough for beaches and well defined strand lines to form. Of course they were hori-zontal at the time, being formed at water level, but they now rise toward the north and give a measure of postglacial warping of the crust. This is discussed on page 444.

STRATIGRAPHY OF THE DRIFT SHEETS

In the region of growth, moving glacier ice scours away the loose mantle and leaves a surface of bare rock, but toward its thinning margin it tends to drop its load and may even override loose mantle and soil. Wherever melting balances forward movement for a time and the ice front remains stationary, a marginal moraine builds up to form a record of the ice margin at that particular time; and when the ice sheet becomes stagnant and finally melts away, it leaves a widespread ground moraine. If a succeeding ice advance were more extensive than the first it would destroy the earlier record, but if less extensive it may leave a second drift sheet superposed on the older one. Fortunately, across the Mississippi Valley the Nebraskan Ice Sheet advanced farthest south, the Kansan not quite as far, and the Illinoian and Wisconsin in turn stopped still farther north (Fig. 18-8). Here, then, four distinct sheets of till are superposed in imbricated fashion as ideally represented in Fig. 18-9, and even the interglacial deposits are locally preserved between successive till sheets (Fig. 18-10).

The **Wisconsin** drift sheet, being the youngest, is the most easily mapped and the best understood. Its end moraines loop across the Central Lowlands in festoons that outline the great southern lobes of the ice. Its surface retains the characteristic features of a glacial deposit—the swells and swales of the end moraine, and the broad undulations and lakes and swamps of the ground moraine, locally diversified by drumlins. A thin soil has formed over it and peat has accumulated extensively in the lakes and swales, but chemical decay is limited to slight leaching to a depth of 2 or 3 feet and most of its boulders are as fresh and solid as quarried stone. Except along the main streams its surface has scarcely been altered by erosion and the outwash plains and valley trains beyond the marginal moraines are still easily recognized. Obviously the Wisconsin till was formed in the recent past.

The **Illinoian** drift extends well south of the Wisconsin drift in Illinois, southern Indiana, and central Ohio, and here it is obviously the older. Its surface retains traces of morainic topography but long weathering and mass wastage on the swells, and deposition in the swales, have softened the relief. In contrast to the freshness of the younger till, the Illinoian drift has been extensively oxidized to a depth of 10 to 25 feet and chemical alteration of the fine particles near the surface has produced a distinctive layer of sticky clay, known as **gumbotil,** about 4 feet thick. Obviously the Illinoian drift has been exposed to weathering

Figure 18-9 Idealized block diagram to illustrate the relations of three imbricated drift sheets. The block represents an area many miles across and the vertical scale is greatly exaggerated.

Figure 18-10 Superposed sheets of till; gumbotil, and loess exposed in a railroad cut southwest of Rhodes, Iowa. The scale is indicated by the man standing at the base of the Wisconsin till. (George F. Kay.)

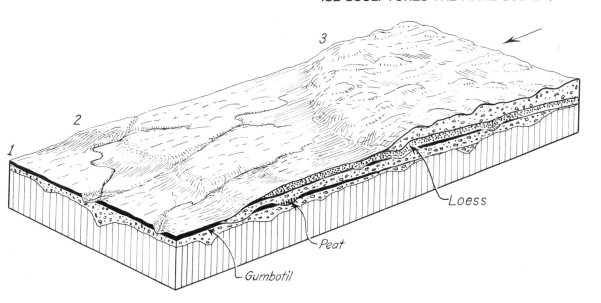

Loess

Peat

Gumbotil

1

2

3

Wisconsin till

Loess

Super-Kansan Gumbo

Kansan till

much longer than the Wisconsin till; and since the gumbotil is present even where overlapped by the Wisconsin drift, it is evident that the long weathering took place before Wisconsin glaciation.

Furthermore, the Illinoian drift is widely covered by a deposit of windblown dust (the Loveland Loess) that passes under the Wisconsin drift and must also have formed before the last glaciation.

Loess plays an important role in glacial stratigraphy in this region. As an ice sheet receded, vegetation followed in its wake but commonly lagged behind, leaving a wide belt of fresh, bare till exposed to the wind. When carried as dust to grass-covered areas the dust settled to form loess. In many places traces of standing grass blades and the shells of land snails confirm this origin. Such beds of loess commonly extend well beyond the till. In this region layers of loess associated with different till sheets can be differentiated by thickness or color or by the contained fauna of land snails and each may serve as a key bed in correlation.

The **Kansan** drift sheet is widely exposed in Iowa, Missouri, and northeastern Kansas where it averages about 50 feet in thickness. It is even more deeply weathered and decayed than the Illinoian, and its capping of gumbotil is much thicker, averaging about 11 feet (Fig. 18-10). In places the Kansan till is covered with outwash gravels and in these the granite pebbles and boulders are so weathered that they crumble and fall to pieces at a blow from a hammer. A thick layer of loess covers the Kansan drift and in several places beds of peat have been preserved between the Kansan and Illinoian till.

The **Nebraskan** drift had about the same distribution as the Kansan, and its exposures are therefore limited, but it lies buried beneath most of the outcropping Kansan till, where it averages 100 feet or more in thickness and rests on an irregularly eroded preglacial surface. The gumbotil on the Nebraskan drift averages about 8 feet thick, and scattered lenses of peat are preserved between it and the Kansan till.

GENERAL EFFECTS OF THE GLACIATION

Migration of Life Zones. As the climate grew colder, and the ice sheets spread over Canada and northern Europe during the last ice age, the mammals of the far north were driven southward and animals of the temperate zone likewise migrated to lower latitudes. At the height of glaciation the reindeer, the woolly mammoth, and the arctic fox ranged widely across Europe, while in North America the reindeer and the woolly mammoth were at home in New England and the musk-ox ranged as far south as Arkansas. With the waning of the ice and amelioration of the climate in postglacial time, these arctic mammals have all retreated to their homeland in the High Arctic barrens as warm temperate species returned northward across Europe and southern Canada.

Fossil mammals found at many places in the interglacial deposits indicate similar mass migrations as the climate changed, and some of these confirm the plant evidence that, at times, the interglacial ages were warmer than the present. In the last interglacial age, for example, such African forms as the lion and the hippopotamus were common in southern Europe.

It is clear, therefore, that the climate was colder than at present during the four glacial ages when ice covered far more than half of North America and approximately half of Europe, and that it was somewhat warmer than at present during at least some parts of the interglacial ages, and that the ice sheets then had retreated far to the north and probably had disappeared both from the mainland of North America and from Europe.

Fluctuations of Sealevel. If the modern ice sheets of Greenland and Antarctica were melted, sealevel would rise about 100 feet, drowning the low coastal plains and transforming the lower courses of many streams into estuaries. If, on the contrary, the former great ice sheets were restored, the water thus withdrawn from the oceans would lower sealevel by at least 300 feet, leaving large parts

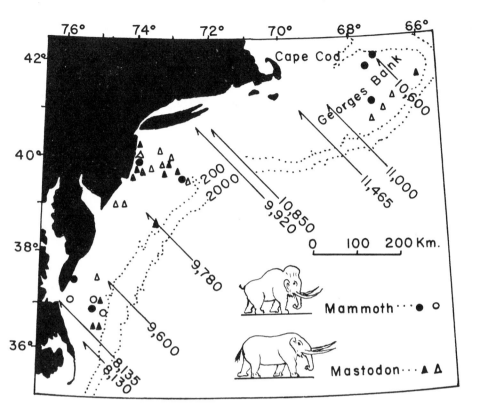

Figure 18-11 Map of localities on the Atlantic Continental Shelf where elephant teeth have been dredged up. Solid dots and triangles mark localities where teeth were found and open dots and triangles mark the approximate location of other finds. Darts point to places where shells or intertidal peat occur; radiocarbon dates of these finds are indicated at the right end of each dart. Obviously at this time the elephants ranged widely over a landscape that was covered with vegetation. (Data from Whitmore et al. [10].)

of the present continental shelves exposed. That this actually happened during the Wisconsin glaciation is demonstrated by recent accounts of the remains of mastodons and mammoths and freshwater peat dredged up off the Atlantic shelf from Cape Hatteras to Georges Bank (Fig. 18-11) [2, 10].

The lowering of sealevel created land bridges that permitted migrations of animals and plants that are impossible at present. England, for example, was united to the continent of Europe so that the hippopotamus crossed the present channel from France; and Sumatra, Java, and Borneo were connected to the mainland of southeast Asia so that elephants, rhinoceroses, and other large mammals crossed

dry-shod the lowlands that are now covered by the Java and Sunda seas. Alaska and Siberia were also united by a land bridge over which the woolly mammoth crossed freely.

Formation of Great Proglacial Lakes. Before the Pleistocene, the basins of the present Great Lakes were lowlands carved by streams flowing northward to Hudson Bay. Some of these streams headed as far south as the upper Ohio Valley and the plains of Minnesota and the Dakotas. As the ice advanced it dammed these streams and thick lobes of the ice followed the basins and deepened them. Figure 18-12 shows three stages in the history of the Great Lakes. For a time the drainage flowed southward into the Mississippi River and from New

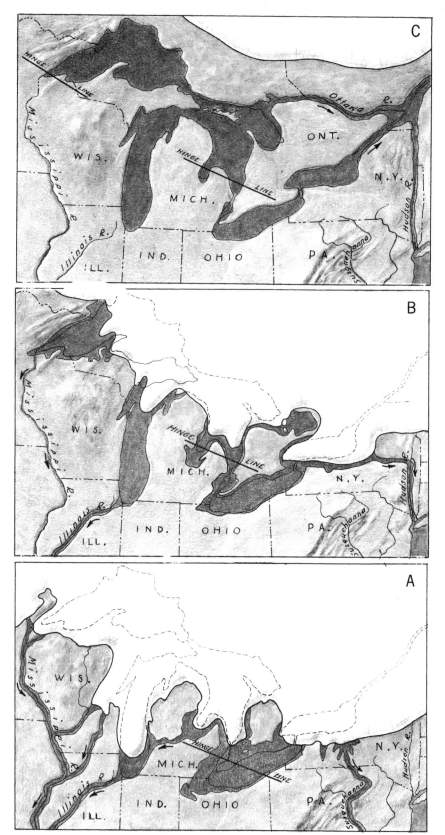

Figure 18-12 Three stages in the development of the Great Lakes during the waning stages of the Wisconsin Ice Sheet. (Adapted from Leverett and Taylor, U. S. Geological Survey.)

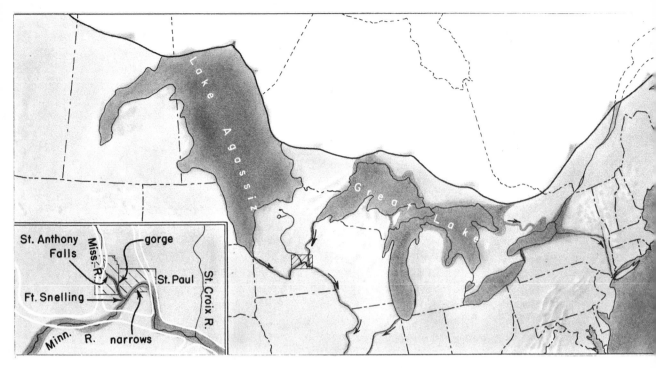

Figure 18-13 Glacial Lake Agassiz and St. Anthony Falls of the Mississippi River at Minneapolis. (Based on data from Leverett and Taylor and from G. M. Schwartz.)

York down the Susquehanna. Then as the ice receded farther the drainage from Michigan eastward found a lower outlet along the front of the ice through the Mohawk Valley of New York and into the Hudson River. Finally when the ice had freed the St. Lawrence Valley two spillways were used, one through the Ottawa Valley and the other over the Niagara Cuesta into the basin of Lake Ontario. It was at this time that Niagara Falls came into existence. At a somewhat later stage the Ottawa Valley was abandoned and all the drainage from the Great Lakes followed its present course.

In the meantime the Red River of the North was still blocked by the ice and the enormous but shallow Lake Agassiz spread widely over Northwest Minnesota, the eastern part of the Dakotas, and much of Saskatchewan (Fig. 18-13). Its outlet was then to the south by way of the Minnesota River, which joined the Mississippi Valley below St. Paul. The preglacial drainage is shown in white on the inset map. The Mississippi River then flowed in a wide valley, its main trunk following approximately the present course of the Minnesota River. When the last ice sheet overrode this area, it filled the old valleys with till, and when the ice finally melted back, the streams at first wandered over the surface of the drift and eventually incised new valleys.

As Minnesota was exposed, three main streams, the St. Croix, the upper Mississippi, and the Minnesota, united near St. Paul. Cutting down through the drift, the trunk stream discovered the old preglacial valley just below St. Paul and proceeded to remove the loose fill below this point with relative ease. But from St. Paul to a point just south of Fort Snelling the new stream was superposed upon the old upland of horizontal limestone beds. Here its downcutting was greatly retarded, and, as it plunged from the surface of the limestone into the old valley, a fall was initiated that gradually migrated upstream until it again cut through into the buried preglacial valley a short distance southwest of Fort Snelling. The recession of this great fall produced narrows in the valley at

St. Paul.

For a time after deglaciation of the region around Minneapolis, all three stream branches carried great volumes of meltwater, but after the ice margin had reached the Canadian border, the drainage through the upper Mississippi diminished to something like its present volume. That of the Minnesota River, on the contrary, increased greatly as it became the outlet of the vast glacial Lake Agassiz, which formed over the Red River Valley while the ice still prevented drainage to the north. Thus, until the disappearance of Lake Agassiz, the stream we now know as the upper Mississippi was but a second-rate tributary to a great glacial stream that followed the present valley of the Minnesota River above St. Paul. The cutting of the narrows at St. Paul was chiefly the work of this great glacial stream and was accomplished while Lake Agassiz

was discharging the meltwater from a great area of the ice. This gives us the clue to the position of the ice margin when the postglacial gorge at St. Paul was cut.

As the ice receded further, Lake Agassiz was finally drained and similar lakes developed farther north in Canada. Along the shores of these wide lakes, beaches and strandlines were formed which can still be followed and mapped. Of course they were horizontal when formed at water level [5].

Depression of the Ice-Covered Regions. The weight of the Laurentide and the Scandinavian ice sheets was sufficient to depress the earth's crust beneath them by large amounts, and since the ice disappeared these regions have been slowly rising again. After the ice had cleared the St. Lawrence Valley shallow sea covered parts of New England and down the St. Lawrence Valley into the

Figure 18-14 Evidence of depression resulting from the weight of the Laurentide ice sheet. Elevation of postglacial strand lines toward Hudson Bay is indicated in feet. The lines are isobases. (Adapted from R. F. Flint [5].)

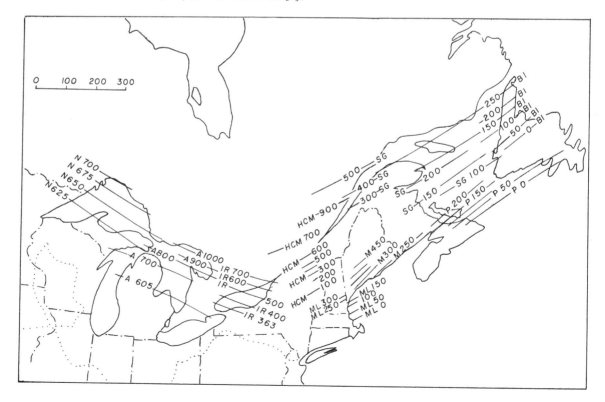

Champlain Valley and the Hudson River Valley so that for a time New England and Maritime Canada were an island. This, the Champlain Sea, spread over the flooded region a layer of clay that yields marine fossils. In this clay at Daveluyville, about 60 miles southwest of Quebec City, the skeleton of a baleen whale was found at an elevation of 275 feet above present sealevel [7] and at the Quebec-Vermont boundary both marine shells and whale bones have been found at 500 feet above sealevel.

Around the southern margin of the ice cap from Lake Superior to Newfoundland the amount of postglacial recovery is measured by the rise of old beaches [5] as indicated in Fig. 18-14. From such data and from theoretical considerations it is suspected that in the Hudson Bay area the depression may have been as great as 3000 feet. To permit such sagging of the crust a vast amount of mantle rock must have flowed laterally from under the basin and the subsequent recovery means that it has since been flowing back. About 1962 evidence of the rate of recovery was discovered near Pelly Bay, Northwest Territory, northwest of Hudson Bay. Here the sea came in while the region was still depressed and cut a conspicuous terrace which is now 540 feet above sealevel (Fig. 18-15). It later cut a lower terrace that is now 290 feet above sealevel and finally a third, now at 175 feet above sealevel. Each of these terraces is covered with beach sand in which fossils are embedded. These have been dated by radiocarbon [9]. They show that the highest terrace was cut about 8370 years ago, the middle one about 7880 years ago, and the lower one 7160 years ago. From these data it is known that during the 490 years between the cutting of the first and second terraces the region rose at an average rate of 5 inches per year. In the 720 years between the two lower terraces the average rise was about 2.15 inches per year; and in the last 7160 years the uplift has averaged only about $\frac{1}{3}$ inch per year. Thus we know that for a time after the disappearance of the ice the uplift was surprisingly rapid and that as the region approached equilibrium the rate decreased progressively. In southern

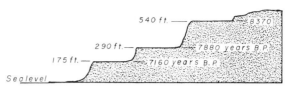

Figure 18-15 Profile section to show the uplifted marine terraces about Pelly Bay, northwest of Hudson Bay.

Sweden where tide gauges have been under observation since 1918 it is found that uplift still continues at the rate of about 1 meter per century.

END OF THE ICE AGES

If our time scale for the Pleistocene Epoch in any sense approaches reality, it is clear that we probably are now in a minor interglacial subage in which the ice fields have shrunk from a maximum of 32 percent of the area of the land surface of the earth to about 10 percent. There are still more than 5,000,000 square miles of ice sheets in Antarctica and Greenland, and the polar seas are choked with floe ice, while lofty mountains the world around bear active glaciers. Furthermore, it is estimated that a decline in the mean annual temperature of not more than 5°C would bring a return of the ice sheets as they were during the last advance [5].

Judged by every criterion we know, the interglacial ages vastly exceeded the time that has elapsed since the last ice sheets began to wane; and fossils preserved between the drift sheets prove beyond doubt that the climate at times during those interglacial ages was appreciably warmer than it is now. Furthermore, there is clear evidence that postglacial world climates reached a maximum of warmth between 6000 and 4000 years ago and since then, with minor oscillations, have become cooler and more moist down to the present time. Whether the ice sheets will spread again or disappear completely during the next few thousands of years is quite impossible to judge; but clearly the **Holocene** is only an age in the Pleistocene Epoch.

Meanwhile the sword of Damocles hangs

over us. If the ice sheets should again spread to the limits they occupied a few thousands of years ago, mass migrations would occur on a scale without precedent in the history of mankind, for the densely populated centers of Europe and the United States, to say nothing of all Canada, would slowly become uninhabitable. And, if, on the contrary, the climate should return to its geologic norm and the last of the ice sheets should disappear, the meltwater would raise sealevel by about 100 feet, slowly submerging all the great seaport cities of the world. In any event, the changes will come too slowly to concern anyone now living, but they may profoundly shape the destiny of civilization within the next few thousands of years.

We can only guess when the end of the Ice Ages will come, as we contemplate some of the problems it will entail for mankind.

REFERENCES

1. Barensen, G. W., Deevey, E. S., Jr., and Gralenski, L. J., 1957, *Yale natural radiocarbon measurements, III.* Science, v. 126, pp. 908–919.
2. Emery, K. O., Wigley, R. L., Bartlett, A. S., Rubin, Meyer, and Barghoorn, E. S., 1967, *Freshwater peat on the Continental Shelf.* Science, v. 158, pp. 1301–1307.
3. Falconer, G., Andrews, J. T., and Ives, J. D., 1964, *Late-Wisconsin end moraines in northern Canada.* Science, v. 147, pp. 608–609.
4. Flint, R. F., and Deevey, E. S., Jr., 1951, *Radiocarbon dating of late-Pleistocene events.* Am. Jour. Sci., v. 249, pp. 257–300.
5. Flint, R. F., 1957, *Glacial and Pleistocene Geology.* New York, John Wiley & Sons, 553 pp.
6. Frye, J. C., Willman, H. B., and Black, R. F., 1965, *Outline of glacial geology of Illinois and Wisconsin.* In *The Quaternary of the United States.* Princeton, N. J., Princeton Univ. Press, pp. 43–61.
7. Laverdiére, J. W., 1950, *Baleine fossile de Daveluyville, Quebec.* Canadian Naturalist, v. 77, pp. 271–282.
8. Urry, W. D., 1948, *The radium content of varved clay and a possible age of the Hartford, Connecticut, deposits.* Am. Jour. Sci., v. 246, pp. 689–700.
9. Trautman, M. A., and Walton, Alan, 1962, *Radiocarbon measurements II,* Radiocarbon, v. 4, pp. 35–42.
10. Whitmore, F. C., Emery, K. O., Cooke, H. B. S., and Swift, D. J. P., 1967, *Elephant teeth from the Atlantic Continental Shelf.* Science, v. 156, pp. 1477–1480.
11. Wright, H. E., Jr., and Frey, D. G., Eds., 1965, *The Quaternary of the United States.* Review Vol. 7th Int. Cong. Intern. Assoc. Quat. Research. Princeton, N. J., Princeton Univ. Press.

Figure 19-1 Little *Eohippus,* the three-toed ancestor of the modern horse, stood only about one foot high at the shoulders. The fossil record of his descendants is one of the most complete in the evolutionary history of the mammals. (From a painting by Charles R. Knight in the American Museum of Natural History.)

CHAPTER **19** *Mammals Inherit the Earth*

Throughout the long Mesozoic Era—for more than a hundred million years—reptiles completely dominated life on the earth; but at its close their dynasty suddenly collapsed. Turtles, lizards, snakes, and crocodiles survived, but they are mostly small and restricted in their range, and few are aggressive. At the dawn of the Cenozoic Era the dinosaurs were extinct, the pterosaurs had disappeared, and the great marine reptiles were gone. Now the way was open for the mammals to begin their conquest of the world. At first they were small and unimpressive, but with a meteoric expansion they soon dominated the earth as completely as the reptiles had done in their day. Thus the Cenozoic has well been called **The Age of Mammals.**

TRENDS OF MAMMALIAN EVOLUTION

Comparative study of early Cenozoic fossils clearly indicates that the first mammals re-sembled the modern hedgehog (Fig. 19-2) in the following respects: (1) they were small; (2) they were short-legged and walked on the soles of the feet; (3) they had 5 toes on each foot; (4) they had 44 teeth of which all but the canines were short-crowned; (5) their brains were small and their intelligence was of a low order; (6) they were long-faced, the jaws exceeding the brain case in size. Unlike the hedgehog, however, they had a long tail. From ancestors of this sort all the modern orders of mammals evolved. In this development four major trends can be detected in most of the groups.

Increase in Size. This was a common trend. The Eocene ancestor of the horse was scarcely larger than a fox, that of the camels no bigger than a jack rabbit, and even the elephants can be traced back to ancestors no larger than a hog. The Eocene forebears of man were tiny half-apes of the size of squirrels. In short, the early Cenozoic mammals, like those of the Mesozoic, were mostly small. As the modern

Figure 19-2 Skeleton of a modern hedgehog, a persistently primitive and generalized placental mammal. Note that the feet are five-toed and plantigrade (i.e., the animal walks on the entire sole of the feet with heel and wrist touching the ground). (Yale Peabody Museum.)

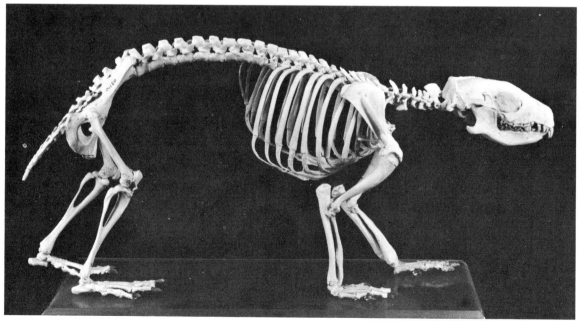

orders evolved, each in time attained larger average size, and several of them produced relative giants. Progress in this regard was not synchronous among the different groups of mammals, however, some having reached their maximum size before the middle of the Cenozoic Era and others much later.

Specialization of the Brain. Unquestionably the most significant trend was specialization of the brain. The earliest mammals, like their reptilian ancestors, had relatively small brains concerned chiefly with the physical senses. Their mentality was probably comparable to that of a modern shrew or an opossum. Progress involved increase in actual size as well as in the ratio of brain to body weight (Fig. 19-3). More significantly, it involved a disproportionate increase in the cerebrum, which is the seat of memory and reason. In this respect, however, progress has been very unequal among the several orders of mammals. At one extreme stand the insectivores (shrews, hedgehogs, and moles) that have changed but little and have survived as stupid, retiring creatures, and at the other extreme stands the dominant primate (man) with skulls grotesquely distended to house a brain capable of pondering the mysteries of time and space and of harnessing atomic energy!

Specialization of the Teeth. This was a third important development. In the primitive placental mammal the cheek teeth had sharp piercing or shearing cusps but were low-crowned and attained full growth early in life. The insectivores and opossums, having stuck to a diet of insects and other delicate food, show little change (Fig. 19-4); but most other groups of mammals have teeth highly specialized according to their feeding habits. Omnivorous feeders such as the swine, the bears, and man show only moderate specialization and have short-crowned jaw teeth with low blunt cusps. Carnivores, on the contrary, show greatly enlarged canines for holding and tearing flesh, and narrow, shearing jaw teeth. The most remarkable specialization is seen in the grazing animals of the prairies, whose teeth must resist the wear of the harsh and commonly dusty grasses while maintaining a rough grinding surface. In these the

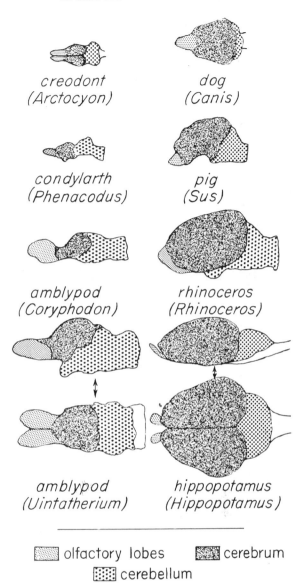

creodont (Arctocyon)

dog (Canis)

condylarth (Phenacodus)

pig (Sus)

amblypod (Coryphodon)

rhinoceros (Rhinoceros)

amblypod (Uintatherium)

hippopotamus (Hippopotamus)

olfactory lobes   cerebrum

cerebellum

Figure 19-3 Brains of archaic mammals on the left paired with modern mammals of comparable body size on the right. (Adapted from H. F. Osborn.)

cheek teeth are large and high-crowned and continue to grow throughout life (Fig. 19-13). Furthermore, the enamel is deeply infolded into the crown, so that even after wear it forms sharp ridges, thus maintaining a good grinding surface. Each order has a distinctive pattern of infolded enamel, making it almost as easy to identify the order of a fossil mammal by its teeth as by a whole skeleton (Fig. 19-5).

Specialization of the Feet and the Limbs. This was a fourth major change. Primitively the feet were short, and the animal walked flat on the soles with the heels touching the ground (Fig. 19-2); but, during the ages, marked specialization occurred according to environment and habits. Tree-dwelling types such as monkeys and squirrels developed prehensile hands with opposable thumb and great toe; carnivores evolved claws to seize and hold struggling prey; and the herbivorous ani-

Figure 19-4 Skull and lower jaw of an opossum showing the character of primitive marsupial mammalian dentition. Note that there are 48 teeth, each jaw including 8 incisors, 2 canines, 6 premolars, and 8 molars, and that each molar is subtriangular bearing 3 blunt cusps. Primitive placental mammals had only 6 incisors in each jaw and the total number of teeth was only 44. (Yale Peabody Museum.)

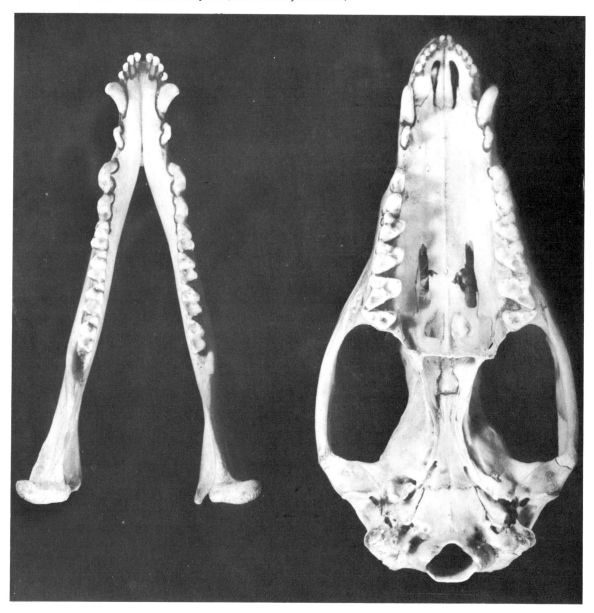

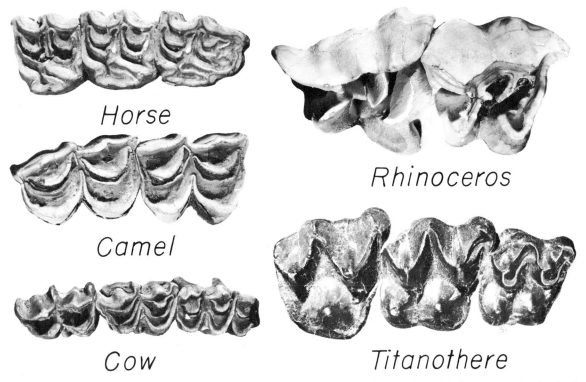

Horse

Camel

Cow

Rhinoceros

Titanothere

Figure 19-5 Crown view of left upper molar teeth of horse, camel, cow, rhinoceros, and titanothere, to show the distinctive patterns of the enamel folds. All on the same scale, about ×½. (Yale Peabody Museum.)

mals of the plains, dependent on fleetness of foot for safety, underwent a remarkable specialization involving a rise onto the very ends of their toes, the development of hoofs, and a reduction and loss of some or all of the side toes.

In swift running, an animal tends to rise up on its toes to attain a longer stride, and those endowed with long, slender limbs have a natural advantage. In the struggle for existence, therefore, many of the plains animals were subjected to an age-long selection in which a great premium was placed on longer limbs and the ability to remain on the toes. The ultimate result is illustrated by the limb of a modern horse, contrasted with the unspecialized limb of a bear (Fig. 19-6). Above the heel there is little difference in size and proportions, the greater height of the horse being due to the elongation of the foot and to the fact that he stands on tip-toes with the heel far above the ground. This affords a great stride and leaves the powerful leg muscles bunched near the body where they can swing

Figure 19-6 Hind limb of bear (left) and horse, showing correspondence of parts.

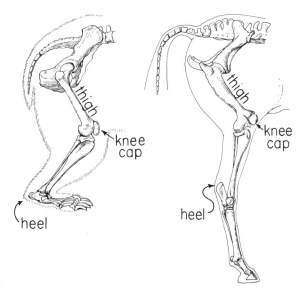

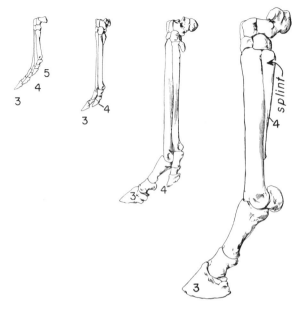

Figure 19-7 Loss of lateral digits in the horse. From left to right, the forelimbs of *Eohippus* of the lower Eocene Epoch, *Mesohippus* of the Oligocene, *Merychippus* of the Miocene, and *Equus,* the modern horse. Corresponding digits bear the same number.

the slender extremity of the limb without sharing in its motion. The hoof developed, of course, as a protective armor for the tip of the toe.

Since the middle toes of a mammal are primitively and normally longer than the side toes, a rise to the ends of the digits lifts the side toes off the ground and leaves them dangling. In this condition they tend to degenerate and disappear (Fig. 19-7). In the odd-toed, hoofed mammals (Order Perissodactyla), the axis of the foot lies in the middle digit and, in the reduction, digits 1 and 5, as a rule, disappeared, leaving a three-toed foot. Heavy-bodied types such as the rhinoceros did not proceed further in this direction, but in the horse, digits 2 and 4 were reduced to mere vestiges, leaving a one-toed foot. In the cloven-hoofed mammals (Order Artiodactyla), the axis of the foot lies between digits 3 and 4. In this group, digit 1 was lost at a very early stage, producing a four-toed foot, as in the hog. Further specialization caused a simultaneous reduction and loss of digits 2 and 5, leaving a two-toed foot, such as that of the cattle, the deer, and the camel.

THE PALEOCENE VANGUARD

At the beginning of Cenozoic time, the mammals expanded like a race delivered from bondage. Although only a few orders are recorded from Late Cretaceous rocks, 14 are now known from the Paleocene Series. Notable among these are the **multituberculates, marsupials, insectivores,** and possibly **primates** that survived from the Cretaceous, and the **rodents, carnivores, condylarths,** and **amblypods** that first appeared in the Paleocene.

The **multituberculates** (Fig. 19-8) were small plant feeders with chisel-like incisors and large jaw teeth crowned with long rows of blunt tubercles. In size and appearance they probably resembled a modern woodchuck; but the resemblance is superficial. Although common in the Paleocene Epoch they died out early in the Eocene Epoch.

The **marsupials** were essentially like the modern opossum and at this time were possibly worldwide in distribution. Outside of Australia and South America, however, they have never competed successfully with more progressive types, and have never risen above the modest and retiring role played by the present opossum.

Insectivores were likewise small and much like the modern shrew and hedgehog. They appear to be the ancestral stock from which the other groups of placentals evolved, but did not themselves share in that evolutionary advance.

Rodents include the modern gnawers, such as mice and rats and squirrels. The oldest representative of this order was found in the Paleocene beds of Montana in 1937.

Primates are represented in the Paleocene by small half-apes (tarsioids and lemurs) scarcely larger than squirrels.

Creodonts were precursors of the modern carnivores. Even in Paleocene time they showed considerable specialization, some being doglike and other catlike. Some had shearing teeth and sharp claws; others strangely blunt teeth and flattened toenails.

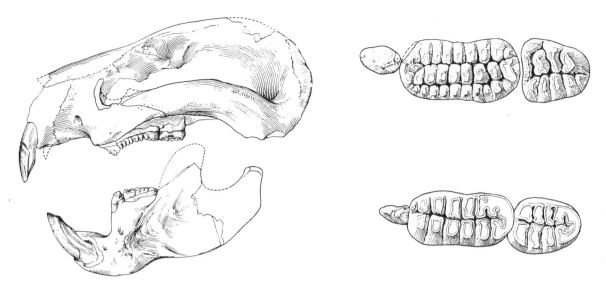

Figure 19-8 Skull and teeth of a Paleocene multituberculate, *Taeniolabis*. Left side view of the skull (×½) and crown view of the lower and upper left cheek teeth (enlarged). (After Granger and Simpson.)

Figure 19-9 Restoration of *Phenacodus*, a condylarth from the Paleocene of Wyoming. The grasslike vegetation consists of sedges, not true grasses. (From a painting by Charles R. Knight, American Museum of Natural History.)

Figure 19-10 Restoration of *Coryphodon,* an amblypod from the Lower Eocene of Wyoming. (From a painting by Charles R. Knight, American Museum of Natural History.)

Their brains, however, were less than half as big as those of modern carnivores of equal stature, and they must have been stupid brutes (Fig. 19-30). Of these the **condylarths** and **amblypods** (primitive ungulates) were the dominant orders of herbivorous animals during Paleocene time and both ranged up into the Eocene. The condylarths were light-bodied and relatively agile, the amblypods stocky and ponderous. *Phenacodus* of the lower Eocene (Fig. 19-9) was a typical condylarth. The slender body, arched back, long tail, and short, five-toed feet give it a super-ficial resemblance to the carnivores; but each toe bore a small hoof, and its teeth were clearly those of a plant feeder. The brain was relatively small, and the teeth primitive and low-crowned. The condylarths appeared early in the Paleocene and ranged upward to the middle of the Eocene Epoch, when they were replaced by the more advanced un-gulates.

Coryphodon (Fig. 19-10) of the lower Eocene, a typical amblypod, was about waist-high to a man. It was thickset and had stout legs and blunt, five-toed feet, each toe bearing a hooflike nail. The canine teeth were tusk-like, but the cheek teeth were relatively small and low-crowned. The amblypods appeared early in the Paleocene and ranged through to the close of the Eocene Epoch. The earliest forms were scarcely larger than a sheep, but they increased rapidly in size and culminated in *Uintatherium*, which had the bulk of a circus elephant and was the largest of the American land animals during the late Eocene time.

In short, the Paleocene faunas would have presented a strange, unfamiliar appearance to a modern, for the dominant forms belonged to groups that are long since extinct, and many of the groups that are now dominant were completely lacking.

EOCENE IMMIGRANTS

At the beginning of Eocene time the ancestors of the modern horse, the rhinoceros, the camel, and other modern groups of mammals appeared simultaneously in Europe and the United States. This sudden advent implies that these modernized stocks had been evolving somewhere in the northern landmass and at this time migrated southward along two

Figure 19-11 Six stages in the structural development of the horse represented by skulls, front legs, and feet. All are shown at the same scale. The forelimb indicates the relative height of each at the shoulder and the skulls are so placed as to indicate the relative height of the head. (Yale Peabody Museum.)

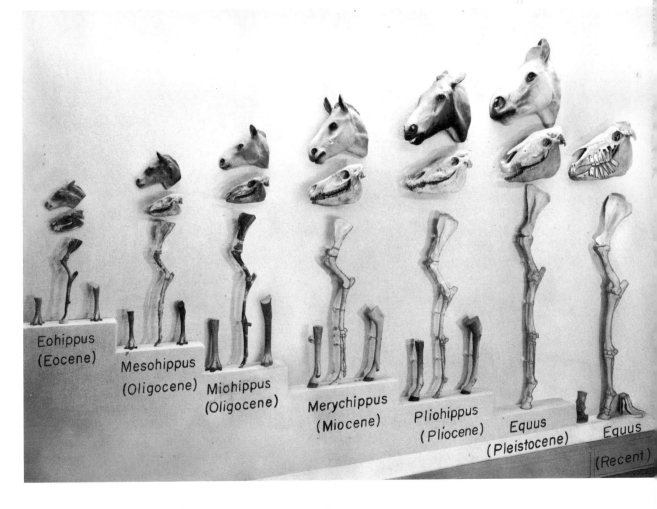

Eohippus (Eocene)
Mesohippus (Oligocene)
Miohippus (Oligocene)
Merychippus (Miocene)
Pliohippus (Pliocene)
Equus (Pleistocene)
Equus (Recent)

Figure 19-12 Restoration of three specimens of *Eohippus* from the Wind River (Eocene) beds. (A painting by Charles R. Knight in the American Museum of Natural History.)

different routes. From this stage on, the history of several of these stocks can be followed in detail; it constitutes one of the most fascinating chapters in the history of life.

CENOZOIC PARADE OF THE MAMMALS

Eohippus and His Progeny. The horse was a native of North America from Early Eocene to Late Pleistocene time and underwent most of its evolution here. Skeletons assembled from successive horizons reveal a gradual development in teeth, limbs, feet,

and size hardly equaled in any other stock of animals. The record is graphically shown in Fig. 19-11.

Eohippus, the "Dawn Horse." The oldest known member of the race was *Eohippus* (or *Hyracotherium*), the "dawn horse." This graceful little animal, scarcely a foot high, had a slender face, arched back, and long tail (Fig. 19-12). Its hind feet bore three toes, and the front feet four toes. Its Paleocene ancestor, we may infer, possessed five toes all around, but no such stage has yet been discovered.

The evolution that followed was long and

complex but may be epitomized by examples from three of the structural levels intermediate between *Eohippus* and the modern horse, *Equus*. The first of these, *Mesohippus* of the Oligocene, was about the size of a sheep (Fig. 19-11), had three toes on each foot, subequal in size and all touching the ground, so as to share equally in the animal's weight. Its cheek teeth were still low-crowned, as were those of *Eohippus*. *Merychippus* of the Miocene grew to the size of a small pony. It possessed three toes on each foot, but the middle toe was much the largest, the others failing to touch the ground and dangling like the dewclaws of cattle. The jaws of this little horse had deepened appreciably, for its molar teeth were becoming high-crowned and prismatic (Fig.

19-13). *Pliohippus*, the first one-toed horse, was somewhat larger than *Merychippus*, and had high-crown teeth and long jaws approximating the condition seen in a modern horse. The side toes were represented only by a pair of splint bones lying alongside of the cannon bone (Fig. 19-7) and invisible externally. The modern horse, *Equus*, appeared about the close of the Pliocene Epoch and survived in America until the last ice sheet of the Pleistocene had disappeared. Wild horses roamed the American plains in great herds until late in the epoch and then for some unknown reason (possibly an epidemic like the modern hoof-and-mouth disease or sleeping sickness) became extinct. Meanwhile, fortunately, they had spread to the Old World

Figure 19-13 Horse dentition. Across the bottom a molar tooth of each of the six genera of horses shown in Fig. 19-11, all at the same scale. That of *Eohippus* had a low crown like that of a modern hog (upper left). Although the teeth increased in size with increase in the size of the body, they remained low through the Oligocene Epoch and then began a rapid increase in height in the Miocene as the horses became adapted to the spreading prairies. The dissection of the skull of a modern horse showing the teeth in position explains why the horse has such an elongate face and deep jaws. (The skull after Chubb, American Museum of Natural History; other specimens from Yale Peabody Museum.)

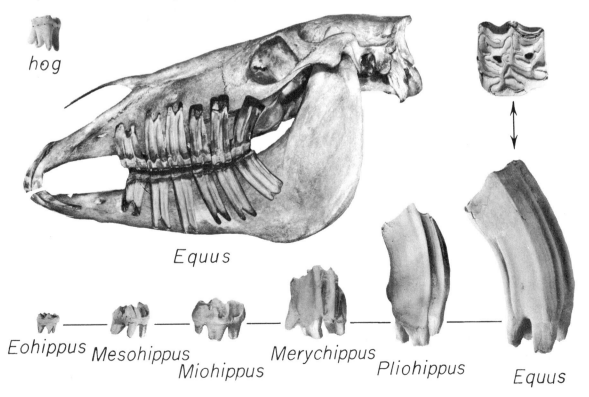

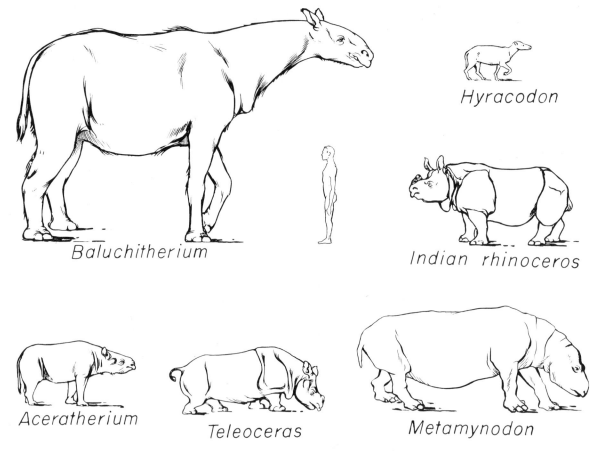

Figure 19-14 Extinct rhinoceroses facing right and a modern rhino facing left and the outline of a man. All are drawn at the same scale to show relative sizes.

(probably via Alaska and Siberia), where they survived to become a servant and a friend of man. The wild American horses of historic times are descendants of those brought over by the Spaniards during their early conquests.

Rhinoceroses. The rhinoceros (Fig. 19-14) also is primarily of North American stock. The group first appeared near the beginning of the Eocene and by early Oligocene time was abundant and specialized into three distinct tribes: (1) the **true rhinoceroses**, which gradually developed into modern types; (2) the **running rhinoceroses**, which were small, light-bodied, and fleet-footed; and (3) the **amphibious rhinoceroses**, which were semi-aquatic and, like the hippopotamus, became thick-bodied and very short-legged. The last two stocks died out during Oligocene time,

but true rhinoceroses were very common in the Great Plains region during Miocene time and were then more varied than they are today in East Africa. One of the striking Miocene forms was *Teleoceras*, a barrel-chested rhinoceros with extremely short legs (Fig. 19-18). In America the rhinoceroses declined to extinction in the Pliocene Epoch, but those that migrated into the Old World survived.

The early rhinoceroses were small and hornless. The true rhinos were still scarcely 3 feet high in Oligocene time, although one of the amphibious tribe then reached a height of 6 feet and a length of 14 feet. In America few of the Cenozoic forms were as large as modern species, but in Asia an aberrant stock of hornless giants developed during Oligocene and early Miocene time and in *Baluchitherium*

(Fig. 19-14) attained the largest size of any known land mammal of any age, standing about 18 feet high at the shoulders, and at least 25 feet long.

Titanotheres. Titanotheres constitute another magnificent tribe of mammals, remotely related to the rhinoceroses and the horses. They were ponderous beasts of rhinoceroslike appearance, many of them with great nasal horns made of bony outgrowths from the skull. Early Eocene titanotheres were scarcely larger than a big hog and were hornless (Figs. 19-15 and 19-30), but the tribe developed rapidly to great size before its extinction about the middle of the Oligocene Epoch. One of the latest was *Brontotherium,* which stood about 8 feet high at the shoulder and far outbulked the largest living rhinoceros. During Oligocene time this was the largest land animal in America.

Titanotheres are known only from the United States, Mongolia, and Europe. They left no descendants, either collateral or direct. Like the rhinoceros, they possessed three toes on each hind foot and four toes on each front foot. The great weight probably prevented further reduction of the digits in either

of these stocks of plains-dwelling mammals.

Chalicotheres. Perhaps the strangest of all the odd-toed mammals were the chalicotheres, a group now extinct, but represented by *Moropus* (Fig. 19-16) in the Great Plains region during Miocene time. The skull of this grotesque creature was shaped like that of a horse, but its body was deep and stocky like that of a camel, and its feet bore narrowly compressed claws. There were three toes on the hind feet and three plus a vestige of the fourth on the front feet. This is the best-known American form, and its remains are not rare in the Miocene beds of Nebraska. Chalicotheres were present also in Europe and Asia and are known to have lived from late Eocene to late Pliocene time.

Camels. Camels underwent a long evolution remarkably paralleling that of the horse. One of the earliest genera, *Protylopus* of the upper Eocene, was a slender, four-toed creature scarcely larger than a jack rabbit. By Oligocene time camels had attained the size of sheep and were associated with *Mesohippus* of the same size. These little camels displayed their true affinities in the enamel pattern of their teeth and in the peculiar carriage

Figure 19-15 First and last of the titanotheres. Right, *Eotitanops* of the Eocene Epoch; left, *Brontotherium* of the Oligocene. The latter stood about 8 feet high at the shoulders. (Adapted from a figure by H. F. Osborn.)

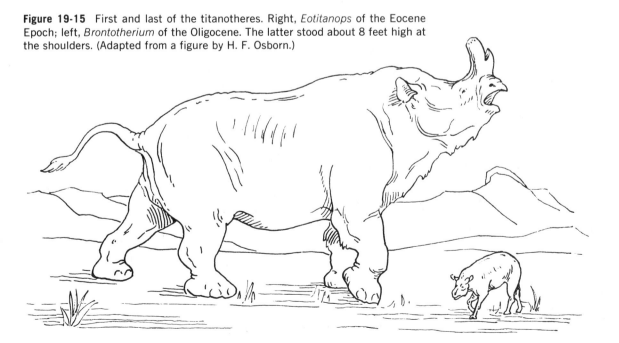

Figure 19-16 Early landscape in Nebraska showing entelodonts, *Dinohyus,* in the foreground, *Moropus,* a chalicothere browsing at the right, and a herd of oreodons at the far left. (A painting by Charles R. Knight in the Chicago Natural History Museum.)

of their heads. The toes had by this time been reduced to two on each foot, but were still free.

In Miocene time the camels diverged into several tribes (Fig. 19-32). The main line continued through *Procamelus* of the Miocene into *Camelops* of the Pliocene and Pleistocene. The latter survived until comparatively recent time in southwestern United States. A larger genus (*Gigantocamelus*) inhabited the Great Plains during Pleistocene time and reached a height of 7½ feet at the shoulders, carrying its head about 9 feet above the ground.

Among the divergent stocks that appeared in the Miocene were the small and very slender "gazelle camels" and the long-necked "giraffe camels."

Like the horses, the camels lived through the several ice ages in America and then for some unknown reason died out before the arrival of the white man. Meanwhile, however, they had migrated to both South America and Eurasia where they still survive (the alpaca and the vicuna of South America and the dromedary and two-humped camel of the Old World).

Oreodons. Among the commonest Mid-Cenozoic animals of western United States were the **oreodons** (Figs. 19-16 and 19-31). These small creatures are not closely allied to any living animals and are therefore difficult to characterize in nontechnical terms. Most of them were of the size of sheep or goats. Although they appeared long-bodied and short-legged like a hog, this resemblance was quite superficial. In a sense they were remote cousins of the camels, for they had similar teeth, they were even-toed, and they were cud-chewers. They were both browsers and grazers, and, if we may judge by the extraordinary abundance of their remains in the Big Badlands of Dakota, they roamed the plains in vast herds. They appeared in late Eocene time, reached a climax in the Oligocene, and persisted into the early Pliocene before dying out, but during all this time they were strangely conservative, retaining four toes (some a small fifth) and short legs, keeping their low-crowned teeth, and failing to increase notably in size. Their short-leggedness was a mark of conservatism; their toes had failed to lengthen as rapidly as those of most other plains animals. So far as is known, oreodons were confined to North America.

Entelodonts. The **entelodonts**, or "giant pigs," were another group of even-toed mammals that assumed a spectacular role for a short time during the Mid-Cenozoic (Figs. 19-16 and 19-32). They were remote cousins of the swine and, like the latter, were adapted for rooting and grubbing in the forest. They are characterized by a large bony extension from the zygomatic arch of the cheek, a structure whose function is entirely problematic. They appeared during Oligocene time, reached their greatest size (6 feet high at the shoulder) in the early Miocene, and then died out.

Bovids. Cattle, sheep, and goats belong to the family Bovidae, which also includes the bison, the musk-ox and the antelopes. In spite of superficial differences, these animals are closely related and have many peculiarities in common. Among other things, they all lack front teeth in the upper jaw, and they possess true horns with an unbranched bony core covered by a horny sheath. Since the beginning of civilization this great family has contributed more than any other to human welfare. To hunting peoples it has been a source of food and of clothing, tents, and thongs. Indeed, the very beginning of civilizations is closely linked with the domestication of cattle, goats, and sheep, and the tending of flocks.

Unlike the horses and camels and rhinos, this family is essentially an Old World stock. The oldest known forms appeared in Eurasia late in Miocene time, having evolved from stocks now extinct. They became highly diversified there in the Pliocene and reached their modern estate during the Pleistocene Epoch, the cattle apparently having developed out of antelopes similar to the gnu. During the Ice Age most of them migrated out of Europe, finding a more suitable environment in the plains of Asia and Africa, but only a few managed to reach America. The buffalo is one of the exceptions. It probably arrived via the Bering land bridge about the beginning of Pleistocene time, soon became enormously abundant, and developed into numerous species, some much larger than present forms. The musk-ox also reached North America in Pleistocene time, having no difficulty in crossing the snowfields of the North.

Elephants and Their Kin. Elephants of many sorts ranged over Europe, Asia, Africa, and North America in Pleistocene time; and, while the last glacial ice was waning, they were still more common in eastern United States than they are now in East Africa.

Compared with the ancestral placental mammal, the elephants show amazing specialization in several respects. Not the least of these is the fusion of nose and upper lip to form the trunk or **proboscis,** from which this order takes the name **Proboscidia.** The earliest known proboscidians are found in North Africa, which appears to have been their ancestral home. Several species of the genus *Moeritherium* (Fig. 19-17) have been discovered in the Late Eocene and Early Oligocene beds of the Fayum Desert not far west of Cairo. When these beds were forming, the climate was humid in North Africa, and the Fayum area was occupied by the delta of an ancient Nile. The moeritheres were thick-set animals scarcely waist-high to a man; they were probably semiaquatic, living like the hippopotamus along the rivers. They had neither tusks nor trunk and showed little superficial resemblance to an elephant, but teeth and skull betray the beginning of specializations that indicate their relations. The head was long and the upper lip was probably prehensile as in a tapir. Among the front teeth the second incisors were enlarged. These were the ones that developed into tusks in later proboscidians. The limbs were quite short and stout but not otherwise specialized. In the same region in the early Oligocene deposits appeared a more advanced type, *Phiomia* (Fig. 19-17, 2), which was about shoulder-high to a man. Its skull was still low and elongate but its second incisors were elongated and directed forward and a short trunk had developed by the fusion of the snout and the upper lip.

Out of this stock developed an amazing variety of proboscidians known as **mastodons** because their enlarged jaw teeth bore pairs of large blunt cusps shaped like human breasts [Gr. *mastos*, breast + *odons*, tooth]. *Phiomia* possessed only 3 pairs of cusps on each tooth but many later relatives added one or more pairs of cusps and in these each transverse pair tended to become united by a ridge. Most of the mastodons bore tusks in both lower and upper jaws. In some the upper pair were larger than the lower, but in others the lower pair became the larger and in some they were extremely specialized. As an example note the long-faced four-tuskers in Fig. 19-18.

Mastodons migrated widely over Eurasia during Miocene time and late in the epoch reached North America over a Siberian-Alaskan land bridge. Among these were four-tuskers and shovel-tuskers, both of which died out late in Pliocene time while the less specialized two-tuskers (Fig. 19-33) lived on in America until about 6000 years ago when they may have been exterminated by primitive

Figure 19-17 Restoration of proboscidian heads, showing stages in the development of tusks and trunks. 1, *Moerotherium*, from the early Oligocene of the Fayum Desert; 2, *Phiomia*, from the later Oligocene of the same region; 3, modern elephant. T, incisor tooth; L, upper lip; N, nostril. (After a drawing by Charles R. Knight, American Museum of Natural History.)

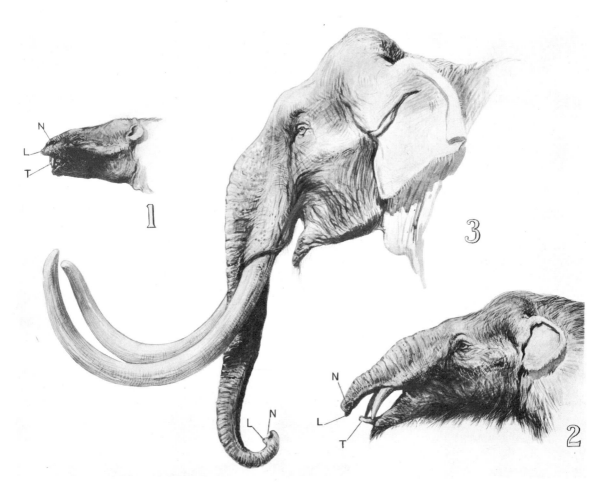

Figure 19-18 A late Miocene landscape in Nebraska showing at the left the short-legged rhinoceros, *Teleoceras,* and at the right a pair of four-tusked mastodons. (A painting by Charles R. Knight, Chicago Natural History Museum.)

Figure 19-19 Molar tooth of a mastodon (left) compared with that of an elephant (right). (Yale Peabody Museum.)

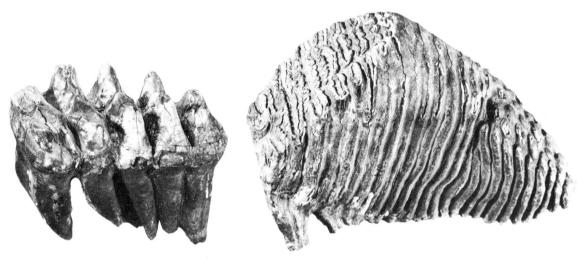

Indians. The mastodons were browsers living in the timbered regions, and their teeth remained relatively unspecialized. The American mastodon, *Mastodon americanus,* ranged widely over the United States during the waning stages of the last Pleistocene ice sheet and its remains are common in the peat and muck of postglacial lakes in which they had become trapped.

The elephants represent a specialized offshoot from the mastodon stock, adapted to a grazing habit which led to an amazing specialization of the jaw teeth used in grinding the harsh grasses. This involved the multiplication of the cross-ridges until they formed relatively thin transverse plates of deeply infolded enamel (Fig. 19-19). So large have these teeth become that there is room for only one on each side of each jaw and as a result the set of six teeth that normally should appear simultaneously came to be inserted in sequence during the life of an individual. As one pair is worn out the next pair grows up to push it out of the jaws and take its place. Of this family, the Elephantidae, the genus *Mammuthus*, reached North America during Pleistocene time and differentiated into several species. The best known of these is the woolly mammoth, which lived on the tundra surrounding the ice fields and ranged across both Eurasia and North America. Frozen carcasses in Siberia show that it bore a thick coat of brownish woolly hair (p. 34 and Fig. 19-34).

Other species inhabited the warmer regions, particularly the plains of the central and southwestern states. One of these, the imperial mammoth of the Southwest, attained a height of 13 or 14 feet at the shoulders and bore tusks as much as 13 feet long. In their present fossil state, a pair of such tusks weighs almost half a ton!

Carnivores. Flesh feeders have a long and complex geologic history (Fig. 19-20). They are intelligent, travel easily, and are highly adaptive. The dog and cat show their typical specializations—clawed feet, enlarged ca-

Figure 19-20 Family tree of the carnivores, on a uniform scale of size. Note that five tribes of creodonts were present in Paleocene and Eocene time, one of which, represented by *Miacis,* gave rise to all the modern carnivores. Only one tribe of the creodonts lived past Eocene time, dying out in the Pliocene. Of the true carnivores, the "beardogs" were the first to attain large size and then became extinct in Pliocene time. (Diagram based on data from Matthew and from Romer.)

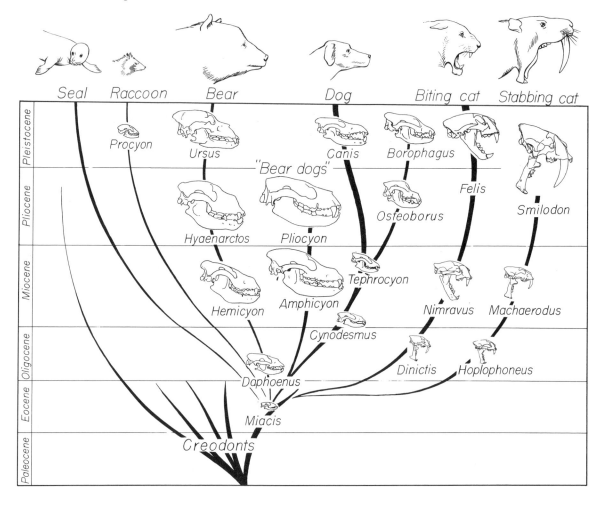

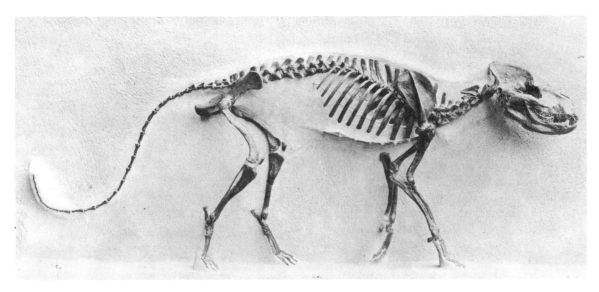

Figure 19-21 Skeleton of a doglike creodont, *Dromocyon vorax,* from the Eocene at Henry's Fork, Wyoming. (Yale Peabody Museum.)

nines, and narrow shearing cheek teeth. These are devices for holding and devouring active prey. Among the modern forms, the dogs, cats, bears, hyenas, raccoons, and seals represent as many well-defined families, but when they are traced back toward the early Cenozoic, their distinctions decrease, and they seem to converge toward a common Eocene ancestor.

Small primitive carnivores appeared in some abundance in the Paleocene, and during the Eocene epoch they diverged into at least five distinct families, adapted themselves to a wide variety of habits, and attained a considerable range of size, a few reaching the stature of a large bear. Some superficially resembled wolves (Fig. 19-30) or cats or other living types, but in all these early forms the brains were very small, as compared with those of modern carnivores, and they must have been a stupid lot. Moreover, certain specializations of teeth or other parts show that four of these families were incapable of developing into any of the modern carnivores. For this reason, they are commonly set off as a distinct order, the **Creodonta** (Fig. 19-21).

Nearly all the creodonts were inadaptive in some respects, and three of the families

died out by the end of the Eocene, the other barely surviving through the Oligocene with one genus ranging up into the Pliocene in India. Thus the first great experiment in carnivore evolution came to an inglorious end!

The fifth family of Paleocene and Eocene flesh-feeders had a higher destiny, even though, at the time, it would have seemed unpromising. It included small slender animals of the size of weasels; but they had better brains than the rest, and adaptive feet; and their teeth were already specializing in the direction followed by the higher carnivores. Among these small Eocene types, the genus *Miacis* appears as a probable ancestor of all modern carnivores.

This second upsurgence of carnivores was under way in the Eocene, and before the close of that epoch the modern families began to emerge. The dogs were represented by *Pseudocynodictis*, the size of a fox, and the cats by *Dinictis*, as large as a small leopard.

The Miocene was a time of rapid expansion, and before its close all the modern families were well defined. In the Pliocene formations of western United States several species of wolves are represented, some large and others small, and by Pleistocene time the species were similar to modern ones. In the asphalt deposits at Rancho La Brea in California a common fossil is the **dire wolf** (Fig. 19-22), which had the stature of a large gray timber wolf.

By Miocene time the cats were diverging into two quite distinct families, the **biting cats** and the **stabbing cats.** In the former, to which all modern cats belong, the lower and upper canines are subequal, and the lower jaw is strong. Such cats kill their prey by biting. In the stabbing cats the lower canines were small, and the upper ones were extended into saberlike blades; the lower jaw was weak and could be opened to a very wide angle to clear the upper teeth, which were then used to stab and tear, bleeding the prey to death. The biting cats are well represented in both the Pliocene and Pleistocene deposits of America, culminating in the panther and lynx. *Felis atrox* of the Pleistocene faunas of southern California was the size of a lion. The true lion, *Felis leo*, ranged widely over Europe during the interglacial ages of the Pleistocene.

The stabbing cats were even more common from Oligocene to late Pleistocene time, culminating in the **saber-toothed tiger** of the Rancho La Brea tar pits (Fig. 19-23). It was the last of its race, becoming extinct during the Pleistocene Epoch.

The Primates: Man's Family Tree. The monkeys and some primitive relatives, the great apes, and man are embraced in the **Order Primates.** For very natural reasons the paleontologic record of this order is tantalizingly inadequate. They were primarily forest dwellers, able to take refuge in the trees and with their increasing intelligence they generally managed to escape floods or other natural catastrophes that would insure mass killings and quick burial. They died individually where carnivores or scavengers almost invariably devoured their flesh and scattered their bones. As a result nearly all their fossil remains are fragmentary — individual teeth, incomplete jaws or skulls, and fragments of the larger limb bones. Even these have had to survive the corrosion of the acid soils of the forest. Comparative anatomy of the living forms provides a firm basis for establishing the main

Figure 19-22 The dire wolf, *Canis dirus,* a common species in the asphalt pits at Rancho La Brea, near Los Angeles, California. Size about that of a timber wolf. Pleistocene. (Model by R. S. Lull, Peabody Museum.)

Figure 19-23 The great saber-toothed tiger, *Smilodon,* common in the asphalt pits at Rancho La Brea, in Los Angeles, California. Height about 3 feet. Pleistocene. (Model by R. S. Lull, Yale Peabody Museum.)

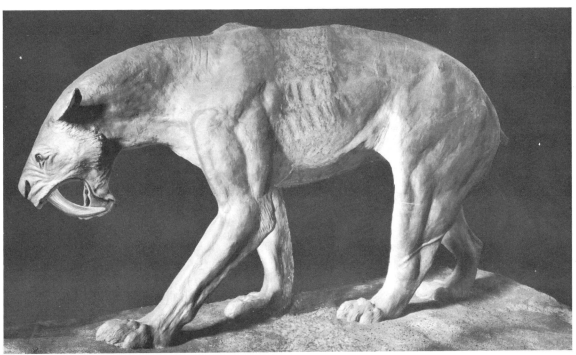

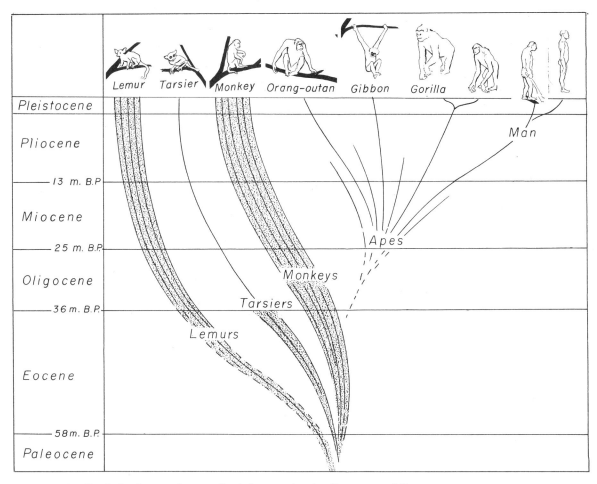

Figure 19-24 Man's family tree. A generalized diagram showing the ranges of the major groups of primates plotted against the Cenozoic time scale.

branches of man's family tree (Fig. 19-24), but the precise time of branching is still the subject of intensive research.

Lemurs. The most primitive branch includes the lemurs (Fig. 19-25), which superficially resemble foxes rather than monkeys for they are distinctly quadrupedal, have long bushy tails, and run on all fours. Their brains are relatively small, their muzzles slender and pointed, and their eyes set far apart on the sides of the head. Their teeth, however, show that they are not carnivores but have descended from the insectivores.

Tarsioids. The living tarsier of the East Indies (Fig. 19-26) is the sole survivor of a group of small, primitive primates that was far more common and more widely distributed in early Cenozoic time. In its nocturnal habits and some other respects the living species is highly specialized, but like its fossil relatives it shows extremely important advances over the lemurs and serves to bridge the gap between them and the monkeys. It has a relatively large brain and a shorter muzzle than a lemur, but the most significant difference is in the position of the eyes, which have migrated forward so that they lie close together and both can focus on the same point. This permits stereoscopic vision, an achievement which has been attained by no other animals save the monkeys and man.

From the start the primates specialized for

arboreal life, finding in the trees a refuge from their more powerful enemies on the ground. Prehensile hands and feet were developed early; and with the depth focus of stereoscopic vision new capabilities for locomotion in the trees were achieved. Instead of running along the limbs like a squirrel, such animals could safely hang by their arms and swing from limb to limb or even from tree to tree. Such free and rapid locomotion through the forest had great selective value. It led first to the evolution of the monkeys and shortly thereafter to gibbonlike apes, in which the arms are longer and more powerful than the legs. This in turn opened up new possibilities when the forests shrank, during late Cenozoic time, and some adventurous apes returned to life on the ground, for now their long arms gave them an almost upright position even when walking on all fours. Bipedal gait was thus easy to achieve and the hands were freed for better uses. Special interest therefore attaches to the tarsioids, whose vision in Paleocene time started us on the highway toward the human estate!

The lemurs and tarsioids were well adapted to the mild, moist climate that prevailed over Europe and the United States during the early part of the Cenozoic Era, and their fossil remains are relatively abundant, though fragmentary, in both these regions in Paleocene and Eocene strata. But during Oligocene time subtropical forests gave way to open plains in the present temperate lowlands, and the primates retreated to lower latitudes. During the Miocene and Pliocene epochs the climate generally became both cooler and drier, and the tropical and subtropical forests gradually shrank to their present distribution. As a result, most of the higher primate evolution took place in parts of Africa and Eurasia that are not yet well known paleontologically.

Lemurs and tarsioids died out in the United States during Oligocene time, and no record whatever is known of monkeys or apes thereafter in all of North America. Small monkeys had reached South America (or had evolved there out of tarsioids) and survive to the present, undergoing an evolution entirely independent of the rest of the primate world;

Figure 19-25 A small lemur, *Galago,* the "bush baby" from Northern Rhodesia, Africa. The body of the animal is about 6 inches long. (Yale Peabody Museum.)

Figure 19-26 The tarsier, a native of the Philippine Islands. About natural size. (Lilo Hess photograph.)

Figure 19-27 The four living great apes drawn to the same scale. They do not live together, but are here posed symbolically to indicate their normal habit. The gibbon and orang-outan inhabit southeastern Asia and the East Indies, whereas the chimpanzee and the gorilla inhabit the tropical forests of Africa. The gibbon travels rapidly by swinging under tree limbs from tree to tree; the orang-outan makes nests in the trees and is a slow, lazy climber; both the chimpanzee and the gorilla spend much time on the ground and are able to walk upright.

they include the marmosets and cebid monkeys, commonly grouped as the New World monkeys.

Until recently the earliest evidence of Old World monkeys was a few fragments, a lower jaw with most of its teeth, found in Lower Oligocene beds of Egypt. This has now been supplemented by more than 100 specimens, mostly teeth but including part of a significant lower jaw. The Miocene record is still very meager, and that of the Pliocene only somewhat better. Tropical forests provide a very poor environment of the preservation of fossils, because the organic acids in the soil cause rapid decay of bones. For this reason we may never have as much evidence for the geologic history of the primates as for most of the other groups of animals. Pliocene monkeys are referred to families still living.

The Apes. The apes include 4 living genera (Fig. 19-27) and a number of fossil genera. They differ from monkeys in significant ways. They are tailless and the arms are longer than the hind legs so that even when walking on all fours the body is almost upright and bipedal gait is not difficult. Their brains are relatively larger than in monkeys.

Although the great apes are more manlike than any other creatures, each is highly specialized for a lazy life in the tropical forest and none of the living genera could be considered the direct ancestor of the human line. The latter must have originated where forests were receding and breaking up into scattered woodlands and some great ape ventured to spend most of its time on the ground, walking upright so that the hands were freed for picking up objects of curiosity and using tools. The gibbon is the most active of the living apes and has relatively longer arms than the others. Its lineage is represented as far back as the Late Miocene in the genus *Pliopithecus* in which the jaw teeth show a significant advance over that of the monkeys—one that persisted in the apes and man. These teeth in the monkeys are subquadrate in section, having four blunt cusps, one at each corner of the tooth; but in the apes and man the lower molars have a fifth cusp.

It now seems probable that the human line diverged from generalized apes such as *Dryopithecus* in Late Miocene or Early Pliocene time. Offshoots of the dryopithecine stock include *Ramapithecus,* an apelike creature somewhat smaller than the chimpanzee but distinguished from all apes by the shape of its jaws and dental arch as well as by the shortening of its face. In apes the jaw is rectangular and the canine teeth are large, rising well above the level of the other teeth, whereas in the human line the canines are small and the dental arch is a gentle curve broadening posteriorly (Fig. 19-28). *Ramapithecus* has the small canines and a dental arch of human form; among the known fossil primates it is the most likely rootstock of the family of man.

Important fossil evidence for early human evolution has been found in limestone caverns near Johannesburg, South Africa, where preserved material of more than 30 individuals includes almost complete skulls and dentition as well as much skeletal material. Although described under four generic names, this material is now believed to represent a single genus, *Australopithecus.* This is a larger animal than *Ramapithecus,* but the two are similar in jaw structure and dentition. They may be in the same line of descent but their remains are separated by a gap in the record that may be as long as 10 million years. The brain of *Australopithecus* was scarcely half as large as that of modern man and the cranium was scarcely as large as the jaw frame (Fig. 20-6). But, as in man, the head was large relative to body size and details of skull and teeth prove it to be in the human family. The pelvic bones indicate clearly that the animal was bipedal, and there is some evidence that it used tools.

Recapitulation

The Paleocene Mammals. As George Simpson has written, "the most puzzling event in the history of life on earth is the change from the Mesozoic, Age of Reptiles, to the Age of Mammals. It is as if the curtain were rung down suddenly on a stage where all the leading roles were taken by reptiles, es-

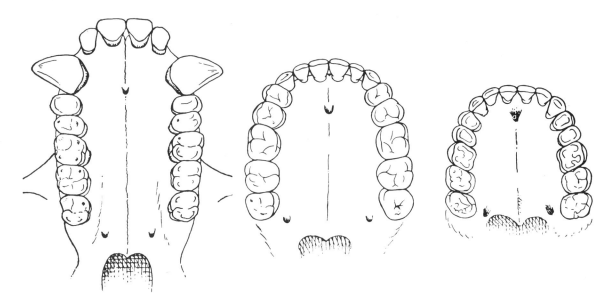

Figure 19-28 Upper jaw and dentition of the gorilla (left), *Australopithecus* (center), and modern man (right). (After LeGros Clark.)

Figure 19-29 Paleocene landscape in the Rocky Mountain region. 1, a boa constrictor hanging from a breadfruit tree; 2, a soft shelled turtle; 3, a crocodile; 4, a flying lemur; 5, a primitive condylarth, *Tetraclenodon;* 6, a primitive amblypod, *Pantolambda;* 7, a larger amblypod, *Barylambda.* (Painting by Rudolph Zallinger from *The World We Live In,* courtesy of Time, Inc.)

pecially dinosaurs in great numbers and bewildering variety, and rose again immediately to reveal the same setting but an entirely new cast, a cast in which the dinosaurs do not appear at all, other reptiles are supernumeraries, and the leading parts are all played by mammals of sorts barely hinted at in the preceding acts."

Clearly these Paleocene mammals (Fig. 19-29) came for the most part as migrants from some region not yet studied where they had been evolving even before the last stand of the dinosaurs. They had been hiding in the wings waiting to take their place on stage. Indeed three groups, the multituberculates, marsupials, and insectivores, had been furtively intruding on the Cretaceous scene. But with the beginning of the Paleocene three new orders—the condylarths, amblypods, and creodonts—appeared suddenly as the dominant land animals. All of these had relatively small brains, large jaws, and relatively unspecialized teeth and feet. They were not nearly as well adapted as the stocks they gave rise to and few survived beyond the Paleocene Era. They are, therefore, commonly referred to as "archaic mammals." Figure 19-29 shows some of these early mammals, and contemporary holdovers from the Age of Reptiles in a Rocky Mountain setting when the mountains were still young and volcanoes were active.

Eocene Land Life. The most striking feature of Early Eocene life (Fig. 19-30) was the appearance in considerable numbers of progressive forms ancestral to the modern orders of mammals. Among these were diminutive **horses**, small hornless **rhinoceroses**, equally small **titanotheres**, tiny **cameloids**, the first **oreodons**, squirrel-like **rodents, bats,** and small **primates**. None of these attained a considerable size, and the largest would hardly have stood waist-high to a man.

With them were associated the "archaic" mammals, some of which were far larger. Creodonts were the carnivores of that time, and of these some were doglike, some hyenalike, and the others more catlike. Common American types reached a maximum size similar to that of a modern timber wolf, though one of the latest Eocene types was as large

as a great bear. The greatest of all Eocene carnivores, however, was the Mongolian *Andrewsarchus*, with a skull about $2\frac{1}{2}$ feet long. The **condylarths** were common during the early half of the epoch but died out before its close. The great animals of this time were the ponderous amblypods, which increased gradually to their maximum bulk in *Uintatherium* of the late Eocene.

In the later Eocene occurred the first mammal adaptation to a marine life, in the form of whale-like animals (zeuglodons), whose fossil bones occur abundantly in parts of the southern United States, in Egypt, and in Europe. One of these, *Basilosaurus*, must have been the "sea serpent" of its time, with 4 feet of head, 10 feet of body, and 40 feet of tail! But even a mammal of this size met its match in the great **sharks** of those seas, one of which (*Carcharodon*) had jaws about 6 feet across.

Oligocene Faunas. By Oligocene time (Fig. 19-31) the modernized types comprised nearly the entire mammalian fauna. A single genus of creodonts remained, but amblypods and condylarths were wholly extinct, and marsupials and insectivores were as inconspicuous as they are today. In western America the **oreodons** roamed in large herds over the plains. Three-toed **horses** (*Mesohippus*) scarcely larger than sheep were common. Rhinoceroses of several kinds were present, the largest of them being amphibious, though not related to the hippopotamus; probably all the Oligocene species were hornless. The **titanotheres** displayed a meteoric evolution, and with the exception of the giant Asiatic rhinoceros, *Baluchitherium*, they became the largest land animals of this time before dying out abruptly about the middle of the epoch. Small camels were present, and so were peccaries and tapirs. The rodents were represented by beavers, squirrels, rabbits, and mice. Among the carnivores there were many small dogs, as well as both **biting cats** and **stabbing cats**. In the Old World the **Proboscidea** were beginning their career, being represented by the first mastodons, which were only about $5\frac{1}{2}$ feet high. Early primates had become extinct in North America, and the only known great ape was repre-

Figure 19-30 An Eocene landscape in the Rocky Mountain states. Here among palms and modern types of deciduous trees lived a strange assemblage of primitive mammals. 1, a lemur, *Notharctus;* 2, a catlike creodont, *Oxyaena;* 3, a condylarth, *Phenacodus;* 4, an amblypod, *Coryphodon;* 5, the "dawn-horse," *Eohippus;* 6, a primitive titanothere, *Palaeosyops;* 7, a creodont, *Mesonyx;* 8, a creodont, *Tritemnodon;* 9, an amblypod, *Eobasileus;* 10, an amblypod, *Uintatherium;* 11, a flightless bird, *Diatryma.* (Painting by Rudolph F. Zallinger, from *The World We Live In,* courtesy of Time, Inc.)

Figure 19-31 An Oligocene landscape in Dakota. 1, a herd of oreodons; 2, an entelodont, *Archeotherium;* 3, a titanothere, *Brontops;* 4, a primitive horse, *Mesohippus;* 5, a deer, *Protoceras;* 6, a creodont, *Hyaenodon;* 7, a rhinoceros, *Subhyracodon;* 8, a tapir, *Protapiris;* 9, a primitive cat, *Hoplophoneus;* 10, a primitive camel, *Poebrotherium.* (Painting by Rudolph Zallinger, from *The World We Live In,* courtesy of Time, Inc.)

Figure 19-32 A Miocene landscape in Nebraska showing animals adapted to the grassy plains. 1, a chalicothere, *Moropus*; 2, an entelodont, *Dinohyus*; 3, a herd of horses, *Merychippus*; 4, a four-horned antelope, *Syndeoceras*; 5, a small camel, *Procamilus*; 6, a herd of oreodons; 7, a giraffe-camel, *Alticamilus*; 8, a pair of four-tusked mastodons, *Gomphotherium*. (Painting by Rudolph Zallinger, from *The World We Live In,* courtesy of Time, Inc.)

Figure 19-33 Scene in the Mississippi Valley during Pleistocene time. Left, a herd of Mastodons, *Mammut americanus;* center, the royal bison; right, a herd of wild horses, *Equus scotti.* (From a mural by Charles R. Knight in the American Museum of Natural History.)

sented in Europe by a single species.

Miocene Faunas. The Miocene was the "Golden Age" of mammals (Fig. 19-32). The spread of the prairies and the change to a more arid climate led to rapid evolution of the grazing stocks, and the formation of a Bering land bridge permitted intermigration between North America and Eurasia.

Within the groups of animals already present there was a rapid expansion into new genera and species and an increase in size in many stocks. The habit of feeding on the harsh prairie grasses resulted in a remarkable change in the teeth of many groups, whereby the jaw teeth became long and prismatic and continued to grow throughout life, thus counteracting the rapid wear at the crowns. Horses now attained the size of small ponies, but the many species all had dangling side toes. Camels were especially abundant and varied, some being little larger than sheep, while others rivaled the modern giraffe in height. **Oreodons** were still very common. Rhinoceroses of several kinds were abundant. At this time the **"giant pigs"** reached their climax in a species (*Dinohyus hollandi*) known from Nebraska that was as tall as an ox and had a skull 4 feet long. *Moropus*, the clawed ungulate, was also most common at this time. Rodents like those of the Oligocene continued through the Miocene. Of carnivores there were numerous wolflike dogs, as well as biting and stabbing cats. There were no North American primates, but in the Old World a great ape (*Dryopithecus*), somewhat related to the gorilla, but much smaller, ranged over Europe and northern Africa. The four-tusked proboscidians arrived in America.

Pliocene Faunas. The Pliocene faunas of North America are still imperfectly known because the terrestrial formations of this age are so sparsely preserved. At this time there was further immigration from the Old World, bringing us the true **mastodons.** Horses continued their rapid evolution and were represented in America by several genera, among which appeared the first single-toed horse, *Pliohippus*. Rhinoceroses were still very abundant. Camels continued to be among the most common animals of the plains. The last straggling survivors of the oreodons were extinct before the close of the epoch.

Pleistocene Faunas. Throughout the Pleistocene Epoch North America and Europe were both inhabited by great animals fully as varied and impressive of those of modern East Africa. In the United States the elephants were perhaps the most impressive, for there were at least four species, two of which exceeded modern elephants in size. The tall, rangy imperial mammoth of the southern Great Plains stood nearly 14 feet high at the shoulders. Another species (*Mam-*

muthus arizonae) was at home in the basins of Arizona and Nevada. Numerous remains of *Mammuthus columbia* and *M. imperator* have been found in the uppermost beds of glacial Lake Bonneville, and these elephants must have been common in the Great Basin region until after the last of the glacial ages. Throughout the forests mastodons (*Mammut americanus*) browsed in great herds (Fig. 19-33); their remains are common in the peat bogs of the eastern states, no fewer than 217 individuals having been discovered in the bogs of New York State alone. In Florida, New York, and elsewhere the remains of this species are associated with human artifacts in such a way as to indicate that mastodons survived the last ice age and may have lived until within the last several thousand years. The woolly mammoths (Fig. 19-34) ranged widely over the glaciated areas, extending northward into Alaska and eastward across Siberia, where their skeletons and tusks are still incredibly numerous in the frozen soil, about half the present ivory of commerce being derived from this source. Siberian ivory was imported into China as early as the fourth century B.C., and began to be extensively transported into Europe early in the nineteenth century. Between 1800 and 1850 the annual sale of tusks at the trading center of Yakutsk averaged about 18 tons, and to date not less than 46,750 pairs of tusks have been recovered in Siberia.

Horses were still common, and at least ten species are known from North America. Most of these were of the size of small ponies, but one fully equaled the greatest modern draught horse. Buffaloes roamed the plains in great herds as they did when the white man first reached America. There were at least seven Pleistocene species, and one of these (*Bison latifrons*) was a colossal beast with a horn-spread of fully 6 feet. Camels also were common. **Wild pigs (peccaries)** now confined to Texas, Mexico, and Central America, then ranged over the United States. **Carnivores** were abundant and varied, including species of such modern types as the wolves, foxes, pumas, lynxes, raccoons, badgers, otters, skunks, and weasels. In addition, there were

extinct types of which the great saber-tooth (*Smilodon*) was perhaps the most striking. Another great cat (*Felix atrox*), also known from the tar pits of Rancho La Brea, was very much like the modern lion in form and size. The great wolf (*Canis dirus*) which was so common in southern California exceeded in size any modern American canines. True bears apparently made their appearance in America at this time as immigrants from the Old World.

One of the striking elements of the Pleistocene fauna was due to the immigration from South America of the **glyptodonts** and the great **ground sloths.** These both reached Texas in the Pliocene, and the latter spread over the United States in the Pleistocene. The ground sloths (Fig. 3-26) were clumsy beasts with the bulk of an elephant, but they were short-legged and curiously club-footed and their ancestors had lived so long in the trees and had developed such long curved claws that it was impossible for them to walk on the bottoms of the feet when they became too heavy to live longer off the ground. These creatures are common fossils in the tar pits of southern California, and at least one genus (*Megalonyx*) ranged eastward to the Atlantic states. A claw of this form was discovered in a cave in Virginia by President Jefferson, who was the first to describe and name the genus but thought it to be a mighty lion.

North and South America were separated throughout Cenozoic time until they became united by the present Isthmus of Panama in the late Pliocene. Just before this connection, South America had 29 families of mammals and North America 27, but with doubtful exceptions they had no families in common, the South American fauna having a large number of marsupials. Shortly afterward, in the Pleistocene, they had 22 families in common, 7 of South American origin, 14 of North American origin, and 1 doubtful. Those emigrating to South America included the mastodons, horses, tapirs, camels, and peccaries; those coming in the opposite direction included the ground sloths, armadillos, and glyptodonts. Most of the South American marsupials did not survive the competition with the placental

Figure 19-34 Pleistocene landscape in Europe during the last ice age, with the woolly mammoth in the foreground and woolly rhinoceroses in the right middle distance. (Picture by Charles R. Knight, Chicago Museum of Natural History.)

mammal groups from the north.

Another striking element of the Pleistocene faunas was supplied by the arctic animals that migrated southward during the glacial ages. For example, the musk ox is recorded as far south as Arkansas and Utah, while in Europe the reindeer, the woolly rhinoceros, the woolly mammoth, and the arctic fox ranged southward into France and Poland.

There were no primates in North America until primitive man arrived from the Old World. Although evidence of the presence of early man in America has been claimed repeatedly, it is still scanty. On the other hand, savage races were present in Europe and in eastern Asia and northern Africa throughout much if not all of the Pleistocene Epoch. The geologic history of man is reserved, however, for Chapter 20.

In Europe, the Pleistocene fauna of the warm interglacial ages included most of the types of great animals now found in Africa. Such, for example, were the lion, the rhinoceros, the hippopotamus, the elephants, antelopes, and lesser animals.

CONTEMPORARY LIFE

Spread of the Prairies. Forests were essentially modern at the beginning of the Age of Mammals. Most of the familiar hardwood trees had appeared during the Cretaceous Period, and their subsequent evolution has been in the main a matter of specific details. However, the development of the grasses during this era was one of the greatest milestones in the history of life, and for the evolving mammals its importance can hardly be overemphasized. It is this stock, for example, that includes not only the forage plants but also the cereals—notably wheat, rice, oats, and corn—that provide the basic food supply for the modern world.

Grass is poorly adapted for preservation, and almost no direct evidence of its early history is known. However, it contains an appreciable amount of silica and tends to wear out the grinding teeth of the grazers, that is, grass-feeders. To compensate for such wear, the plains mammals have high-crowned cheek teeth that grow at the roots throughout

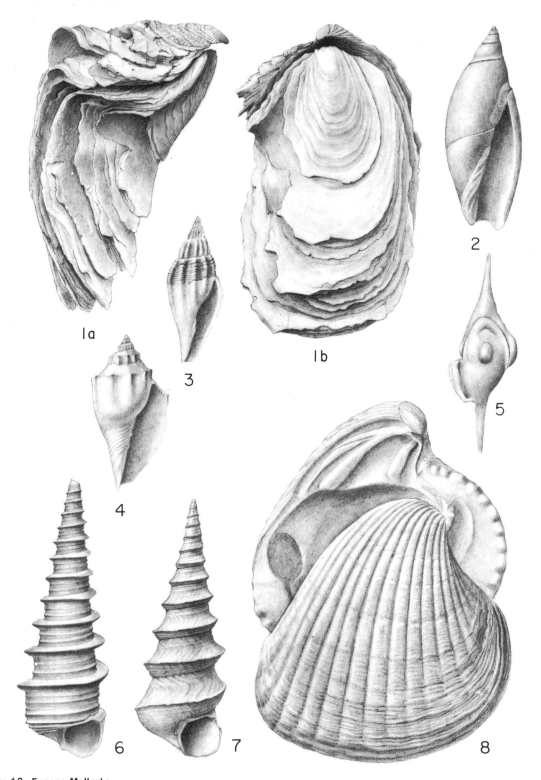

Plate 18 Eocene Mollusks.
  Figures 1a, 1b, *Ostrea sellaeformis;* 2, *Oliva alabamiensis;* 3, *Plejona rugata;* 4, *Volutilithes savana;* 5, *Calyptrophorus trinodiferus;* 6, *Turritella praecincta;* 7, *Turritella mortoni;* 8, *Venericardia planicosta.*

Plate 19 Miocene Pelecypods.

Figures 1–4, *Glans decemcostata;* 6, *Chlamys decemnaria;* 7, *Lyropecten ernestsmithi;* 8, 9, *Venus berryi;* 10, *Glycimeris americana.* Figure 5 is an Oligocene species, *Pecten poulsoni.* (From Gardner, U. S. Geological Survey.)

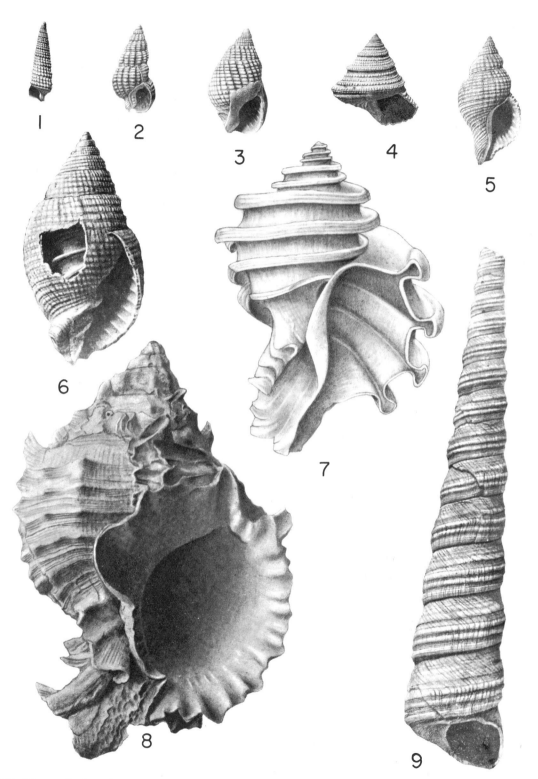

Plate 20 Miocene Gastropods.

Figure 1, *Triphora bartschi;* 2, *Utzia neogenensis;* 3, *Illyanassa grandifera;* 4, *Calliostoma mitchelli;* 5, *Urosalpinx trossula;* 6, *Cancellaria rotunda;* 7, *Ecphora quadricostatus;* 8, *Murex pomum;* 9, *Turritella pilsbryi.* (All except Fig. 7 after Gardner, U. S. Geological Survey.)

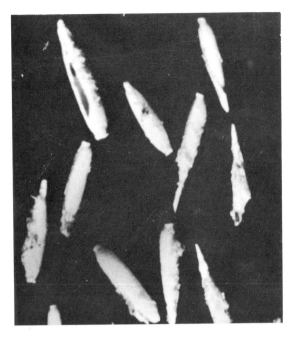 

Figure 19-35 Seeds of grasses and other herbs. Left, the spear grass *Stipidium* from the Valentine Formation (Pliocene); right, the borage herb *Biorbia* from the Ash Hollow Formation (Pliocene). (Maxim K. Elias.)

life (Fig. 19-5). Pre-Miocene representatives of each of these groups, like the modern forest-dwellers that browse on more succulent leaves, had low-crowned cheek teeth. It appears evident that the high-crowned grinding teeth are a direct adaptation to a grazing habit; and since this specialization began early in the Miocene, it is inferred that prairie grass had then, for the first time, become widespread. About 1940, fossil grass seeds were discovered in some of the sandy beds of western Nebraska, and it was soon found that they are abundant over the High Plains region in beds ranging from Mid-Miocene to Late Pliocene date (Fig. 19-35). Other herbaceous plants are still imperfectly known, but roses with characteristic leaves and thorns have been found in the Oligocene, and unmistakable leaves and seeds of the grape occur in Eocene and later beds in both Europe and North America. Petrified grapevine is known from Miocene beds in Nevada.

Marine Invertebrates. The marine invertebrates (Plates 18 to 20), like the forest plants, were essentially modern (and how different

they were from the Cambrian faunas with which we started!). Special note must be made, however, of the large discoidal foraminifera such as the **nummulites** (Fig. 19-36) and **orbitoids** that accumulated on the shallow Eocene sea floors to contribute largely to extensive limestones and provide some of the most useful zone fossils for correlating Eocene to Miocene formations. Great diversification of the smaller planktonic **foraminifera** also marks the Cenozoic.

Decline of the Reptiles. With the extinctions at the end of the Mesozoic, the reptile dynasty collapsed. Turtles, crocodiles, lizards, and snakes lived on about as they do today except that large land turtles and crocodiles were more widely distributed during the warmer epochs. Fossil turtles are abundant in the badlands of Oligocene and Miocene age in Nebraska and the Dakotas. Enormous species of land turtles big enough to stand waist-high to a man were common in Florida in Pleistocene time. Snakes, first recorded in late Cretaceous beds, must have survived, but their retiring habits and their delicate

Figure 19-36 A chunk of nummulite limestone from the Pyramid of Gizah, Egypt. Natural size. Herodotus, about 450 B.C., alluded to this stone as an example of petrified animals. Eocene. (Yale Peabody Museum.)

Figure 19-37 A giant flightless bird, *Phororhacos,* from the Miocene of Patagonia. (From Lucas, *Animals of the Past.* American Museum of Natural History.)

skeletons make them rare fossils. Nevertheless, an Eocene species found in Patagonia related to the boa constrictor is estimated to have been 35 feet long.

Birds. Modern types of birds appeared before the end of the Cretaceous and by Eocene time most of the modern orders were represented.

Nearly all the continents at one time or another during the Cenozoic Era had large flightless birds. One of these, *Diatryma*, known from the Eocene of Wyoming, stood nearly 7 feet tall and had a skull almost as large as that of a horse. Another (*Phororhacos*), found in the Miocene of the Argentinian Pampas, stood 7 to 8 feet high and had a very massive skull 23 inches long with a strongly hooked beak (Fig. 19-37). It was undoubtedly the greatest of all birds of prey. The largest bird, however, was an ostrichlike form, the moa (*Dinornis*), which lived into historic time in New Zealand and was exterminated by the Maoris only a few centuries ago. This enormous bird stood about 10 feet high and was therefore more than 2 feet taller than the greatest living ostrich. Still another giant (*Aepyornis*) of Madagascar laid the largest known eggs, which normally measured 13 inches long and 9 inches across. Discovery of the eggs of this bird by early navigators inspired the thrilling tales of the roc told by Sinbad the Sailor in the *Arabian Nights*.

It was long believed that the toothed birds were all extinct before the end of the Mesozoic. It was a great surprise, therefore, when a bed in the Pliocene near Shasta Maria, California, was found to bear tracks as well as impressions of the beak, jaws, skull and leg bones of a giant toothed bird, estimated to have had a wing spread of 16 feet. The remains are now in the Mission Canyon Museum.

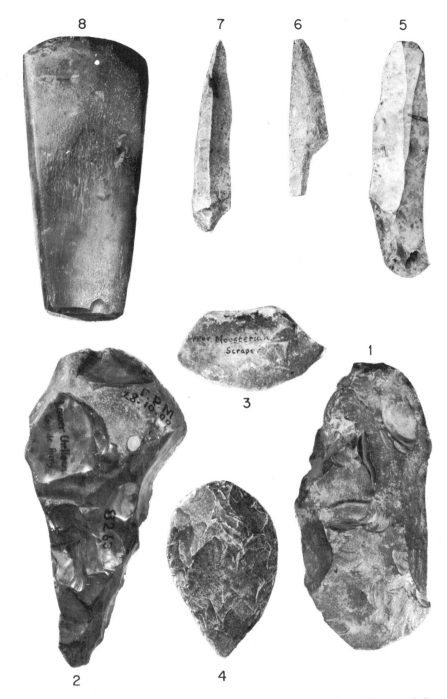

Figure 20-1 Stone artifacts. 1, an *eolith,* used but not shaped by man; 2–7, *paleoliths;* 8, a *neolith.* No. 2 is Chellean, nos. 3 and 4 are Mousterian; nos. 5 and 6 are Solutrean, and no. 7 is Aurignacean.

CHAPTER **20** *The Coming of Man*

During Early and Middle Cenozoic time the climate was mild over most of the present temperate zone and the widespread forests of deciduous trees and herbaceous plants provided abundant food for the primates in the form of foliage and fruits and nuts. But during the Pliocene Epoch the climate gradually became cooler and the hardwood forests of Eurasia retreated gradually toward the tropics and were replaced by evergreens. As the forests retreated and broke up into open woodlots scattered among the grasslands, the great apes had a choice to make. They could either retreat to the tropics (where they now live) or adapt to the changing environment. Nearly all of them took the easy choice. The gibbons and ourangs and the gorillas and chimpanzees found life easy in the tropics and changed but little. The more intelligent and venturesome apelike early humans stayed with the woodlots, spending more and more time on the ground, searching for berries, insects, and small animals that gave them a varied diet and improved their ability to walk and run on the hind limbs. This freed their hands for more important uses — seizing small prey, picking berries, and picking up objects that caught their fancy, and it stimulated their curiosity. Thus they learned to pick up sticks to use as clubs, which certainly were the first weapons of the chase. Stones could also be used for killing small game. Sharp-edged pieces of flintstone could also be used for cutting and scraping skins. For a long time such progressive animals probably stayed near the trees in which they could escape the ever-present carnivores; but as they learned to use clubs and stones for defense, a great stimulus was placed on the ability to shape better weapons, both for offense and defense. This marked the beginning of man the hunter.

Human fossils of this early stage are extremely rare for two natural reasons. First, these most intelligent of all the primates of the time were able to avoid the natural catastrophes by which they would be quickly buried. Second, they were intelligent enough to be conscious of death and to hope for some kind of renewal after death. This was the beginning of religious customs and of reverence

and care of the bodies after death. The cultures of the most primitive living peoples indicate the probable forms adopted for this purpose. The American Indians chose a high place and built scaffolds on which the dead would be out of reach of scavengers. Other people bury the dead in graves covered with stones. Some cremate the bodies. In no case are they ever placed in low ground where they would soon be covered by accumulating sediment. They were all doomed to be destroyed by weathering and erosion within a few centuries at the most. As a result early human fossils are extremely rare.

To this generalization there is one notable and fortunate exception. Where limestone caverns or overhanging ledges were available along the river bluffs, primitive men eventually found them to be a safe place in which to live, for here they could protect themselves from inclement weather, from the large carnivores, and from other tribes. Some of these people learned to bury their dead in the floor of the caves and commonly surrounded them with weapons, food, and charms for use in the next life. Here we find entire and well-preserved skeletons along with the artifacts used in life, and clear indication of primitive religion.

Much of the earliest human history, however, has been gleaned from the stone implements which are virtually indestructible whether lost or abandoned about camp sites in the open or in the refuse heaps before permanent shelters. We shall turn our attention to these first.

Artifacts. Stone implements shaped by prehistoric man are widespread and locally abundant in western Europe. During and after the Renaissance they were commonly collected as curios and became the subject of strange superstition. In 1655, for example, Olaf Worm, a Danish student of such objects, wrote that "they are commonly supposed to fall with the lightning from the sky," though "opinions differ as to their origin, since some believe they are not thunderbolts but petrified iron implements, seeing they resemble the latter in shape so closely." And as late as 1802 Thorlacius, writing of stone artifacts discovered in burial mounds, concluded that "the objects found in the mounds are nothing else than symbols of the weapons employed by the Gods of Thunder in chasing and destroying evil spirits and dangerous giants. They could not be ordinary tools and weapons as these have been made of metal since the earliest times."

It now seems almost incredible that beliefs so fantastic could have persisted in Europe for 300 years after the American pioneers were in contact with the Indians, who were using such artifacts. It must be remembered, however, that European thought at that time was completely dominated by the belief that man was created only about 6000 years ago and that the whole of human history was recorded in the Scriptures. This left no place for prehistoric man in Europe, and there was a tendency to regard the American Indians and other primitive peoples as degenerates who had wandered far from the center of civilization and had lost the art of working metal. Occasional thinkers, far ahead of their time, had indeed realized the true meaning of archeological remains since before the beginning of the Christian Era, but it was not until within the last century and a half that these objects were generally accepted as evidence of prehistoric races. Even then, no one considered the possibility that they might represent the work of extinct **species** of man until after Darwin's *Origin of Species* had paved the way for a belief in the gradual evolution of man from the lower animals.

The appreciation of the true meaning of stone artifacts was first developed in Scandinavia shortly after 1830. It began with the creation by the Danish Government of a scientific commission to study the refuse heaps and shell mounds that had already attracted attention in the region. As a result of this project, extensive collections were assembled at the Royal Museum in Copenhagen and were studied with respect to their stratigraphic occurrence in the mounds. On this basis, Thomsen, the director of the Museum, in 1837 proposed a chronology of human culture divided into the **Stone Age**, the **Bronze Age**, and the **Iron Age.**

Man undoubtedly advanced through this sequence of cultures on the way to civilization, but in western Europe the making of bronze and the smelting of iron had been mastered while aborigines in many parts of the world were still using crude stone implements. Thus the Iron Age in Europe was contemporaneous with the Stone Age in many other parts of the world. Thomsen's chronology is therefore applicable only locally.

The first tools used by man were doubtless those accidentally shaped by nature to fit his hand, such as sharp-edged chips of flint that he could use to scrape skins or to fashion wooden tools. Such stones, which he picked up and used without modification (Fig. 20-1), are known as **eoliths** [Gr. *eos*, dawn + *lithos*, stone]. Showing evidence of wear but not of conscious shaping, they represent the lowest stage of human culture and are found in deposits as old as the Late Pliocene.

Eventually man learned to flake off pieces of stone and to shape them, by chipping, into scrapers, hand axes, spearheads, and other useful tools. This was an art slowly acquired through countless generations of trial and experiment by primitive peoples whose lives often depended on the quality of their weapons. Such artifacts, shaped by chipping alone, are known as **paleoliths** [Gr. *palaios*, ancient + *lithos*, stone], and cultures represented by such implements characterized the **Paleolithic Age,** which endured in Eurasia until a few thousand years ago.

The American Indians and other primitive peoples in Africa, Australia, and the Pacific Islands never progressed beyond the paleolithic stage of culture; but in Europe some of the prehistoric people went a step further when they learned to shape and sharpen stone implements by grinding and polishing them against natural abrasive stones. Such objects, known as **neoliths** [Gr. *neos*, recent + *lithos*, stone], are found only in deposits younger than the last glacial drift in Europe and mark the highest type of Stone Age culture, attained shortly before the discovery of the use of copper.

Cultures. The assemblage of material objects used by a people reflect and represent its

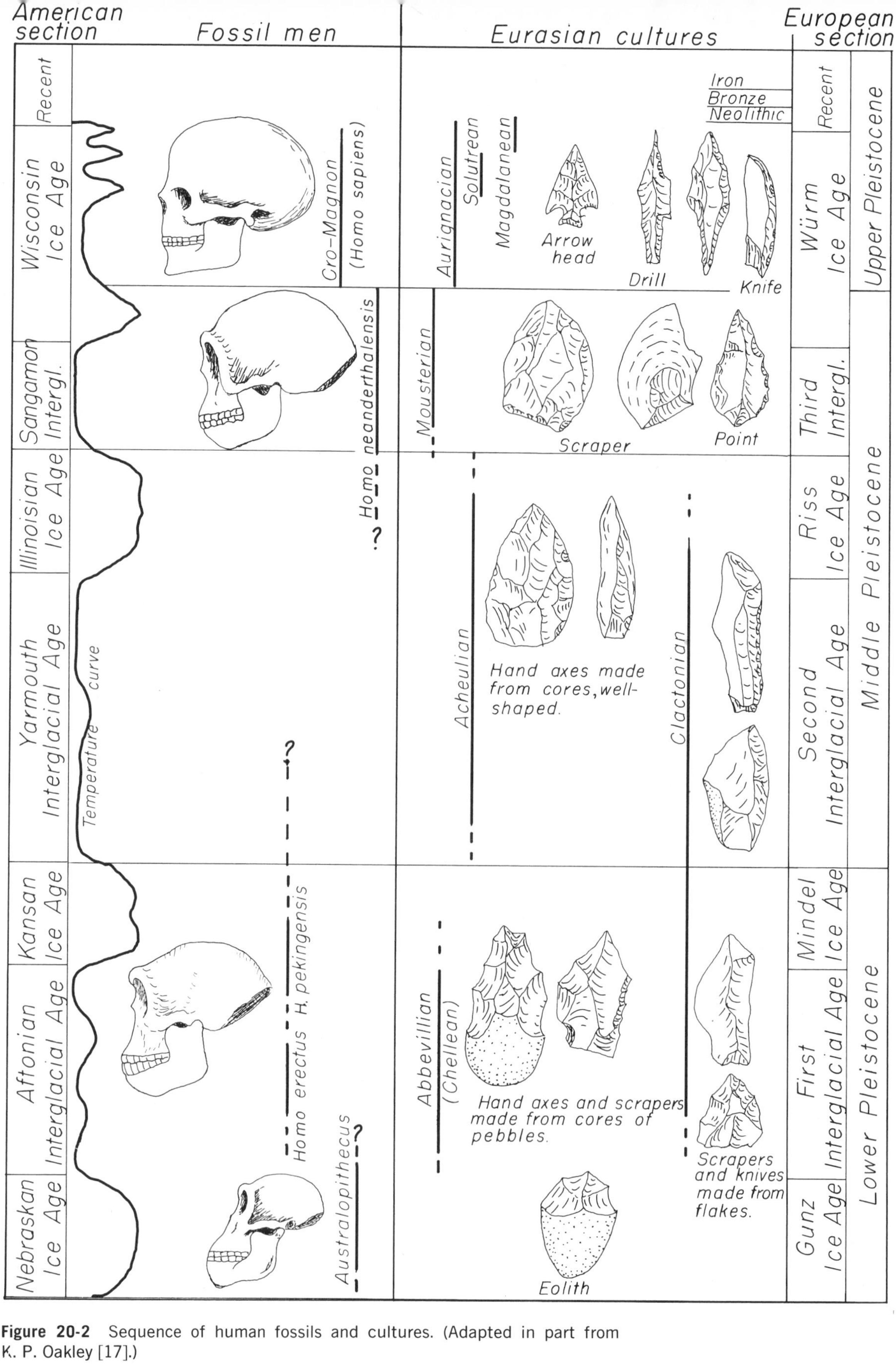

Figure 20-2 Sequence of human fossils and cultures. (Adapted in part from K. P. Oakley [17].)

Figure 20-3 Rock shelter at Les Eyzies in the Dordogne Valley in France. The reentrant across the middle of the cliffs was occupied by Paleolithic man and camp refuse accumulated on the slope below, which is now occupied by houses. (G. G. MacCurdy.)

culture. For primitive people such objects include weapons, utensils made of stone or pottery, and clothing and adornment such as amulets or beads.

During the Paleolithic Age in Europe the character and variety, and the functions, of the stone implements changed from time to time as new techniques were discovered for shaping flint objects and as climatic fluctuations induced changes in living habits and dress and in available game for food. Thus it has been possible to recognize in the burial mounds and the refuse heaps of Europe a series of distinct cultures, the chief of which are indicated in Fig. 20-2.

The sequence of cultures is most clearly recorded in the stratified deposits that accumulated about camp sites that were repeatedly occupied, especially the limestone caverns along the river bluffs that were a favorite home of primitive man since they offered protection against inclement weather, wild beasts, and warlike neighbors (Figs. 20-3 and 20-5).

The sequence in time of these cultures has been determined in numerous places by simple stratigraphic principles, as illustrated in Fig. 20-4. Here at Laussel, France, a fine rock shelter was used repeatedly by Paleolithic man, who discarded the refuse from his camp onto the slope below, where it accumulated layer upon layer. The sediment that accumulated while the shelter was occupied includes discarded or broken artifacts as well as the bones of animals on which he fed. Layers of sediment formed when the shelter was abandoned are barren. Careful excavation here has revealed five layers of culture separated by barren layers [14]. Of course the order of superposition shows their sequence in time, the Acheulian being the oldest and Solutrean the youngest. In similar stratified refuse heaps before other shelters or caves, the sequence is always the same, though in some, for example, the Cave of the Kids (Fig. 20-5), more subdivisions are recognizable.

Finally, when objects distinctive of a par-

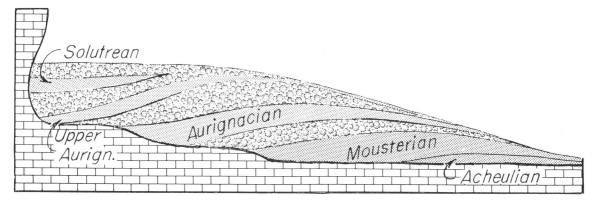

Figure 20-4 Diagrammatic section of the stratified deposits below a rock shelter at Laussel, France, showing a sequence of five distinct cultures. The layers bearing cultures are stippled and named. (Adapted from G. G. MacCurdy [14].)

ticular culture are found in stratified deposits associated with one of the drift sheets, or in interglacial deposits, it is possible to correlate the human history with the glacial history of Europe as shown in Fig. 20-2. Thus it is known that the Abbevillian culture belongs to the first interglacial age, the Acheulian spans the time from early in the second interglacial to near the end of the third interglacial age, and that Mousterian, Aurignacian, and Magdalenian cultures are related to the last ice age.

The time relation of these last three to the advance and retreat of the ice is indicated by the mammal bones associated with cultural objects in the refuse heaps and burial mounds, for they record the chief animals of the chase upon which these peoples were feeding. It is well known, for example, that as the ice advanced over northern Europe an arctic fauna spread southward and the reindeer became the chief source of food, but during the last interglacial period, as in postglacial time, the reindeer retreated to the far north and wild horses became the chief object of food. A refuse heap near one of the Solutrean camp sites, for instance, has been estimated to include the remains of some 70,000 wild horses.

Finally, for these later cultures, less than 50,000 years old, radiocarbon dating based on charcoal from the ancient camp sites, or bones or wooden artifacts, gives us an absolute chronology in years.

Early Flint Workers of England. The most ancient relics of mankind in Europe are found in England. Near the city of Ipswich, about 65 miles northeast of London, the upper Pliocene beds include seven layers that have yielded abundant eoliths, now accepted by archeologists as the implements of primitive man. About the turn of this century there was found among these stone implements a stiletto made of deer horn that must have been shaped by human hands. Here, in the coastal margin of England, the beds carrying flint implements are overlain by a thin marine deposit with Late Pliocene fossils, thus proving that tool-using humans were living in western Europe a little before the beginning of the Pleistocene Epoch.

Slightly younger deposits (Early Pleistocene) of similar nature are found at Foxhall, a few miles northeast of Ipswich, and there the flints are associated with charred wood, suggesting that the ancient flint workers had already discovered the use of fire. At Cromer, still farther northeast of London, there is another bed (the Cromer Forest bed) of Early Pleistocene age, from which large but crudely chipped flint implements have been recovered. In none of these localities have any skeletal remains been found, but the crude implements indicate clearly the presence of man in England at the close of the Pliocene and early in Pleistocene time.

BEGINNINGS OF THE HUMAN FAMILY

Early human history is now based on a very few significant discoveries of primate fossils made in Africa and India. These seem to indicate that the human family (*Hominidae*) extends back into the Late Miocene, at which time it first became distinguishable from the family of apes (*Pongidae*).

Ramapithecus. As early as 1910 primate teeth and jaw fragments were discovered in the Miocene and Pliocene strata of the Siwalik Hills of northern India. The remains were attributed to a group of generalized apes called **dryopithecines** (p. 471). These animals were smaller than chimpanzees and gorillas but so much like them that they are considered to be the stock from which these modern apes evolved. In 1934 a new Siwalik primate fossil, *Ramapithecus*, was described by G. E. Lewis, who included it with the dryopithecine apes but pointed out the resemblance of its dentition to that of humans [22]. Modern restudy by Elwyn Simons of the Siwalik fossils, together with a recently discovered African jaw fragment from Kenya, has shown that *Ramapithecus* is much more like *Australopithecus* (see below) than it is like the dryopithecine apes and it is now regarded as the earliest representative of the human family [23]. The African specimen of *Ramapithecus* has been fairly reliably dated at about 14 million years by isotopic analysis of mica from an overlying bed; this places it within the Miocene [23].

Australopithecus. The African discoveries began in 1924 when quarrying in a limestone near Taungs, about 100 miles north of the

Figure 20-5 Cave of the Kids in the Valley of Caves (Wady el Mughara) in Palestine. The stratified refuse before these caves is about 70 feet thick and includes a sequence of 11 cultures ranging in age from pre-Acheulian to the Bronze Age. (G. G. MacCurdy.)

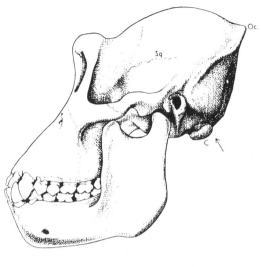

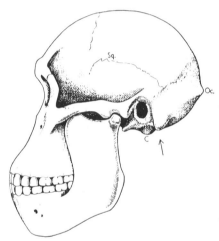

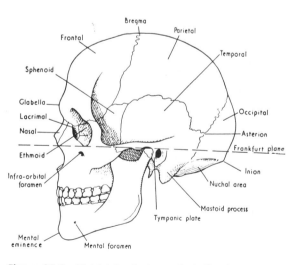

Figure 20-6 Right lateral views of skulls of the gorilla (top), *Australopithecus* (center), and modern man (bottom). (After LeGros Clark.)

494

Kimberley diamond mines in the Union of South Africa, brought to light a nearly complete and well-preserved skull covered with travertine. It came into the hands of a trained anthropologist, Raymond Dart, who, supposing that it was the skull of a great ape, proposed for it the name *Australopithecus africanus* [6], meaning the southern ape of Africa.

During Late Pliocene and Early Pleistocene time the area was an open veldt much as it is now. Trees were sparse but extensive limestone caverns about Taungs provided a safe refuge for *Australopithecus* and since 1924 they have yielded remains of no less than 30 individuals, including 25 skulls and a variety of limb bones. With these were associated an abundance of remains of other animals, which it is now believed were game brought into the cave by *Australopithecus* [6]. If so, this large primate was a hunter and a user of tools. As study of all these fossils progressed it gradually became evident that *Australopithecus* was more man than ape.

The evidence that he was a hunter, using tools and bringing his game into the caves, seems critical. No stone implements have been found, and he may not have made tools, but merely selected large limb bones to use as bludgeons and sharp splintered bones for stabbing and teeth for scraping skins. He was also selective in the game he hunted. The vast majority of kills were antelopes and baboons. Horses, giraffes, and wart hogs were less common. Dart [6] has pointed out that most of the baboon skulls had been crushed by a blow in the temporal region, a condition that could hardly have been due to accidental death. Moreover, the majority of these indicate a right-handed blow with a bludgeon such as a large limb bone, and some of the latter have the head of the bone broken down as though it had been so used.

But one problem remained. *Australopithecus* still had a prognathous face and a brain only intermediate in size between that of a gorilla and man (Fig. 20-6). Still it shows marked progress in other respects [4, 5]. The canine teeth were reduced in size and the dental arch was curved as it is in man (Fig.

Figure 20-7 Exposures in Olduvai Gorge.

19-28), the brow ridges were much reduced, and the head was more nearly balanced over the occipital condyle (the articulation with the spine). Furthermore, the thigh bones were straight, indicating that he walked upright, as does man, instead of slouching as do the apes. With this stance the powerful neck muscles attached to the back of the skull were reduced in size. In a recent summary of all that is known about *Australopithecus*, Professor LeGros Clark has made a convincing case for accepting him as a representative of the human line [5]. It may be noted that since all the apes have dark skins, the early humans were almost certainly not white men.

The geologic age of *Australopithecus* is still not well established. He was too old to be dated by radiocarbon and there are no associated igneous rocks from which other radioactive isotopes can be used. The rather large fauna of associated animals is our only basis for age determination, but since South Africa is so far from the classical sections of Europe on which our chronology is based, close comparisons are not yet possible. The date is tentatively placed at the base of the Pleistocene, approximately 2,000,000 years ago.

The Olduvai Gorge. Probably the most informative of all known localities for the record of early human history is Olduvai Gorge in Tanzania, south of Lake Victoria and west of the great volcano Kilimanjaro in East Africa (Fig. 20-7). During Late Pliocene and Early Pleistocene time this locality was a shallow, arid basin to the west of a chain of active volcanoes. The center of the basin was occupied repeatedly by a shallow lake along the margins of which primitive man lived.

The stratigraphy of the basin, complicated by the intertonguing of volcanics and fluvial beds from the east with the lake deposits in the center, has recently been carefully worked out by Hay [7] (Fig. 20-8). Since discovery of evidence of human camp sites Dr. and Mrs. L. S. B. Leakey [10, 12] have spent many years in the Gorge and in 1959 were rewarded by the discovery of a magnificent human skull (Fig. 20-9) to which they gave the name *Zinjanthropus boesei* but which is now generally conceded to be a species of *Australopithecus*. It was found at the spot numbered 2 in Fig. 20-8, near the top of Bed I, which also contains stone artifacts [7]. The associated volcanic rocks have been dated by the potas-

495

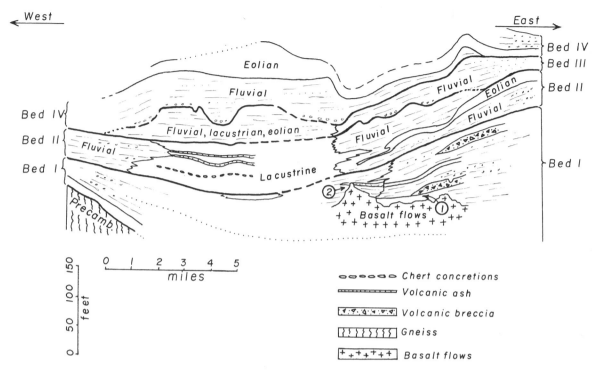

Figure 20-8 Section across the basin along Olduvai Gorge, showing the relations of Beds I to IV. The approximate position of the *Zinjanthropus* skull is marked by no. 2. (Courtesy of National Geographic Society.)

Figure 20-9 The skull of *Zinjanthropus.* (Courtesy of the National Geographic Society.)

sium/argon method as approximately 1,750,000 years old [11]. Since this is far older than had been expected, a large number of tests have been run on many rock samples both below and above the skull and they are relatively consistent and average at the date stated above, which now appears to be firmly supported. By 1951 Dr. Leakey had discovered a succession of 11 zones of human culture, each an old camp site with abundant stone implements. Of these, 4 are in Bed II, 2 in Bed III and 5 in Bed IV. It is evident therefore that primitive man occupied this basin repeatedly for a very long time and the artifacts are identified with several stages of the Acheulean cultures of Europe. In addition to the record of early man the beds at Olduvai Gorge have yielded spectacular faunas of other fossil mammals.

The significance of *Australopithecus boesei* is unparalleled because he is the only one of the very early human fossils that has been accurately dated. If it be granted that this is an australopithecine, it throws light on the age of the related species found in South Africa and confirms the judgment that they were tool users and were humans.

EARLY MEN OF THE FAR EAST

Homo erectus (Java Man). The most discussed of all human fossils was discovered in 1891 by Eugene Dubois, a Dutch army surgeon stationed on the island of Java. He had opened a quarry for vertebrate fossils in a 3-foot bed of gravel exposed in the bank of the Solo River, and there he came upon several human bones—a skull cap, a left thigh bone, fragments of nasal bones, and three teeth. Although each bone was isolated, and the thigh bone was found almost 50 feet from the skull, Dubois assumed that they belonged to one species if not to one individual, and recent application of the fluorine test confirms his inference that they are at least of the same age.

The skull cap was remarkably thick, the brow ridges very massive, and the forehead low and receding. The brain of this skull, estimated to have had a volume of 900 cubic centimeters, is intermediate in size between that of the largest apes (about 600 cubic centimeters) and the average for the lowest types of living men (about 1240 cubic centimeters). Moreover, the scars of attachment for the great neck muscles at the base of the skull clearly imply that the head was carried forward, as in the apes, instead of being well balanced on the neck, as in modern man.

Soon after discovery, this find was hailed as a "missing link" between the apes and man and was given the name *Pithecanthropus erectus* [Gr. *pithecos,* an ape + *anthropos,* a man]. Almost at once it became a subject of controversy. Skeptics argued that it was an abnormal individual, perhaps an idiot; but statisticians pointed out the extreme improbability of an abnormal individual being the sole survivor of a population to be preserved and discovered. All uncertainty was cleared up by the extensive and careful restudy of the area by Koenigswald [8] between the years 1935 and 1940, which brought to light three additional skulls. The last and most important of these (Fig. 20-10) includes the upper jaw, part of the lower jaw, and several teeth, along with the posterior and basal part of the braincase. It is somewhat larger and more massive than the original skull and is believed to be that of a male, whereas the original was female. These skulls fully confirm the interpretation previously made of the brain size and the shape of the head and face of *Pithecanthropus,* and prove beyond possible doubt that this is a well-defined but primitive human type.

The small brain, low forehead, heavy brow ridges, protruding mouth, and receding chin give the skull a striking resemblance to that of a great ape, as shown in Fig. 20-11; yet the brain is far larger than that of any great ape, the toothline is even, the canine teeth are relatively small and the dentition is in all respects human rather than simian; moreover, the straight thigh bone proves that he walked upright. Volcanics associated with the fossils indicate an age of about 500,000 years [9]. There is no longer any doubt that *Pithecanthropus* was human and he is now placed in the genus *Homo.*

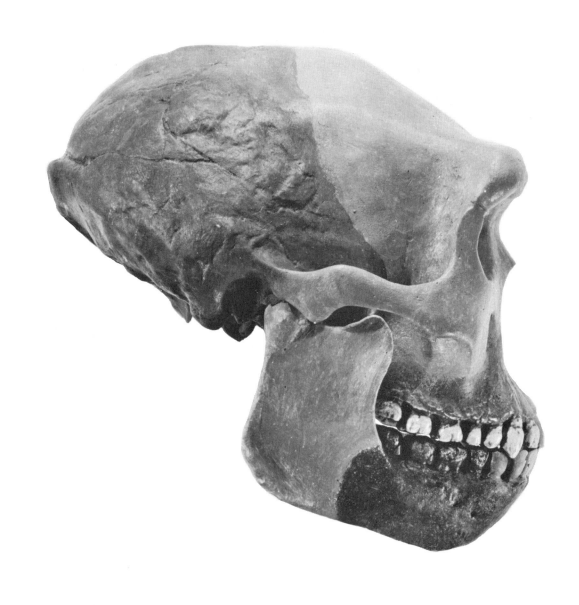

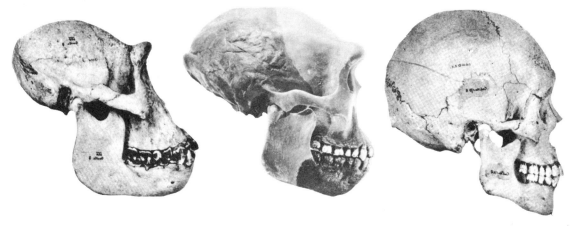

498

Six faunal zones are now known in the Pleistocene deposits of Java, and all the remains of *Pithecanthropus* are from a single one of these, the so-called **Trinil horizon.** Other human remains of more modern type are found in some of the higher zones.

Homo pekingensis (Peking Man). A series of discoveries in 1928 and 1929 near Peking, China, brought to light another species, commonly known as Peking Man. The remains were found amid cave deposits of Chicken Bone Hill (Chou Kou Tien) about 30 miles south of Peking [4]. At the time of habitation, the site was a spacious limestone cavern, but it has since been filled with debris fallen from the walls or washed in from above and cemented in part by travertine.

Once the great significance of this primitive human race was perceived, systematic exploration of the deposits was undertaken with the joint support of the Geological Survey of China and the Rockefeller Foundation, and for more than a decade, 50 to 100 technicians and laborers worked continuously at the excavation. As a result, about 40 individuals were recovered, including men, women, and children. In addition, a large fauna of contemporary mammals was found, many of which represent the prey that Peking man brought home from the chase. About seven-tenths of these are deer, suggesting that this was the chief animal hunted.

The human remains are nearly all skulls and lower jaws (Fig. 20-12), though a few limb bones have been found. The absence or rarity of other skeletal parts, as well as evidences that the base of each skull had its base broken away in a definite manner, suggests strongly that the heads had been severed from the bodies and that the brains had been eaten. Professor A. C. Blanc [3] of the University of Rome has advanced this interpretation, citing the observations of Wirz [21] on the Marind Anim tribe of New Guinea. This tribe opens the base of the skull in exactly this manner in order to extract the brain, which is then baked in a pie with sago and eaten as part of a ceremonial rite concerned with the naming of a child. The child is then

Figure 20-10 *Homo erectus.* Skull No. IV found by Dr. von Koenigswald in 1939, as restored by Franz Weidenreich. The darker parts are actual bone; the lighter parts are restored by comparison with other skulls of the same species. (Courtesy of Dr. Koenigswald.)

Figure 20-11 Skull of gorilla (left), *H. erectus* (center), and modern man (right). The apelike appearance of *H. erectus* may be seen in the low forehead, heavy brow ridges, protruding jaws, and chinless face. (After Weidenreich.)

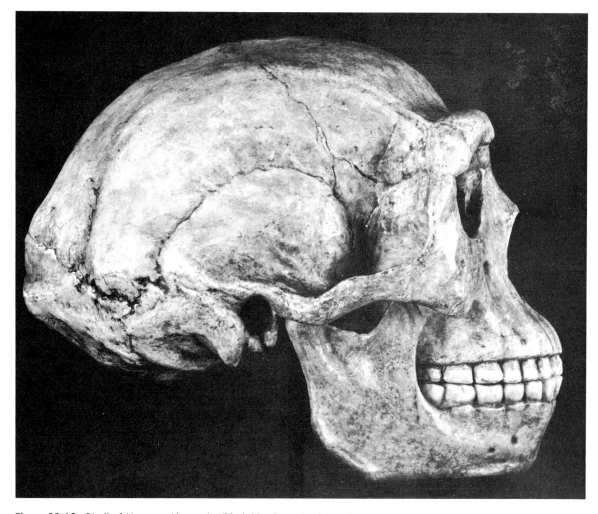

Figure 20-12 Skull of *Homo pekingensis*. (Model by Franz Weidenreich.)

given the name of the one whose brain has been eaten, the skull being dried and painted and preserved as a sacred household object during the life of the child.

It is a tragedy that during World War II all the hard-won remains of Peking Man disappeared. It is believed that they were packed by the Japanese for shipment to their homeland and were lost in transit. Fortunately, however, they had been critically studied and well described and illustrated by Davidson Black and Franz Weidenreich, and excellent casts of the most important skulls are still extant.

Associated with the skeletons have been found charred animal bones and layers with charcoal debris, ranging through a thickness of 20 feet of deposits. It is therefore clear that these people used fire. With them have also been found more than 2000 crude artifacts of the Chellean cultural type, made from greenstones and vein quartz. These stone implements include choppers, scrapers, gravers, and awls. Some of them were evidently used also for fashioning weapons from animal bones, such as the daggers made from deer antlers. In other words, Peking Man was already human, and was able to organize his life so as to select intelligently the materials useful for fuel, weapons, and tools, besides being a successful hunter of animals.

The associated mammals indicate that Peking Man was approximately contemporaneous with *Homo erectus* of Java. Indeed, there is some uncertainty whether the two are specifically distinct.

NEANDERTHAL MAN

Best known of all the extinct species of man is *Homo neanderthalensis,* who inhabited the caverns of western Europe during the last interglacial age and part of the last glacial age. The original discovery of this race was made in the Neander Valley near Düsseldorf, Germany, whence the name. Although found the previous year, this remarkable skeleton was first described in 1858, the year before the publication of Darwin's *Origin of Species.* Its striking characteristics and the timeliness of the discovery led to an immediate appreciation of the significance of the Neanderthals as a species far more apelike than any living men.

Since 1858, several entire skeletons have been recovered, and incomplete remains of many men, women, and children of the Neanderthal race have been found in the caves and rock shelters of Belgium, France, Italy, Spain, Croatia, Iraq, Crimea, and Palestine. Their stone implements (the Mousterian culture), moreover, are found scattered throughout western Europe and farther eastward in Asia Minor, North Africa, Syria, northern Arabia, and Iraq. Among the striking Neanderthaloid discoveries may be mentioned the skeleton at Broken Hill, Rhodesia; that near Galilee in Palestine; and more recently others at the Cave of the Kids near Mount Carmel, Palestine, and in caves in Shanidar Valley in the Zagros Mountains of Iraq [19]. From these abundant remains it is possible to present an adequate picture of the racial characteristics and the culture of these interesting people.

The Neanderthals (Fig. 20-13) were stocky and short of stature, rarely exceeding 5 feet 4 inches. Although they stood upright, their carriage was more like that of a great ape than is that of living man, because the spine lacked the fourth or cervical curvature and the thigh bones were sigmoidally curved in compensation. The head accordingly was carried far forward, and the body had a slouched appearance (Fig. 20-14). Both hands and feet were large, and the great toe was offset against the rest, as in the great apes.

The head differed from that of modern man in the very low forehead, heavy brow ridges, and receding chin (Fig. 20-15). The face was undoubtedly coarse-featured and brutal. Nevertheless, the brain was approximately equal in size to that of modern man (1400 to 1600 cubic centimeters), the brain case being low at the front but large in the back and lower part. It has been inferred from the proportions of the brain that the species was deficient in the higher qualities of reasoning and association, and probably less capable than modern man in social organization. It must be remembered, however, that Neanderthal man dominated all Europe during the last interglacial age and the early part of the last glacial age, a period estimated to exceed 100,000 years.

The Neanderthals made fairly good stone implements, and they also knew how to kindle a fire, for hearths have been found in their cave abodes. In at least two instances the skeletons have been found in their original burial places, where they had been laid away with implements, paints, and food, indicating that the race held a belief in immortality and buried the dead with ceremonial rites.

Among the stone implements of these people, the hand ax, scraper, and point are most characteristic. Flint-tipped spears were used, but there is no evidence that Neanderthal man used the bow and arrow. In view of his relatively feeble weapons, it is remarkable that he was a successful hunter of big game, including the bison, cave bear, horse, reindeer, and mammoth, all of which inhabited Europe during his reign.

Recent excavation in a cave in Shanidar Valley in the Zagros Mountains of northern Iraq [19] has revealed four stratigraphic levels bearing archeological remains, and in the next-to-the-lowest two Neanderthal skeletons were

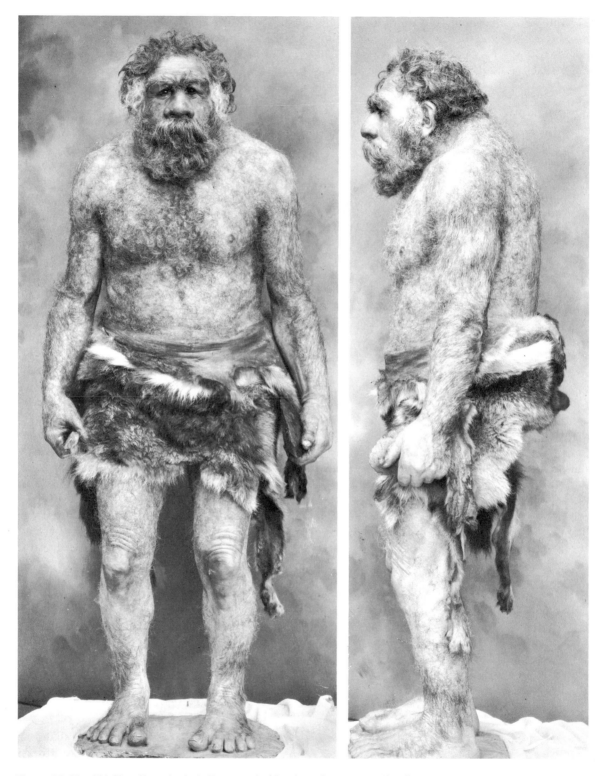

Figure 20-13 Old Man Neanderthal. Front and side views in a restoration by Blaschke in the Chicago Museum of Natural History.

Figure 20-14 Right, skeletons of Neanderthal man (left) and of a modern man showing contrast in posture. (After Boule.)

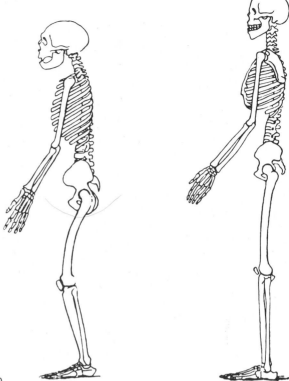

Figure 20-15 Below, Neanderthal skull from the grotto of Monte Circeo, south of Anzio, Italy. (A. C. Blanc.)

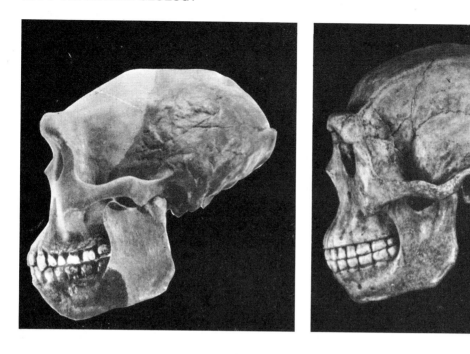

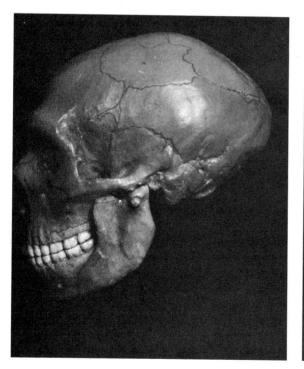

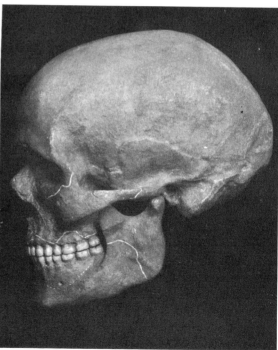

Figure 20-16 Skulls of *Homo erectus, H. pekingensis, H. neanderthalensis,* and Cro-Magnon man. (The first two are restorations by Franz Weidenreich, the last two by J. H. McGregor. All in the American Museum of Natural History.)

found along with artifacts of the Mousterian culture and hearths where fires had been tended. This layer is estimated to be between 45,000 and 60,000 years old.

The fine skull shown in Fig. 20-15 was found in the inner recesses of a former sea cave, now uplifted, at Monte Circeo on the coast south of Anzio, Italy [3]. The circumstances here prove that the cave was occupied late in the last interglacial age. The skull was surrounded by a ring of stones indicating some sort of burial ceremony, and the base had been broken away as were those of Peking Man, indicating that the brain had been eaten.

The Neanderthaloids resemble *Homo erectus* in their prognathous jaws, receding chin, massive brow ridges, and low forehead, in all of which they stand in contrast to modern man. They show a distinct advance over *H. erectus*, however, in having a larger brain, about equal in volume to that of modern man, though differing in shape.

In one important respect the Neanderthal people differ from both *H. erectus* and modern man, namely, in the curvature of the thigh bone, which gave them a slouched stance. It is now suspected, therefore, that they represent a specialized sideline in which the curved thigh bones, the very massive brow ridges, and thick skull bones became accentuated, whereas some of the less specialized remains currently attributed to *Homo neanderthalensis* are in the direct line to modern man and should be regarded as *Homo sapiens* [4, 20].

HOMO SAPIENS ARRIVES

The Cro-Magnons. Early in the last glacial age, about 35,000 years ago, modern men appeared in southern Europe and soon replaced the Neanderthalers. They have been called the Cro-Magnon race for the original discovery of five skeletons at the rock shelter of Cro-Magnon in the French village of Les Eyzies in the Dordogne Valley. They brought with them the distinctive Aurignacian culture.

The original find included the skeleton of a grown man, two young men, a woman, and a child. Numerous other remains have since come to light, so that the race is well known from many skeletons and from associated stone implements and other evidences of culture and art.

Unlike the Neanderthaloids, the Cro-Magnons were tall and straight, with relatively long legs, straight thigh bones, and the complete double curvature of the spine that permitted the balance of the head as in modern man. The chin was prominent, the jaws not protruding, the forehead high, and the brain fully as large as in modern races. In physical development the Cro-Magnons were essentially modern (Fig. 20-16).

In mental development, also, they were superior, for they had abundant and well-formed implements. They used bone for awls and ivory for skewers and ornaments; they made spears and bows and arrows; and they dressed themselves in fur. Their bodies they ornamented with sea shells derived from the Mediterranean and Atlantic coasts, with fossil shells from places far inland, with the teeth of mammals and even of human beings. Toward the close of the last ice age they made beads and bracelets and other objects of shell and ivory.

Endowed with more intelligence and better weapons than their predecessors, the Cro-Magnons lived better and enjoyed leisure that enabled them to develop a type of art and culture that has excited the wonder and admiration of all anthropologists. Besides personal adornment and the use of clothing, this artistic development was expressed in picture writing on the walls of their caves, in sculpture on fragments of stone or on bone, and finally in the polychrome paintings left on the walls of certain caverns in southern France and Spain (Fig. 20-17).

The Cro-Magnons were the last of the Paleolithic peoples in Europe. Their history is recorded in four cultural stages, the Aurignacian, the Solutrean, the Magdalenian, and the Azilian. It is probable that there is no break in the lineage, however, between them and the people who introduced the Neolithic culture into Europe about 5500 years ago and modern man.

Figure 20-17 The cave of Gargas in southwest France as restored by Frederick Blaschke showing a prehistoric artist of the Cro-Magnon race at work. Engravings of bison and elephants may be noted. (Chicago Natural History Museum.)

A significant recent find was that at Shanidar Cave in Iraq, mentioned earlier, where the Aurignacian culture is recorded in layer number 3 and material from near its base gives a radiocarbon date of 34,000 years ago [1]. The Magdalenian culture near Hamburg, Germany, gives radiocarbon dates ranging from about 9300 to 10,500 years ago [2].

When the remarkable cave art of the Cro-Magnons was discovered it seemed puzzling that the pictures were drawn on the walls of the most remote chambers where they could be seen only by artificial light. It now seems obvious that these chambers were holy places where religious and other ceremonial rites were performed (Fig. 20-18). Some of the finest paintings are on the walls

of the cavern at Lascaux northeast of Les Eyzies where charcoal from a hearth gives a radiocarbon date of about 15,500 B.P.

NEOLITHIC PEOPLES AND THE BEGINNING OF CIVILIZATION

As the last ice sheet disappeared from Europe, the climate moderated and became moister. The reindeer, which had been the chief source of food and clothing for Cro-Magnon man while the ice still occupied northern Europe, now vanished from most of the continent. These changes in climate and food were accompanied by human migrations as man spread northward in the wake of the vanishing ice. About this time the art of finishing stone implements by grinding and polishing was developed in southern Europe, and a new culture, the Neolithic, spread quickly over the continent. This was accompanied by the development of the art of making pottery, the domestication of animals, and the adoption of habits of communal life. Later on, permanent habitations in the form of stone or wooden huts or tents of skins became general, and agriculture was pursued. In order to secure protection, villages were commonly built on piles over lake shores, swamps, or streams.

An important factor in the development of communal life was the development of agriculture, and in Europe this centered around the cultivation of the wild einkorn that grows naturally in Armenia, Georgia, and Turkey, and from which some 14 modern species of wheat have been developed [15].

The oldest known association of einkorn with human culture is at Jarmo in Iraq, where radiocarbon dates the association at 4750 B.C., or 6700 years ago. It was domesticated at about the same time in the eastern Caucasus, Asia Minor, and Greece. Carbonized grains of einkorn have been found in the neolithic deposits of the Lake Dwellers in central Eu-

Figure 20-18 Idealized restoration of a cavern in France suggesting initiation ceremony of youths into the clan at puberty. (Painting by Rudolph F. Zallinger from *The Epic of Man.* Courtesy of Time, Inc.)

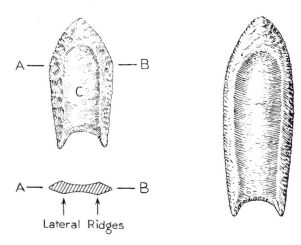

A — B

C

A — B

Lateral Ridges

Figure 20-19 Folsom points from the Lindenmeier ranch in Colorado. The longitudinal fluting of these finely chipped points is distinctive. (After F. Roberts.)

rope, and a modern type of wheat has been found at Mohenjo-Daro, in the Indus Valley, where it is dated at 2500 B.C. [1].

EARLY MAN IN NORTH AMERICA

All the Paleolithic races of men described above lived in the Old World. It is generally accepted that man evolved in that hemisphere, for no remains of the higher primates are known in America. The date of man's first migration to this continent is a problem that has been long under discussion, but one on which numerous recent finds have thrown light. It now appears certain that he arrived before the extinction of several of the characteristic Pleistocene mammals, notably the Columbian elephant, the American mastodon, a large extinct species of bison, a native camel, three species of horse, and the giant ground sloth. Indeed, it has been suspected that he may have contributed to the extinction of some of these great animals. All the American human fossils are attributable, nevertheless, to the modern species *Homo sapiens*.

A notable discovery of human bones was made at Vero, south of Daytona Beach, Florida, in 1916, and at Melbourne, about 40 miles farther south. Numerous fragments of human

remains occur at the latter place in a bed of sand that also contains bones of the giant ground sloth, two extinct species of horse, the mastodon, the Columbian elephant, and a saber-tooth tiger.

In 1926 stone implements were found near Folsom, New Mexico, associated with an extinct species of bison, *Bison taylori*. Later excavation revealed an arrowhead between two bison ribs, and eventually the remains of 40 to 50 specimens of the extinct bison were found within a small area, representing a kill and barbeque. Among these were 16 arrow points of a distinctive type (Fig. 20-19) now known as Folsom points. In recent years the Folsom culture has been found to be widely distributed in the United States, and at several places it is associated with extinct animals. A notable example is the Lindenmier site north of Fort Collins, Colorado, where at a depth of 14 to 15 feet below the present surface the Folsom culture is associated with a "bison kill" similar to that at Folsom. Here the remains of an extinct camel were found with the bison. From this ancient camp site some 2000 stone implements have been recovered, including scrapers, drills, gravers, and blades [18].

A similar occurrence was found near Plainview, Texas, in 1944, where extensive quarrying in a gravel bed led to the recovery of skeletons of between 50 and 100 bison of an extinct species larger than the modern buffalo [18]. With these were found 19 projectile points and 8 stone scrapers. It is believed that the bison were stampeded into falling from the river bluff. The artifacts resemble those at Folsom. At this quarry the only other animal found was a large wolf, but, near by, the same bed yielded the Columbian elephant and a fossil horse.

Burned bison bone associated with the Folsom culture near Lubbock, Texas, has given radiocarbon dates of about 9700 to 9883 years B.C.

Folsom points found in the northern foothills of the Brookes Range in Alaska tend to confirm the generally accepted belief that man reached North America from Asia by way of Bering Strait.

Artifacts have also been found deeply buried in river-terrace gravels at numerous localities in southwestern United States, and in several of these they are associated with remains of extinct mammals.

Terraces along Blanco Creek about 100 miles southeast of San Antonio, Texas, for example, have yielded six sites in which flint artifacts are associated with bones of extinct animals including elephants, mastodons, horses, bison, camels, and glyptodons.

The valley about Mexico City has yielded several bits of evidence of early man. A skeleton and associated artifacts were found beneath a lava flow in the suburban village of San Angel, and the same culture was found beneath 10 to 12 feet of sediments northwest of the city. The lava flow is believed to have occurred at least 2000 years ago, and possibly as much as 10,000 years. Hence these people long preceded the Aztecs, who date from about A.D. 500. In 1946 more remains were found near Tepexpan. Here a human skeleton was discovered in a layer that also yielded several artifacts (three gravers, a scraper, and a bone point), as well as bones of the imperial mammoth, bison, horse, and glyptodons.

These are but samples of a large number of occurrences of either artifacts or human fossils associated with the large extinct mammals that were common in North America during the Pleistocene Epoch.

One of the most remarkable archeological sites in America is Russell Cave [16], discovered about 1953 and now a national monument presented to the Government by the National Geographic Society in 1958. It is a limestone cavern in a hillside in Jackson County, Alabama, and was repeatedly occupied by prehistoric man since about 10,000 years ago. Stratified deposits more than 20 feet thick show a succession of cultures that have been dated by radiocarbon. The deepest of these includes a hearth from which charcoal gives a date of 9020 years, and the youngest is dated as A.D. 1650. Some 2½ tons of artifacts have been recovered from the cave, along with several skeletons.

MAN'S DAY ON EARTH

We might put human history in perspective by plotting it as a 12-hour period on the face of a clock. Assuming that our ancestors attained the human estate about two million years ago, each minute on the dial will represent about 2800 years, each second about 470 years; we begin at midnight.

From the appearance of *Australopithecus* in Africa to the arrival of Cro-Magnon man in Europe, our ancestors were nomadic hunters using implements of chipped stone and of bone, and searching for game and wild fruit. For long ages these tools were held in the hands—hand axes, choppers, scrapers, and bludgeons of wood or bones. It was about 6:00 A.M. when the Chellean people learned to make bolas by tying stones together with leather thongs. By 11:30 A.M. Neanderthal man was using flint-tipped spears, but it was 11:57 A.M. when Cro-Magnon man invented the bow and arrow and began making neolithic tools and adorning the walls of his caves with magnificent paintings. It was about this time that he began to domesticate animals and plants and to live in villages and tend his flocks. This marks the beginning of civilization; but our industrial revolution began only a split second before high noon!

REFERENCES

1. Anonymous, 1957, *Archeological discoveries in Iraq.* Science, v. 126, p. 834.
2. Barensen, G. W., Deevey, E. S., Jr., and Gralenski, L. J., 1957, *Yale natural radiocarbon measurements, III.* Science, v. 126, pp. 908–919.
3. Blanc, A. C., 1939, *L'uomo fossile de Monte Circeo e la sua posizione cronologica nel quadro del pleistocene glaziale.* Soc. Toscana Sci. Nat. Atti, Pr. verbali, v. 48, no. 2, pp. 19–20.
4. Clark, W. E. LeGros, 1955, *Fossil evidence for human evolution.* Chicago, Univ. of Chicago Press, 181 pp.
5. ——, 1967, *Man-Apes or Ape-Men?* New York, Holt, Reinhart, and Winston.

6. Dart, R. A., 1955, *Cultural status of the South African man-apes.* Smithsonian Ann. Rept. for 1955, pp. 317–338.

7. Hay, R. L., 1965, *Preliminary notes on the stratigraphy of Beds I–IV, Olduvai Gorge, Tanganyika.* In Leakey, L. S. B., *Olduvai Gorge 1951–1961,* V. 1, Cambridge Univ. Press.

8. Koenigswald, G. H. R., 1937, *A review of the stratigraphy of Java and its relation to early man.* In G. G. MacCurdy, Ed., *Early Man,* Philadelphia, Pa., J. B. Lippincott, pp. 23–32.

9. ——, 1961, *Age of the basalt flows at Olduvai, East Africa.* Nature, v. 192, pp. 720–721.

10. Leakey, L. S. B., 1951, *Olduvai Gorge.* Cambridge Univ. Press.

11. Leakey, L. S. B., Evernden, J. F., and Curtis, G. H., 1961, *Age of Bed I, Olduvai Gorge, Tanganyika.* Nature, v. 191, pp. 478–479.

12. Leakey, L. S. B., 1965, *Olduvai Gorge 1951–1961,* v. 1, *Geology and faunas.* Cambridge Univ. Press, 108 pp., 97 figs.

13. MacCurdy, G. G., Ed., 1937, *Early man.* Philadelphia, Pa., J. B. Lippincott.

14. MacCurdy, G. G., 1924, *Human origins,* V. 1, New York, D. Appleton Co., 440 pp.

15. Mangelsdorf, P. C., 1953, *Wheat.* Scientific American, v. 189, no. 1, pp. 50–59.

16. Miller, K. F., 1958, *Russell Cave: New light on Stone Age life.* Nat. Geographic Magazine, v. 113, pp. 426–437.

17. Oakley, K. P., 1952, *Man the tool-maker.* British Museum (Nat. History) Handbook, 98 pp.

18. Sellards, E. H., Davis, G. L., and Mead, G. E., 1947, *Fossil bison and associated artifacts from Plainview, Texas.* Geol. Soc. America Bull., v. 58, pp. 927–954.

19. Solecki, R. S., 1963, *Prehistory in Shanidar Valley, northern Iraq.* Science, v. 139, pp. 179–193.

20. Weckler, J. E., 1957, *Neanderthal man.* Scientific American, v. 197, no. 6, pp. 89–96.

21. Wirz, P., 1925, *Die Marind-Anim von Holl, Neu-Guinea.* In Univ. Abh. aus dem Gebiete der Auslandeskunde. Hamburg III, p. 59.

22. Lewis, G. E., 1934, *Preliminary notice of new man-like apes from India.* Am. Jour. Sci., v. 27, pp. 161–179.

23. Simons, E. L., 1968, *A source for dental comparison of Ramapithecus with Australopithecus and Homo.* South African Jour. Sci., v. 64, no. 2, pp. 92–112.

CHAPTER 21 *Epilogue*

For I dipt into the future
As far as human eye could see,
Saw the vision of the world
And all the wonders that would be.
TENNYSON

The long history of the past impels one to wonder what the future holds for our world and especially for mankind. No answers will be found by gazing into a crystal ball, but the geologic record may set guidelines for our speculation.

OUR ENDURING WORLD

For more than 4000 million years the earth has serenely pursued its measured course about the sun. Meanwhile the unflagging energy of our guiding star has bathed it in life-giving warmth and light. We can feel confident that our physical world will persist thus for the forseeable future — at least for many millions of years.

In the meantime regional changes in the surface environment have been endless, but such changes have been episodic rather than cumulative. Mountains have risen to impressive height only to be worn low within a few million years. And although all the lands have been under the constant attack of erosive forces, the continents have been renewed from beneath and now stand as high as ever. The climate has varied regionally from frigid to warm and from arid to humid, but extremes in one direction have been reversed after a time. The recent widespread glaciation can be matched at several levels in the remote Precambrian record. Tall corn now grows in Illinois and Iowa where summer temperatures occasionally exceed 100°F, and the vast wheat fields of the Dakotas stretch far to the north on the plains of Alberta; yet less than 20,000 years ago this entire region lay beneath the great Laurentide ice sheet. The widespread Permian tillites of South Africa are overlain by Mid-Permian formations that include much coal and a prolific fauna of reptiles that could scarcely have endured a mild winter freeze. Strongly arid conditions have been followed by extreme humidity in various parts of the world and vice versa. The Silurian deserts of New York and Michigan, for example, preceded the humid conditions of Pennsylvanian time when vast subtropical swamps covered much of the central and eastern United States. Conversely, the Fayum

desert southwest of Cairo now overlies Oligocene deposits of a humid swampy lowland. The causes of climatic change are complex and in part they are not yet understood, but the record of the past shows no evidence of a cumulative change in world climate, and we can feel confident that changes similar to those of the past will recur throughout the forseeable future.

THE CONSTANT CHANGE OF LIVING THINGS

In contrast to that of our physical world, the history of life has been one of ceaseless and **irreversible** change. Dynasties of animals (and of plants) have held the center of the stage for a time only to pass and be forgotten. Trilobites and ammonoids and dinosaurs are gone forever. Every living creature faces the competition of others and the problem of adjusting to changes in the physical environment. Species that cannot adjust fail to survive; those that are sufficiently adaptive are eventually molded by natural selection into new species. Some of the less specialized invertebrate animals have relatively long ranges, but among the vertebrates few genera have survived for as long as an epoch of time and the duration of most species is measured in tens of thousands of years. What, then, are the prospects for the future of mankind?

Our ancestors appeared in Africa about two million years ago as a single genus, *Australopithecus*. These were half-apes, walking upright and probably using tools of wood or bone, but their brains were only about half as large as ours and their jaw frame was rather larger than the braincase. Early in the Pleistocene Epoch they were replaced by the living genus, *Homo*, whose successive stages of development are marked by three dominant species. The first of these was *Homo erectus* (*Pithecanthropus* and **Peking man**). In these men the forehead was low, the face prognathous and chinless, and the brain considerably smaller than in modern man. Next came the Neanderthalers (*Homo neanderthalensis*) who spread over Europe during the last glacial age some 50,000 years ago. They had brains as large as modern man but they walked in a slouched position and lacked his high forehead and prominent chin. Nevertheless, they used a variety of stone implements and were competent hunters. Late in the Ice Age some population of the Neanderthalers evolved into the living species, *Homo sapiens*. The first well-defined tribe of our species were the Cro-Magnon men, who spread over Europe in the wake of the receding ice cap some 15,000 years ago. They were tall men, fully our equals in physique and brain size, and their high foreheads and prominent chin gave their faces the familiar profile of modern man.

Unlike all other species, modern man is no longer threatened by competition with other creatures. With modern firearms he can exterminate any large enemy, and with modern drugs and medical techniques he has already gained control over the microorganisms that once spread pestilence and disease. Advances in agriculture and animal husbandry have assured him of more and better food than the earth has previously known. Furthermore, he can effectively insulate himself from the physical environment as no other creature can. Clothing and permanent shelter protects him where the climate is cold and air conditioning and refrigeration shield him where it is hot. He can drain swamps and exterminate their noxious insects, and he can make the deserts bloom by irrigation. He is the only creature in all the history of life with the capacity to control his environment and to assure his survival. As the master of all he surveys, **he has only himself to fear.**

But there are danger signs ahead and it is the burden of this chapter to discuss them.

THE HAZARD OF ATOMIC WAR

The stockpiles of thermonuclear weapons developed during the last two decades place in jeopardy not only the human race but the whole animal kingdom. **No such power of death has existed before in all the history of the earth.** While the two most powerful nations face each other in a balance of terror other nations strive to join the Atomic Club. It is inconceivable that the present leaders of the great powers would willfully launch an all-out atomic war. But as long as military leaders contemplate the use of atomic warheads as tactical weapons in a local war, the danger is appalling that man may unwillingly drift into a major world war.

Only a madman would launch an atomic war under present conditions, but history records a number of seemingly insane rulers, one of whom undertook to exterminate an entire race and to conquer the world by launching the most powerful rockets at his command on unprotected cities. And if the ability to produce atomic bombs spreads to many nations, the danger is compounded. A single bomb launched by a madman, or by an error, anywhere might start a chain reaction that would ignite the world. It should be the most urgent objective of all the nations to work for peace and understanding that would make possible the reduction of the present stockpiles of atomic weapons and prevent their spread.

THE POPULATION EXPLOSION

In the process of evolution nature has endowed every species with the capacity to produce far more offspring than can survive to sexual maturity. This capacity is most spectacular in the lower animals that do not care for and protect the young. A salmon, for example, may lay 100,000 eggs in a single season and a marine turtle will bury scores of eggs in a single clutch, yet on the average less than two of each species will survive to maturity. In the higher animals where the young are nurtured and protected, fewer are born each year; but, in any case, if more are born than die, the population tends to increase exponentially.

This presents a problem well illustrated by the oft-told allegory of the little girl who once begged her father for a penny and promised that if he would double it each day for a month she would never ask for more. The unsuspecting father readily agreed and gave her a penny and on the next day 2 pennies. But on the seventh day he had to pay her $1.28; on the fifteenth day $327.68; on the twenty-first day $20,972.32 and on the thirtieth day

$10,737,827.84. The total for the month amounted to more than $21,000,000!

Where natural checks are removed and an inflationary increase in the population ensues the results can be equally spectacular. For example, about the middle of the last century a litter of rabbits was introduced into Australia; in the absence of natural enemies, they multiplied so fast that within a few decades they were stripping the range grass and threatening the sheep and cattle industries. A massive campaign of poisoning and hunting was necessary to keep them in check.

About the same time an immigrant family brought a few cactus plants for their yard and later threw them away. Another family introduced blackberries for their garden. In the absence of the insects that keep them in check in their native land, both plants went wild and within a few decades had transformed thousands of square miles of virgin prairie into bramble patches of blackberries and cactus. To save the grazing lands the Australian government finally sent its distinguished economic entomologist, R. J. Tillyard, to America to find and bring back live insects that keep both plants in check in their native homeland.

Humans do not multiply as fast as rabbits, but they live much longer, and it is the ratio of births to deaths that determines whether the population will expand or decline.

Nature maintains a balance in the wild kingdom by ruthless slaughter of the young and by famine and disease. "Little fish have larger fish upon their backs to bite 'em, and on these fish are larger fish and so ad infinitum." A rapid increase in the population of a herbivorous mammal is followed shortly by an increase in the carnivores that prey upon it. And if by chance this is an inadequate check, overgrazing results eventually in famine. In the forest most any tree will produce hundreds of thousands of seeds each year, but nearly all of them fall in the shade where young trees cannot survive. Only as the old trees die is room provided for the young to replace them.

When man appeared there was a wide world to populate and it was not until a few thousand years ago when he ceased being a roving hunter and settled down to domesticate animals and plants and to build cities that congestion became a problem. Then the Four Horsemen of the Apocalypse began to play their cruel role: Conquest, War, Pestilence, and Famine provided an effective population check.

But since the Renaissance, better standards of living and, more recently, spectacular advances in medicine have upset the balance. The dread diseases—yellow fever, smallpox, scarlet fever, polio, and the rest—have almost disappeared, and life expectancy steadily increases. Human population is expanding out of control. It is estimated that in 1770 the world population was about 700,000,000. By 1900 it had doubled to 1,600,000,000; and by 1965 it had doubled again to 3,300,000,000; and at the present rate of increase it will reach 7,000,000,000 by the end of this century [2]. In short, the time required to double the population has decreased progressively since 1770, from 150 years to 65 years and then to 35 years. The increase is most rapid in the underdeveloped countries—India, China, Latin America, and parts of Africa. For example, at the present rate of increase the population of Costa Rica will double within the next 10 years [4]!

Where the population is already dense and the increase is most rapid, many people live in ignorance and squalor and famine is an ever-present threat as it is now in India and China. Even in affluent America, where the population growth is less rapid, our cities are growing into vast megalopolises that are decaying at the centers into slums. Traffic congestion outruns our capacity to provide transportation, and more people are being killed on the streets and highways than in all our wars. Sewage disposal has become a critical problem and we are polluting our rivers and lakes so that many of our cities face a crisis in the search for adequate clean water for domestic use. Air pollution has become a growing menace to the health of millions of our people.

Meanwhile the intense competition for wealth and status in the advanced nations has made heart attacks man's great killer, mental disorders grow like a cancer in his society, and the steady increase in crime is a shocking menace. In the most affluent society the world has ever known, millions of our citizens are denied social justice and live in slums where ignorance and degradation spawn unwanted children and crime. At the current rate of reproduction the world population will exceed 14,000,000,000 people by the year 2050 and people now living may see a world of hungry men fighting for room and for food. We may be sure that if we drift into such chaos nature will strike a balance by means of wars and famine. This is the most awesome problem faced by man.

But surely there is a more humane method. By family planning and birth control man can stabilize and even reverse the population explosion. The means are already available and will undoubtedly be improved within the next few years. Great obstacles stand in man's way—ignorance, inertia, and dogma—but the time bomb is ticking away and the crisis is upon him. Will we take action in time?

PROBLEMS OF CLIMATIC CHANGE

Climatic changes have been slow but inexorable throughout the past, and they will undoubtedly continue. We cannot foresee what direction they will take next, but it may be wise to contemplate some of the possibilities.

During the Pleistocene Epoch glaciation occurred four times and the three interglacial ages were each much longer than postglacial time. We may be living in a fourth interglacial age. If the ice should return again, it would force mass migrations on a scale without parallel in human history. If North America were still sparsely settled as it was two centuries ago it could accommodate the migrations from the north, but the crowding in southern Europe would be most critical. The population explosion has already made the prospects frightening. If the present ice caps of Greenland and Antarctica should melt away, sealevel would rise to drown all the major seaports of the world. On the other hand, a return to more humid conditions in Central Asia, the Sahara, the continent of Australia, and the western United States would expand the useful land. None of these regions has always been as arid as it is now.

DEPLETION OF OUR NATURAL RESOURCES

Before the Industrial Revolution began nearly all the world's people still lived close to the land, on farms or in rural villages. Food was grown locally, homes were simple, and clothes and household conveniences were made by handicraft, either in the home or in small shops in the villages and towns. The world's work was done by beasts of burden or by human toil. To be sure, the royalty and landed aristocrats commonly enjoyed wealth and luxury, but the great masses of mankind led simple and frugal lives. Pestilence and disease were a permanent threat, and life expectancy was less than it is now. All the past civilizations shared these characteristics and even today millions of people in the underdeveloped countries still live in this simple way.

In the advanced nations industry is made possible by the use of machines powered by the stored-up energy of fossil fuels, coal, and oil, recently supplemented by atomic fuels. Industrialization began in Europe about 1750, quickly reached to North America, and is now spreading like a tidal wave to the ends of the world. In all the history of mankind the industrial culture stands unique. Machines have lifted the burden from human shoulders and, by multiplying our productive capacity, have enriched our lives with comforts and luxuries such as the world has never known before. Moreover, in the industrial nations these benefits now reach the great masses. The conveniences in a middle-class kitchen would have been the envy of royalty in any previous civilization.

Within living memory, advances in science and technology have outrun even the imagination of Jules Verne and we have come to think that we can accomplish anything we can imagine. It is small wonder that our political and industrial leaders are committed to the philosophy of an ever-expanding economy. It may seem like apostasy, therefore, if we suggest that this is an impossible dream, and point to some of the problems that lie ahead. We do so only because problems must be understood if they are to be solved or avoided; and massive social changes cannot be made quickly or without turmoil.

The Sinews of Industry. Industry depends on several factors. Science and invention first made it possible. Without these the Indians had inhabited this rich continent for at least 10,000 years and were still living in the Stone Age when the white man arrived. The invention of the steam engine, the harnessing of electricity, and the techniques of mass production made industry come alive. Capital and business organization have played a vital role and the lack of these is one of the major obstacles hindering the underdeveloped nations in their struggle to achieve industrialization. But, above all else, industry depends on the use of metals and fuels, both of which lie buried in the earth's crust. They have been concentrated into usable deposits by geological processes that operate so slowly that none will be renewed or appreciably increased within the next million years. They are man's heritage from the long geologic past, and in exploiting them he is using not the interest but the principal of his inheritance. Fortunately they are vast and in large part they are not within easy reach, **but they are finite.** Man's industrial societies have already depleted all of the more readily accessible rich deposits, and in an ever-expanding economy his demands upon them grow at an ever-increasing rate. As Abelson recently observed, we in America "have had a cavalier attitude toward our environment and toward our resources. We have acted as if we had to consume everything in sight within a generation or sooner." This is in the nature of *all* expanding industrial economies. Their appetite is insatiable. The case histories of petroleum and iron will put the problem in perspective.

Petroleum. As of today petroleum is our most important source of energy, but its use began only a little more than a century ago when the first oil well in the world was drilled near Titusville, Pennsylvania. The year was 1859 and the well produced 10 barrels of oil daily. Two years later a larger producer was drilled near Paola, Kansas. But for several years the chief petroleum products were lubricants, and kerosene for use in "coal oil" lamps and domestic stoves. The invention of the internal combustion engine in 1867 soon created a demand for gasoline to drive the machinery in factories. The automobile came into use about the turn of the century and the airplane followed about 15 years later. Trucks, powered by diesel engines, followed shortly, and during recent decades the coal-burning locomotives have been replaced by diesel engines. Meanwhile millions of our homes are now heated by fuel oil. With the pyramiding of all these technical advances the demands for petroleum have skyrocketed to astronomic heights. By 1950 the United States was using 6,500,000 barrels of oil daily. By 1960 our demands had jumped to 7,000,000 barrels daily and the Standard Oil Company has estimated that by 1975 we will be using 13,000,000 barrels daily. In 1966 the American Association of Petroleum Geologists published the estimate that by 1985 the need will be 17,500,000 barrels daily [3]. This will amount to 6,387,500,000 barrels annually. But our domestic production is already on the decline and we are importing large amounts of oil from foreign fields, chiefly in Venezuela and the Middle East.

Our petroleum reserves are still great but they are not boundless. During the 1920s flowing wells and gushers were still common and the average depth of production was not over 4000 feet; but today we seldom hear of a flowing well and the average depth of production exceeds 5000 feet. Meanwhile the percentage of dry holes increases. The number of exploratory wells increased from 6000 in 1945 to a peak of 16,000 in 1956 but had declined to 10,000 by 1965. In 1945 21.6 percent of the exploratory wells were rated as producers but by 1965 the percentage had dropped to 15.4 and many of these were too small to be commercially valuable. In 1959, for example, when 9000 exploratory wells were drilled, about 1 in 10 were rated as producers but after 6 years only 121 of these were still being pumped [3]. And year by year as the percentage of dry holes increases the cost in capital and labor goes up. For example, in 1965 we drilled a total of 9,000,000 feet to get producing wells, whereas 40,000,000 feet of drilling was unproductive.

The land areas of the United States are now so extensively explored that there is little probability that new major oil fields will be discovered. The greatest promise now is in submarine oil pools along the Gulf shelf, and prospective fields in the inhospitable coastal plain of extreme northern Alaska.

In the mid-1950s when we were still at the peak, one of the major oil companies published an attractive brochure entitled *Oil for the Foreseeable Future*. Shortly afterward the petroleum engineer of the company confided to one of us that the data for the report were secured as follows: He had written to each of the large producing companies asking for an estimate of their reserves for the next 5, 10, and 15 years, respectively. We submit that 15 years is a poor gauge of the foreseeable future.

In contrast, a report of the National Fuels Study Group [4], published in 1962 under the auspices of the U. S. Senate, stated: "By 1980 our energy needs will double. If that proves true, we will use as much energy of all kinds in the next two decades as we have used in all our previous history dating back to the American Revolution." And the Atomic Energy Commission recently quoted from the same report as follows: "At today's rate of fuel consumption, our total recoverable resources of fossil fuel (coal, oil, and gas combined) would last only 800 years. But when projected increases in the rate of consumption are taken into account the estimate of 800 years shrinks to 200 years more or less. And if low grade sources such as lignite and oil shale are left out of the calculation, the estimate shrinks to 100 years or less."

Now if we look backward, 100 years would take us to the close of the Civil War, 200 years to the signing of the Declaration of Independence, and 800 years to the Crusades. These are short steps in the history of Western civilization!

Perhaps the estimates quoted above are too low. The oil shales will someday be used; and atomic power will supply an increasing share of our energy. If we succeed in harnessing controlled atomic fusion (which we have yet to do), we shall literally have energy for the foreseeable future. But our study thus far should indicate some of the problems involved in the use of finite resources in an ever-expanding economy. Among other things our automobile industry may sooner or later have to shift to electric instead of gasoline engines. And our petroleum geologists and engineers and our coal miners may then have to find new employment.

Iron. Above all else industry depends on machines made of metals. A bit of make believe will prove the point. Let us suppose, then, that some evil genie should suddenly turn all the metals into common rock. If so, every automobile, every truck, every airplane, and every train would suddenly stop. The machinery in every factory would cease to function. We can neither generate nor transmit electricity. If this happens suddenly, every city will face mass starvation, for only locally grown food will be available. Our tools will be made of flint or bone or wood and our clothes will be made of skins until our women can learn again how to card and spin cotton and wool and weave cloth. In short, we will suddenly be back in the Stone Age.

But this bit of whimsy is not so fanciful after all, for industry destroys the metals upon which it feeds. They are soon scattered and left to rust and waste away. Fortunately, the loss is not sudden and this gives us time to plan for the future. One hundred years ago we in the United States imported most of our iron and steel from England. The fabulously rich deposits of iron ore in the Lake Superior Region had been discovered several years earlier but they lay in the wilderness far from the centers of industry and their exploitation had to await the arrival of a railroad. In 1864 they yielded 250,000 tons of iron. Within the next 9 years the output had quadrupled to a million tons a year, and in the next 75 years it had increased a hundredfold, since in 1948 we produced 108,000,000 tons of iron. The cumulative production from colonial days to 1948 amounted to 1,836,000,000 tons; but if it had all been available to use at the rate current in 1948 it would have lasted only 17 years!

World War II made enormous inroads on our reserves and by its close deep concern was felt because the rich hematite ore of the Lake Superior region was fast approaching exhaustion. In 1946 C. K. Leith reported that if the war had lasted two years longer we would have been hard pressed to produce enough iron to meet our urgent needs.

Our national concern was temporarily relieved by the almost simultaneous discovery of vast foreign deposits in the tropical jungles of Venezuela and in the barrens of central Quebec. We immediately prepared to exploit both. Bethlehem Steel opened an outlet from the Venezuelan field to the coast and built an enormous processing plant at Fairless, Pennsylvania. Simultaneously our government, in collaboration with the Canadian government, began construction of the St. Lawrence Waterway whereby ore from the Quebec field can be brought in barges through Lake Erie to our great inland steel mills and coal fields. Thus our production of iron has continued to rise, reaching 127,000,000 tons in 1964; but in the meantime 33 percent of the iron was coming from foreign fields and our domestic production had fallen by almost 25 percent in 16 years.

In view of our expanding economy it is evident that if the deposits in Venezuela and Quebec were each as vast as that of the Mesabi Range of the Lake Superior region, and if we had the exclusive use of each, they would probably be exhausted within the next 50 years.

It is likely that other rich deposits will yet be discovered in some of the undeveloped and only partially explored countries, but since all of these are now striving to attain industralization, the competition for such ores will certainly increase.

Fortunately iron is one of the most abundant elements in the earth's crust and lower grade ores, not now profitable, will eventually be used. Recently man has discovered how to beneficiate and utilize taconite, the low-grade refractory silicate from which the rich hematite ores of the Lake Superior region were derived by millions of years of weathering. This is our largest potential reserve and it has recently been estimated that if it were completely mined out, to a depth of 3000 feet, it may last for 250 years. When this last great reserve is exhausted, what then?

Of course we will not suddenly reach the "bottom of the barrel." Thus far we have been skimming the cream of the richest deposits, most of which lie at the surface where they can be mined in open pits with great steam shovels. As these are exhausted and we turn to leaner ores, the cost in both labor and capital will increase and our productive capacity will gradually decline.

The substitution of other metals for iron is hardly a promising long-range solution for our problem since all the metals are already being exploited and none are as abundant as iron. At present aluminum is the major substitute, but the chief aluminum ore, bauxite, is essentially a superficial accumulation formed by the weathering and leaching of granitic rocks under warm and humid conditions during the present erosion cycle. In North America bauxite deposits are quite limited and our supply comes chiefly from foreign fields in tropical and subtropical regions of slight relief. Unless we can develop techniques to extract aluminum cheaply from fresh rocks, the world reserves of aluminum are more vulnerable than those of the iron ores.

If our reasoning is valid, the extravagant industrial culture that has become our way of life may reach a crest within the next few centuries and then begin a slow decline. If we are more frugal in our use of metals, we may preserve the best in our culture and reduce some of its harmful aspects. *Our most important natural resources are not expendable.* We can preserve and enrich the soil, we can renew our forests and improve wood technology, we can extend irrigation, and, by improved agriculture and animal husbandry, we can increase the food supply. Our future will depend on our imagination and on the continuing advances in science and technology. We will probably have fewer automobiles and more bicycles and we will walk more.

OLD PATTERNS AND NEW PROBLEMS

The prospect of major social and cultural changes will not be surprising to historians; such changes have happened before in human history. Each of the great civilizations of the past has risen to a crest within a period of several centuries and then declined as another took the center of the stage. People survived, of course; only the culture and human institutions changed. But this simple pattern has little relevance to man's present predicament. In contrast to the localized influence of earlier civilizations, the impact of the present industrial civilization is global.

Perhaps the major cumulative effect of modern civilization and its burgeoning population has been the gradual spoiling of man's natural environment; this has gone essentially unheeded until the present decade. Overpopulation together with uncontrolled industrialization and indiscriminate use of chemicals has introduced pollutants into the air man breathes, the water he drinks, and the food he eats. The poisoning of his environment is most obvious about his industrial centers, which are, ironically, also the showplaces of his culture and learning. But the effects of this pollution have proved to be far-reaching and man's ability to influence the environment of the entire surface of the earth has been established beyond question.

For example, it is estimated that over the past several decades the carbon dioxide content of the earth's atmosphere has increased nearly 10 percent chiefly as a result of the burning of coal and petroleum products. We also know that poisons from certain common pesticides, directed at one or two species, survive in the environment to harm many other organisms. Detectable amounts of some of these poisons are now found throughout the oceans of the world.

The evident dangers of overpopulation and the concomitant pollution and depletion of the natural environment are crises that man has not previously faced. They transcend his national and factional groupings and call for levels in the control of his own destiny that he has never before attained, and he is unlikely to attain them without far-reaching changes in the structure and objectives of his present institutions. The crucial problems lie clearly before him, the task is immense and the challenge to his intelligence, humanity, and ingenuity has never been greater.

REFERENCES

1. Anonymous, 1966, Population Reference Bureau press release of February 7.
2. Cook, Robert C., 1965, *World Population Projection 1965–2000.* Population Bulletin, v. 21, no. 4, p. 92.
3. Dillon, E. L. and L. H. VanDyke, 1966, *Exploratory drilling in 1965.* Am. Assoc. Petroleum Geologists Bull., v. 50, pp. 1114–1138.
4. Paley, W. S. et al., 1952, *Resources for Freedom,* U. S. Government Printing Office, Washington, D. C.

APPENDIX

An Introduction to Animals and Plants

Scientific Names. The beginner in natural history is usually dismayed by the scientific names of unfamiliar types of animals and plants, and is inclined to wonder why common names are not used instead. The answer to this is very simple: a name is common only because it is familiar. Such words as **boa constrictor**, **Gila monster**, and **rhinoceros** are common, but a small child finds them fully as difficult as *Homo sapiens* (man), or *Equus caballus* (the horse). Moreover, only a few thousand kinds of animals are at all commonly known, and the rest, already exceeding 825,000 kinds, can therefore have no really "common" names.

Furthermore, so long as we have a diversity of languages, there can be no really common names of general and worldwide usage. To the Germans, the common name of the horse is **Pferd**; to the French, **cheval**; to the Italians, **cavallo**, etc. Moreover, a common name such as **bear** has many different meanings: to a New Englander it implies one species, to a Montanan another, and to an Alaskan a quite different kind of bear. Therefore, if naturalists of all countries are to share in scientific studies, it is obviously necessary that each species should bear a name that applies to one kind alone and is recognized in all languages. For such names, naturalists have turned to the classical languages, Greek and Latin.

The early naturalists did, indeed, give each kind of animal but a single name. Thus in the Roman Empire, the cat bore the name *felis* and the horse was called *equus*. This scheme sufficed so long as only a few hundred kinds of animals were known, and the scholars of the civilized world all lived in a relatively small area about the Mediterranean. But as culture spread during the Middle Ages, and animals from other regions were studied, it became necessary to accompany each name with a short diagnosis or description in order to distinguish it from some similar, previously known kind. Latin scholars at first used the name *felis* for the domestic cat, but they later became acquainted with the great tawny cat of Africa, as well as the spotted cat of the tropics.

When it became necessary to accompany the common name with sufficient adjectives or descriptive phrases to indicate clearly which kind was meant, the great Swedish naturalist Linnaeus devised the plan of giving each species a double name, the first representing the group (genus), the second a special or specific name for that particular kind (species). The species name is generally a descriptive or qualifying adjective standing in place of a descriptive paragraph, and, in harmony with the Latin custom, it follows the word which it modifies. Thus the house cat became *Felis domestica*, and the great spotted cat *Felis leopardalis* [L. *pardalis*, spotted]. The scientific name is therefore in reality a nickname, or an abbreviation of the longer diagnosis that would otherwise be required.

It should be noted that the generic name is capitalized, and that the specific name is not.

Classification of Animals and Plants. In dealing with any large or complex group of objects, some scheme of classification or orderly grouping is required. To appreciate this fact, we need only contemplate an army of individuals without organization, a great library with the books placed at random on the shelves, or a dictionary with the words arranged by chance! Nowhere is the need for organization more keenly felt than in the study of the enormously varied forms of animal and plant life. Here, obviously, the most useful basis of classification is blood kinship, and a **biologic** classification has therefore been adopted which aims to group creatures according to their degree of actual relationship, regardless of superficial resemblances or differences.

In this biologic scheme, the animal and plant kingdoms are divided, first, into **phyla** [Gr. *phylon*, stock or race], each phylum including organisms that are alike in some fundamental anatomical characters. For example, animals with backbones form the phylum **Vertebrata** [L. *vertebratus*, having a backbone]; those with jointed legs and bodies, such as insects, spiders and crabs, form the phylum **Arthropoda** [Gr. *arthron*, joint + *pous*, foot].

Each phylum in turn is divisible into **classes**, within which the resemblances are still closer. For example, among the vertebrates the fishes constitute one class, birds another, and mammals a third. Classes are further subdivided into **orders.** Thus the class **Mammalia** includes the orders **Carnivora** (flesh-eating types), **Rodentia** (gnawers like rats and squirrels), etc. Orders are divisible into **families,** and these in turn into **genera,** each genus including one or several **species** of animals that are very closely related and structurally alike. For example, the cats form the genus *Felis*, and the dogs and wolves the genus *Canis*. The species is the next smallest unit, including individuals very closely alike.

ANIMALS

The following table presents in simple form the major subdivisions of the animal kingdom; (see also Fig. 6-1).

A Simple Classification of the Animal Kingdom

Phylum **Protoza** — single-celled, generally microscopic animals. Examples: amoeba, foraminifera, radiolaria.

Phylum **Porifera** — sponges.

Phylum **Coelenterata** — coral-like animals, lacking viscera.

    Class **Hydrozoa** — hydroids, stromatoporoids.

    Class **Anthozoa** — corals and sea anemones, tetracorals, hexacorals, honeycomb corals.

Phylum **Platyhelminthes** — flatworms (never fossil).

Phylum **Nemathelminthes** — threadworms (never fossil).

Phylum **Trochelminthes** — rotifers, all microscopic (never fossil).

Phylum **Brachiopoda** — brachiopods.

Phylum **Bryozoa** — moss animals.

Phylum **Echinodermata** — echinoderms.

    Class **Asteroidea** — starfish, brittle-stars.

    Class **Echinoidea** — sea urchins, heart urchins, sand dollars.

    Class **Crinoidea** — sea lilies or feather-stars.

    Class **Blastoidea** — sea buds or blastoids.

    Class **Cystoidea** — cystoids.

Phylum **Mollusca** — mollusks.

    Class **Pelecypoda** — bivalves, clams, oysters, scallops.

    Class **Gastropoda** — snails, conches, etc.

    Class **Cephalopoda** — squids, octopuses, nautiloids, ammonoids and belemnoids.

Phylum **Annelida** — segmented worms, earthworms, beach worms, etc., fossils rare but traces common.

Phylum **Arthropoda** — invertebrate animals with jointed legs.

    Class **Trilobitoidea** — trilobites.

Class **Crustacea**—lobsters, crabs.
Class **Myriapoda**—centipedes, millepedes, etc.
Class **Chelicerata**—spiders, scorpions, eurypterids.
Class **Insecta**—insects.
Phylum **Vertebrata** (Chordata)—animals with backbones.
    Class **Pisces**—fishes (actually four classes).
    Class **Amphibia**—salamanders, frogs, labyrinthodonts.
    Class **Reptilia**—crocodiles, turtles, dinosaurs, ichthyosaurs, plesiosaurs, mosasaurs, pterosaurs, snakes.
    Class **Aves**—birds.
    Class **Mammalia**—milk-feeding, warm-blooded animals (including man).

Phylum Protozoa

The so-called single-celled animals constitute the phylum **Protozoa** [Gr. *protos*, first + *zoön*, animal]. Although widely distributed and extremely numerous, protozoans are nearly all microscopic and therefore seldom seen.

Each protozoan is a tiny droplet of fluid living matter enclosed in a membrane. Unlike higher animals, it has no special visceral organs for digestion, circulation, reproduction, etc. Food consists of other microscopic creatures, commonly plants, which are swallowed whole. Some lack a mouth and take food through a temporary rupture in the cell wall, and later void the indigestible residue by the same means. Once in the body, the food particle is attacked by fluids that digest it, and the single-celled animal assimilates this food without the need of a circulatory system. When fully grown, the tiny animal reproduces by simply splitting into two or more young, each of which is like the parent but smaller. Since the parent passes completely into its offspring, **there is no death** in the normal course of events for these simple creatures. Individuals may be killed, of course, by unfavorable environment or by other animals; this, however, may be considered accidental, for death is not the inevitable fate of each individual, as it is with all higher animals.

Although protozoans probably exceed all other types of animal life combined, both in number of individuals and in total bulk, the vast majority are soft-tissued and incapable of fossilization. There are, however, two prolific groups of them that form delicate shells of calcium carbonate or silica, and have left an imposing record. These are the orders Foraminifera and Radiolaria.

The **Foraminifera** (Fig. A-1) build tiny chambered shells, commonly of calcium car-

Figure A-1 Protozoan shells. Left, globigerina ooze (×10) dredged from a depth of 2898 feet about 100 miles west of Martinique, West Indies. Right, radiolarian ooze (×35) from the Miocene beds of the island of Barbados, West Indies. (Yale Peabody Museum.)

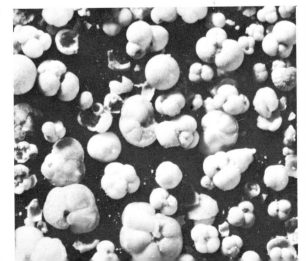
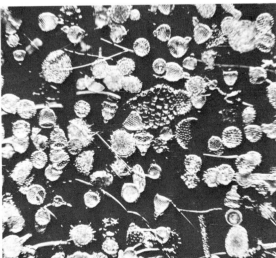

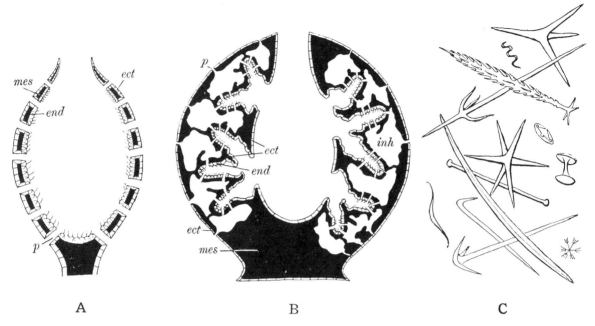

Figure A-2 Sponges. A, diagrammatic vertical section of a very simple sponge; B, similar section of a more complex sponge; C. various types of sponge spicules, greatly enlarged. *ect*, ectoderm; *end*, endoderm; *inh*, inhalant canal; *mes*, mesoglea; *p*, wall pore. (From Brooklyn Museum of Natural History.)

bonate. They inhabit all the oceans but are rarely found in fresh waters. The majority live on the bottom or cling to seaweeds, but about 20 kinds float near the surface of the open oceans, whence their shells, abandoned at time of reproduction, rain down like a snowfall to cover the sea floor. The commonest of these is *Globigerina,* and the soft, fine-grained, limy deposit made mainly by its shells is known as **globigerina ooze** (Fig. A-1). Approximately 50,000,000 square miles of the sea floor are now covered to an unknown depth by these deposits. At various times in the geologic past, foraminiferal shells have accumulated in shallow water to form extensive beds of chalk or limestone. The pyramids of Gizeh, for example, are made of a limestone that is widely spread in the Mediterranean region and is largely made of coin-shaped shells known as nummulites (Fig. 19-36). Still older limestones of the late Paleozoic are formed of **fusulines,** a tribe having shells about the size and shape of wheat grains (Fig. 12-21).

The **Radiolaria** (Fig. A-1) make their shells of silica. Differing from the capsulelike shell of the foraminifer, these are of a loose, open texture, like a delicate glass sieve. They form deposits of radiolarian ooze on parts of the deep-sea floor.

Phylum Porifera

The Porifera or sponges are multicellular animals in which there is little specialization of tissues. The "bath sponge" is but the silken skeleton of one highly specialized type. The essential features of the group are better displayed by a very simple sponge (Fig. A-2). Such an individual has the form of a slender vase; there is nothing to it but a living wall surrounding a large hollow space. This wall is made up of three layers like a jelly sandwich, the outer layer being formed of protective cells (ectoderm), the inner layer of feeding cells (endoderm), and the middle layer of a noncellular jellylike substance (mesoglea). The endodermal cells feed as do protozoans, each capturing and swallowing other microscopic organisms; the ectodermal cells do not take food but absorb what is needed from the nearby endodermal cells. Thus no digestive or circulatory organs are required.

To strengthen this delicate wall, either mineral spicules or threadlike fibers of spongin, an organic substance allied to silk, are formed in the gelatinous layer. These are secreted by specialized cells and are united to form a loose meshwork. In the bath sponge, the spicules are all made of spongin, but in many sponges the spicules are of silica or calcium carbonate (Fig. A-2C). The mineral spicules are commonly preserved, and are among the oldest records of Phanerozoic life.

Phylum Coelenterata

The third phylum of animals includes the **hydroids, corals,** and **jellyfish.** These, like the sponge, consist essentially of a body wall, lacking any internal organs, whence the name **Coelenterata** [Gr. *koilos,* hollow + *enteron,* inside cavity].

Hydroids. *Hydra* is a simple representative of the phylum, and particularly of the class *Hydrozoa.* A vertical section through its slender, subcylindrical body (Fig. A-3) shows a wall of three layers, as in the sponge. But in three respects it is vastly ahead of the sponge. First, it has muscular tissue that permits change of shape and even locomotion, and makes possible a circlet of muscular

tentacles about the mouth with which to capture food; second, in the inner layer are special gland cells that excrete digestive juices into the central cavity; and, third, it bears stinging cells like those of the jellyfish that can paralyze other animals coming into contact with it. The hydroid is thus able to capture, swallow, and digest animals almost as large as itself (Fig. A-3).

Hydras live as solitary individuals, but many hydroids reproduce by budding and thus form colonies of individuals organically

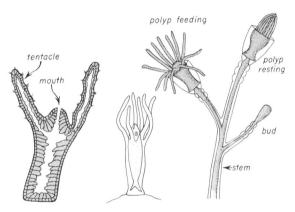

Figure A-3 Hydroids. Left, diagrammatic vertical section of *Hydra,* much enlarged; center, *Hydra* devouring a young trout (×2); right, a colonial hydroid, *Obelia,* with two adult individuals and a young bud.

Figure A-4 Left, portion of a colony of a modern coral with living polyps (a–d) and an exposed coral showing the septa (e); right, a horn coral.

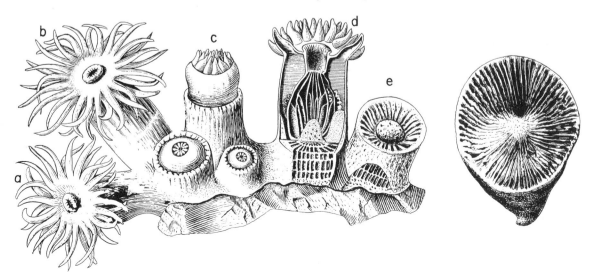

united (Fig. A-3). Many of these secrete a bell-like or vaselike sheath (hydrotheca) about their body as a protection. These sheaths are formed of chitin, a substance similar to finger-nails.

Many hydroids secrete calcium carbonate about the base of the colony, forming lami-nated coral-like structures.

Corals. The corals and sea anemones form another class of this phylum, the **Anthozoa** [Gr. *anthos*, flower + *zoön*, animal], so-named because of their bright colors and flowerlike symmetry (Fig. A-4). The coral animal re-sembles *Hydra* in essentials, but has in addi-tion thin radial partitions (mesenteries) that extend from the wall part way into the central cavity, subdividing it into a series of alcoves.

The coral animal secretes about its side and base an external skeleton of calcium carbon-ate. The animal is correctly termed a **polyp**, and its skeleton, **coral**.

For some unknown reason the base of the coral polyp is invariably marked by radial in-foldings which alternate in position with the internal mesenteries. The skeleton secreted against this base is marked by radial ridges or plates known as septa. Coral polyps may live singly, but many kinds reproduce by budding new polyps from the margins of the older, and thus develop colonies in which many indi-viduals cooperate to build a complex stony skeleton.

The skeleton of a solitary coral may be cushion-shaped or horn-shaped or subcylin-drical, and usually has a cuplike depression at the summit, in which the base of the polyp is housed. In colonial forms the skeleton is commonly branching or massive, with depres-sions for the bases of the individual polyps. The radiating septa are absolutely distinc-tive of the coral skeleton and serve to distin-guish it from all other types of shells.

It is well known that corals make reefs in the sea, but only where it is warm and shallow. Accordingly, the coral reefs in the rocks of past geologic ages are regarded as evidence of mild temperature and of shallow water. Corals have been important agents in rock for-mation at various times in the past. Coral reefs now occupy an area of about half a million square miles of the shallow seas, and their limy debris spreads over a vastly greater area.

Nearly all the modern corals belong to the subclass **Hexacoralla,** so named because their septa are introduced in cycles of six or mul-tiples of six. In these the septa are equally spaced, so that they seem to radiate with regu-lar symmetry in all directions from the center. This group has been the dominant one since the beginning of Mesozoic time. The subclass **Tetracoralla** (Fig. A-4, right), on the contrary, which was dominant in the Paleozoic Era, shows more or less conspicuous bilateral sym-metry, with septa introduced in cycles of four.

Phylum Brachiopoda

The brachiopods constitute a phylum of rather small marine animals that invariably bear an external shell of two pieces, known as **valves** (Fig. A-5). Although between 200 and 300 kinds are now living, they are seldom seen on the beach and are hardly known ex-cept to specialists; but they are extremely abundant fossils, especially in Paleozoic rocks, and so challenge the interest of geolo-gists.

The body is much more complex than that of the coral, having a digestive system, kid-neys, a nervous system, reproductive organs, and well-developed muscles. In life, the animal is attached to the bottom by a fleshy stalk (pedicle) at the posterior end, its mouth facing upward. The two valves of the shell are hinged together at the posterior end of the body and can be opened slightly at the front to permit circulation of water, which car-ries food particles to the mouth. The shells are borne on the back and front surfaces of the body (not on the sides), and thus a plane of symmetry passes through the middle of each valve. Special muscles open and close the shell, and a pair of interlocking teeth and sockets at the posterior end forms a hinge. A hole (pedicle foramen) for the passage of the stalk is usually present near the apex of the ventral valve.

Insofar as the brachiopod shell consists of two valves, it resembles the shell of a clam, but the likeness is purely superficial. The

Figure A-5 Brachiopods. Left, a shell in position of growth, attached to a point of rock by the pedicle. Center, dorsal view of same, showing the pedicel foramen and the bilateral symmetry (i.e., each side is a mirror reflection of the other). Right, side view showing that the valves are unequal in size and in shape.

structure of the body is unlike that of a clam, and the shell valves are borne in a wholly different position. In the clam, the valves are on the sides of the body and are hinged along the back; in the brachiopod, they are borne on the back and front surfaces and are hinged across the posterior end. As a result, the symmetry of the shell is quite dissimilar in these two groups, the brachiopod shell being **inequivalved** although each valve is **equilateral**, each half being a mirror reflection of the other, like the left and right sides of a coat. The clam shell is **equivalved**, right and left valves being mirror reflections of one another (except in deformed types like the oyster), but each valve is **inequilateral**, the front and hind ends being normally different in shape.

Brachiopod shells vary greatly in shape (Plate 4, Figs. 10 to 22), p. 204, some being strongly biconvex, others planoconvex, and others concavoconvex, the space between the valves in the last type being so thin that the animal must have had the proportions of a flatworm. In many brachiopods the hinge is short, and the posterior end of the shell pointed or "beaked" as in Fig. A-5; in others, it is long and straight, and the shell is "square-shouldered." Some shells are smooth, many are ribbed, and some are spiny. In spite of the diversity of form, the **brachiopod shell is easily recognized by its symmetry.**

Brachiopods are rather small animals, the average length of shell being between 1 and 2 inches. A few attained a diameter of 3 or 4 inches, and the largest that ever lived had a breadth of about 1 foot.

Phylum Bryozoa

The Bryozoa or moss animals form another important phylum little known to the general public, in spite of the fact that living forms are commonly attached to the rocks and seaweeds everywhere along the seacoast (Fig. A-6). Anatomically the bryozoan is very simple and in many respects more like a brachiopod than other animals, but, unlike the brachiopod, it is invariably minute and always grows in colonies. The individuals rarely attain a diameter much greater than that of a period on this page, but thousands of them living together may form a colony some inches across. Locally they combine to make reef limestones.

Unlike the brachiopod, the bryozoan forms a simple skeleton in the form of a slender tube or a boxlike cell, with an opening at or near one end through which the front end of the body can be thrust out while feeding. Many bryozoans have only a soft, delicate covering of chitin, but the majority secrete a skeleton of calcium carbonate.

The form of the colonial skeleton varies enormously with different species. It may be branching and mosslike, whence the name [Gr. *bryon*, moss], or stemlike, leaflike, massive, or encrusting. In spite of all this diversity, the bryozoan skeleton is easily recognized because it is made up of minute tubules or cells.

Phylum Echinodermata

The echinoderms are peculiarly different from all other animals. This great phylum

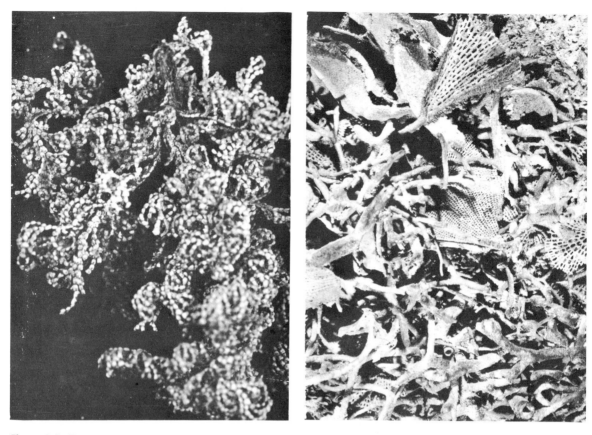

Figure A-6 Bryozoa. Left, *Menipea,* a living colony of mosslike form. Right, fossil bryozoans of many kinds weathering from a piece of Lower Devonian limestone. Natural size.

Figure A-7 Three views of the echinoid *Cidaris.* Left, a young individual as it appears alive; center, upper surface of shell with spines removed, showing the five ambulacral areas (with rows of pores) and five interambulacral areas (with large bosses for spine bases); right, side view showing the arrangement of the plates to be in vertical columns. About ½ natural size.

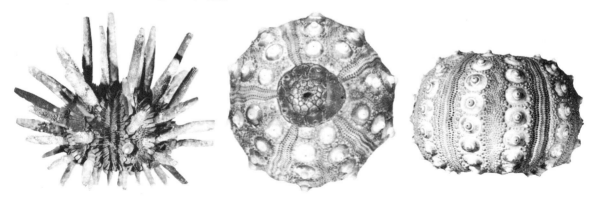

includes the starfishes, echinoids, crinoids, blastoids, and cystoids. Their bodies are short and commonly globular. Almost all have a radial and five-rayed symmetry. Nearly all develop a shell in the form of limy plates that are secreted in the body wall and fit edge to edge like the pieces in a mosaic. These types represent five distinct classes, which will be discussed in order.

Echinoids or Sea Urchins. A typical echinoid (Fig. A-7) has a globular or bun-shaped body bristling with slender, movable spines, whence the name [Gr. *echinos*, hedgehog]. The mouth is at the center of the lower side, and the axis of the body is vertical. Stripping away the spines, we find the body wall of the animal to be a rigid, boxlike shell of polygonal plates arranged in 20 vertical columns. Upon these plates are scattered small rounded nubs, each of which was the pivot for a spine.

Radiating from the summit of the shell to the mouth are five paths along which the plates are thickly perforated with small double-barreled pores. These paths are the ambulacral areas. In life each pair of pores bears a slender muscular organ known as a tube-foot. The tube-feet are part of a remarkable hydraulic system, found only among the echinoderms, which serves for feeble locomotion, for the gathering of food, and for respiration. The body cavity is spacious and includes a well-developed digestive system, a nervous system, and reproductive organs.

We have described a typical sea urchin as a radially symmetrical animal, but there are some specialized types (heart urchins and sand dollars) in which a secondary bilateral symmetry has modified the primitive, pentameral form. In all echinoids, however, a five-rayed symmetry is clearly evident even though somewhat irregular.

Starfishes. The starfishes are distinguished by their star-shaped form (Fig. A-8), the body being depressed and extended at the sides into tapering rays. The skeleton is made of small limy plates, articulated by fleshy tissues to permit some flexibility, as in a coat of chain mail.

Starfishes probably have been abundant since Early Paleozoic time, but they are

Figure A-8 A primitive starfish, *Devonaster eucharis.* Artificial cast from a natural mold in Middle Devonian sandstone of New York. Natural size. (Yale Peabody Museum.)

rarely found as fossils, because their loosely joined plates fall apart with the decay of the flesh, and such small irregular plates are not easily recognized.

Crinoids. The crinoids or sea lilies (Fig. A-9) look more plantlike than animal-like, for their globular bodies are supported by flexible stalks which anchor them to the sea floor, mouth upward. The animal consists of three chief parts, a stem, the body proper, and a series of branching arms.

In spite of its plantlike appearance, the crinoid is an animal, essentially comparable in its structure to a starfish, though in making this comparison we must turn the starfish with its mouth upward.

The body of the crinoid is covered by a series of limy plates which fit edge to edge like those of the echinoid. These plates are arranged in several horizontal cycles, one above another, beginning at the upper end of the stem. Normally the plates in each cycle number five or some multiple of five, so that five-rayed symmetry is the rule here, as in the sea urchins and starfish. The stalk is strengthened by the secretion within it of a

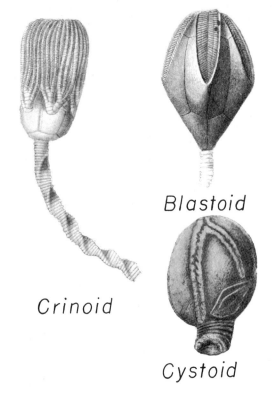

Blastoid

Crinoid

Cystoid

Figure A-9 Stalked echinoderms. A crinoid, *Platycrinus*, with part of its stem; a blastoid, *Pentremites*, without its stem; a cystoid, *Sphaerocystites*, with most of its stem missing. Natural size. (After Wachsmuth and Springer, Romer, and Schuchert.)

series of disk-shaped limy plates which are superposed like buttons on a string. These "stem joints" are united by muscular tissue, so that the stem has flexibility enough to let the animal swing with the currents. The mouth is at the summit of the body, and from it food grooves radiate along the upper sides of the arms. Indeed, the arms are structures developed merely to extend the food grooves.

The living crinoids are brilliantly and beautifully colored with shades of lavender, purple, red, lemon-yellow, or brown, and it is fitting that they should be called sea lilies. They tend to grow in patches on the sea floor and where present are commonly very abundant, so that they present much the appearance of submarine flower beds as they sway gracefully with the bottom currents. Upon the death of the animal its limy plates

commonly fall apart. Crinoid remains are among the commonest fossils in some of the Paleozoic formations, and they give distinctive character to "crinoidal limestones," some of which have wide extent.

Blastoids. The blastoids or sea buds [Gr. *blastos*, bud] form another group of stalked echinoderms (Fig. A-9). Their bodies are globular or bud-shaped and are encased in a shell of 13 chief plates of which 3 form a basal cycle, while 2 succeeding cycles have 5 plates each. There are no arms, the food grooves lying upon the surface of the body. These food grooves are always simple, 5 in number, and arranged in perfect five-rayed symmetry about the mouth, which is at the summit. The 5 ambulacral areas are submerged a little below the level of the chief body plates in such a fashion as to give the entire body a superficial resemblance to a flower bud in which the sepals are just beginning to part.

Blastoids are all extinct and are known only from Paleozoic rocks. They were very abundant during only one geologic period, the Mississippian.

Cystoids. The cystoids [Gr. *cystis*, bladder] are primitive echinoderms with globular or almond-shaped bodies but differ from both crinoids and blastoids in that their plates are irregularly arranged, so that the body shows no definite symmetry (Fig. A-9). They are extremely varied in details. Some had arms, and others had none; many were stalked, but some apparently were attached directly to the sea floor or were free. They were among the most primitive echinoderms.

Phylum Mollusca

This great phylum includes the bivalves, the snails, the octopuses, the squids, the pearly nautilus, and the extinct ammonoids. These commonly possess solid, limy, external shells, and they are generally known as "shellfish." The phylum is an enormous one, with probably no fewer than 50,000 species now living.

Pelecypods (Bivalves). A typical bivalve has a laterally compressed body encased in a bivalved shell, the two halves of which lie on the sides of the body and are hinged along its back.

The body is generally elongated, and the mouth is at the front end, but there is no distinctly marked head. Lining each valve there is a thin, fleshy extension of the body wall, the mantle, which hangs freely about the body like a loose garment. The most conspicuous organs are the great gills, which hang as a double pair of thin plates between the mantle and the sides of the body.

The shell (Fig. A-10) is opened by an elastic ligament at the hingeline, which is placed under tension (or in some cases under compression) when the valves are closed. The shell of most clams is closed by a pair of heavy transverse muscles which run through the animal's body from side to side. Normally one muscle is near the front, and the other near the back end.

The typical clam is free to creep slowly about by thrusting out its foot, the muscular ventral edge of its body wall. Secondarily, many clams have given up this freedom and lie on one side and adhere to the bottom, as does the oyster, or attach themselves to the rocks by silken threads, as does the blue clam. Still others burrow in the mud or even in hard rock.

The clam shell, like that of all other mollusks, consists of three layers. The outer one is a film of organic material to protect the limy part of the shell from solution. The second layer is of calcite and commonly has a white, porcelainlike appearance, while the inner layer is made of aragonite or mother-of-pearl and in some has the iridescence of pearl. In fact, the precious pearl is a secretion formed between the mantle and the shell, usually in an attempt by the animal to protect itself against a parasite or other irritant.

Gastropods (Snails). Though fundamentally like the clam in many of its anatomical structures, the snail is in several ways more highly specialized. It has an elongate muscular body with a distinct head bearing a pair of eyes and a pair of tentacles or feelers. Its mouth is provided with a flexible rasping tongue whereby it can shred either plant or animal food; it is therefore not dependent on microscopic ob-

jects. As a result its gills are small and plumelike, since they are used only for respiration. Internal organs resemble those of the clam in most respects.

The shell of the snail is coiled spirally and consists of a single valve. As in the clam, the shell is secreted by a mantle, so that the body may be measurably free.

When disturbed, the snail can withdraw completely into its shell, but normally it extends most of its body and creeps about, carrying the shell upon its back (Fig. A-11). The ventral surface of the body has developed into a muscular creeping sole, whence the group is known technically as the **Gastropoda** [Gr. *gaster*, stomach + *pous*, foot].

Most commonly the gastropod shell is coiled in a helicoid (corkscrewlike) spiral, but many fossil forms are bilaterally symmetrical like a watch spring. Rapidly expanding shells have few volutions, but slowly expanding ones commonly have high, slender spires. The shell may be ornamented with spines, ribs, or nodes.

Cephalopods. The squids and octopuses and cuttlefish represent a class of mollusks that has been abundant and important in all the seas since Early Paleozoic time. Unlike the sluggish snails and clams, these are active and aggressive predators. With their keen eyesight and strong powers of swimming, they alone of all invertebrates are able to compete actively with the vertebrate animals of the seas. The living giant squid is the largest invertebrate animal of all time.

The name cephalopod [Gr. *cephale*, head + *pous*, limb or foot] was suggested by the fact that all members of the class bear a circlet of

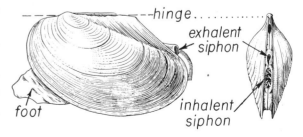

Figure A-10 Bivalve shell. The common river clam, *Anodonta*, seen from the left side and from the posterior end.

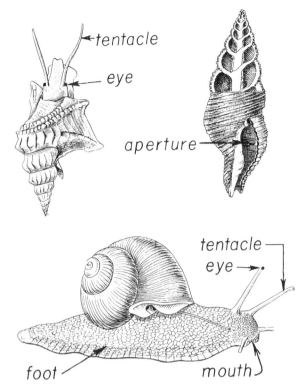

Figure A-11 Gastropods. Upper right, empty shell with front side of spire cut away; upper left, another shell with the snail crawling upward on the page, the head, with eyes and tentacles, showing above the shell; below, a common land snail in crawling position with creeping "foot" extended and shell coiled over the back. (After Hatschek and Cori.)

fleshy limbs about the mouth, which is at the front of the head.

The squid and octopus are shell-less, but the rare pearly nautilus (Fig. A-12) bears an external chambered shell, as did a host of forms known only as fossils. Here, then, are living examples of both shell-less and shelled cephalopods. The first is represented by the squids, whose only shell is vestigial and internal; the second by the nautilus with its external, chambered shell. Obviously the shelled cephalopods are of chief interest to the geologist, since they alone are ordinarily preserved as fossils.

The cephalopod shell has essentially the form of a slender cone, which may be straight or coiled (Fig. 9-20). If coiled, it is almost invariably in a flat spiral like a watch spring. The animal's body occupies only the larger end of the shell, the rest having been partitioned off into a series of chambers by transverse plates known as septa. In the living nautilus (and presumably in extinct types) the chambers may be filled with gas or fluid and the buoyancy of the shell regulated.

The chambers represent successive portions of the shell that were vacated as the growing body moved forward; and the partitions or **septa** are walls secreted by the bluntly rounded posterior end of the body. These partitions are attached to the inner surface of the shell, and the line of junction is known as a **suture.** The form of the sutures and the course which they take in the shell are of importance in the classification of the cephalopods. Since the septa are formed within the shell, the suture is not visible externally, but in fossil forms where the chambers have been solidly filled with mineral matter and the outer shell then dissolved away, the sutures show clearly as sharp lines on the fossil (Fig. 9-20). The animal retains connection with the abandoned chambers by a slender tube, the **siphuncle,** which runs back through all the septa. Septa and a siphuncle are absolutely distinctive features of the cephalopod shell and serve to distinguish it readily from that of the snail.

Nearly all the primitive cephalopods had straight shells and met problems of buoyancy by secreting limy deposits in the siphuncle and chambers. Curved, loosely coiled, and tightly coiled shells were probably all developed to cope with buoyancy and straight shells eventually became extinct. Besides being compact, the coiled shell brings the center of buoyancy of the chambers directly above the center of gravity, so that the animal can float at ease in the water.

The shelled cephalopods are divided into a number of subclasses. In most the septa were simple, saucer-shaped plates and the sutures run directly around the shell as simple lines without marked flexures (Fig. 9-20), as in the nautilus (Fig. A-12). The pearly nautilus is the only living representative.

One group of externally shelled cephalopods, the **ammonoids,** had septa that were fluted or ruffled along their edges, and as a result the sutures form strongly crenate lines around the shell. In the earliest ammonoids the fluting of the septa was very slight, and each suture showed only a few simple bends (Plate 8, Figs. 9 and 12), but fluting became highly complex by the close of the Paleozoic and in the Mesozoic most all ammonoids had complicated sutures (Plate 14, Fig. 11).

Belemnites. The belemnites (Fig. 15-22) were squidlike cephalopods that appeared in Mississippian time but were common only during the Mesozoic Era. They possessed a conical chambered shell like that of a primitive straight-shelled nautiloid, but it was internal, having been overgrown completely by flesh (Plate 14, Fig. 8) and weighted by a sheath of calcium carbonate.

Squids are closely related to the belemnites, but in the former the shell (also internal) is reduced to a mere cartillaginous vestige (the "pen").

Phylum Arthropoda

The insects represent a phylum of animals characterized by jointed walking legs, whence the name **Arthropoda** [Gr. *arthron*, joint]. Other examples are the spiders, scorpions, lobsters, and crabs, and some important fossil groups such as the trilobites and eurypterids. All these have segmented bodies and jointed limbs. Their bodies are protected by a neatly fitting jointed armor made of chitin. In some, like the lobster and crab, this skeleton is strengthened by the addition of calcium carbonate. This is undoubtedly the largest and most diversified phylum in the animal kingdom.

Insecta. The insect possesses an elongate, bilaterally symmetrical body distinctly divided into head, thorax, and abdomen. The sharp constriction separating thorax and abdomen, as in the wasp, has suggested the group name [L. *insectus*, cut into]. The abdomen is without limbs, but the thorax bears three pairs of walking legs and commonly two pairs (in one order, one pair) of wings. The head bears a pair of compound eyes and a pair of slender feelers or antennae. The mouth is provided with specialized biting or sucking devices.

The physical senses are rather highly developed.

Insects are too easily destroyed to be commonly preserved as fossils, but locally they are found as far back as the Late Paleozoic (Fig. 12-26).

Crustacea. The lobster and crab represent another great class of the arthropods, but, unlike the insects, these and nearly all their kin live in the water and breathe by means of gills. The name **Crustacea** [L. *crusta*, crust] refers to their hard, crustlike armor, but it must be confessed that many examples of the class do not have a hard shell.

In most Crustacea each of the segments of both thorax and abdomen bears a pair of jointed appendages (in some forms, part or

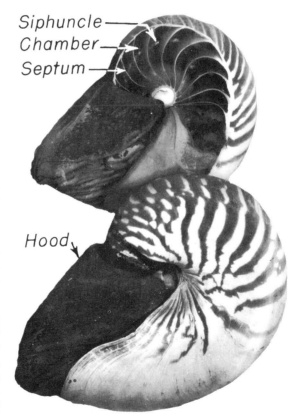

Siphuncle
Chamber
Septum

Hood

Figure A-12 Cephalopods. The pearly nautilus in its shell. The lower shell is intact, but the upper one has the left side cut away to reveal the hollow chambers separated by curved septa and connected by the tubular siphuncle.

all of the abdominal segments lack append-ages). Commonly those on the thorax are walking legs (as in the lobster), and those on the abdomen are for swimming, but in many of the primitive Crustacea all the limbs are flattened swimming paddles. The gills are generally plumose and are attached to the legs.

The most common crustaceans in the geologic record are the **ostracods**, generally tiny, bivalved forms whose shells are locally common in both marine and nonmarine strata. Today they are widespread in the seas and lakes of the world and also have terrestrial forms that live in forest soils and leaf mold.

Chelicerata. Spiders and scorpions belong to a class of arthropods known as the Chelicerata. The head of a spider is fused to the thorax to form a cephalothorax, which is separated from the abdomen by a deep constriction and bears four pairs of walking legs. The abdomen is not segmented. The head bears several simple eyes but nothing like the compound eyes of the insect. Also it does not bear antennae. The ability to spin a web is a characteristic feature of the spider. Spiders, being soft-bodied, are rarely preserved as fossils, but specimens found in the Lower Devonian rocks of Scotland are among the most ancient records of land animals.

The eurypterids or "sea scorpions" (Figs. 10-21 and 10-22) were a remarkable race of large aquatic chelicerates closely resembling the scorpions in bodily form. Indeed, they were probably the direct ancestors of the scorpions. The eurypterids were confined to the water, and chiefly marine waters, and their limbs were partially modified into swimming paddles. They were relatively large, the average length being several inches. One form, *Pterygotus*, from the Silurian rocks of New York State, had a length of about 9 feet and ranks as the largest arthropod of all time. The eurypterids are common and striking fossils only in certain Middle Paleozoic formations, but the race died out before the close of the Paleozoic era.

Trilobita (pronounced trī′ lo bī′ ta). These animals formed a primitive but exceedingly important group of arthropods which is now extinct and is known only as fossils from Paleozoic rocks. The body was depressed and distinctly divisible into head, thorax, and tail (Plate 1). Head and tail were each covered by an unsegmented shield, but the thorax was jointed. The entire body was longitudinally trilobed by reason of a pair of grooves that separated a rounded central axis from the lateral areas. This trilobation is at once the most distinctive feature of the group and the one which gave it the name **Trilobita**.

The trilobites usually possessed a pair of compound eyes and a pair of antennae or feelers. Each body segment bore a pair of legs, and these were essentially alike from head to tail. Each leg consisted of two branches, the lower one of which was a jointed limb for crawling, while the upper and outer branch was a delicate, featherlike structure, commonly regarded as a gill.

Phylum Vertebrata

The most advanced of all animals are those possessing a vertebral column or backbone. So distinctive is this phylum that it is often contrasted with all the other animals, which are known collectively as **invertebrates**. The vertebrates are characterized by the highly organized character of their nervous system, with the spinal cord running along the dorsal side of the body, and by many other important details, but the possession of a backbone is the most obvious and distinctive character. There are eight distinct classes, as follows: **fishes** (four classes), **amphibians, reptiles, birds,** and **mammals**.

Fishes. Fishes are primitively aquatic, cold-blooded vertebrates that breathe by means of gills. Most of them possess paired lateral fins as well as a tail fin.

This is an enormous and highly diversified group with a long geologic record and, although commonly treated as a single class, is now subdivided into four distinct classes. Many are scaled, but several of the extinct groups bore an armor of bony plates and some (for example, the catfish) are not protected by either scales or bone. The skeleton is made of bone in the majority of modern fishes, but in most of the early groups it consisted of cartilage, as in modern sharks and sturgeon.

It is important to distinguish between the fishes, which are primitively adapted to life in the water and breathe by means of gills, and, on the other hand, certain fishlike animals that have returned from the land to become secondarily adapted to the water. Among the latter are whales, porpoises, and seals, and certain extinct reptiles (ichthyosaurs, plesiosaurs, and mosasaurs). These secondarily aquatic animals breathe only by means of lungs and must come to the surface for air. They mimic fish but are not closely related to them.

Amphibia. The frogs and salamanders constitute a class of vertebrates that are only partially adapted to life on the land and have many features to remind us of a fishlike ancestry. Indeed, they are certainly the most primitive class of land vertebrates, and they clearly evolved from fishes. The salamander is a typical representative of the class (the frog being a modern and extremely specialized form). Because its members live partly in water and partly on land, the class has received the name Amphibia [Gr. *amphi*, on both sides + *bios*, life]. The metamorphosis from an aquatic youth (tadpole) to a terrestrial adult life and the need to return to the water to reproduce distinguish amphibians from all other land animals.

Reptiles. The reptiles constitute a very large class of vertebrates and one that for long geologic ages completely dominated the earth. The group includes the alligators (Fig. A-13), crocodiles, lizards, turtles, and snakes, and the extinct dinosaurs and pterosaurs and 3 great groups of marine reptiles of Mesozoic seas, the mosasaurs, plesiosaurs, and ichthyosaurs. The crocodile is a very typical reptile.

Reptiles are cold-blooded, egg-laying animals. In shape, many of them closely resemble amphibians, but they all differ from the amphibians in the way the young develop. The reptile lays its eggs on the land, that is, out of the water. These are provided with stored-up food in the form of yolk, so that the young can develop fully enough to crawl about and care for themselves immediately upon

Figure A-13 A typical reptile, the Florida alligator.

hatching. In other words, the reptiles are completely adapted for terrestrial life.

Although united by such features as their cold blood, their egg-laying habit, and various details of skeletal structure, the many different stocks of reptiles specialized greatly; some returned secondarily to aquatic life, and one group, the extinct pterosaurs, developed wings.

Birds. The birds constitute a very well-defined class of vertebrates characterized by the presence of feathers and warm blood and by the egg-laying habit. In them the front limbs are specialized as wings. The birds diverged from one group of reptiles at about the same time that mammals were developing from another.

Mammals. The mammals are warm-blooded. They bring forth their young alive and nourish them with milk. All of them bear hair, though in some, as the whale and the elephant, the hair is almost lost through specialization. They constitute the dominant group of land animals of the Cenozoic and modern worlds, and some (seals, porpoises, and whales) have secondarily returned to the sea.

One small aberrant group, the **monotremes,** lays eggs. Two living genera are believed to represent an ancient, primitive stock of the mammals, but no paleontological record of this group is known.

The mammals did not appear on the earth until Mesozoic time, and there is clear evidence that the ancestral types descended from one of the primitive groups of reptiles.

PLANTS

Plants do not lend themselves to a simple, inclusive classification. The groups listed below include those most commonly found in the fossil record, but it is not intended to serve as a comprehensive classification.

Nonvascular plants.
    Algae, fungi, bacteria.
    Phylum **Bryophyta** — mosses, liverworts.
Vascular plants (Phylum **Tracheophyta**).
    Class **Psilopsida** — primitive tracheophytes.
    Class **Lycopsida** — scale trees, club mosses.
    Class **Sphenopsida** — horsetails.
    Class **Pteropsida** — ferns and seed-bearing plants.
        Subclass **Filicinidae** — ferns.
        Subclass **Gymnospermidae** — seed-ferns, cycads, ginkgos, conifers.
        Subclass **Angiospermidae** — flowering plants.

Nonvascular Plants

Algae. The term algae covers a wide variety of very simple plants that are commonly arranged in seven different phyla. Nearly all are aquatic and they range in size from microscopic filaments to seaweeds that may attain a length of 100 feet. They have no woody tissue and no vascular tissue such as that used to circulate sap in higher plants. All reproduce by means of spores.

Most algae are not likely to be preserved as fossils. Some blue-green algae (Cyanophyta) cause precipitation of calcium carbonate which settles over them to form a limy deposit. Figure A-14 shows such a deposit (a "water biscuit") formed by a moldlike colony of microscopic blue-green algae that covered a pebble in a stream bed. The deposit has a finely laminated texture due to the addition of concentric films of calcium carbonate. Sim-

ilar limy deposits, called **stromatolites** (p. 159) occur in rocks of all ages and are the only common fossils in Cryptozoic rocks.

Diatoms are single-celled, microscopic algae, included in the phylum Chrysophyta. They live in vast numbers near the surface of the seas and lakes and form one of the chief sources of food for all the marine animals. They secrete delicate siliceous shells which accumulate over large areas of the modern ocean floor as **diatomaceous ooze.** Fossil oozes are commonly called **diatomite.**

Bacteria and Fungi. Two other kinds of simple plants have left a very sparse fossil record. The bacteria and fungi probably have existed from far back in Cryptozoic time (p. 160). Fungi lack chlorophyll and live by absorbing nutrients from other organisms, living or dead. Bacteria are single-celled and microscopic. Both may occur among the interesting Cryptozoic fossils from the Gunflint Chert (Fig. 7-23).

Phylum Bryophyta. This small phylum, including the mosses and liverworts, represents the simplest stage of adaptation to land life. Like the algae, these plants lack vascular tissue, hence remain very small and thrive only in moist places. Almost nothing is known of them as fossils.

Phylum Tracheophyta (Vascular Plants)

Psilopsids. These are generally small, branching vascular plants bearing spore cases on the tips of stems. They lack roots and true leaves but commonly have a scale-like foliage. This highly varied but primitive group represents the first radiation of land plants in Silurian and Devonian time. It did not survive the Devonian.

Lycopsids. Scale trees constitute another well-defined tribe of plants that occupy a humble place in the modern world but were very important in the Paleozoic forests (Figs. 12-24 and 12-25). The ground pines or club mosses are modern examples. Like the **sphenopsids,** these plants bear spores in strobili at or near the tips of the stems.

Many of the Paleozoic species attained the size of forest trees. In these the leaves, when

Figure A-14 Calcareous algae. A "water biscuit" from the bed of Little Conestoga Creek, near Philadelphia, Pennsylvania. Left, external view of the colony; right, a median section showing the concentric laminae of algal deposit about an angular pebble. Slightly less than natural size. (Yale Peabody Museum.)

shed, left prominent scars regularly spaced over the bark of the trunk and limbs (Fig. 12-25).

Sphenopsids. Horsetail rushes and their kind constitute a tribe characterized by regularly jointed stems. The existing horsetail or scouring rush (*Equisetum*) grows abundantly in moist places in many parts of the country. In all horsetails the stem has only a thin cylinder of woody tissue around a large center of pith. Reproduction is by means of spores which are borne in **strobili** (cones) at the tops of the stems.

Modern horsetails are mostly small, and the race passed its climax in the Late Paleozoic, when giant scouring rushes (calamites) grew to the height of trees and had stems as much as 12 inches in diameter (Figs. 12-1 and 12-24).

Ferns. The great tribe of the ferns are the simplest of plants to be well adapted to land life. They have well-developed vascular tissue and are differentiated into roots, stem, and leaves. Whereas many of the ferns are small and herbaceous, the tree ferns of the tropics have woody trunks and commonly reach a height of 20 to 50 feet, bearing a crown of large fronds. Ferns are distinguished above all else by the fact that they reproduce by means of spores borne on the under sides of the leaves or on slightly modified leaves, never in cones.

Ferns are among the oldest fossil land plants known, being recorded first in the Lower Devonian rocks. They have been a prolific tribe in all subsequent ages.

Gymnosperms.

Seed Ferns. The oldest and most primitive seed-bearing plants were fernlike in everything but their fruit, whence their formal name, **Pteridospermales** [Gr. *pteris*, fern + *sperma*, seed].

The distinction between the spore-bearing and seed-bearing plants is comparable to that between the egg-laying and the viviparous animal, both in its nature and in its significance. The egg is a simple cell, deposited to hatch into an embryo that must look out for itself at a very immature stage of its development; the spore, likewise, is a single cell cast free to generate and grow as best it can. On the other hand, the viviparous animal retains the egg in the mother's body until it has developed into an embryo of considerable com-

Figure A-15 Stems, foliage, flowerlike parts of the cone of the cycadeoid *Williamsonia.* Lower Jurassic of Mexico. Natural size. (G. R. Wieland.)

plexity before birth liberates it to shift for itself. Similarly, the seed is an embryo plant formed after the fertilization of the ovum, which is retained and nourished by the mother plant until considerable size has been attained and food is stored up to give the new plant a good start in life. The development of seed is a very considerable specialization. That it was an obvious advantage is suggested by the dominance of seed-bearing plants on the modern lands.

The seed ferns were common from the Late Devonian to the end of the Paleozoic, when they died out, probably having given rise meanwhile to other phyla of seed-bearing plants.

Cycads. The living sago palms or cycads represent an extensive tribe, mostly extinct. It is a large group, divisible into two well-marked orders, the one, **Cycadeoidales,** en-

tirely extinct, and the other, **Cycadales,** represented by the modern cycads. The cycads (Fig. 14-15) possess short trunks and large pinnate leaves, and bear seeds in loose cones. The trunk is heavily armored with persistent leaf bases.

The **cycadeoids** (of Mesozoic age) bore conspicuous cones, some superficially flowerlike in appearance. These plants were of two chief types, one stocky, like the modern cycads, and the other slender and branching. The branched cycadeoids possessed rather slender stems bearing cones at forks (Fig. A-15). These plants probably developed from seed ferns and evolved into the thick-bodied cycadeoids. Cycadeoids are abundantly represented in the Triassic, Jurassic, and Cretaceous periods, and probably were present in the Late Paleozoic (Fig. 16-14).

Conifers. These constitute another tribe of

gymnosperms characterized by the development of cones and, generally, by evergreen foliage. The leaves are as a rule needlelike or straplike and have parallel veins. There are several orders. The pines are typical.

Cordaites. A related group is the **cordaites,** which were common in the forests from Devonian to Permian time. Unlike the conifers, they had large straplike leaves, and their seeds were loosely arranged in racemes instead of cones.

Ginkgos. These were a common Mesozoic gymnosperm group that has survived to the present. The trees have distinctive fan-shaped leaves.

Angiosperms. The most advanced of all plants are the hardwood trees and the true flowering plants. At least 125,000 kinds have been described. They are all, of course, seed-bearing, but show a great advance over the other seed-bearing phyla in that their seeds are protected in a closed capsule or ovary. For this reason they are known as angiosperms (covered seed) in contradistinction to the gymnosperms (naked seed), which include the seed ferns, conifers, and cycads.

The angiosperms are commonly characterized as **the flowering plants,** but this is hardly justified, since even the conifers and cycadophytes have flowers of a sort.

The angiosperms are first certainly identified in the Lower Cretaceous rocks, where the characteristic net-veined leaves of deciduous trees make their appearance.

AUTHOR INDEX

SUBJECT INDEX

*Asterisks refer to illustrations.

+Includes names and ideas that appear in the text, but not formation names shown on the correlation charts or biologic
  names in the legends of the plates of fossils.